Foundations of
Modern Cosmology

John F. Hawley
Katherine A. Holcomb

University of Virginia

New York   Oxford
OXFORD UNIVERSITY PRESS
1998

To our parents
and other teachers

Oxford University Press

Oxford   New York
Athens   Auckland   Bangkok   Bogotá   Bombay
Buenos Aires   Calcutta   Cape Town   Dar es Salaam
Delhi   Florence   Hong Kong   Istanbul   Karachi
Kuala Lumpur   Madras   Madrid   Melbourne
Mexico City   Nairobi   Paris   Singapore
Taipei   Tokyo   Toronto   Warsaw

and associated companies in
Berlin   Ibadan

Copyright © 1998 by Oxford University Press, Inc.

Published by Oxford University Press, Inc.
198 Madison Avenue, New York, New York 10016

Oxford is a registered trademark of Oxford University Press

Library of Congress Cataloging-in-Publication Data
Hawley, John Frederick., 1958–
Foundations of modern cosmology   /   John F. Hawley, Katherine A.
Holcomb.
p.   cm.
Includes bibliographical references and index.
ISBN 0-19-510497-8 (cloth)
1. Cosmology. 2. Astrophysics.   I. Holcomb, Katherine A., 1957– .
II. Title.
QB981.H378   1998
523.1—dc21   97–3338

9 8 7 6 5 4 3 2

Printed in the United States of America
on acid-free paper

Contents

Preface

Recent discoveries in astronomy, especially those made with data collected by satellites such as the *Cosmic Background Explorer* and the *Hubble Space Telescope*, have brought cosmology to the forefront of science. New observations hold out the tantalizing possibility that the solutions to some especially elusive mysteries might be found in the near future. Despite an increase in public interest in black holes and the origins of the universe, however, the unavoidable lack of context with which discoveries are reported prevents most people from understanding the issues or appreciating the true significance of the new data. Popular books on cosmology abound, but often they present the subject as a series of "just so" stories, since some basic physics is a prerequisite for comprehending how cosmology fits into modern science. The lay reader may well have trouble distinguishing knowledge from speculation, and science from mythology. Furthermore, the popular literature often emphasizes the more exotic aspects of the field, often at the expense of the firmly grounded achievements of modern cosmology.

Cosmology holds an intrinsic interest for many college students, who are granted, as part of their general education, the time and opportunity to learn more about the scientific discoveries they see described in newspapers and magazines. Most colleges and universities offer a comprehensive introductory astronomy course, with the primary objective of offering science to as broad a population of students as possible. Topics such as relativity, black holes, and the expanding universe are typically of particular interest, but they are covered in a cursory fashion in most introductory courses and texts. In our experience, there is always a sizable number of students who find astronomy sufficiently interesting that they wish to continue their study of the subject at a comparable technical level but with greater depth. With little but astronomy-major or graduate-level courses available, however, such students often have no such opportunities. These students, who are genuinely interested in learning more about these topics, deserve the opportunity to further their learning and to do so in a serious way.

The course from which this book grew is intended for upper-division liberal arts students at the University of Virginia. Most of the students who take it have some basic science background, such as would be provided by a general introductory astronomy course; however, well-prepared students can and do take the course in lieu of general astronomy. Students from wide-ranging areas of study have taken this course. Their relative success is not necessarily correlated with their major. Some exceptionally strong students have come from the ranks of history and philosophy majors, while occasionally an engineering or astronomy major has floundered. Extensive experience with math and science are not prerequisites; interest and willingness to think are.

This text is intended to fill the gap between the many popular-level books that present cosmology in a superficial manner, or that emphasize the esoteric at the expense of the basic, and the advanced texts intended for students with strong backgrounds in physics and mathematics. The book is self-contained, appropriate for a one-semester course, and designed to be easily accessible to anyone with a grasp of elementary algebra. Our goal is to present sufficient qualitative and quantitative information to lead the student to a firm understanding of the foundations of modern theories of cosmology and relativity, while learning about aspects of basic physics in the bargain.

The level of mathematical detail is always a concern for instructors of undergraduate astronomy. We have aimed for a middle ground; some may regret the lack of calculus and accompanying derivations, while others may recoil from the appearance of any equation. The real difficulty with a topic like cosmology is not the mathematics per se, but the challenging concepts and the nonintuitive way of thinking required. However, without some understanding of the mathematical basis for cosmology, the student may find it difficult to distinguish science from mythology; without data and quantitative analysis, science becomes just another narrative. Thus, while we have tried to keep the level of mathematics consistent with minimum college-level algebra, we have not shied from including some equations within the text, rather than relegating them to an appendix or omitting them altogether. The resulting level is comparable to some of the more comprehensive introductory astronomy texts. Of course, more or less mathematical detail may be included or required by the instructor, depending upon the backgrounds and wishes of the students.

The book contains more material than can usually be presented in one semester. The instructor has a good deal of flexibility in designing their particular course. Depending on the background of the students, various sections can be given more or less emphasis.

The text is divided into five major sections. Since many students are unaware of the historical background from which modern cosmology grew, we begin with an overview of historical cosmology, from ancient myths to present scientific theories. The history of cosmological thought demonstrates that the universe is not only knowable to the human mind, but that the modern physical universe, constructed in the light of our new understanding of physics, is far grander than the constricted heavens of the ancients. This section also lays out the important cosmological questions and introduces the ideas of natural motion, symmetry, and the relation of physical law to the structure of the universe. For students who have just completed a typical introductory astronomy course, the historical and review sections could be covered quickly, with an emphasis on Newton's laws.

The second section exists primarily to make the book self-contained; it quickly reviews points that are likely to have been covered in an introductory astronomy or physics course. We do not assume or require introductory astronomy as a prerequisite; a motivated reader can find all the necessary background material here. While this section can be discussed briefly, or skipped entirely, even those students who have previously studied astronomy might find the review beneficial.

The theories of special and general relativity are presented in the third section, with emphasis on the fundamental physical consequences of these theories. Many textbooks, particularly at the graduate level, de-emphasize relativity, since it is true that little knowledge of the theory is required for the study of cosmology. However, readers who form the intended audience of this text often find relativity particularly fascinating, since it is so drastically different from anything they have previously learned or thought. Relativity is the setting upon which much of modern cosmology takes place, but professional astronomers often take this worldview so much for granted that they do not appreciate the point of view of students who have never encountered this material. Class surveys have consistently shown that relativity makes the greatest impression upon most of the students. In any case, portions of this section are indispensable. Chapter 6 presents the cosmological principle, a concept that is obviously required for the remainder of the book. Chapter 7 introduces several essential concepts including the space-time interval, light cones, and the metric. Chapter 8 on general relativity includes the necessary introduction to the non-Euclidean isotropic and homogeneous geometries. General relativity is highlighted by a chapter on black holes (chapter 9), which includes some of the latest astronomical ideas and discoveries. While this chapter on black holes can be omitted, students often find that topic to be the most interesting of all.

The theory of relativity provides the background for the next section, which presents basic modern cosmology. Chapter 10 discusses the discovery of the external galaxies and the expanding universe, and the theoretical interpretation in terms of Einstein's theory of relativity. This leads into chapter 11 which presents the simplest mathematical models of the universe itself, and the standard big bang models. Chapter 12 deals with the discovery and interpretation of the cosmic background radiation, as well as other modern cosmological observations. The history of the universe, starting from the "bang," follows as the next topic. Throughout, emphasis is given to the standard models, with some discussion of the most likely variants.

The final section covers topics that are the subject of current ongoing research. In this section, we emphasize that the standard model of cosmology has been spectacularly successful as a scientific theory; it simply does not yet provide all the answers. We consider the possibility of dark matter in the universe and the formation of large-scale structure in chapter 14. Inflationary models have been advanced as a possible solution to some of the quandaries of the big bang; they are presented in chapter 15 with an explanation of how they might answer these questions. We end in chapter 16 with the most speculative topics: the unification of gravitation and quantum mechanics, the two great triumphs of twentieth-century physics, the enigma of the arrow of time, time travel, and the fate of the universe itself. Any of the chapters in this final section can be used independent of the others, as time permits. Instructors may wish to supplement this material with additional information from current research, or from their own notes, as appropriate.

As an aid to the students, each chapter includes a list of key terms and review questions. A glossary of terms is provided at the back of the book. A brief description of scientific notation, units, and physical constants is given in the appendices.

We wish to acknowledge those colleagues and friends who provided comments, criticism and advice during the preparation of this book. We thank Steven Balbus, Jane Charlton, Marc Davis, Dorothy James, Hannu Kurki-Suonio, Karen Kwitter, Michael Norman, Christopher Palma, James Stone, John Tonry, David Weinberg, Mark Whittle, and the many students from Astronomy 348.

PART I

History

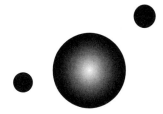

CHAPTER 1

In the Beginning

On a clear, moonless night, in a field far from city lights, the sky might be the cabinet of some celestial jeweler, displaying glittering points of light on a field of black velvet. A faint, irregular band meanders overhead, like a river of cosmic milk. On any particular night, noticeably bright stars might stand out among the others; on subsequent evenings, an observant watcher would find that these wandering lights had shifted their positions against the backdrop of stars. As the seasons change, so does the sky; some groups of stars visible in summer disappear during the winter, whereas others remain above the horizon all year. In the morning, the Sun appears on the eastern horizon. It climbs upward into the sky, then descends, and vanishes beneath the western horizon. As the Sun disappears, the stars rise, retracing the Sun's motion from east to west across the sky. The Moon rises as well, but keeps its own schedule, independent of the stars. At times, the Moon appears as a silvery disk marked with gray splotches; the imaginative may see a man, a rabbit, or even a beetle in the face of the Moon. At other times, the Moon shows us a crescent, or half its disk. Sometimes, it never appears at all.

Today most people pay little attention to the sky, its contents, and its motions. Electric lights and mechanical clocks have dethroned the celestial sphere from its classical importance in human affairs. The inhabitants of brightly lit cities may never even have seen the stars clearly, much less watched the motions of the planets. Some would also argue that modern science has removed the wonder from the sky; the planets, the Sun, the Moon, and the stars have all been "explained." Yet how many among us understand what those explanations are or what they mean?

Romantics often declare that understanding a phenomenon somehow takes away its beauty, reducing it to a desiccated specimen, like a stuffed bird in a museum case. But it is not the scientific understanding that is at fault. The failure to observe, and to ask the questions posed by the observations, shows that the beauty was never truly appreciated in the first place. To those who take the time to look, the sky is still a marvel, and its wonder is only magnified by the extraordinary discoveries of modern astronomy. The Milky Way retains its grandeur, but now we know that this faint, diffuse light is the combined glow of some of the billions of stars filling the unimaginably huge galaxy in which we live, an awesome contemplation on a dark evening.

The heavens still pose many questions to those who take the time to ponder them. What are the stars? Where are they? What makes the Sun rise, and what carries it across the sky on its daily journey? Where does it go at night, and where are the stars during the day? Why do the "wanderers" roam among unshifting stars? Such questions follow immediately even from casual observations. From there, the study of the cosmos leads us toward even more profound mysteries. How did it all begin? Was there a beginning at all, or have the heavens and Earth existed forever? Will the universe come to an end? What is the nature of the universe, and what role might humans play in it?

Many of these questions puzzled the ancients and have long since been resolved; but for the modern observer of the night sky, astronomy has deepened some of the old mysteries and added new ones. Many literate persons have heard such expressions as "the big bang" and "expanding space." They may be aware that astronomers debate whether the universe is "open" or "closed," infinite or finite, eternal or doomed. But what does it mean to say that the universe expands? Is the universe really expanding? Into what? When astronomers say that most of the mass of the universe is "missing," what do they mean? Where could it have gone? What are space and time, and why does time move in only one direction? What is the big bang? How did elements originate? What happens to stars when they die? What is a black hole? What will be the ultimate fate of our Sun and even of the universe itself? Were there other universes before this one, and will others follow ours?

Questions such as these fall within the domain of **cosmology**, the study of the universe. Today we regard cosmology as a modern science, but cosmological yearnings have been part of humanity throughout history. All cultures have a cosmology, for such questions have been asked by all peoples for as long as we have wondered at the stars. The explanations have varied from culture to culture, and from time to time, but all seek to impose an order upon the cosmos, to make it accessible to the human mind. This is just as true of scientific as of prescientific cosmologies.

Prescientific cosmologies generally interpret the universe in strictly human terms. Early cosmologies certainly began with basic observations; the connection between the changes in the skies and the days and seasons is difficult to miss. Mythological models of the universe sought to render such observations intelligible and to fit them into a theory of existence. However, in the mythological worldview, observations were, for the most part, of secondary importance. Scientific cosmologies, in contrast, are based upon and judged by data, the measurements obtained by direct, objective observations of the universe. The better the data, the better the cosmologies we can develop.

Cosmology can lay a defensible claim to the title "the grandest science," for no other field can have so vast an object of study: the universe in its entirety. But what do we mean by the "universe"? We might define the universe as "the sum of all that exists," but this is insufficient, for existence draws its meaning from the universe. The universe exists independent of any, or all, of its contents. A complete definition of the "universe" may not be possible, for it may be that some aspects of the cosmos are forever beyond our limited understanding. Here we will define the **universe** as that which contains and subsumes all the laws of nature and everything subject to these laws; that is, all that is physical. Is cosmology, then, the study of everything? Are all sciences cosmology? Such a definition would be too broad to be useful. We restrict our definition of cosmology to the study of the formation, structure, and evolution of the universe as a whole. This will prove more than sufficient.

Cosmology is sometimes regarded as a subfield of **astronomy**, but this is not an accurate division. Astronomy is the study of the contents of the universe. Although modern cosmology is intimately linked with astronomy, as our only way to observe the universe is to observe the objects it contains, it is also closely tied to physics. The universe consists not only of bodies, but also of forces and laws which govern their interactions. Indeed, we shall find that physics plays a much greater role than astronomy in describing the earliest moments of the universe. Cosmology draws upon many fields, and itself contributes to other sciences, sometimes in unexpected ways.

At the close of the twentieth century, cosmology can take pride in its accomplishments. A coherent view of the universe has emerged, the "hot big bang" model, which successfully explains a remarkably broad range of observations. While the big bang model has never claimed to represent the final truth, it nonetheless provides a framework for understanding the cosmos from the earliest few fractions of a second of its existence till the present; the model even predicts how it all might end. Surely this must count among the greatest of human achievements, even though this model cannot yet explain all the unknown.

In this text we will present the foundations of modern cosmology. The history of the development of scientific cosmology shows keenly how our intuitions and common sense continually mislead us, from our perception of an unmoving Earth, to our persistent belief in the absoluteness and inflexibility of space and time. Nevertheless, we can transcend these human limitations and arrive at a picture of the universe that is much closer to the way it truly is.

Cosmological Roots: Mythology

Although modern cosmology is scientific, and based upon highly detailed observations of great sensitivity and precision, the standard big bang model has a long lineage of human explanations of the cosmos. Most of these ancestral models have much to do with human hopes, desires, and preoccupations, and precious little to do with observations. To some extent, this was due to the limited capabilities of the unaided human senses. Much more important, however, were the philosophical prejudices that prevailed for millenia. Only slowly have humans learned to understand what our senses tell us.

Young children tend to interpret their worlds as reflections of themselves; so it was with humanity for most of its history. The earliest cosmologies were anthropomorphic cosmologies. An **anthropomorphism** is the interpretation of what is not human or personal in terms of human or personal characteristics. Attributing human motivations and emotions to the cosmos as a whole, or phrasing the existence of the universe in terms of a literal birth and death, are examples of anthropomorphisms. One form of an anthropomorphic worldview is *animism*, the belief that all things are animated by spirits, all of which hold some opinion toward humans, and any of which may actively aid or frustrate human plans. Less purely anthropomorphic cosmologies may hold that some portions of the universe are inanimate but are created and affected by animate beings, perhaps by a pantheon of gods as well as humans.

The trend to anthropomorphism comes from the quite natural tendency of human cultures to describe the universe with imagery from familiar, and necessarily human, experiences. When a cosmology is expressed in the form of a narrative tale that explains or illustrates the beliefs of a culture, it is a said to be a **myth**. Cosmological myths make the culture's ideas of the origin and structure of the universe generally intelligible and broadly accessible. Some of these myths were interpreted quite literally and anthropomorphically, while others were understood to be only analogies that could make the incomprehensible more familiar. Mythology tends to reflect what is known or important to the culture in which it arose. The myths of agricultural societies typically revolve around the imagery of the seasons, of planting and harvesting, while the myths of hunting and gathering peoples often involve animals which take on human characteristics.

Ancient mythologies still hold so much power that even modern cosmologists sometimes inappropriately blend objective data and mythological leanings. This is illustrated by the special fascination that cosmological beginnings and endings continue to hold. There is no particular reason to believe that the universe must have a beginning or an end, based solely upon our immediate observations; on the scale of human life, the Earth and sky seem eternal and unchanging. Yet it is also true that seasons begin and end, plants develop and wither, and animals and humans are born and die. Perhaps, then, the universe too has a beginning and an end. Not all mythologies assume this, nor do all modern scientific models. Even among scientists of the twentieth century, preferences for one model over another have sometimes been based more upon philosophical beliefs than upon data. A distaste for the big bang or for an infinite expansion is an emotional choice based upon a personal mythology. When the big bang model was first introduced, many of the most prominent scientists of the day reacted quite negatively; such an abrupt beginning for the cosmos was uncomfortable for the older generation. Still others interpreted the big bang as scientific vindication for the existence of a creator. Today the big bang is well accepted on its own objective merits, but now some discussions of the possible end of the universe carry an echo of the aesthetics of the cosmologist. Regardless of such intrusions of human wishes, the major difference between science and mythology stands: in science, the cumulative evidence of data must be the final arbiter.

The recognition of familiar concepts makes even a cursory study of the mythologies of many cultures an enjoyable pursuit. It is worth remembering that many of our unexamined cosmological ideas, including some most firmly embedded within the human psyche, have mythological origins. Many genesis myths can be seen to share common themes. There are three categories of imagery commonly invoked to explain the beginning of the universe. One is the action of a supreme craftsman, mirroring the image of a human craftsman at work. Another is generation from a seed or egg, reflecting biological generation. The third is the imposition of order onto chaos, as in the development of human society. These three are not mutually incompatible, and many myths incorporate two or more of these motifs. Nor are these the only possible themes; the early Hindu creation epic, the Rig-Veda, makes no explicit claims about the creation of the universe, suggesting only that perhaps some highest god knows and hinting that it is beyond mortal comprehension.

Another recurrent theme of great importance to humans, although of less significance to the universe as a whole, is the origin of imperfection in the human condition. Many cultures have believed that humans were originally close to the gods, but sinned and were punished. The origin of death is often attributed to human misbehavior; in a number of traditions, women take the brunt of the blame. Death is not always a punishment, however; some see death as an active choice, specifically, the choice to have children. (From a biological perspective, this is rather accurate.) Such myths seek to understand humanity's place within the universe, an issue with which we still struggle today.

A few specific examples will serve here; they are not intended to be comprehensive, but are fairly typical of the range of themes in cosmological mythology. The myths of a society spring from that society, and these examples vividly illustrate this, but myths, once established, can also mold societies long after they cease to function literally.

The *Enuma Elish* is the "Babylonian Genesis." Babylon, a great city of ancient Mesopotamia located near present-day Baghdad, was originally settled by the Sumerians around 3500 B.C. The Sumerians irrigated the desert and developed the cuneiform writing system, but they were conquered by Akkadians and Amorites from the north, and their culture was assimilated and eventually forgotten. The great king Hammurabi, famous as the first ruler known to have written down a code of laws, was an Amorite ruler of Babylon in the eighteenth century B.C., during the height of Amorite power in the region. A few centuries later, however, Babylon and its possessions came under the control of the Kassites, a tribe which may have originated in central Asia. The *Enuma Elish* dates from the Kassite regime, perhaps around 1450 B.C., though only later copies of it now exist. The second millennium B.C. was a peak period in Babylon's history, and this creation myth was probably composed at least partially with the motive of justifying the city-state's political power by making its patron deity the chief among the gods.

The story incorporates ancient Sumerian themes, as well as contributions from the later conquerors of Mesopotamia; like many Babylonian myths, it is evocative of later stories indigenous to the Middle East. In this tradition, the tumultuous sea is identified with disorder. The Sumerians and their successors believed that the cosmos began with a chaos of fresh water, sea, and mist. From this confusion, pairs of gods were created representing the silt, the horizon, and the sky, as well as embodying male and female aspects; this echoes the creation of new land in the delta region of Mesopotamia, between the Tigris and Euphrates Rivers, in what is now Iraq. Following the initial creation, there was conflict between "order" and "evil," in the form of two deities, Marduk, the protector of Babylon, and Tiamat, the sea-goddess, representing chaos. Marduk killed Tiamat and created the Earth from her body, then created humans from the blood of another rebel god, Kingu. The struggle between two powerful deities mirrors the development and nature of human societies: both good and evil are present, while custom and authority create order, backed by the application of force, if necessary.

A Tanzanian myth, although quite different in detail from the *Enuma Elish*, similarly reflects the lives of the people who created it. In the beginning was "the Word," which was the creative force; there was also air and sky, a single tree, and some ants who lived on the tree. One day a great wind blew away a branch of the tree, carrying some ants with it. The ants continued to eat from the branch, but soon they ran out of food and were forced to eat their own excrement. The excrement grew into a huge ball, which became the Earth. The Earth eventually enveloped the tree, at which point the Word sent wind and water to Earth; the ants were subsequently destroyed in a flood. But the tree continued to grow, and its roots gave rise to plants on the Earth. The atmosphere then created animals and humans, each kind with its own voice. Fighting over food led to war between humans and animals. The war became so terrible that parts of the Earth broke away to form the stars, the Moon, and the Sun. The Sun glows because it came from a part of the Earth that was on fire when it was separated, while the Moon and stars are transparent disks through which the Sun's light shines. Some of the animals became the slaves of humans, while others remained wild and attacked people. At the end of the war, a sheep kept by humans leapt to the sky, where it killed the Word and ruled the cosmos, bringing thunder and lightning to the Earth. Because of this transgression, humans were punished by the gods when they dared to ask for help after their sheep caused the death of the Word. Humans were made lowly and warned that Earth shall be eventually consumed by fire.

Chinese cosmologies were diverse, but all emphasized that the doings of humans, particularly the mandarins of the court, were reflected in the heavens, a preoccupation not surprising in a country which was highly and hierarchically organized from very ancient times. Evil works would show themselves by disruptions in the sky, so the Chinese were keen observers, seeking auguries in the stars. Because of this, and because of the antiquity of the Chinese writing system, the Chinese annals constitute the longest unbroken records of the sky, which has proven important for some aspects of modern astronomical research. For example, the Chinese recorded in detail the supernova of 1054. This "guest star," as they called it, should have been sufficiently bright to have been visible for a short while during the day, yet it is completely absent from European chronicles. Perhaps the Europeans of the time, with their ironclad belief in a perfect, immutable heaven, ignored this strange phenomenon. The Chinese believed that the universe was huge, possibly infinite; the Sun, Moon, planets, and stars consisted of vapor and were blown about by a great wind. Many Chinese also accepted that the Earth moved, although like most peoples they did not realize that it rotated; they envisioned, rather, a smooth oscillation they believed to cause the seasons.

A fairly widespread Chinese creation myth tells of the giant Pan Gu. In the beginning, the cosmos was a great egg. For 18,000 years, Pan Gu slept within the egg. Finally he awoke and broke free, shattering the cosmic egg which had contained him. The lighter, purer elements rose and became the heavens, while the heavier, impure elements sank to form the Earth. Pan Gu maintained the separation of Earth and heaven with his body, supporting heaven with his head while his feet rested on the Earth. As the distance between heaven and Earth grew, Pan Gu grew to equal it. Finally heaven and Earth seemed securely in place, and Pan Gu died. His breath became the wind, his voice the thunder, and his perspiration the rain; his left eye was transformed into the Sun and his right eye into the Moon. His four limbs became the four directions, his trunk the mountains, while his blood ran as the rivers and his veins laid out roads and paths. His flesh created fields and soil, his skin and hair became the plants of the Earth, while his bones went into rocks and his marrow became the precious gems. After the sacrifice of Pan Gu, the Earth was a pleasing place, but the goddess Nu Wa, who had the face of a human but the body of a dragon, found it lonely. Stooping by the bank of a pond, she fashioned some amusing little creatures from mud. It was too tiring to create them constantly, so she endowed them with marriage and the capability to reproduce on their own. Later, a great battle between the spirit of water and the spirit of fire resulted in a catastrophe when the fleeing spirit of fire struck the great mountain that supported the western part of the sky. The heavens tilted and ripped apart, while the Earth fissured. Nu Wa melted the prettiest stones from the riverbeds to repair the holes and cracks, then killed a giant turtle, and cut off his legs to form the four pillars that support the sky. But the tilt to the west remained, and thus the Sun and stars slide down it, while on the Earth the water runs eastward into the ocean.

A common theme among many peoples indigenous to the Americas, both North and South, is the primacy of the four directions and of the number four in general. In some Native American cosmologies, these directions correspond to the four cardinal directions also utilized by Europeans, whereas in others, the major directions are those of the rising and setting of the Sun at summer and winter solstices. Some Native American cosmologies add the idea of the center to the directions, making five the principal mystic number. In many American cultures, particularly in the Southwest of the United States and in Mesoamerica, the world has been destroyed and recreated four or five times. Each world consists of layers, typically three; an Upper World of spirits and pure birds, a Middle World of humans and animals, and a Lower World of evil creatures.

The most elaborate cosmologies of this kind were found in Mesoamerica. In Aztec and Maya belief, the cosmos passed through four ages in the past. At the end of each "Sun," a great disaster destroyed the world. The current era, the Fifth Sun, began with the self-sacrifice by fire of Nanahuatl, the ugliest god, who was reborn as the Sun Tonatiuh. After this, the god Teucciztlan, whose courage had faltered at the great bonfire, threw himself into the flames and became the Moon. But the new Sun sullenly refused to rise until it was placated with the sacrifice of hearts and blood. Xolotl, a twin and aspect of the serpent god Quetzalcoatl, performed the sacrifices of all 1,600 deities present, then sacrificed himself; after this, the Sun rose. Quetzalcoatl ("plumed serpent") was, in many Mesoamerican traditions, a creator god who took several aspects, including the wind god Ehecatl and the monster Xolotl. He was a very important deity in Mesoamerica, especially to the Maya, who called him Kukulkan and associated him with the planet Venus, which as the Morning Star rises in the east just before the Sun, and as the Evening Star was thought to plunge sacrificially into the Sun just after it sets. Mayan astronomers kept meticulous records of Venus, and their observations enabled them to compute the length of the solar year to within a few seconds.

The importance of Quetzalcoatl played a pivotal role in the conquest of Mexico by the Spanish. Legends told of a great battle between Quetzalcoatl and his rival, the jaguar god Tezcatlipoca, after which Quetzalcoatl disappeared, promising to return from the east. The year 1519 corresponded roughly with the year Ce Acatl (One Reed) in the Azteco-Mayan calendar, the date-name associated with Quetzalcoatl as the Morning Star. When Hernán Cortés and his men appeared on the east coast of

Figure 1.1   El Caracol temple at Chichén Itzá in Mexico. Built by the Mayans around A.D. 1000, this temple was used as an astronomical observatory to record celestial events such as the rising and setting of Venus.

Mexico, the Aztec emperor Montezuma II took him for a representative of the returning Quetzalcoatl, and sent treasures of gold and silver from his capital of Tenochtitlán. The riches merely whetted the Spaniards' appetites for conquest, and they quickly made alliances with tribes held vassal by the Aztecs. The Conquistadores, hardly the salt of their own society, soon enough demonstrated by their behavior that they were not gods but merely an unfamiliar and especially rapacious kind of human. Yet Montezuma persisted in his delusion until Cortés appeared at Tenochtitlán and threw the pious emperor into prison. Montezuma was stoned by his own people for his failure to resist the invaders, and he died a few days later.

The example of Montezuma should make it apparent that cosmological considerations are not idle speculations but have significant consequences for the individual and society. Creation myths reflect the values and observations of the cultures that created them. Culture shapes the world view of its society, and conversely. The actions of the society's leaders, for good or ill, can be dictated by the prevailing cosmological mythology.

Even in our modern, industrialized societies, many unspoken cosmological assumptions mold our thinking. One of the most significant is the belief that the bounty of the universe is without limit. Although rarely articulated explicitly, this principle pervades many cultures, encapsulating the view that resources and opportunities are infinite. This point of view fits nicely with the attitude that the Earth is here for the benefit of humanity. As has become increasingly apparent in the twentieth century, such an outlook has important, and perhaps disastrous, consequences. Much of economic theory is founded upon the postulate that growth can continue indefinitely; that if we run out of some resource, a substitute can always be found. Yet it is clear that the illusion of boundless resources occurs only because the Earth is much larger than a human being, and geological timescales are much longer than a human lifespan. Our perceptions of the Earth, its history, and its contents, are skewed by our human limitations.

The perception of bounty continues to affect modern thought, sometimes in unexpected ways. Even those who recognize the limitations of Earthly resources often argue that space exploration and colonization can provide the materials and living space for a human population that grows without bound. Since the dawn of civilization, the human population has grown at an exponential rate, that is, at a rate in which the increase in population is always approximately proportional to the current size of the population. If humanity continued to reproduce at such a pace, eventually expansion into space could not occur fast enough to accommodate the new population. Indeed, in a relatively short time, by astronomical standards, we would reach the point at which all the particles in the observable universe would be required just to make up the physical bodies of people. Obviously, this is absurd. Nature will take care of our numbers, by its own methods, if we choose not to do so ourselves. As we achieve greater control over our immediate environment, we require an increasingly better assessment of how we fit into the greater world. We may be just as self-assured as Montezuma and ultimately just as surprised when we find that the way the world *is*, is quite different from how we *believe* it to be.

The "I" in the Center of Universe

Mythology casts ideas and aspects of the universe into human terms. In some respects, this is essential to our comprehension; we can deal with such issues only in terms we can understand, which must, by necessity, be of human construction. The universe is required for us, but we are mistaken if we invert this assertion and assume that humanity is essential to the universe itself. Yet the attitude that humankind occupies a special place in the universe is an overriding theme in almost all mythology. This is **anthropocentrism**, the belief that humans are important to the universe, which may well have been created especially for their purposes. To early peoples, observation seemed to support this viewpoint. The Earth is big, while the Sun and planets and stars seem small. All celestial objects appear to revolve around the Earth. Humans have power over plants and animals. The Earth provides the things that make human life possible, so it must have been created for us. (Early peoples did not generally consider the alternative, that humans require for life what the Earth was able to provide. That is, humans are adapted to the Earth, rather than the Earth being adapted for humans.) In contrast, some phenomena, such as the weather, remain beyond our power. Such things are important, both blessings and curses to humans. Weather brings rain for crops but also storms that destroy. Since anthropocentric cosmologies assume that humans are cardinal, these natural powers demonstrate that a still greater power exists, which is inflicting upon us the good and bad; if we may not be in charge, at least we occupy much of the attention of the powers that are. The aspect of "punishment" is often central, sometimes almost an obsession; humans did wrong and were punished, hence bearing forever the responsibility for death, decay, and imperfection.

Anthropocentrism is still a powerful concept in popular thought. Among the many possible examples, the most familiar may well be astrology—the belief that the planets and stars themselves relate to personal actions and destiny. Astrology is one of the oldest systems of belief known. The version that is common in Western countries is based upon a systematization of ancient lore by the Greek scholar Ptolemy, whose *Almagest* still forms the basis for the casting of horoscopes. Astrology is based upon the supposition that the stars influence our lives in mysterious ways, or foretell our destinies through their motions and configurations. Before there was any understanding of gravity or of the orbits of planets, some explanation had to be devised for the regularity of the celestial motions. In the prevalent anthropocentric view, those motions must surely have something to do with human events. In Greek and Roman belief, the planets were explicitly associated with specific gods and goddesses, whose names they still bear. The five planets known to the ancients, those that are visible with the unaided eye, are Mercury, Venus, Mars, Jupiter, and Saturn. The Sun and Moon were also considered "planets," making seven in all. The days of the week correspond to these seven planets. Sunday is the Sun's day, Monday the Moon's, Tuesday is ruled by Mars.[1] Wednesday is the day of Mercury, Thursday is governed by Jupiter, Friday corresponds to Venus, while Saturday belongs to Saturn.

The gods and goddesses of ancient Rome may have faded to amusing anachronisms, but astrology still holds the attention of many people. Who has not had the experience of reading the appropriate horoscope in the newspaper and finding that it applies perfectly? This is an example of a phenomenon well known to psychologists. People are much more likely to believe very general statements about themselves than they are to believe genuine specific psychological assessments. Moreover, there is the universal tendency to interpret vague descriptions in terms appropriate to the individual reading them. Finally, there is the phenomenon of *selective memory*, in which "hits" are remembered vividly, while "misses" are forgotten. Even so, if astrology had never been developed, it seems likely that people would be drawn to some similar system, such as "paranormal" phenomena, unidentified flying objects, "channeling" of spirits, past lives, and so forth. Many humans are unwilling to believe that their lives are subject to random occurrences; the wish to seek order in the cosmos is powerful.

Astrology may be easy to ridicule, but other common viewpoints are no less anthropocentric. For example, many believe that the land, the sea, the air, the animals, and plants exist primarily for our benefit, to be used as we see fit. Even if we do not believe in astrology per se, we frequently believe that we must "deserve" our fates; our goodness or badness determines the vicissitudes that befall us in life. We believe in cause and effect, but even more, we have a strong desire to believe that the causes of events are purposeful, not due to chance. If they are purposeful, they are understandable, predictable, and controllable. However, if the behavior of the universe were controlled or dictated by the needs and actions of some 6 billion humans, with their conflicting motives and desires, then we might as well return to the ancient myths of unpredictable gods.

The triumph of scientific cosmologies over the anthropocentric worldviews has not always been welcome; many people mourn the ancient universes in which humans played a clear and important role. The new universe seems, to some, a bleak and sterile place, while the ineffable universes of the past seemed awesome and meaningful. But this attitude often results from a confusion of the knowledge of a thing, or, more precisely, the model that allows us to know it better, with the thing itself. Science knows that crystals are highly ordered arrangements of atoms; quartz, for example, is simply a chunk of a common mineral, a major component of sand, which happens to have an ordered structure. It is the unusual large-scale symmetry of crystals, compared to most objects, that accounts not only for their rarity in nature, but also for their beauty.[2] But this leaves many people dissatisfied; they feel that the ability of polished crystals to refract light, which sometimes even makes the light appear to originate within the crystal itself, must mean that these humble rocks possess mysterious powers. Others, while not so extreme, still find the description of a diamond as a tightly bound collection of carbon atoms repugnant, as though this knowledge somehow takes away from the beauty of the gem. In reality, a diamond's sparkle depends mostly upon human knowledge and artifice to find its expression. A rough diamond is hardly more than a dull, gray pebble, with perhaps a bit of sheen. The trial-and-error experience gained over the centuries by diamond cutters has now been augmented by technology; a diamond to be cut is often subjected to a micrograph to determine planes along which it will most readily fracture. The various standard cuts must be carefully prepared in order that the stone show its greatest fire. It is knowledge that elicits the greatest beauty of a diamond.

Thus the knowledge that we acquire need not preclude awe. Rather than the constricted, unchanging universe imagined by our ancestors, we now find ourselves in a dynamic and evolving universe too large for any real comprehension of its size. If some people might be distressed that humans now seem so small and insignificant, science can only respond that we are nevertheless a part of this grand cosmos, and we should feel privileged to have the ability to appreciate its true majesty. If we have been forced to abandon our anthropocentric models, in return we have gained a far grander home.

A New Explanation

In the beginning there was neither space nor time as we know them, but a shifting foam of strings and loops, as small as anything can be. Within the foam, all of space, time and energy mingled in a grand unification. But the foam expanded and cooled. And then there was gravity, and space and time, and a universe was created. There was a grand unified force that filled the universe with a false vacuum endowed with a negative pressure. This caused the universe to expand exceedingly rapidly against gravity. But this state was unstable and did not last, and the true vacuum reappeared, the inflation stopped, and the grand unified force was gone forever. In its place were the strong and electroweak interactions, and enormous energy from the decay of the false vacuum. The universe continued to expand and cool, but at a much slower rate. Families of particles, matter and antimatter, rose briefly to prominence and then died out as the temperature fell below that required to sustain them. Then the electromagnetic and the weak interaction were cleaved, and later the neutrinos were likewise separated from the photons. The last of the matter and antimatter annihilated, but a small remnant of matter remained. The first elements were created, reminders of the heat that had made them. And all this came to pass in three minutes, after the creation of time itself. Thereafter the universe, still hot and dense and opaque to light, continued to expand and cool. Finally the electrons joined to the nuclei, and there were atoms, and the universe became transparent. The photons which were freed at that time continue to travel even today as relics of the time when atoms were created, but their energy drops ever lower. And a billion years passed after the creation of the universe, and then the clouds of gas collapsed from their own gravity, and the stars shone and there were galaxies to light the universe. And some galaxies harbored at their centers giant black holes, consuming much gas and blazing with exceeding brightness. And still the universe expanded. And stars created heavy elements in their cores, and then they exploded, and the heavy elements went out into the universe. New stars form still and take into themselves the heavy elements from the generations that went before them.

And more billions of years passed, and one particular star formed, like many others of its kind that had already formed and would form in the future. Around this star was a disk of gas and dust. And it happened that this star formed alone, with no companion close by to disrupt the disk, so the dust did condense and formed planets and numerous smaller objects. And the third planet was the right size and the right distance from its star so that rain fell upon the planet and did not boil away, nor did it freeze. And this water made the planet warm, but not too warm, and was yet a good solvent, and many compounds formed. And some of these compounds could make copies of themselves. And these compounds made a code that could be copied and passed down to all the generations. And then there were cells, and they were living. And billions of years elapsed with only the cells upon the planet. Then some of the cells joined together and made animals which lived in the seas of the planet. And finally some cells from the water began to live upon the rocks of the land, and they joined together and made plants. And the plants made oxygen, and other creatures from the seas began to live upon the land. And many millions of years passed, and multitudes of creatures lived, of diverse kinds, each kind from another kind. And a kind of animal arose and spread throughout the planet, and this animal walked upon two feet and made tools. And it began to speak, and then it told stories of itself, and at last it told this story. But all things must come to their end, and after many billions of years, the star will swell up and swallow the third planet, and all will be destroyed in the fire of the star. And we know not how the universe will end, but it may expand forever, and finally all the stars will die and the universe will end in eternal darkness and cold.

Is this a myth? If we define a myth as a narrative of explanation, it would qualify. How does this myth differ from others? For one thing, it is highly detailed. The fanciful description above is extremely condensed; the complete version of this story occupies the remainder of this book. In addition, it is not overtly anthropocentric. People play only a very limited role, even though this description was developed by humans. Nevertheless, if all you knew of this explanation was a tale such as that written above, you might have difficulty in distinguishing it from a story of ants in the tree of life. But this story differs fundamentally from the earlier myths. The most important distinction is the way in which this explanation was developed. It was based upon many centuries of observations of the universe and its contents. It draws upon the experience and thoughts of generations of thinkers, but always the most significant factor has been the accumulation and interpretation of observations. The story is held to a set of stringent constraints; it must explain known facts, and it must hold together as a coherent narrative, all the parts fitting like pieces of a grand jigsaw puzzle. How humans have arrived at this narrative, what it means, which aspects of it are more certain and which less so, and how it is to be judged, are the subject of this book. It is a lengthy story that will unfold over many chapters, but let us begin with the most fundamental basis: the establishment of criteria by which our narrative of the universe can be evaluated.

The Scientific Method

Over the past 400 years, a new viewpoint has come to fruition: the scientific viewpoint. At first glance, this may seem to be no better than the mythology of our ancestors; it is just another belief system. However, there are significant differences between scientific and mythological explanations. In science, the ultimate judge is the empirical data, the *objective* observations. The truth, whatever it may be, is independent of humanity; but it can be known and understood, at least in approximation. The results of a set of observations, that is, of an **experiment**, must not depend upon who makes the observations. The validation of any theory lies in its ability to make predictions that can be tested by further experiments. Regardless of the internal consistency of a theory, or its philosophical or aesthetic appeal, it is the data that judge the success or failure of that theory.

The realization that the universe is, at least in a practical way, knowable, developed only slowly in human thought. Although many cultures contributed to this dawning, it appeared in the first coherent way among the ancient Greeks, during the age of the philosophers some 3,000 years ago. The Greeks incorporated into their system of logic the formal connection between a cause and its effect, introducing the concept, novel for the time, that a phenomenon could have a natural, consistent cause, and that cause could be identified by rational thought. The Greeks were eventually conquered by the Romans, who held the Greek philosophy in the greatest esteem but had little interest in furthering it themselves. After a lengthy decay, the Roman Empire itself finally collapsed in the fifth century, ushering in the Dark Ages in Europe and extinguishing nearly all the memory of the achievements of the ancients. During the Dark and Middle Ages rational thought was almost entirely absent in Europe. The knowledge gained by the Greek philosophers was preserved primarily within the Islamic world until the Crusades and increased travel and trade brought Europe into contact with other cultures once again. The Greek writings were rediscovered early in the thirteenth century, beginning with Aristotle. Although, as we shall see, Aristotle did much damage to scientific inquiry, both in his own time and during the late Middle Ages, his texts did bring the concepts of logic and inference back into European thought, helping to pave the way for the Renaissance. During the Renaissance, and the Enlightenment that followed it, European science took shape and matured.

Science gradually became systematized. The British philosopher Sir Francis Bacon developed a procedure for scientific inquiry during the last decades of the sixteenth century. Bacon's methods were further refined and codified in the nineteenth century by a subsequent British philosopher, John Stuart Mill. Mill provided a formal approach to establishing inductive inferences of any kind, but science is one of its most important applications. It should be emphasized that almost all scientific hypotheses are *inductive*, not deductive. Induction is the drawing of general conclusions from an examination of particular instances, whereas deduction is the inference of particulars from general principles; the distinction between the two is often ignored in popular usage, but the difference is significant if we are to understand clearly what our observations can tell us. Since we cannot inspect every particle of matter in the universe, and our scientific laws must necessarily be based upon the data available, we must generalize from our limited experience to all the universe. Unlike deductive conclusions, which proceed from the general to the specific and can be rigorously and decisively proven, inductive hypotheses go from the specific to the general, and if the number of possibilities is too large for us to examine all of them, as is usually the case, an inductive hypothesis cannot be conclusively proven. We can, however, use deduction to test repeatedly these general hypotheses by developing specific predictions for comparison with observation.

Despite the fundamental limitations of the inductive process, science has made great progress in building a consistent and comprehensible picture of the universe. The occasional failure of established hypotheses has never overturned the scientific edifice completely; instead, such failures lead to new and better knowledge of the way in which the universe works. Methodology can guide the construction of a valid (in the inductive sense) hypothesis from the known data but cannot give a blueprint; often the great scientific hypotheses are the result of genius, hard work, or even simple luck.

Over the past 200 years, a systematic approach has developed for the treatment of scientific problems. This so-called **scientific method** is a method for testing and verifying scientific hypotheses; it proceeds, at least in principle, by several steps. First comes the gathering of **data**. We cannot build scientific explanations without careful objective observations of the phenomenon in question; this is one of the most important distinctions between scientific and unscientific explanations. Study of the data enables the scientist to look for patterns, for similarities with other phenomena, and so forth. Once some unifying concept has been found, it may be phrased as a **hypothesis**, a working explanation for the phenomenon that can lead to further observation.

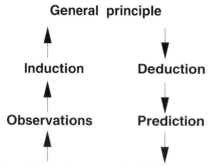

Figure 1.2 The process of induction moves from observations of specific events to a general principle. The general principle can never be "proved" since all specific instances cannot be observed. However, the principle can be tested through deduction, by which particular instances following from the general principle can be inferred.

In order to be scientific, a hypothesis must have five characteristics. First, it must be **relevant**. This may seem self-evident, but it is significant. The hypothesis should be related to some observed phenomenon, not merely something invoked because the theorist happens to like it.

Second, the hypothesis must be *testable* and potentially **falsifiable**. That is, it must be possible to make observations that could confirm or, even better, refute the hypothesis. The importance of this characteristic cannot be overemphasized; indeed, it may be regarded as *the* distinguishing feature of a scientific explanation. The hypothesis that the planets are controlled by spirits was accepted for centuries, but it is not scientific because it cannot be tested; there is no observation that could disprove it. The Newtonian hypothesis, that planets are controlled by a force emanating from the Sun that causes them to move in specific ways, is falsifiable; if a new planet, or other orbiting body, were discovered and found not to obey the laws that Newton had discovered, his hypothesis would be disproven. On the other hand, if the new body were found to obey Newton's laws precisely, it would add evidence for the validity of the hypothesis but would not prove it.

Falsifiability unambiguously distinguishes scientific from nonscientific explanations. The philosopher of science Karl Popper put forward the proposition that the criterion for the scientific status of a theory is the potentiality that the theory may be falsified. Pseudoscience is often based on observations and may cite much "confirming evidence" but never permits refutation. Either the contrary data are ignored, or new details are continually added to the theory in order to explain all new observations. Seen in this light, the scientific status of a theory is granted not so much by its explanations but by its prohibitions: the theory says what cannot happen, and if those things are observed, then the theory is wrong.

A scientific hypothesis must also be **consistent** with previous established hypotheses. If a known hypothesis explains a phenomenon well and has passed many experimental tests, we would be ill advised to abandon it merely because a newer and shinier explanation might appear. This principle is often little appreciated by the public or by pseudoscientists who cite Einstein and Galileo as iconoclasts who refuted established science. In fact, Einstein's theory of relativity would not have been accepted had it not been consistent. Newton's laws of motion were well established even during his lifetime as a very good explanation of mechanics. Over the next three centuries, they were verified time and again. Yet there remained one nagging problem, which Albert Einstein set out to solve. In doing so, he was forced to give up notions about the universe that had been cherished for centuries but which were not essential to understanding the Newtonian observations. The special theory of relativity revolutionized our conceptual view of space and time and showed itself to be a more *complete* theory of motion, in that, unlike Newton's laws, it was applicable at all speeds, and it made electromagnetics consistent with mechanics. The special theory of relativity is, nevertheless, fully compatible with Newtonian mechanics and can be shown to reduce to the Newtonian theory for all material motions at speeds well below the speed of light. This is precisely the regime in which Newton's laws were known to be valid to within the accuracy of the data available. Einstein did not *refute* Newton, but rather he modified and *extended* the laws of mechanics into previously unexamined and untested domains.

The criterion of consistency is important but not absolute. It *is* possible for an old theory that is well accepted to be simply wrong, and a new one replaces it completely; but such incorrect theories survive only in the absence of data. A good example of this is the caloric theory of heat. For many years, heat was believed to be some sort of invisible fluid, which flowed from a hotter to a colder body. The caloric theory was able to explain many common properties of heat reasonably well. It was not until more careful measurements were made and better data were collected, beginning with Count Rumford's observations of cannon boring at the end of the eighteenth century, that the theory was called into doubt. In 1799, Sir Humphrey Davy conducted a **crucial experiment**, one which has the power to decide between two competing theories on the basis of a single incompatible prediction. Unfortunately, his experimental design was somewhat lacking, and his results were not convincing. But the way was shown, and within 50 years several scientists, most especially Sir James Joule, developed the kinetic theory of heat, which is accepted today. The new theory was incompatible with the old, and the caloric theory was discarded.

A fourth criterion for a scientific hypothesis is **simplicity**. This is a rather vague and subjective criterion, to be sure, but it has guided the development of many theories. All other things being equal, the simpler explanation is favored, an assertion often known as **Occam's razor** for the medieval English philosopher William of Occam (or Ockham), who asserted that "entities must not be needlessly multiplied." A good theory does not require a special rule for each observation.

The fifth important criterion for judging a scientific explanation is its **predictive power**. Predictive power is not quite the same thing as falsifiability, although the two are interrelated. Predictive power refers to the ability of the hypothesis to predict new, previously unobserved phenomena. Similar to this, and part of the same criterion, is the **explanatory power** of the hypothesis, which is a quantification of the number of facts the hypothesis can encompass and explain. Given two otherwise similar hypotheses, the one with greater explanatory power is generally preferred. Predictive power is even better, for then the hypothesis can be bolstered if the new phenomenon is observed, or discredited or even disproved if the phenomenon is not observed, or is observed but behaves contrary to the prediction of the hypothesis.

In order to be accepted, any new hypothesis must represent an improvement. It must explain more facts, or provide a better explanation of those existing, than the older theory. Although great theories are often advanced by individuals, science as a whole is a social activity. It is not the brilliance or authority of one person that forces the acceptance of a hypothesis. Hypotheses, like clothing, may come into fashion or fall from favor for all-too-familiar human reasons, such as dominance by one powerful individual or a scientific fad, but it is inevitable that over time, only those explanations that can win the acceptance of the scientific community prevail. And by the communal nature of science, such hypotheses must fit in with the overall picture in order to win any such contest.

If a hypothesis becomes especially well established and survives tests that could have refuted it, it may be elevated to the status of a **theory**. A theory, in strict scientific usage, is a hypothesis that is sufficiently accepted and which shows enough explanatory power to be strongly confirmed by experiment. It is *not* a conjecture, as the word "theory" often connotes in popular usage, where it has little more import than an opinion. Occasionally, an especially well confirmed theory is called a **law**, but this usage has diminished considerably in the past century. The terminology is by no means consistent, and in any case, most scientific explanations, being inductive, are necessarily hypotheses with lesser or greater degrees of verification. However, in no case is a scientific hypothesis or theory a mere guess. It is always founded upon a careful methodology for correct inductive inference, and it is judged by the criteria we have described. Ultimately, the data decide. No matter how beautiful the theory, its success or failure is determined by how well it explains our observations, both those already known and those that will come from experiments yet to be made.

The progressive nature of science should also make clear that it does not seek a revealed or absolute truth but instead looks for models of reality. A **model**, in this context, refers to the coherent description established to explain a phenomenon. It is more or less equivalent to a theory; that is, it is a well-confirmed hypothesis or set of interrelated hypotheses. For example, the big bang model of the universe is a mathematical construction that provides illumination and interpretation for the data we collect. This does not mean that a model is a fiction having nothing to do with reality; on the contrary, in modern science, a model represents the best description of the phenomenon which we can devise, and insofar as it succeeds at reproducing the observations, it surely must touch some facet of reality. It does mean that a model never claims to *be* reality. If better data invalidate part or all of our model, we must replace it appropriately. The failure of a model does not represent a failure of science; science fails only when scientists cling stubbornly to a model that has clearly ceased to be the best possible.

A model must never be confused with the entity it represents. No matter how good a cosmology we may eventually develop, it is still a product of the human mind, yet no one would claim that the universe is a human construction. Humans have strong intellectual gifts, especially with our unique ability to consult with one another, but our brains are still finite; it may be that some aspects of reality are beyond our grasp. Even if physicists develop an ultimate theory that explains all that can be known about elementary particles, this will tell us little about how consciousness arises, or about a host of other complex problems. Reality may be a fleet runner we can never overtake, but which we can approach ever closer.

Despite the grandeur of its subject, modern cosmology is a science and obeys the rules of scientific method. Cosmologists formulate hypotheses and appeal to data to test them. Cosmology is primarily an observational science. We cannot arrange to perform our own experiments on the universe, controlling them as we like, but must be content to observe what we happen to see. Cosmologists attempt to tie those disparate observations together with physical theory to create the best cosmological hypotheses possible. These new hypotheses may then suggest new observations, as a good scientific hypothesis should do, and from those observations we may strengthen or discredit the explanation. Thus we humans make cosmological progress despite our confinement to the immediate vicinity of a small planet orbiting a modest star in a run-of-the-mill galaxy. Yet even from our restricted vantage point, we shall find a universe more wondrous than our ancestors, with their capricious gods and their preoccupations with geometrical or mystical perfection, could ever have dreamed.

KEY TERMS

cosmology	universe	astronomy
anthropomorphism	myth	anthropocentrism
experiment	scientific method	data
hypothesis	relevant	falsifiable
consistent	crucial experiment	simplicity
Occam's razor	predictive power	explanatory power
theory	law	model

Review Questions

1. For at least one myth, either one from the text or one of your own choosing, identify the major theme(s) and explain how the myth fitted the social and political circumstances of the people who developed it.

2. Give an example of how one or more cosmological assumptions have influenced the behavior of modern political leaders in an industrialized nation.

3. Find your horoscope for one particular day in a newspaper. Keep track of your activities for the day, observing any occurrences which could appear to be fulfillments of the horoscope. Did anything happen that was explicitly contrary to the predictions?

4. Repeat the activity in question 3, but for a horoscope that is *not* yours, and is chosen randomly from the horoscopes separated from yours by at least two "houses." (Recall that the ordering of the houses is circular; Aquarius follows Capricorn.) Ideally, this and the preceding exercise should be done with the help of a friend, so that you do not know whether the horoscope you are given corresponds to your birthdate or was randomly selected.

5. Describe at least two examples of anthropocentric beliefs that are still widespread.

6. What is an experiment, and what is its role in science?

7. Explain the distinction between inductive and deductive reasoning.

8. Describe the five major criteria for evaluating scientific hypotheses. Which are most important? Why?

9. Define the word *theory* as it is used in science. How does this usage differ from a common everyday meaning of the word?

10. Choose an example of a pseudoscientific theory and explain how it fails to be testable or refutable.

11. What is the ultimate arbiter of "truth" in science? How does this distinguish science from other systems?

Notes

1. In English, most of the names of the days of the week come from Norse gods and goddesses who played roles similar to those of the Graeco-Roman deities.

2. All true solids are, in fact, crystalline, but usually they consist of aggregates of many tiny crystals. Only occasionally does a crystal naturally grow large enough for us to appreciate its symmetry without a microscope.

CHAPTER 2

Cosmology Becomes a Science

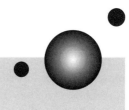

For thousands of years, the universe that occupied human minds was small, limited by human senses and abilities. The world seemed to end at the horizon, and few traveled far from the towns of their births. The heavens were the realm of gods, beyond the understanding of mortals. From such a narrow perspective, it is not surprising that the universe appeared to be dominated by the Earth. The stars were held to be eternally fixed in their positions on the celestial sphere. The "wanderers," or planets, known from ancient times and by nearly all cultures as entities distinct from the stars, were thought to be under the control of, if not literally the embodiment of, gods or spirits. The Earth was apparently motionless, while the sky and planets, including the Sun and Moon, revolved around it. But if the Earth is still and everything else moves, is it not perfectly reasonable to conclude that the Earth is the center of the universe?

This was the dominant cosmology in Europe from ancient, perhaps prehistoric, times until the close of the Middle Ages. Then, over an astonishingly brief span of less than two centuries, the prevailing worldview changed dramatically and irrevocably, bringing about what is often called the *scientific revolution*. Over the past 300 years, further elucidation of the new cosmology has continued, bringing us to our modern models. The new universe that has emerged might seem as strange to Isaac Newton, as would his to ancient philosophers.

Greek Cosmology

More than 2,000 years ago, Greek philosophers developed a sophisticated system of rational thought, establishing the basic rules of deductive logic that are still followed today. Some of the early philosophers were also scientists, performing feats of astronomy that, in light of their extremely limited ability to make quantitative observations, seem impressive even now. When Greek culture was temporarily forgotten, European thought degenerated into the superstition and fear of that dismal period known as the Dark Ages. The rediscovery of Greek culture, as well as the discovery of the achievements of other Mediterranean cultures, led ultimately to the Renaissance.

Today we acknowledge that Western science has its roots in Greece. The Greeks did not invent their system from nothing, but were influenced by neighboring peoples; however, it was they who were chiefly responsible for establishing the basic principles of scientific inquiry. Among their accomplishments was the identification of cause and effect. This may seem obvious to us now, but it was an important conceptual advance and an essential prerequisite for scientific thought. The Greeks realized that it was possible to observe a natural phenomenon and to seek an explanation for the observation. It was even possible to understand nature in precise, mathematical terms, which, to the ancient Greeks, meant geometry. To move from an understanding of Earthly phenomena to a grasp of the universe is then merely a matter of scale. If we can measure the size and shape of the Earth, we can do the same for the heavens. With the concepts of cause and effect in place, the world is no longer random and capricious; instead, it is ordered and predictable.

The predominant feature of the mainstream Greek cosmology was the centrality of an unmoving Earth. As remarkable as it may seem, the spherical shape and the size of the Earth were well known to the Greeks. Despite the restricted ability of the ancients to travel, they were aware that the view of the constellations, at the same time of year, changes as one moves north or south. More evidence was found in the fact that ships with tall masts disappear as they move away from the coast, but not in a proportional manner; first the hull drops from view, then later the mast. This would not happen on a flat plane, as geometers could appreciate; thus they concluded that the surface of the Earth must be curved. Furthermore, the Greeks had also deduced the cause of lunar eclipses and realized that the shadow of the Earth on the Moon was curved. Once the shape of the Earth was determined, it became possible to ascertain its size. The Greek geometer Eratosthenes (ca. third century B.C.) computed the diameter of the Earth by measuring the altitude of the Sun in the sky at two different locations on the Earth at noon on the summer solstice. With the reasonable assumption that the Sun's rays were parallel to one another, he was able to use these measurements to obtain a result that historians believe to be quite close to the correct figure. To surround this spherical Earth, the ancient Greeks supposed that the sky too was a physical sphere; they believed that it hung overhead, relatively close to the Earth.

An important factor in establishing the Greek cosmos was the conclusion that the stars moved, whereas the Earth did not. Motion had long been recognized in the heavens; the patterns of stars changed with the seasons, while the planets, including the Sun and Moon, moved among the stars. But for observers confined to its surface, all available evidence indicates that the Earth itself does not move. If the Earth, rather than the celestial sphere, were turning, then near the equator, a point on the surface would have to be moving at the incredible speed of nearly 1600 kilometers per hour. Surely such speeds would be perceptible! Would not such a great motion generate winds with enormous velocities? Moreover, how could someone jumping from the surface of the Earth land in the same spot from which he leaped? When you stand upright, you feel no sensation of motion. Drop an object; it falls straight down. The concept of a moving Earth also seems to conflict with the observation that moving objects tend to come to rest and to remain at rest unless impelled. How could the Earth sustain movement, when all other Earthly motions rapidly come to a halt?

To the ancients, arguments such as these established unambiguously that the Earth was motionless. The next task was to describe and explain the heavenly motions. The model of the celestial motions must do more than provide a general description. It should make detailed predictions that would be as accurate as the observations. The difficulty, as any casual student of the heavens rapidly comes to appreciate, is that the motions in the sky are intricate. The prejudices of the ancient Greek geometers for certain figures further complicated their construction of a model. The scientific process has always required an interaction between ideas (theory) and observations (data and experiment). Today we regard accurate observations as supreme; theory must give way if need be. The Greeks felt the opposite to be true. Theory, which sprang from pure rational thought, was considered to be supreme over observations, which were sullied by the unreliability of human experience and senses. Thus the early Greek scientists felt no qualms about forcing the universe to conform to their philosophical ideals.

The first systematic cosmology we shall consider was developed by the philosopher Plato and his student Eudoxus in the third century B.C. Eudoxus set out to create a system that adequately agreed with real observations while preserving accepted ideas about motion and geometry. The resulting cosmological picture, later considerably refined by Aristotle and others, was based upon the sphere, which, according to Greek philosophy, was the most perfect of solid geometric forms; correspondingly, the circle was the most perfect curve. The sphere has the appealing property that it encloses the largest possible volume for a given surface area, an aesthetic much appreciated by the Greeks. Justification for the spherical universe was also found in the recognition that the Earth itself was a sphere; surely this shape was no accident but reflected the geometrical design of the universe. We should not criticize the Greeks for relying so heavily upon their notions of symmetry, for even modern physicists profess a great appreciation for symmetry in their theories. In modern science, however, symmetry is a guide and not an arbiter, and the symmetries invoked are often quite subtle.

The obvious approach to constructing a cosmos based upon spheres, centered on a spherical Earth, requires separate spheres for the Sun, the Moon, and each of the planets; only one sphere is required for all of the fixed stars, which move as a unit. Unfortunately, the motions of everything but the fixed stars are more complex than can be accommodated within such a straightforward model. Even at the time of Eudoxus, observations were adequate to rule out a simple circular motion of the planets about a stationary Earth. Multiple spheres are needed to account for the various observed motions of even one celestial body. For example, the Sun exhibits its familiar daily motion through the heavens, for which a single sphere, rotating on a 24 hour schedule, can account; but the Sun also has a longer seasonal motion as it moves north and south of the equator. The seasons, therefore, require a second sphere. The more complex motions of the Moon and the planets required even more spheres. In order to fit the observations, Eudoxus was obliged to introduce 27 different celestial spheres, each with a different rate of rotation and orientation of its axis. The result was less geometrically beautiful than it was practical; it fitted the observations of the day reasonably well. Eudoxus' model set the pattern for future refinements.

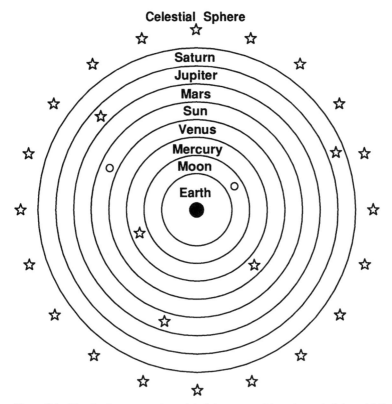

Figure 2.1 The simplest geocentric model of the cosmos. The universe is finite, with Earth at the center, surrounded by the spheres of the Sun, the Moon, and the planets. The sphere of the stars lies at the outer edge. The motions of the Sun, Moon, and planets cannot be adequately described by a single sphere for each. The models of Eudoxus and his successors postulated multiple spheres for each of these bodies.

Aristotle

The most famous of Plato's students was not Eudoxus, nor was Eudoxus the only one who pondered cosmology. By far the most influential of the ancient Greek philosophers was another pupil of Plato, Aristotle (384–322 B.C.). Aristotle wrote widely and voluminously on nearly every conceivable philosophical issue of his day. Much of what we know of Greek cosmology comes from his writings, and from later members of his school, who edited and revised his texts. Although he made many original contributions to a number of fields and was one of the first to develop a theory of biology of any kind, most of the elements of Aristotelian cosmology are common to other cosmologies of the era. The writings of Aristotle became particularly influential because he justified his cosmology on rational grounds.

Aristotle's cosmological model is particularly important because he developed it within a general physical theory. This remains the basic approach in modern scientific cosmology. If the universe has certain properties and the objects within it behave in certain ways, then there must be principles behind these behaviors, specifically, natural laws. The universe embodies these laws, and they can be discerned by humans if enough observations are made. It is sometimes claimed that Aristotle developed his theories by thought alone, without regard to observations, but this is not true. He made observations, to the best of his ability, then attempted to reason from those observations. In this respect, his work represented a break with the earlier Platonic school of thought, which held that truth lay only with ideas. To Plato, observations were misleading because the physical world was at best a pale manifestation of the truth; only pure geometry could claim to represent the ultimate reality.

An important element of Aristotle's cosmology was his theory of motion; today we call this branch of physics **mechanics**. Motion, taken mostly for granted before his time, presented many questions to Aristotle. Why do objects on the Earth have the tendency to move as they do with respect to the Earth? Why do objects fall when dropped, and why do stones sink in water, while bubbles rise? It must be, thought Aristotle, because it is in their fundamental nature to move so. In many ancient theories, all objects are composed of the four basic "elements" of earth, water, air, and fire. In the Aristotelian view, each of these elements was believed to move differently: earth toward the center, fire away from the center (flames rise), while water and air occupy the space between. Air bubbles up through water, but rocks sink. Consequently, objects of different compositions fall at different rates. An object containing a higher proportion of the lighter elements air or fire, or both, would fall slowly, whereas an object consisting mostly of earth would fall quickly. The conclusion that various bodies fall at different rates was consistent with casual observation. The composition or "nature" of the object thus determined its mechanical behavior. All things sought to move to their natural place in the cosmos. Because of this, all motion must be with respect to the basic structure of the cosmos; specifically, Aristotle proposed that Earthly objects move in a straight line, that is, *linearly* with respect to the center of the universe. The Earth is a sphere, and so objects falling straight down are actually moving toward the center of the sphere, as Aristotle realized. Therefore, the center of the cosmos must lie at the center of the Earth.

A cosmology which places the Earth at the physical center of the universe is said to be **geocentric**. Quite apart from Aristotle's rationalization on the basis of a physical theory, the geocentric model of the universe fitted perfectly with the anthropocentric attitudes that dominated most human thought. Indeed, the geocentric model remained the mainstay of cosmology until scarcely 300 years ago.

Natural motion is thus defined. But what of other motions? Aristotle's law of motion incorporated the idea that **force** caused a deviation from natural motion, a significant advance in understanding. The concept of "force" is intuitive; it is a push or a pull, an action by one thing on another. This concept still remains a fundamental part of modern mechanics, although in a much more quantitative form. Aristotle observed that a force is required to initiate motion in a stationary object and that Earthly motions tend to die out soon after they are so initiated. For example, a rock thrown, no matter how energetically, soon falls to the ground and stops. Aristotle proposed that a force is required to make an object move in any manner different from its natural motion. A horse must continually pull on a cart to move it. Similarly, an arrow shot from a bow must experience a sustained force during its flight, or so Aristotle thought. Aristotle believed that objects in flight, such as an arrow, were somehow pushed along in their paths by the air, with a kind of highly localized wind.

Although ground-breaking, Aristotle's law of motion was erroneous. Viewed with the hindsight provided by modern physics, we can see where he went astray. Aristotle's difficulties arose because he only partially grasped the concept we now call **inertia**, the tendency of a body to resist changes in its state. He realized that a body at rest will remain at rest unless impelled by a force, but he missed the other, equally important, part of the law of inertia: a body in motion in a straight line will remain in that state unless a force is exerted. From modern physics, we know that a force is required to produce a *change* in a state of motion. To Aristotle, continuous motion required the continual application of force; he could not conceive of the possibility that an Earthly object might travel forever on its own.

But what of the heavenly motions? In contrast to Earthly motions, celestial motions *do* continue indefinitely. The motions of heavenly objects cannot follow straight lines, since straight lines would end at the edge of the universe, and thus all such motion would ultimately be finite. Hence there must be two separate types of natural motion: straight-line (linear) limited motion in the Earthly realm and continuous circular motion in the heavens. This was one of Aristotle's most influential axioms: the *primacy of circular motion* in the heavens. It made a certain geometric sense: lines are of finite length, whereas a circle closes back upon itself and has neither a beginning nor an end. Because the heavenly bodies had a different natural motion, circular and eternal, they could not be composed of earthly materials, which could move only linearly toward their proper place in the cosmos, as determined by their composition. Instead, celestial objects were composed of *ether*, a fifth element. Since they were already in their proper place with respect to the center of the cosmos, they moved in perfect circles. Heavenly bodies would thus continue to move indefinitely without the action of any force.

In the ancient cosmos, the heavens consisted of distinct objects: the Sun, the Moon, the five planets, and the stars. These were all accommodated in the cosmology of Aristotle. The Earth was surrounded by the nested, crystalline (transparent) spheres of the heavens, to which were attached the celestial bodies. Whereas Eudoxus apparently thought of the spheres as mathematical entities only, useful for description but not to be taken literally, Aristotle gave them physical reality and a composition. These spheres rotated around the Earth, carrying the heavenly bodies with them. Aristotle argued that the ethereal heavens were eternal and unalterable, perfect in their structure and unchanging. The Earthly world below changed, but the heavens did not. Any apparent change in the heavens must therefore be linked to the Earth. Aristotle argued on these grounds that meteors and comets were manifestations in the upper atmosphere of the Earth.

The spherical universe of Aristotle was consistent with the physical and philosophical reasoning of his time, but the final model lacked much of the aesthetic quality that had originally motivated the Greek philosophers. Alas, the geometrical beauty of spheres and perfect circular motion encountered the obstacle that plagues all theories: better observations. In order to meet the challenge of the observations of the day, Aristotle was obliged to postulate 55 separate spheres to account for the motions of a far smaller number of bodies.

How large was Aristotle's grand construct? The size of the universe was limited by its fundamental geocentric property. The heavens were moving, not the Earth.

Consequently, the universe must be finite, for an infinite universe rotating around a center would necessarily travel an infinite distance in a finite interval of time. In Aristotle's cosmology, the distance to the stars is very small, so as to prevent them from moving at unreasonable speeds. The entire Aristotelian universe would fit comfortably into a region smaller than that defined by the Earth's orbital radius around the Sun. This finite universe had an edge, but it could never be reached because any motion toward the edge would shift from linear to circular as the traveler approached the heavenly realm.

Even though space was assumed to have an edge, Aristotle apparently could not imagine an edge to time, and so he took the point of view that time must be infinite, without beginning or end. The Greeks were aware that recorded history did not stretch back to infinity and that things changed. This fitted into the philosophy of the Earth as imperfect, made of four base elements, while the heavens were composed of eternal, perfect matter. The Earth changed, while the heavens did not. Conversely, the Earth did not move, while the heavens did. Few philosophers of the time, or even much later, seemed inclined to question why the center of a perfect universe would be located on an imperfect, woeful planet.

Aristotle's cosmology was very much a product of its time, and of its author. While it cleaved tightly to the ancient view of the Earth and sky in its insistence upon an unmoving, central Earth and a perfect heaven, it still contained important, original contributions. As we have suggested, the supreme accomplishment of Aristotelian cosmology was the argument that the universe could be described in terms of natural laws that could be inferred through rational thought. Aristotle founded the science of mechanics and developed the concept of "force" into something that was at least vaguely systematic. In Aristotle's cosmology, the structure of the universe is inextricably linked to physics and to the definition of natural motion. Remarkably, this is true for modern cosmology as well. Cosmology cannot exist as a science without physics; the general structure of physical theory affects the underlying cosmology, and vice versa. As humanity's understanding of physics improved, first from Newton and later from Einstein, the universe changed as well.

Unfortunately, Aristotle's work was also fundamentally flawed. As we have mentioned, the Aristotelian laws of motion did not include the correct concepts of natural motion or inertia. Also influential, but quite wrong, was his separation of the universe into Earthly and celestial realms, governed by separate laws and composed of separate elements. These misconceptions, especially the demand that celestial motion be circular, would confuse and confound physics, astronomy, and cosmology for seventeen centuries.

We cannot blame Aristotle too much for developing a physics that was largely incorrect. The fault in his method was that his observations were often misleading. For example, he did not understand phenomena such as air resistance, nor was he able to recognize that if all objects on a surface are moving with that surface, the motion will be difficult, perhaps even impossible, to detect locally. Perhaps most important, he did not recognize his own limitations, both as an observer and a theorist. Aristotelian physics matched the intuitive beliefs of most people and suited their philosophical leanings as well. Consequently, the geocentric theory was retained and enshrined, eventually reaching the point of religious dogma during the Middle Ages.

Not only the Aristotelian cosmology was venerated during the Middle Ages; the corresponding Aristotelian physics of motion was further elaborated into the *impetus* theory. In this view, objects moving on the Earth are propelled by an "impetus," a vaguely defined, traveling, generalized force. For example, the impetus theory holds that a rock shot from a catapult is endowed with some amount of impetus that continues to propel the rock forward. The rock falls back to Earth when it has consumed all the impetus provided by the catapult. Similarly, in the case of an arrow shot from a bow, the medieval theory took the arrow to be pushed not by any vortex of air, as Aristotle had believed, but by impetus imparted to it by the bow. In this picture, air resistance is a factor acting on bodies to exhaust their impetus; the more massive the object, the faster the resistance dissipates its impetus. Impetus was also hypothesized to follow the form of the original motion; if a ball was whirled in a circle and then released, it would carry circular impetus with it, and thus would continue to execute curved motion.

Modern researchers into science education have found that many people hold an "intuitive" view of the world that is very similar to that at which Aristotle arrived, and that contains many elements of the later impetus theory. Most people consistently misinterpret observations of motion; a notion of something like impetus still governs the way in which many of us think about motion. For example, when shown a ball traveling along a spiral track toward an exit, and asked to describe the ball's motion after it leaves the track, many people believe that the ball will continue in a circular path. As we shall see when we study Newton's laws, this is incorrect. When we observe the flight of a ball or arrow, effects such as air resistance, or aerodynamic lift provided by a ball's spin, alter the trajectory in complicated ways. It is very difficult to derive the true laws of motion from our observations of such everyday occurrences.

Perhaps even more remarkable has been the survival of the Aristotelian distinction between the Earthly and the celestial realm. Despite the great gains in understanding over the past few centuries, this viewpoint lingered even into the modern era. Newton demonstrated in the seventeenth century that celestial motion was governed by the same laws as Earthly motion, yet space remained a mysterious realm. Prior to any manned spaceflights, exaggerated scientific and medical concerns about grave dangers were voiced. Fears that astronauts might go insane merely from being exposed to "outer space," for example, might well have been a relic of the Aristotelian cosmology. Prominent scientists expressed opinions that a moon landing would be extraordinarily dangerous because of deep seas of dust, or "moon germs," or highly reactive compounds in the lunar soil that would burst into flame when first exposed to oxygen. These concerns were put to rest in a most decisive way: humans visited the Moon, and in just a few years transformed it from the exotic to the mundane. Television transmissions of astronauts bouncing about on the Moon showed it to be a real physical object, made of rock, covered with fine dust, interesting but also familiar. Going beyond the Earth's immediate vicinity, photographs sent back by the *Viking* landers from the surface of Mars resembled scenes of terrestrial deserts. Later spacecraft, especially *Voyager I* and *II*, have visited most of the worlds of our inner solar system. We have found that each planet and moon is unique, with its own history and geology, yet each is a physical world, obeying the same natural laws as does the Earth.

Heliocentrism ahead of Its Time

Aristotle was by far the most influential of the ancient Greek thinkers, especially among later Europeans. Part of the reason for this is undoubtably because many of his observations seem intuitively correct. Nevertheless, his theories were not the only ones developed by Greek scholars. Ancient scientists who belonged to competing philosophical schools, especially the Pythagoreans, were making remarkable progress with the very limited tools, both mathematical and observational, that were available to them. One of the most outstanding of these Greek scholars was Aristarchus of Samos (ca. 310–230 B.C.). Aristarchus came close to the modern description of the solar system, a millennium and a half before Copernicus.

Aristarchus set out to calculate the relative sizes of the Earth, the Sun, and the Moon, using geometry and eclipse data. The relative sizes of Earth and Moon can be determined by comparing the shadow of the Earth with the angular size of the Moon during a total lunar eclipse. From these data, Aristarchus was able to conclude that the Moon had approximately a third the diameter of the Earth, very close to the correct ratio. He also obtained a very accurate value for the distance from the Earth to the Moon.

Obtaining the distance from the Earth to the Sun is more difficult. This measurement was carried out to good accuracy for the first time only in 1769, after dramatic improvements in knowledge and technology made it possible to exploit for triangulation the rare passage of Venus directly across the face of the Sun. Aristarchus instead used a method that was extremely clever, although difficult to make work in practice: he attempted to triangulate on the Sun by using the phases of the Moon. When the Moon is in its first or third quarter, i.e., half its surface is illuminated, the angle defined by the lines from the Earth to the Moon, and the Moon to the Sun, is a right angle. The other angle required for the triangulation is proportional to the ratio of the time elapsed between first and third quarters, and third and first quarters. The closer the Sun is to the Earth, the shorter is the time elapsed between the third and first quarters of the Moon, in comparison to the corresponding interval between the first and third quarters. Unfortunately, Aristarchus could not have carried out an accurate determination with this technique, as he had neither a precise method of detecting, by naked-eye observation, the moment at which the Moon is exactly half illuminated, nor did he possess accurate clocks to measure the time intervals required. Nevertheless, Aristarchus obtained a distance to the Sun of 19 times the distance to the Moon. This number is much too small, by about another factor of 20 (the correct result is that the Sun is 390 times as far as the Moon), but the principles were sound, and Aristarchus was led to an incredible conclusion. He knew that the Sun and Moon had the same apparent size in the sky, from the remarkable fact that the Moon precisely covers the Sun during a solar eclipse. By his measurement, the Sun was roughly 20 times as distant as the Moon; therefore it must be 20 times the diameter. Since the Moon was about one third the size of the Earth, the Sun must be much larger than the Earth. This led Aristarchus to propose the first **heliocentric** cosmology, in which the Sun, not the Earth, was the center of the universe.

Aristarchus' heliocentric model was never accepted by his contemporaries, who raised what they considered to be sound objections against it. First, it required that the

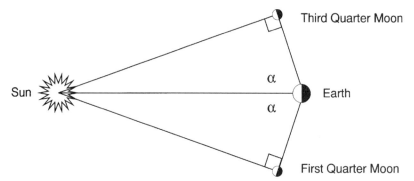

Figure 2.2 Method proposed by Aristarchus to measure the distance from the Earth to the Sun. Assuming that the Moon's orbital speed is constant, the angle α can be determined from the ratio of the time interval from third to first quarter, to the interval from first to third quarter. At the quarter lunar phase, the Earth, Moon, and Sun form a right angle, as shown. Simple geometry, plus the value of α, yields the Earth–Sun distance.

Earth move, in violation of both sensory evidence and the prevailing physics of the day. Second, a moving Earth has definite observable consequences. Since the Greeks believed that the stars were located on a relatively nearby celestial sphere, the Earth's orbital motion should bring different regions of that sphere noticeably closer at certain times of year; no such stellar brightening was seen. Moreover, over the course of a year the stellar positions should shift as the stars are viewed first from one side of the Earth's orbit and then from the other. This phenomenon, known as **stellar parallax**, had never been observed. The only way in which the absence of parallax could be explained within the context of the heliocentric models was to demand that the stars be at enormous distances from the Earth (which, of course, they are). Aristarchus' cosmos was, by the standards of his day, fantastically huge, with a radius comparable to the distance we now call a light year. We now know that this is barely a quarter of the distance to the nearest star, but at the time this immense size could not be accepted by most people. Aristarchus, who was probably one of the most brilliant of the ancient scientists, was too far ahead of his time. His theory probably also did not win favor because people were not yet ready to accept that their Earth was not the center of the universe and the sole preoccupation of its gods.

Ptolemy

The work of the Greek astrologer and geographer Claudius Ptolemaus, called Ptolemy (ca. A.D. 100–170), brought the Aristotelian system to its pinnacle. Ptolemy worked in an observatory near Alexandria, the great seat of learning of ancient Egypt. His principal work, the result of his years of study, is generally known by its Arabic name of the *Almagest* (*The Great System*). This opus brought together all the refinements of Aristotelian cosmology to better describe the observed motions of celestial objects. Ptolemy was not only a theorist but spent time charting the movements of the stars and refining his system to fit his observations. Furthermore, by his time a long history of observations had accumulated, exposing the inadequacies of earlier models. Slight inaccuracies in predicting conjunctions of Jupiter and Saturn might not be noticed over the course of a few years, but in a few hundred years these errors would become substantial. To provide an accurate *predictive* model for projecting future motions of the known celestial bodies, an essential prerequisite for the practice of astrology, Ptolemy developed an elaborate system of multiple circular motions. Such a complex scheme was required to match the observations, which were, by then, becoming quite refined.

Of course, the Greeks had known long before Ptolemy's time that the simplest possible geocentric system, in which each planet describes a circular orbit around the Earth, could not fit the data. Eudoxus' 27 spheres and Aristotle's 55 were the consequence of this celestial complexity. Ptolemy's model continued this tradition. The actual details of Ptolemy's system are rather complex, and they are of interest today mainly to historians. However, a few examples of the observational challenges, and the way they were answered, are instructive.

By observing the planets over the course of several months, it can easily be seen that they vary in brightness. This is difficult to accommodate within a philosophy that expects that the heavens are perfect and unchanging, unless the distance between the planet and the Earth changes with time. Another interesting planetary behavior is known as **retrograde motion**; this occurs when a planet reverses its usual direction with respect to the fixed stars and moves "backward" for a while before

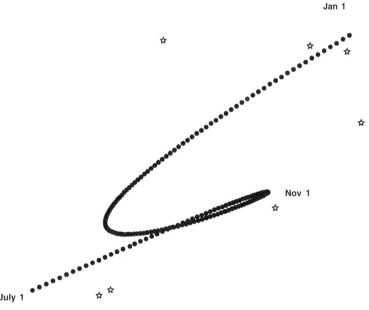

Figure 2.3 Plotting the position of a planet against the background stars, here for Mars during the interval from July 1, 1988 to January 1, 1989, reveals that the planet slows, stops, and reverses itself, traveling "backward" for a while. It then reverses again and continues in the forward direction. This cosmic pirouette is known as retrograde motion.

resuming its "forward" motion. In addition to such directional changes, the speed with which a planet moves with respect to the background of fixed stars varies with time.

Like Eudoxus and Aristotle before him, Ptolemy was obliged to construct a hierarchical system of circles in order to account for the observations. The major circles, which carried the planets around the sky, were called the *deferents*. Superposed on each deferent was a smaller circle, the *epicycle*. With the addition of epicycles, the planets no longer executed strictly circular motion, although the net motion was still a sum of circular motions. Ptolemy shifted the center of the deferents away from the center of the Earth so as to account for the apparent changes in brightness and speeds of the planets. The net center of motion of each planet was also moved away from the center of the Earth, to a point called the *equant*. As viewed from the equant, the rate of rotation of the planet was constant. However, this new feature meant that the center of motion no longer corresponded with the supposed center of the universe.

The resulting model described planetary motions well, but it fell prey to the same failings as earlier cosmological systems: the accumulation of error over time and improved observations. It became necessary to tinker further with the system, adding epicycles upon epicycles, the "wheels within wheels," in an attempt to achieve the elusive perfection. Accuracy was obtained at the expense of simplicity, a fact that was not lost upon even adherents of the system. Alfonso, a fifteenth-century king of Castile and Leon, is said to have remarked upon learning the Ptolemaic system, "If the Lord Almighty had consulted me before embarking upon Creation, I should have recommended something simpler." In retrospect, we can see how this increasingly elaborate and cumbersome construction continued to succeed. The true motions of the planets are not circular, but elliptical, and are centered upon the Sun, not the Earth. Nevertheless, any arbitrary closed curve can be approximated by a sequence of circles. But perfect accuracy requires an infinite number of circles, so ultimately the Ptolemaic system was bound to fail.

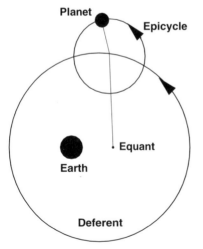

Figure 2.4 Components of the Ptolemaic model for planetary motion. The planet moves on a small circle, the epicycle, which itself moves around the Earth on a larger circle called the deferent. The center of this motion is the equant.

Ptolemy and his successors probably did not intend for their system to be taken literally, although ultimately its fate was to be taken all too literally. Their original purpose was a model which would serve as a mathematical tool to predict the positions of the planets. In that, the Ptolemaic system was quite successful for hundreds of years. It was eventually rejected not because it was inaccurate or incapable of correction, but because the heliocentric model proved to be much simpler. Moreover, the Ptolemaic model had no underlying, unifying predictive principle. If a new planet were to be discovered, the model could not describe its motion in advance, but only after many observations had been made to fit the required deferents and epicycles. The scientific method gives preference to the simpler theory with greater predictive power. The heliocentric model that ultimately arose has taken us far, yet all models must be constantly tested by observations. Indeed, the Newtonian model of the solar system has a tiny, but significant, discrepancy in the orbit of Mercury, which eventually contributed to the acceptance of the general theory of relativity.

The Renaissance

With the decline of Greek culture, scientific cosmological modeling came to a halt. Greek learning was preserved by the Arabs, who added further observations to the growing volume of data and made additional refinements to the Ptolemaic system. Some Arab scholars were dissatisfied with Aristotelian physics and wrote detailed critiques of it, but no new theory arose in the Middle East. Aristotle's writings, along with further elaborations by his successors and by Ptolemy, were rediscovered in Europe at the beginning of the thirteenth century. The Greek/Ptolemaic cosmology eventually became incorporated into medieval European philosophy, with sufficient modifications to be compatible with Judaic and Christian theology. One important alteration was the change from a universe of infinite duration to one with a creation from nothing at a finite time in the past. The Earth remained at the center of the cosmos, although not because the Earth was considered to be an especially wonderful place. Indeed, in this cosmology the center of the Earth was the lowest, basest point of the cosmos, the location of Hell. The celestial realms were the domains of angels, with God beyond the outermost sphere. In this form, Thomas Aquinas and other medieval theologians elevated the pagan Ptolemaic cosmology and Aristotelian physics into a cornerstone of Christian doctrine.

The supremacy of Aristotelian authority throughout the Middle Ages may well have occurred because, in essence, he told Europeans what they wanted to hear at the time. It was an authoritarian era, when the control of the Church in matters of belief was absolute, and dissent, whether in theology or science, was not tolerated. Aristotelian physics and especially his cosmology fitted the prevailing attitudes. It was believed that all that could be discovered had already been discovered. The search for new knowledge was regarded as a pointless enterprise, since Aristotle had anticipated and resolved all questions.

Human curiosity could not be suppressed forever, however. The rediscovery of Greek scientific thought began a transformation in Europe that led eventually to the

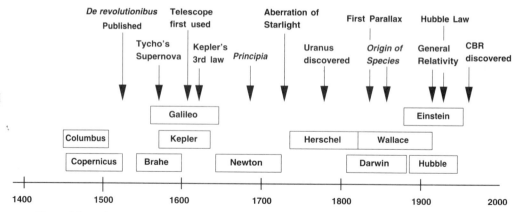

Figure 2.5 A timeline for cosmological discovery, from the Renaissance to the present.

Renaissance. By the fifteenth century, every educated European was versed in Greek learning. Astronomy, which was more like what we would today call astrology, was one of the original "liberal arts." For example, the English poet Geoffrey Chaucer wrote a treatise on the use of the astrolabe, an instrument for measuring the positions of stars. Educated Europeans also were well aware that the Earth was a sphere and even knew its diameter to fairly good accuracy.[1]

The intellectual community of Europe in the sixteenth century was in a ferment. The increased level of literacy and education, the rediscovery of ancient scholarly works, and the development of printing raised the intellectual standards and dramatically altered the political climate. This new environment made possible such changes as the Reformation, which directly challenged the prevailing doctrinal authority of the time, the Roman Catholic Church. It is ironic, then, that the man who was to set in motion the coming cosmological revolution should have been a canon, a cathedral officer, in the Church. This man was Nicholas Copernicus.

Copernicus

Nicholas Copernicus (1473–1543) is the Latinized name of the Polish scholar Mikolai Kopernik, who is credited with the introduction of the proposal that the Earth revolves around the Sun. This is called the **Copernican revolution**, and it was a revolution in more than one sense of the word: the revolution of the Earth and the revolution in thought. Copernicus was not the first to propose such a *heliocentric*, or Sun-centered, system; Aristarchus of Samos had anticipated him by 1,700 years, and Copernicus apparently learned from one of his teachers about the work of Aristarchus. Copernicus, however, introduced his system into a world that was more receptive to new ideas, although it still was many years before heliocentrism was generally accepted. Indeed, Copernicus released his work *De revolutionibus orbium coelestrium* (*On the Revolution of Heavenly Spheres*) for publication only near the end of his life. It appeared in 1543 and immediately created a sensation among the literate scholars of the day.

Figure 2.6   Nicholas Copernicus (1473–1543), the Polish scholar whose Sun-centered model of the cosmos marks the beginning of modern astronomy. Courtesy of Yerkes Observatory.

Why did Copernicus propose such a radical change? We can only speculate, as he left no specific explanation for his reasoning, but he apparently had several motivations. First, he was dissatisfied with the complexity of the Ptolemaic system. The continued addition of epicycles and eccentrics had made a mockery of the original goal of geometric purity in the celestial motions. Copernicus may well have hoped that by shifting the center of motion to the Sun, he could restore the heavens to simple circular motion. He was also aware of the inaccuracies in the predictions of planetary

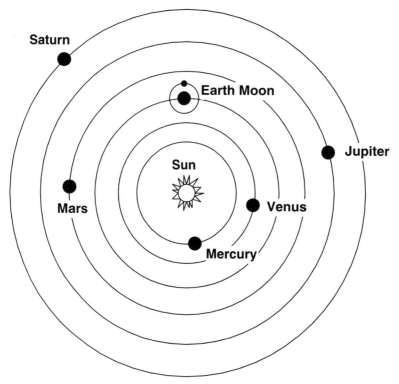

Figure 2.7 The heliocentric model of Copernicus (not to scale). Copernicus was able to arrange the planets in their proper order, as well as give accurate relative spacings between them. The stars remained on a fixed sphere, now further removed so as to explain the lack of observable parallax.

positions and must have expected that his model would make better forecasts. It also appears that he might have been attracted to the model by aesthetic considerations; where better to light the worlds than from the center of the universe?

The new theory had some immediate successes; most prominently, it explained retrograde motion in a very natural way. The planets are like sprinters running around the lanes of a circular track. The innermost racers run faster, with outer racers lagging ever slower. We are the sprinter in the third lane out. As we overtake and pass the slower outer runners, they appear to move backward with respect to the distant background, resuming their apparent forward motion after we are well past. Similarly, the inner runners are moving faster and pass us; as they turn the corner, we see them briefly move backward.

The Copernican system also made it possible to compute the relative spacing of the planets in their orbits. The two inner planets, Mercury and Venus, never travel far from the Sun in the sky. Simple geometry, combined with measurements of the angle of their maximum elongation away from the Sun, provides the size of their orbits, relative to that of the Earth. A similar, albeit slightly more complex, calculation gives the relative sizes of the orbits of Mars, Jupiter, and Saturn. In the Ptolemaic system, the diameters of the various spheres were arbitrary and were usually computed by assuming that they nested so as just to touch one another.

The heliocentric model also had the advantage of explaining the daily motions of the Sun and stars in terms of the simple rotation of the Earth. The seasonal changes in the patterns of the fixed stars were comprehensible as a consequence of the Earth's journey around the Sun, thus dispensing with the deferent that carried the fixed stars around the Earth in the Ptolemaic system.

These successes made the Copernican model the subject of much interest and discussion, well before its formal release as a printed work. With regard to improved accuracy for observed planetary positions, however, the model failed. Copernicus placed the center of the cosmos at the Sun, but he still relied upon uniform circular motion. In the end, to fit the model to the known observations, Copernicus was obliged to include many of the same complexities as the Ptolemaic system: equants, epicycles, and so forth. What Copernicus did not know was that circular motions would not suffice. Planets move on ellipses, not circles, but the true elliptical nature of planetary motions had yet to be discovered.

Parallax was still another sticking point for a heliocentric model, just as it had been for Aristarchus so many centuries before. Because stellar parallax had never been observed, Copernicus was obliged to expand greatly the size of the cosmos, although he himself continued to regard it as finite, with a fixed sphere of stars removed to a great distance. But once the intellectual wall was breached, and the heavens no longer hung close to the Earth, others grasped that the distances might be enormous, perhaps even infinite.

Although Copernicus did not produce a "better" cosmology, in the sense that Copernican calculations of celestial motions were not as accurate as those of the well-refined Ptolemaic model, his model did have an appealing simplicity. In one particular respect the Copernican model had a clear advantage over the Ptolemaic system: it made a *prediction*. In arranging the planets in their proper order from the Sun, Copernicus discovered that the inner planets moved faster than the outer ones. Thus, if a new planet were to be discovered farther from the Sun, it should be found to move more slowly than the known planets. However, he proposed no law to explain *why* the planets moved as they did; this explanation had to await the arrival of Newton.

For Copernicus, the inability of his model to make precise forecasts of planetary positions meant failure and may represent part of the reason that he did not publish his work until the end of his life. His book, published well after his theories were already widely discussed, was a highly technical work, read by few. Why, then, was Copernicus so revolutionary? By abandoning the geocentric model, Copernicus struck at the philosophical underpinnings of the prevailing cosmology. In the Copernican system the Earth is not the center of the cosmos; it is just another planet. This development, with further elaboration, is now embodied in what is often called the **Copernican principle**, which, in its most elemental form, states that the Earth is not the center of the universe. This principle is the most valuable legacy of Copernicus.

The Copernican system was obviously a much more severe challenge to medieval theology than were any of the Greek models. Many passages in the Christian scriptures support the model of a stationary Earth, including the command by Joshua that the *Sun* should stand still. Belief in the Copernican system came to be regarded as heresy and was suppressed by both the Roman Catholic Church and the renegade Protestants. The Catholic Church still wielded formidable political power with which to back its damnations, and at the time it was fighting the ultimate challenge to its authority, the Protestant Reformation. Dissension from accepted theology was thus especially dangerous. This alone was ample reason for the timid Copernicus to avoid publication as long as possible. At this he was quite successful; the page proofs for his book arrived as he lay dying.

It may be that Copernicus developed an idea whose consequences ran away from him. He intended to save the phenomenon, to restore the Platonic purity of the circle, and to recreate the geometric beauty of the heavens as they were originally conceived. Instead, he set in motion a revolution that would not be complete until both the cosmos, and the very foundations of physics, had been overturned.

Tycho Brahe

The intellectual climate of the Renaissance was receptive to the new Copernican ideas, but the most important driving force leading eventually to their adoption was increasing dissatisfaction with the Ptolemaic tables. With the development of the printing press, the tables were widely and accurately disseminated. Errors of a few days in the prediction of an important conjunction of planets could be blamed only upon the tables and not on the stars or on transcription errors. Although the telescope had not yet been invented, increasingly accurate observations made the faults of the Ptolemaic model all too apparent. This astronomical trend reached its peak in the work of the Danish astronomer Tycho Brahe (1546–1601), the last of the great naked-eye observers.

Tycho is memorable both as a methodical scientific observer and as a remarkable personality. He was a member of the aristocracy, yet he devoted himself to the decidedly unaristocratic art of astronomy. In this pursuit, he benefited from his association with King Frederick II of Denmark, whose financial support enabled Tycho to build Uraniborg, a lavish observatory on an island just offshore from Copenhagen. Here Tycho lived the life of a self-indulgent noble, while still devoting both his own efforts and those of a considerable staff to gathering his detailed observations of the heavens. Tycho's personality stood in marked contrast to his careful scientific work. He was a flamboyant and fiery man who sported a metal nose, the original having been cut off in a sword duel in his youth. He loved parties, which in his time were often lengthy binges, involving much heavy drinking. He may have met his end as a result of such customs. Legend has it that he imbibed excessively at a royal banquet in Prague in 1601, but the protocol of the day prohibited guests from leaving the room when royalty was present. Tycho died shortly after this banquet, possibly as the result of a ruptured bladder.

It is easy to focus on such interesting details of Tycho's personal life, but he should be remembered instead for his exceptionally careful and systematic observations of celestial motions. Tycho repeated his measurements, and used the additional information to estimate his errors, a revolutionary idea at the time. In this he was one of the first investigators who could be called a scientist in the modern sense of the word. His amassed data provided a record of unprecedented accuracy and detail and clearly showed the deficiencies in the Ptolemaic tables. Better observations do not

Figure 2.8 Tycho Brahe (1546–1601). Tycho's meticulous naked-eye observations of the heavens revealed the inadequacies of the Ptolemaic tables and provided the essential data that enabled Kepler to formulate the laws of planetary motion. Courtesy of Yerkes Observatory.

simply destroy old theories; these observations were also accurate enough to allow Johannes Kepler finally to determine the correct planetary orbits, thus laying to rest forever the Ptolemaic system and establishing the basis for Newton's laws of motion.

Although primarily an observer, Tycho was not above trying his hand at cosmological modeling. Tycho was no Aristotelian; he knew particularly well the failings of the Ptolemaic system. Yet neither was he a Copernican. He ultimately rejected the heliocentric model because he was unable to detect stellar parallax, the shift in the apparent position of a star due to the change in our vantage point as the Earth orbits. He knew that the lack of observable parallax could be explained by only two hypotheses: either the stars were so far away that their parallaxes were smaller than his measurement error or else the Earth did not move. Tycho believed that the stars were near because he thought he was able to detect their apparent sizes. He did not realize that the finite disks of stars are an optical illusion, caused by the shifting of parcels of air in the Earth's atmosphere (stellar "twinkling"). If the stars had the sizes he measured, such great distances as were required by their lack of parallax implied them to be enormously large objects. Hence he concluded that the Earth could not be in motion.

Tycho was a true scientist; he proposed a test of the heliocentric theory, the stellar parallax. The theory seemed to fail his test, and so he rejected it. But even though he was not a Copernican, he did appreciate the simplicity of the heliocentric theory. Faced with conflicting observations and philosophical leanings, he proposed his own model in which the Sun and Moon revolved around the Earth, but everything else revolved around the Sun. In essence, he recreated the Copernican model but shifted the center back to the Earth. Aside from differences in the frame of reference, the two systems were nearly equivalent. Like most compromises, however, Tycho's model pleased no one, except, possibly, himself.

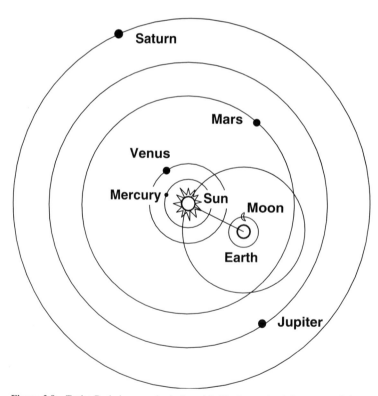

Figure 2.9 Tycho Brahe's cosmological model. Earth remained the center of the cosmos, and the Sun circled the Earth, but the other planets orbited the Sun.

Stellar parallax is an important prediction of the Copernican theory, and Tycho's objection was taken seriously. But the true distances to the stars are so great that Tycho could not possibly have detected any parallax without advanced telescope technology. If we were to shrink the radius of the Earth's orbit to a meter, the distance to the nearest star, Alpha Centauri, would be 274 kilometers (km) (around 170 miles)! Measuring the parallax of this star amounts to determining the smallest angle of a triangle whose short side has a length of 1 meter (m), and whose two long sides are 274,000 m long. This angle works out to be less than 1 arcsecond, approximately 100 times smaller than the unaided eye can resolve.

Parallax is so difficult to observe for even the nearest stars that the first proof of the Earth's motion was indirect and came as late as 1728, more than a century after the deaths of Tycho and Kepler. The English astronomer James Bradley was attempting, unsuccessfully, to measure parallaxes when he noticed that *all* stars he observed showed a systematic shift with the seasons. At last the explanation came to him while he was boating; watching a vane turn with the winds, he realized that the Earth was traveling through a "wind" of starlight. An even better analogy is a sprint through the rain. If you are caught outside without an umbrella in a sudden downpour, you must tilt your body forward in order that the newspaper you are holding over your head can be oriented perpendicular to the raindrops, even when the wind is perfectly still and the rain is falling straight down. The apparent direction of the source of the rain shifts because of your motion. The analogous phenomenon discovered by Bradley is called the *aberration of starlight*.

It was not until 1838 that F. W. Bessel, F. G. W. Struve, and T. Henderson independently detected the parallaxes of the stars 61 Cygni and Vega, in the Northern Hemisphere, and Alpha Centauri, visible only from the Southern Hemisphere, thus proving once and for all the heliocentric model. Observations of stellar parallax retain

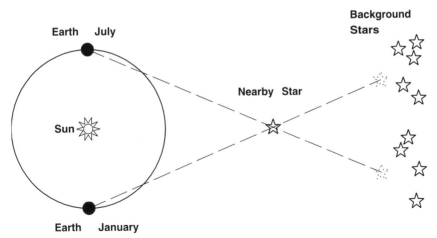

Figure 2.10 Earth's annual orbital motion produces an apparent shift of a nearby star's position on the sky with respect to the background stars. This shift is known as parallax; observation of the parallax angle determines the distance to the star via triangulation.

their cosmological importance even today, as they are still the fundamental basis for all stellar and galactic distance measurements.

In addition to the catalog of accurate stellar and planetary positions, Tycho made several important discoveries. In 1572, he observed what was, at the time, an unbelievable sight, the sudden appearance of a new star in the constellation of Cassiopeia. This was what we now call a supernova, a stellar explosion. When Tycho was unable to measure a parallax for this object, he realized that it could not be merely a brightening in the atmosphere of the Earth, but must belong to the realm of the fixed stars. This showed that the heavens were not immutable, a stunning revelation at the time. In the same vein, Tycho demonstrated, again by means of parallax, that the orbit of a comet lay beyond that of the Moon. Until that time, comets had been believed to be vapors in the atmosphere of the Earth. Suddenly, the Aristotelian view of a perfect, changeless, unblemished heaven was untenable. New stars appeared and then disappeared. Unpredictable, rapidly moving comets belonged to the celestial realm. Indeed, the Aristotelian physics then accepted required that the crystalline spheres be real, physical entities; Aristotle believed that "nature abhors a vacuum," and thus he asserted that the spheres must fill all space. Yet now it seemed that the comets followed a path which must take them *through* the planetary spheres.

The old model of the universe was disintegrating; there yet remained the task of building the new. Tycho's observations did as much as anything to chip away the foundations of the prevailing cosmology, but his own attempt at a new cosmological model met with indifference. Clearly he was not the man who could create the new synthesis. It happened, however, that Tycho became embroiled in a dispute with King Christian, the new sovereign of Denmark, who ascended to the throne after the death of Tycho's exceptionally generous benefactor, King Frederick. Tycho packed up his instruments and records in 1597 and moved from his private island off the coast of Denmark to central Europe. Tycho's misfortune was the great fortune of science, for there he took a new assistant named Johannes Kepler.

Kepler

Tycho's schizophrenic cosmology was characteristic of a transitional era; the established model was rapidly failing, but its successor, the heliocentric model, had not yet been established. It fell to Johannes Kepler (1571–1630) to develop the new paradigm. Kepler, a reserved Bohemian Protestant, came to Prague to work with the temperamental and outgoing Tycho in 1600 and set about interpreting his data. After Tycho's death in 1601, Kepler absconded with Tycho's vast collection of observational data, and its study occupied him for the rest of his life.

Kepler settled upon the objective of explaining the motion of Mars, a project suggested by Tycho, apparently because Mars shows the most irregularities in its motion. (We now know that this is because of its unusually eccentric orbit and its proximity to Earth.) Kepler spent years considering all manner of epicycles. Nor did he find better luck from a different philosophical approach, in which he fitted the observations to his pet geometrical objects, a class of figures called Platonic solids. He felt that these objects had just as much right to be perfect as did a sphere, since they can be precisely surrounded by a sphere, but they yielded no improvement. Kepler did eventually hit upon a traditional Ptolemaic scheme that fitted the observations better than any existing model of the time. He could have stopped at that point and continued in his modest employment as a fairly successful man. But Kepler was ruthless with himself and strictly intellectually honest. He was aware that none of his models, even his best, could describe the planetary motions to within Tycho's stated errors, and he was confident that Tycho had estimated his errors accurately. Therefore, Kepler struggled onward.

Finally, in 1604, he achieved success. Some inspiration caused him to abandon the ancient philosophical prejudices and to consider the motion of Mars as seen from the Sun. He found that he was able to fit the data to within the observational errors with an **ellipse** rather than a circle. The ellipse is the curve representing a constant

Figure 2.11 Johannes Kepler (1571–1630). Kepler's three laws of planetary orbits provided the first simple, predictive description of celestial motion. Courtesy of Yerkes Observatory.

sum of the distance from two fixed points, called *foci* (singular: focus). Because of its oval form, the ellipse has not a single diameter, but two perpendicular axes, the major (longer) axis and the minor (shorter) axis. The shape of an ellipse depends upon the separation of its foci. As the foci move farther apart, the ellipse becomes increasingly elongated, or *eccentric*; conversely, a circle is a "degenerate" ellipse whose foci coincide. Thus an ellipse is really a generalization of the circle, so the ancients were not quite so far wrong after all. The eccentricity, or deviation from circularity, of the orbits of almost all the planets is very small; for Earth's orbit, the major axis is a mere 0.014% longer than the minor axis. However, these relatively small differences from circular motion were more than sufficient to confuse astronomers for many centuries.

Kepler discovered that the Sun was located at one focus of the ellipse. (The other focus is empty.) Each planet moved on its own elliptical path, with its own eccentricity. This insight was to unlock the secret of the heavens, although the work had only begun. It was not until 1621, after laborious calculations using the only mathematical tools available at the time, that Kepler finally arrived at his three laws of planetary orbits.

Kepler's first law: Planets orbit the Sun in an ellipse, with the Sun at one focus.

Kepler's second law: The line from the Sun to the planet sweeps out an equal area in an equal time. Thus planets move faster when they are nearer the Sun.

Kepler's third law: The square of the period of the orbit is equal to the cube of the semimajor axis (half the major axis) of the ellipse.

If the period, symbolized by P, is measured in years, and the size of the semimajor axis of the ellipse, R, is measured in terms of the *astronomical unit* (AU), where the AU is defined as the mean distance of the Earth from the Sun, then this law can be expressed mathematically as

$$P^2 = R^3. \tag{2.1}$$

Kepler had strong mystical leanings and always hoped to find deep meaning in the cosmos. His third law, often called the *harmonic law*, was probably the most personally satisfying discovery of his life. Kepler went so far as to assign musical notes to the planets, based upon his third law. Today the mathematical beauty of the harmonic law is understood to be a direct consequence of more fundamental, and perhaps even more beautiful, laws of physics. In Kepler's time, however, this achievement was a great triumph. It is fair to say that Kepler was the first to hear the true music of the spheres.

With Kepler's laws in place, simplicity swept away complexity. There was no need for circular motion; the Copernican system, freed of its epicycles, finally revealed the elegant simplicity of the travels of the planets around the Sun. Now it could be shown that the new model agreed with observations to a far better precision than even the carefully elaborated Ptolemaic system. Geocentrism was dead.

The complexities of the geocentric systems were due not only to their inappropriate frame of reference, but, in retrospect, to the impossibility of fitting an ellipse with any finite sequence of circles. The data were forcing a change, but the idea of the primacy of circular motion was so strong in European thought that the correct solution could

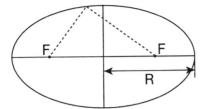 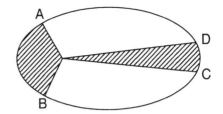

Figure 2.12 Kepler's first and second laws of planetary motion. (*a*) An ellipse is defined as the curve traced by a constant sum of distances from two arbitrary points, the foci, whose locations are marked by *F*. Planets orbit in an ellipse, with the Sun at one focus. The semimajor axis of the orbit is indicated by *R*. (*b*) The law of equal areas. The planet is observed to move from A to B in the same time as it requires to move from C to D. The lines connecting the Sun to those points, along with the curve of the ellipse between them, indicate the area swept out during that time (*hatched regions*). Since the time from A to B is the same as from C to D, the two indicated areas would be equal by Kepler's second law.

not been seen. It was Kepler's great achievement that he was able to break through this mindset. And it was not so much that the old theory was demolished as it was a crystallization of what was already known, now seen in a new light. The old theory had reached the end of its possibilities.

Kepler's laws provide a correct mathematical description of planetary motion. Unlike the Ptolemaic model, the Keplerian model has considerable predictive power. If a new planet were discovered, not only could we predict whether it would orbit faster or slower around the Sun, but, from only a few observations to determine the length of the semimajor axis of the orbit, we could predict the period of that orbit. However, Kepler's laws alone do not provide the reason why the motion occurs as it does. Kepler recognized that the third law provides a clue. If planets orbit more slowly the greater their distance from the Sun, then their motion must be related to some influence from the Sun. Sunlight also diminishes with distance from the Sun, so perhaps there is some force emanating from the Sun that sweeps the planets along in their orbit; this force must decrease with distance, just as does the intensity of sunlight. Unfortunately, Kepler still labored in the shadow of Aristotelian mechanics. Kepler lacked the proper definition of inertial (natural) motion, so he was not quite able to grasp the law of gravitation; the correct formulation had await the arrival of Newton. Perhaps it is too much to expect a single individual to do more than to overthrow the cosmology accepted for two thousand years.

Kepler was a quiet and unassuming man who might not have seemed destined for the greatness he achieved. He was not highly regarded in his day, yet he was persistent, mathematically gifted, and intellectually honest. While he never completely abandoned his philosophical prejudices, continuing to think about his Platonic solids even after his success with ellipses, he was able to put them aside rather than allow them to twist his theories away from their observational roots. His achievements are eloquently summarized by Kepler himself, in his own epitaph. The original was written in Latin, the scholarly language of the day:

I measured the heavens, now I measure the shadows,
Skyward was the mind, the body rests in the earth.

Galileo

Kepler was the scientist who discovered the mathematical laws of the celestial motions, and it was he who made the bold leap from circles to ellipses that finally vindicated the Copernican heliocentric system. Yet the name most popularly associated with the championing of this new world view is that of the Italian astronomer Galileo Galilei (1564–1642).

Galileo was one of the great Renaissance scientists. He made significant contributions in many areas of research, although he is most remembered for his astronomical discoveries, which he made by putting the newly invented telescope to its first celestial use. It is often believed that Galileo invented the telescope, a misconception common even during his lifetime and one that Galileo himself made no attempt to dispel. However, credit for the invention of the telescope is usually assigned to Hans Lippershey, a Dutch lens grinder, although earlier lens makers may have discovered the basic principles. In any case, as soon as Galileo heard of this new instrument in 1609, he immediately built one and turned it toward the sky.[2]

One of Galileo's first observations was of craters and mountains on the Moon. This showed that the Moon was not a smooth sphere, but a world with its own detail, much like the Earth. He also turned his telescope to the Sun. Although he was not the first to discover sunspots (they had been, and still can be, observed by the unaided eye at sunrise or sunset), he was the first to conclude correctly that the spots were associated with the Sun itself and were not foreground objects. Galileo also recognized that the Sun carried the spots around as it rotated on its own axis, and this enabled him to estimate the rotation rate of the Sun. Observations such as these pounded away at the Aristotelian concept of the perfection of celestial bodies. As Tycho had discovered around the same era, the skies were not the abode of perfect, immutable objects. The Earthly and celestial realms were not distinct, but might obey the same laws and be made of the same substances.

Galileo made another surprising discovery when he turned his telescope toward the Milky Way, which to the unaided eye appears only as a diffuse glow spanning the sky. He resolved the glow into a myriad of stars too faint to see without the new device. But if these stars were too dim to see, while others were visible without the aid of the telescope, how could they reside upon the same crystalline sphere, as required by the ancient cosmology? Under magnification, the new, faint stars had the same apparent size as all the others. This suggested that the apparent disks seen by earlier observers, including Tycho, were an illusion. Even today, a sweep through the Milky Way with a simple pair of binoculars gives a distinct sensation of vast depth to the skies. The Copernican model and the lack of observable parallax required the stars to be at a great distance; the telescope made such a heresy believable.

Although the stars remain unresolved points even to modern instruments, Galileo found that the planets *did* present disks to his telescope; in fact, Venus went through phases, just like the Moon, and its phases accounted for some of its dramatic changes in brightness. The gibbous and full phases of Venus observed by Galileo could not be explained by the Ptolemaic model, which could produce only crescent and new phases. The Ptolemaic model made a testable prediction about the phases of Venus, which it failed when the observation was made. The Copernican system, on the

Figure 2.13   Galileo Galilei (1564–1642), ardent champion of the heliocentric model. Courtesy of Yerkes Observatory.

other hand, predicted a full range of phases; hence Galileo's observations are an example of a crucial experiment, providing strong evidence in favor of the heliocentric model.

Perhaps Galileo's most dramatic observation was that Jupiter commands its own miniature system. Galileo discovered the four largest moons of Jupiter, still known as the *Galilean moons*. It is one thing to observe new details of known objects; far more sensational is to discover completely new objects. Galileo's careful charting of the motions of these objects demonstrated unequivocally that the moons orbited Jupiter. The Earth was not the only center of motion, refuting one of the basic tenets of Aristotelian cosmology.

The impact of Galileo's findings was widespread. When he wrote of his observations in his book *The Starry Messenger*, he wrote in Italian rather than in the Latin of

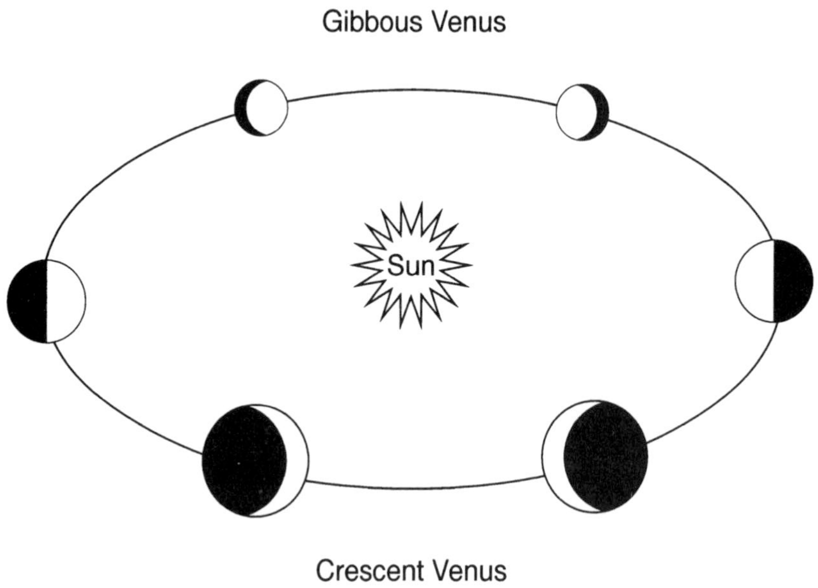

Gibbous Venus

Crescent Venus

Figure 2.14   Galileo's observations of Venus revealed a full ensemble of phases, from crescent to full. This is consistent with the Copernican model in which Venus circles the Sun, but not with the Ptolemaic model, in which Venus always lies between the Earth and the Sun.

scholars, so that everyone could read about his discoveries. Soon many people were turning telescopes skyward to share in these new wonders.

Although Galileo began his career teaching the standard Ptolemaic model, he apparently was not satisfied with Aristotelian cosmology. He had little patience with his fellow scholastics, who unquestioningly repeated Aristotle's laws of physics. Galileo was not content to accept the word of even so venerated an authority as Aristotle and often put the Aristotelian precepts to the test. When his astronomical observations converted him completely to the Copernican model, he was faced with the problem of reconciling his findings with physics. Aristotle's physics explicitly denied the motion of the Earth, which seemed to be perfectly consistent with the observations of our senses. Yet the skies supported the Copernican model. How was physics to be modified to explain this apparent contradiction?

Fortunately, Galileo had devoted much of his career to the physics of mechanics. In particular, he was intrigued by the motion of falling bodies. Aristotle held that the rate of fall depends upon the composition of the falling body, and of the medium through which the body fell. Galileo recognized that this idea could be put to the test, as several other scholars of the time had done. He carried out his own experiments (none of which, apparently, involved dropping any objects from the Leaning Tower of Pisa) and made measurements in support of his conclusion that all objects fall at the same rate, contrary to the Aristotelian claim. But the limitations of the technology of his time forced him to appeal for many of his arguments to **thought experiments**, that is, mental experiments that could, in principle, be performed if the technology were available. As an example, consider a stone falling from a height. Now imagine cutting the same stone into two equal pieces, then dropping them together. Would the severed halves fall at different rates from the whole? What if the two pieces were connected by a short string? It should be clear that a boulder will not suddenly fall at a different rate if a crack appears in it. From such reasoning, Galileo concluded that all objects must fall at the same rate in a vacuum. This important observation, that in the absence of air resistance or other complicating factors, all objects fall at the same rate in a gravitational field, is now called the *equivalence principle*; Galileo was one of the first to articulate it clearly. Yet even Galileo could not have realized how profound was this observation, as much later it became the basis of general relativity; more immediately, it formed a foundation of Newton's theory of gravity.

A key rule of mechanics, with which Galileo struggled, is the law of inertia. Galileo's knowledge of contemporary experimental results, plus his own experiments with pendula and with balls rolling on an inclined plane, convinced him that impetus was not lost but was *conserved* in freely moving bodies. Hence not only does an object at rest remain so unless a force acts upon it, but a body in motion in a straight line remains in that motion unless a force acts. The essential break from Aristotelian mechanics to modern mechanics is to recognize that force is responsible not for motion, but for *changes* in motion. From this realization, the relativity of uniform motion follows. Galileo understood the experimental fact that if everything moves together uniformly, such as the furniture and lamps in the interior of a moving ship, then it will seem no different from when the ship is at dock. To take a more modern example, imagine you are in the supersonic Concorde airplane, flying faster than the speed of sound. You have dinner; the flight attendant pours coffee. There is no more sensation of speed than is felt while sitting in one's living room.

Constant-velocity motion is not necessarily perceptible if you and your surroundings are moving together. The Earth could thus be moving through space, yet this not be directly noticeable by the humans moving along with it. This was the critical conceptual breakthrough that made the heliocentric model plausible. However, Galileo never completely worked out the laws of motion that would replace those of Aristotle. That task fell to Isaac Newton.

Galileo summarized his cosmological conclusions in 1632 in a new book *Dialogues Concerning Two Chief World Systems*, in which he showed how his discoveries supported the Copernican system. The book caused a sensation throughout educated Europe and paved the way for the new paradigm of the universe. It also set the stage for Galileo's later troubles with the Church. His outspoken advocation of the Copernican model had earlier discomforted Church authorities, and this new book provided further provocation. One of his political missteps, perhaps a natural consequence of his arrogance, was to place the defense of the Aristotelian cosmology into the mouth of Simplicio, an obvious fool. Galileo was brought to trial for heresy in 1633, was forced to recant his scientific beliefs, and was confined to his home for the rest of his life. Only in 1980 did the ecclesiastical authorities finally exonerate him.

Galileo was a vain, arrogant man; in the end, he came to regard himself as much of an authority as Aristotle had considered himself. He deliberately provoked the Church and was, in fact, given an unusually light penalty at his celebrated trial, partly because of his fame and partly because of his advanced age and infirmity at the time he was brought before the Inquisition. Galileo certainly promoted himself and was not above claiming credit, or allowing credit to be assigned to him, for nearly every discovery in astronomy during his lifetime. Despite such human character failings, he was an important figure in the history of science. He was one of the first to understand fully how critical is the role of experiment. Both he and Kepler realized that *data*, not our philosophical wishes, must be the final arbiter of science. One modern school of thought in the philosophy of science holds that great discoveries are more the products of an era than of individual genius. If Galileo had not made his discoveries, someone else would have done so. There is probably much truth in this idea, as it is clear from history that important discoveries are often made simultaneously and independently by more than one researcher. Yet there must be some due given to individuals. Perhaps it is the combination of the right person at the right time. Kepler and Galileo were the right people at the right time; between them they irrevocably changed our view of the world.

KEY TERMS

mechanics	geocentric	force
inertia	heliocentric	stellar parallax
retrograde motion	Copernican revolution	Copernican principle
ellipse	Kepler's laws	thought experiments

Review Questions

1. From what evidence did the ancient Greeks (and others) conclude that the Earth was immobile?

2. From what evidence did the ancient Greeks deduce that the Earth was a sphere?

3. Why did Eudoxus demand spherical motions for the planets? What were the consequences for his model of this assumption?

4. According to Aristotle, what caused motion on the Earth? In the heavens? What type of motion was appropriate to each realm?

5. What was the impetus theory of motion?

6. (More challenging.) While stationed on the planet Zorlo, you decide to replicate the calculation of Aristarchus for the Earth and the Sun. Zorlo's moon, Crastig, completes one revolution (360°) in 42 Zorlo days. You observe that Crastig requires 20.985 days from third to first quarter. What is the ratio of the distance from Zorlo to its moon Crastig, to the distance from Zorlo to its sun? (Hints: first compute the number of degrees traveled by Crastig in one day. From Figure 2.2, note that the desired ratio of distances is given by the cosine of the angle α.)

7. How did Ptolemy account for the retrograde motion of the planets?

8. Describe two major weaknesses of the Ptolemaic model of planetary motions.

9. The imagery of Hell existing "down below" and Heaven having a location above the clouds is still common, at least metaphorically. How is this connected to medieval European cosmology?

10. Was the original Copernican model simpler than the Ptolemaic? What phenomena were more easily explained by the Copernican theory than the Ptolemaic? What is the most valuable legacy of Copernicus?

11. What Aristotelian belief did the observations of Tycho Brahe most seriously challenge? Why did Tycho reject the Copernican model?

12. A new asteroid orbits the Sun at a mean distance of 40 AU. What is the period of its orbit in Earth years? Does the answer depend on how eccentric the orbit is?

13. Describe three observations of Galileo which supported the Copernican model. State also why they falsified the Ptolemaic/Aristotelian system.

Notes

1. It is an interesting historical tidbit that, regardless of what some legends might claim, Christopher Columbus certainly was not waging a lonely battle against ignorance by contending that the Earth was spherical. On the contrary, Columbus had carried out his own erroneous calculation of the diameter of the Earth; he argued that it was a much *smaller* sphere than others believed and maintained that the great Ocean was traversable by the small sailing ships of the era. In this case, conventional wisdom was correct and the supposed iconoclast was wrong. Others asserted, quite correctly, that a journey in a small sailboat across the distance proposed by Columbus was impossible. Columbus would have vanished, both from his countrymen and from history, had not an unknown (to Europeans of the time) continent intervened. The myth that Columbus was fighting the ignorant scholars of the time who insisted that the Earth was flat is pure fiction, apparently invented from whole cloth a few hundred years after his voyages, and popularized by the writer Washington Irving. Columbus himself refused to accept that he had found a new land, believing to his dying day that he had discovered a route to Asia. Sometimes, it would seem, it is more important to be lucky than to be right.

2. Like many new technologies before and since, one of the initial applications of the telescope was for military purposes. Galileo demonstrated the military possibilities to the local authorities in Padua, impressing them sufficiently that they provided Galileo with funding and status. After this, Galileo became the first to apply the telescope to scientific inquiry.

CHAPTER 3

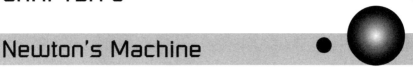

Newton's Machine

I think Isaac Newton is doing most of the driving right now.

—ASTRONAUT BILL ANDERS, ABOARD THE *APOLLO 8* COMMAND
MODULE DURING ITS RETURN FROM LUNAR ORBIT

Isaac Newton

If modern physics and cosmology were to be assigned a birthday, it would coincide with that of Isaac Newton (1643–1727). Born prematurely in Lincolnshire, England, on Christmas Day of 1642 (according to the calendar then in use in England), the infant Newton barely survived. His father had died before his birth; when his mother remarried a few years thereafter, he was given over primarily to the care of his maternal grandmother. He distinguished himself scholastically even as a child, and his family decided that he should enter a university; he began his undergraduate study at Cambridge University in 1661. Newton set about the study of mechanics, a science which at that time was still dominated by the theories of Aristotle. The University was not immune to new ideas, however, and Newton also acquainted himself with the more recent work of Kepler, Galileo, and the French philosopher René Descartes.

Newton left Cambridge in 1665 when the university closed during an epidemic of the plague. Returning home, he began the independent research that was to revolutionize science. During his 18 months in Lincolnshire, Newton developed the mathematical science of calculus; performed experiments in optics, the science of light; and carried out his initial derivations of the laws of mechanics and gravity. In 1669, at the age of 26, Newton was named the Lucasian Professor of Mathematics at Trinity College, Cambridge, in recognition of his accomplishments with calculus; he held this position for the rest of his scientific career. In 1672 Newton was elected to the Royal Society of London on the basis of his work in optics, particularly his invention of the reflecting telescope, a device which uses a mirror rather than a lens to focus light. This is still the basic design of all large astronomical telescopes.

Newton was an embodiment of the eccentric genius. In addition to his work in physics, he dabbled in alchemy and theology and considered his efforts in those fields every bit as important as his physics, although today they are regarded to be at best of no consequence and at worst completely wrong. Newton shared with Aristotle and

Figure 3.1 Isaac Newton (1643–1727). His master work, the *Principia,* established the science of mechanics and provided the law of gravitation that explained Kepler's laws. Courtesy of Yerkes Observatory.

Galileo the conviction that he was always right. He once ended a friendship because the friend dared to disagree with Newton's interpretation of the Old Testament's Book of Daniel.

Newton's pathological personality almost denied the world the benefit of his insights. Newton was neurotically protective of his privacy, becoming greatly disturbed when his first published work, a report of his discoveries in optics, drew some moderate attention to him. The paper was an important one: Newton showed that a prism separated light into colors because the index of refraction, a measure of the bending of light in a medium, differed for each color. Newton proved that white light was a mixture of these colors by showing that a second prism could recombine the spectrum into white light. Previously, it had been believed that a prism somehow manufactured the colored light internally. Although the paper was generally well received, the publication of this early optical work dragged Newton into a dispute with his chief rival, Robert Hooke. Hooke attacked the paper viciously because it made some sweeping statements about Newton's corpuscular theory of light, on the shaky basis of some rather crude experiments. Stung, Newton withdrew for a while even further into his shell. His full optical researches were not published until after Hooke's death.

Robert Hooke made important contributions of his own; among other things, he discovered the rotation of Jupiter. Yet he is also remembered as a thorn in Newton's side, an egotist who claimed for years that Newton had stolen "his" theory of gravitation. In fact, Hooke and some other intellectuals of his day had independently arrived at the hypothesis that gravity obeyed an inverse-square law. (Kepler himself had suspected that the Sun's influence over the planets obeyed such a relation, so the idea was hardly new.) In January of 1684, Hooke boasted to Edmund Halley and Christopher Wren that he could prove this assertion easily, but he failed to produce a demonstration after several months. Later that year Halley, in Cambridge for other reasons, stopped to see Newton. Halley asked Newton what would be the orbit of a planet obeying an inverse-square law of gravity, and Newton replied immediately that it would follow an ellipse. When the astonished Halley asked Newton how he knew this, Newton replied, "I have calculated it." Halley requested to see the work, but Newton, rummaging through his stacks of papers, claimed he could not find it. (Most likely, the proof was incomplete and contained an error, and Newton did not wish to expose himself to criticism.) Although Newton had carried out the basic calculations years earlier, he had delayed publishing his findings because he had difficulty in proving an important result in his theory of gravity. Now, goaded by Halley's request, Newton turned again to the problem. Three months later, Newton had worked out the proof in detail and sent a copy to Halley, who immediately urged Newton to publish a full description of his work. Two years later Newton delivered a manuscript that laid down the fundamental laws of mechanics and gravitation, laws that are still today the basis of mechanics in the usual limiting case of speeds that are not too extreme and gravitational fields that are not too strong.

Newton published this great work, the *Philosophiae Naturalis Principia Mathematica (The Mathematical Principles of Natural Philosophy)*, in 1687. This book, issued with the imprimatur of the Royal Society, is usually known simply as the *Principia*. Halley paid for the publication from his own pocket; without the intervention and support of Halley, Newton's discoveries might never have reached the world. Had such a calamity occurred, at best the progress of science would likely have been much delayed. The laws of mechanics and gravity might well have trickled out slowly, attributed piecemeal to the work of others, rather than emerging, as they did, as a unifying whole. Edmund Halley, who had the patience to remain Newton's friend for years, despite Newton's tantrums and quirks, might have been one of the few who could have persuaded him to publish.

Newton's scientific career ended only a few years after the *Principia* was published. In 1693, he suffered an unmistakable mental breakdown, possibly from years of exposure to mercury, a very toxic heavy metal, during his alchemical studies. He recovered, but never made any further contributions to science. By then, however, Newton's renown was so great that the British government arranged for him to receive a comfortable position. He spent the last 30 years of his life managing the British Mint, an office which brought his considerable eccentricities into the public eye. He became a common object of ridicule, lampooned in popular plays and pamphlets as a pompous, overblown martinet. History has been more generous. Newton is now recognized as one of the greatest scientists in history, and the *Principia* as possibly the greatest scientific work ever published.

Newton's Laws

The science of Newtonian mechanics, as elaborated in the *Principia*, is summarized in Newton's three laws of motion. These three laws can be stated quite briefly and in simple language, yet they are of overarching importance, transcending this apparent simplicity.

The first law of motion returns to the question of "natural motion," or how things move if left on their own. Aristotle believed that things moved only if acted upon by a force, that is, a push or a pull exerted by one thing upon another. An arrow flies only because it is pushed along by the air through which it moves. Otherwise, all things move to their proper location within the cosmos. This implicitly assumes that there is a universal standard of rest, relative to which all things move. For Aristotle, this standard of rest was the center of the unmoving Earth, the center of the universe. By Newton's time, however, scientists and philosophers were well aware of the problems with these ideas. The Copernican system requires the Earth to orbit around the Sun and to rotate on its axis. Hence the Earth is partaking in some stupendous motions, yet without the obvious application of a force or any sensation of motion. Galileo developed the idea that motion is *relative*: if all things move uniformly together, sharing in a common motion, there is no discernible effect. The Earth's motion through space is imperceptible because we take part in that motion. This suggests that there is no absolute standard of rest; the state of "rest" is relative.

We can retain the aspect of Aristotelian physics which asserts that an object at rest will remain at rest unless acted upon by some force. But if what is meant by "at rest" is relative, then motion per se cannot require the continual application of a force. Consider two persons, each moving uniformly with respect to the other. Each feels at rest, and no force is required to remain at rest. Each feels no force, even though the two are moving with respect to one other. To initiate some other motion does require a force. It is not motion per se that requires a force, but a *change* in the state of motion. Hence we arrive at:

> **Newton's first law of motion:** A body at rest or in a state of uniform motion will remain at rest or in uniform motion, unless acted upon by a net external force.

In this law, also called the **law of inertia**, Newton grasped what others had failed to see: not only would a body remain at rest, but a body in **uniform motion**, that is, traveling in a straight line with constant speed, would also remain in that state unless a force acts. This law also clarifies and defines what is meant by a **force**: a force is that which causes a body at rest or in uniform motion to change its state. Note that the first part is really just a special case of the second part of the statement, since a body at rest has a velocity (both speed and direction) of exactly zero, which surely is uniform. In the absence of force, a body in uniform motion will remain in motion forever.

We often have difficulty in grasping intuitively Newton's first law because the motions we commonly experience are always affected by forces. The arrow, flying through the sky, is slowed down by the act of *pushing* the air out of its path. The force of air does not keep the arrow flying; it is the force of air that eventually brings it to a halt. Similarly, an automobile will come to a stop if the engine shuts off. This is a result of the resistive force exerted on the car's tires by the ground, a force we call *friction*. In the absence of friction an automobile, once started, would travel down a straight road without the need for any motive power whatsoever. More realistically, if the friction between the tires and the road is reduced, by driving on glare ice, for example, the driver will quickly discover the difference between Aristotelian and Newtonian physics. Regrettably, uniform straight-line motion will continue until acted upon by the force of the collision with the tree.

Thus we conclude that a force is required to produce a change in velocity. A change in velocity means a change of speed or direction, or both, with respect to time. This is an **acceleration**, and mathematically it is expressed as the change in velocity per unit of time. But what is the relation between force and acceleration? Is force simply equal to acceleration, or is it more complicated? We know that our arms can exert a force, for instance, when throwing a ball. Presumably there is only so much "push" we can exert upon the ball with our arms, so the force we can produce is limited. And we know that the same force, when exerted upon a hollow rubber ball, produces a much greater velocity than when exerted upon a bowling ball. Therefore, the amount of acceleration generated by a force is linked with how "massive" something is. The exact statement of this idea is:

> **Newton's second law of motion:** The acceleration of an object is equal to the net force applied to it, divided by its mass.

Mathematically, this law can be expressed in the form

$$F = ma, \tag{3.1}$$

where F is the symbol for the force, including both its magnitude and its direction, m is the mass, and a is the acceleration, both in magnitude and in direction. This simple law contains most of the science of mechanics. The second law also provides us with a formal definition for **mass**: mass is the source of inertia; it is that property by which an object resists a change in its state of motion. The greater the mass of an object, the larger the force must be to produce a given acceleration. A change in either direction or speed, or both, is an acceleration and requires a net force. Moreover, an acceleration can be either positive or negative; either the speeding up, or the slowing down, of a body is an acceleration. A negative acceleration, that is, the slowing down of a body, is often called a "deceleration," but in physics the word "acceleration" can cover both cases.

The force which appears in Newton's second law is the *net* force, the sum of all forces acting upon the body. If you pull a wagon over a rough surface, the horizontal forces on the wagon are your pull in one direction, and friction, which occurs whenever one object moves over another, in the other direction. The net horizontal force is the sum of these two forces, and this net force determines how successfully you accelerate the wagon. When you are pulling the wagon at a constant speed in a fixed direction, the force you exert is exactly the same in magnitude as, and opposite to the direction of, the frictional force, so the net force is zero, as Newton's first law requires.

The second law can also be written in terms of *linear momentum*. In Newtonian mechanics, the linear momentum of a body is simply its mass times its velocity, i.e., $p = mv$, where p is the usual symbol for (linear) momentum in physics. (In physics, the word "momentum," when unqualified, always refers to linear momentum, i.e., that momentum which is attributable to straight-line translations in space.) From the definition of acceleration, it follows that the change in momentum with time is equal to ma. Force, then, can also be defined as *that quantity which causes a change in the momentum* of a body. The expression of the second law as a change in momentum is more general than its formulation involving acceleration, since a change in momentum can occur because of a change of mass as well as a change in velocity. Of course, the mass of an isolated object never changes in Newtonian physics, but the concept of momentum enables systems to be treated in which mass can change. A favorite example in many physics textbooks is an initially empty boxcar rolling under a hopper while being loaded with coal. It is possible to compute the force on the boxcar much more easily by the application of the momentum law than by attempting to calculate the accelerations of all parts of the system involved.

Since a force causes a change in momentum, it follows that if no force acts, the momentum of a system does not change. The law of inertia is thus generalized to the law of **conservation of momentum**, which states that *the linear momentum of a system never changes as long as no external force acts.* The law of conservation of momentum is considerably more powerful than the law of inertia alone, since it permits such complicated systems and interactions as collisions, compound objects, and so forth to be handled elegantly by relatively simple mechanics. More important, the conservation of linear momentum is more fundamental to the laws of physics than is the bare law of inertia. Unlike the force-acceleration form of Newton's laws, the momentum laws can be readily extended to more advanced physics, such as special and general relativity and quantum mechanics. Even deeper, the law of conservation of momentum can be shown to arise from fundamental symmetries of space. Momentum is one of the basic quantities of the physical universe.

Force, acceleration, momentum, and velocity are all **vectors**; that is, both the magnitude (size) and direction are important. An acceleration can occur if just the direction of motion, and not the speed, changes. When you drive around a curve, you feel yourself accelerated toward the side of the vehicle, even if the needle of your speedometer never moves. There is a special, essentially geometrical, way to add vectors which we will not treat here. It is sufficient to realize that vertical forces cause only vertical motions, while horizontal forces create horizontal motions. A force that is neither strictly vertical nor strictly horizontal can be broken into *components* along those directions; its vertical component can be added to any other vertical forces, and similarly for the horizontal component.

As an example of the vector nature of forces, consider a cannonball shot straight from a cannon. By Newton's first law, a force is required to start the cannonball into horizontal motion; that force is supplied by the explosion due to the gunpowder, and the cannonball is then accelerated in obedience to the second law. After the cannonball exits the barrel, there is no further horizontal force upon it. Therefore, again by Newton's first law, the cannonball should continue to move at a constant horizontal speed in a straight line. But if the cannon is fired in a gravitational field, there is always a vertical force upon it. How does that affect its motion? By the previous paragraph, the vertical force of gravity cannot affect the horizontal motion of the cannonball. It does affect the vertical component: it causes the cannonball to fall to Earth. The combination of straight-line falling to the ground, with acceleration, and straight-line motion horizontally, with constant speed, creates the net curved motion of the cannonball, which is a *parabola*. The rate of fall of the cannonball is *exactly* what it would be had it been simply dropped from its initial height. That is, if you were to drop one cannonball at the same instant and from the same height as you fired a second one from a horizontal cannon, both balls would hit the ground at exactly the same time! The distance traveled during this time interval by the second ball would depend upon its muzzle speed, of course. This effect accounts for the difficulty in observing the fall of a fast projectile such as a bullet, since it travels a great distance and generally strikes a target before it has time to fall far.

If, rather than pointing the barrel horizontally, you aimed the cannon upward, you would gain a little more time, because the time of travel would now be that interval required for the ball to rise to some maximum height and then fall back to Earth. But by firing upward you may lose some horizontal distance, because the ball's initial velocity now has a vertical component, which does not contribute to crossing the horizontal distance to the castle you are bombarding. It turns out to be possible to find the best compromise between these two competing effects, and, if air resistance is negligible, the maximum range is obtained when the cannon is pointed at an angle of 45° to the ground. Bombardiers and artillery gunners, as well as pilots of all kinds of airplanes, as a few examples, are well aware of the vector nature of forces and velocities and must accommodate them constantly.

Now we can understand why circular motion is not a "natural" motion. It is, in fact, accelerated motion. It has a uniform speed, indeed, but the direction of motion changes constantly. Without the force of gravity, planets would not orbit but would travel forever through space in a straight line. Gravity causes them to bend constantly, deviating from the straight line they would otherwise follow. We can also now see how

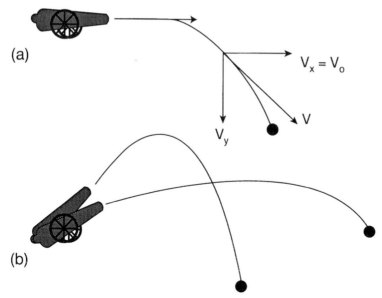

Figure 3.2 Trajectories of cannonballs in a constant gravitational field. (*a*) The projectile has only a horizontal initial velocity, $V_x = V_0$. At later times, gravitational acceleration creates an increasing downward velocity, and the net velocity is the vector sum of these two, V_x and V_y. (*b*) Tilting the cannon up produces both horizontal and vertical components in the initial velocity.

to correct the misconception many people hold about circular motions. Suppose you attach a ball to a string and whirl it around your head. The tug you feel from the string tells you that there is a force. The force is exerted by your hand and is transmitted through the string to the ball. What if you did not tie the string securely, and the ball slips out? What will be its subsequent motion? According to Newton's first law, once it is freed of the force from the string it will fly off in a straight line, *not* continue its circular motion. This is an easy experiment to try. (You can simply let go of the string in order to remove the force.) Careful observation, ignoring any preconceptions you might have, will show that the ball does move away in a straight line.

We now know how to create changes in a body's motion. Push it, pull it, or exert some other kind of force on it, and it accelerates. But if you act on something—for example, you push against a stalled automobile with its transmission in neutral—are you yourself unaffected? Are you able to exert forces on objects without any back reaction? Obviously, this is not the case; applying forces to physical objects has consequences for you. If you push on something, it pushes back on you. The exact relation is one of equality, leading to:

> **Newton's third law of motion:** For every action, there is an equal and opposite reaction.

This law is easy to misunderstand and probably causes more confusion than the other two put together. The action and reaction forces always act on different bodies. Body A exerts a force on Body B, and Body B exerts an equal and opposite force on Body A. If you do not understand this law, you might wonder how a horse can pull

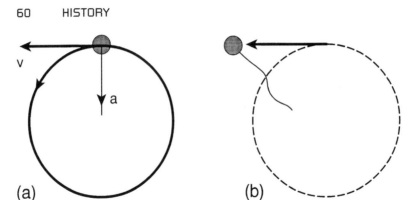

Figure 3.3 (*a*) A ball tied to a string and whirled about one's head moves in a circular path due to the force exerted by the string. The force, and hence the acceleration **a**, is directed from the ball toward the center. This is known as a centripetal force. (*b*) If the string breaks, there is no force and the ball moves in a straight line with constant speed.

a wagon. The horse exerts a forward force on the wagon, but the wagon exerts an equal backward force on the horse. How can they move? It is correct that the wagon pulls backward on the horse. If you have ever pulled a wagon, you are aware of the stretching of your arm, caused by the backward pull of the wagon on you. But we are not asking the right question here. The horse's hooves push against the Earth, and it is the reaction of the *Earth* upon the *horse* which ultimately moves the wagon.

A familiar example of Newton's third law is the "kick" of a gun or cannon. Not everyone has ever fired a gun, but those who have, have experienced this phenomenon

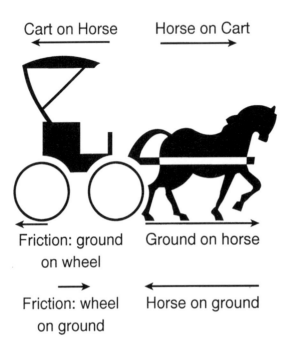

Figure 3.4 Forces on a wagon being pulled by a horse. The net force is the (vector) sum of all the forces. In this case, the net force on the horse and cart is the difference between the pull of the horse and the friction of the ground. The direction of frictional forces is always opposite to the direction of the motion.

first-hand. The explosion of the powder within the gun exerts a considerable force on the bullet. By Newton's third law, there is an equal and oppositely directed force upon the gun. This force causes the gun to accelerate, in the opposite direction from the acceleration of the bullet. Hence a force must be exerted to bring the gun to rest after its recoil. The object against which the gun is braced, often the shooter's shoulder, produces a force against the gun that, again by Newton's third law, exerts an equal and opposite force against the shooter, producing significant effects such as a pain in the shoulder. The amplitude of this force depends upon the acceleration (deceleration, if you will) of the gun. The more slowly the gun is brought to a halt, the smaller the deceleration and hence the less the force; conversely, a rapid deceleration requires a large force. Padding on the shoulder of a shooting coat helps to slow the deceleration and thus to reduce the force.

Does Newton's third law also imply that if the Earth attracts a brick, then the brick attracts the Earth? Indeed it does. Why, then, does the brick fall to the Earth, and the Earth not rise to the brick? The answer is found in Newton's second law. The mass of the Earth is so much larger than that of the brick that, although in principle the Earth does move, in practice, its acceleration due to the brick is unmeasurably small. The brick, on the other hand, acquires from the same magnitude of force a very large acceleration, and crashes to the ground, to the hazard of anything in its path.

The Law of Universal Gravitation

Now that we understand the laws of motion, we may see how they apply to gravity. According to his reminiscences, the basic ideas came to Newton when he was home in Lincolnshire during the plague in 1665. Many years later, a younger friend reported that the aged Newton told him over tea how his thoughts turned to gravity when, upon watching an apple fall, he began to contemplate that the same force that caused the fall of the apple might account for the orbit of the Moon. (He did not mention being hit upon the head by the apple, and that detail is probably just a bit of legend embroidered upon this account, if indeed there is any truth to the story at all.) It was perfectly well understood at the time that there was some force that causes objects to fall to the ground, a force called "gravity." Newton's bold leap was to imagine that the force extended not only at the surface of the Earth, but to the distant Moon.

If the Moon were moving according to Newton's first law, it should travel in a straight line. Since its path is curved, its velocity changes, and hence there must be a force causing this acceleration. The force must be directed toward the Earth, or, more precisely, along the line joining the center of the Moon to the center of the Earth. Newton calculated the acceleration required to keep the Moon in orbit and found it to be about 1/3600 as great as the acceleration due to gravity at the surface of the Earth, a quantity which had been measured by Galileo. Newton knew, from Kepler's third law, that the force had to decrease as the distance between the bodies increases. He conjectured that the force varied as the *inverse square* of the distance. Since the distance to the Moon is close to 60 times the radius of the Earth, the inverse square law is consistent with the observed acceleration. Newton later wrote that he "...thereby compared the force requisite to keep the Moon in her Orb with the force of gravity at the surface of the earth, and found them to answer pretty nearly." Thus he was able to conclude that gravity is, in fact, described by an inverse-square law.

Once Newton had determined that the force law was an inverse square, he was able to show that Kepler's laws followed necessarily. There was, however, one stumbling block. It is not obvious what the "distance" between the Earth and the Moon should be. Newton had assumed that the Earth, an extended body, attracts the Moon as if its mass were concentrated at a point at the center. This seems like a reasonable approximation for the Earth-Moon system, but what about the Earth-apple system? Yet Newton's estimate indicated that even for the apple, the Earth attracted it as if all its mass were concentrated at the center. In order to prove why this should be the case, Newton was forced to invent a new system of mathematics, integral calculus.[1] With integral calculus in hand, Newton was able to prove that the gravity of a spherical body is the same as that produced by the equivalent amount of matter concentrated in a point at the body's center.

We can explain Newton's proof qualitatively by exploiting the symmetry of the sphere. Imagine a small "test particle" located just above the north pole of a spherical object. Consider a small volume element on the left side of the sphere, as seen from the position of the particle. The gravitational pull of this element on the test body is directed downward and toward the left. For each such element, there is an equivalent volume element on the right that pulls the particle down and to the right. For the case of a spherical distribution of mass and an inverse-square gravity law, the left- and right-directed forces cancel, leaving only the attraction toward the center. This

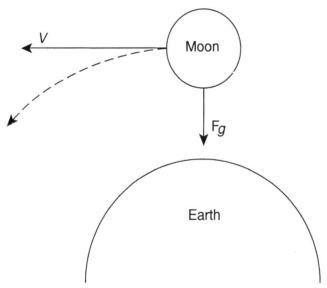

Figure 3.5   The direction of the gravitational force F_g acting on the Moon is toward Earth. The Moon's instantaneous velocity V is perpendicular to the direction toward Earth, but the constant centripetal gravitational acceleration produces a circular orbital path (*dashed line*).

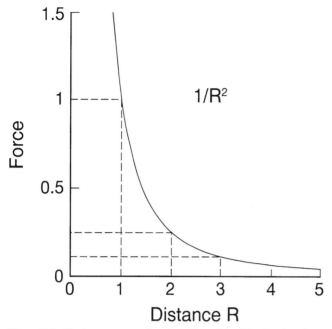

Figure 3.6 The inverse-square function. The value of the function decreases very rapidly as the distance increases.

argument relies on the assumption that the distribution of matter within the sphere depends only on the radial distance from the center, which is quite a reasonable approximation for a planet or star. With some further reflection along these lines, you should be able to see that for *any* point *outside* the sphere, the gravitational attraction of all the matter in such a sphere cancels for all directions except toward the center. Therefore, the spatial extension of the sphere does not matter; an object with perfect spherical symmetry behaves gravitationally as if all its mass were concentrated at the center.

After proving that the gravitational force was proportional to the inverse square of distance, Newton was obliged to evaluate its explicit form; ratios alone would not be adequate to develop the full formula. Newton was aware of Galileo's demonstration that all masses fall with the same acceleration in the gravity of the Earth. In fact, Newton repeated Galileo's work, using pendula whose bobs were of different masses; he was also able to take advantage of advances in the technology of timekeeping in order to time the periods of oscillation of the pendula. Newton found no difference in the period for a wide variety of bobs, a confirmation of Galileo's original experiments with inclined planes. The only way in which the acceleration due to gravity could be independent of the mass of the falling object would be if the force of gravity itself were proportional to the mass of the falling object. Let us write Newton's second law with a subscript to indicate that we are referring specifically to the force of gravity; the gravitational acceleration shall be denoted with the conventional lowercase g:

$$F_g = mg. \tag{3.2}$$

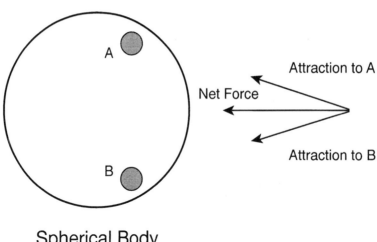

Spherical Body

Figure 3.7 The symmetry of a sphere means that those components of the inverse-square gravitational force directed away from the center cancel, while those components directed toward the center add. The net force is the same as if all the sphere's mass were concentrated at its center. This essential simplification greatly aided Newton in his search for the basis of Kepler's laws. Although real objects deviate from this idealization, the sphere is an excellent first approximation to very many astronomical objects.

But g is independent of mass, as determined from experiment; hence F_g/m must also be invariant with mass. Therefore, the formula for F_g must contain m, the mass of the falling body.

The next step employs Newton's third law. The mutuality of force, as required by the third law of motion, requires that the gravitational force also be proportional to the mass of the "attracting" body, which we will symbolize M. Both masses must be involved in a symmetric way, if the force of mass A on mass B is to be equal in magnitude to the force of mass B on mass A. Thus Newton arrived at the conclusion that gravity was proportional to the product of the masses, divided by the square of the distance separating them. This is known as Newton's **law of universal gravitation**, and it can be written mathematically as

$$F_g = \frac{GMm}{R^2},$$ (3.3)

where M is the mass of one body, m is the mass of the other, and G is a constant called the **gravitational constant**. The quantity R is the distance from the center of one object to the center of the other.

Newton indicated G symbolically because he could not compute its numerical value; it had to be determined experimentally. It is sufficiently difficult to measure G that Newton was long dead before anyone was able to find its value. Henry Cavendish, working in the last decade of the eighteenth century, invented a very sensitive balance, with which he was able to measure the extremely weak gravitational attraction between two objects of known mass. Since the gravitational acceleration can be measured relatively easily, Cavendish made it possible to "weigh" the Earth, or for that matter, the Sun, or indeed any other object whose gravitational force can be determined by one means or another, since G, g, and R together suffice to compute the mass of a planet. Even today, G is the most poorly determined of all the fundamental constants of nature, perhaps an appropriate situation in light of the fact that gravity remains the least-understood of the fundamental forces.

We can now understand the meaning of the **weight** of an object. The weight is nothing more than the force of gravity upon that object. At the Earth's surface, the radius R is very nearly constant, and so the force of gravity seems to be the same everywhere. (The Earth is not quite a perfect sphere, and its rotation introduces additional effects, but the corrections are small.) We know that if we drop an object, it accelerates toward the center of the Earth. We can compute the acceleration due to gravity at the surface of the Earth by combining Newton's second law in the form of equation (3.2) with his law of universal gravitation, equation (3.3). We obtain:

$$g = GM_{\mathrm{E}}/R_{\mathrm{E}}^2, \tag{3.4}$$

where M_{E} is the mass of the Earth and R_{E} is its radius.

Notice that the weight of an object varies with distance from the center of the Earth. We do not ordinarily notice this effect because we hardly ever travel far enough from the Earth's surface for g to be perceptibly different. At the greatest height to which most of us will ever travel, the 30,000 feet (9144 m) of a cruising jetliner, the acceleration due to gravity still has 99.7% of its value at the surface of the Earth. Thus for most practical purposes, g is a constant. However, delicate instruments called gravimeters can detect tiny changes in g. They can even detect the effects due to a local mass concentration in the crust of the Earth, such as a nearby hill or perhaps an accumulation of a relatively massive mineral or ore.

Equation (3.4) also shows how weight depends upon the mass of the planet. If we were to travel to the Moon, both the mass and the radius of the "planet" would be much smaller. The effect of the lower mass of the Moon, which is only 1.2% the mass of the Earth, is greater than the effect of the reduced radius, which has 27% of the Earth's value; hence the gravitational acceleration on the surface of the Moon has about one-sixth the value on Earth. Conversely, Jupiter is so much more massive than the Earth that the gravity at the top of the cloud layers of Jupiter is 2.5 times as great as at the surface of the Earth.

The MKS[2] unit of force is called, appropriately, the *Newton*. Since weight is a force, the correct MKS unit of weight is also the Newton. The kilogram, which is the MKS unit of mass, is often used for "weight." This convention is possibly due to the small size of the Newton relative to the weight of most "everyday" objects. In the very nearly constant gravity at the Earth's surface, the distinction is not of much practical importance, although, of course, one should keep the conceptual difference clear. On the other hand, the British unit *pound* is a unit of force and so is correctly used for weight. (The British engineering unit for mass is not very well known, except perhaps to fans of crossword puzzles; it is called a *slug*.) In MKS units, the acceleration due to gravity at the surface of the Earth is 9.8 meters per second per second, or 9.8 m/s². That is, if an object falls from rest, and air resistance can be neglected, at the end of 1 second, it will be traveling 9.8 meters per second; at the end of another second, it will attain a speed of 19.6 meters per second; and so forth, until it hits the ground or air resistance balances the force due to gravity.

Escape Orbit

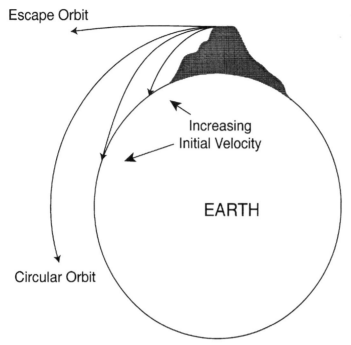

Figure 3.8 A planet's gravitational acceleration determines the speed required to reach orbit. Consider possible orbits for an object shot out in a direction parallel to the surface of Earth from the top of a high mountain. If the initial velocity is too low, the object falls back to Earth. At a velocity of about 7 km/s, the object moves so rapidly that the surface of the Earth curves away at a rate that exactly matches the fall of the body, resulting in a circular orbit. At about 11 km/s, the object exceeds the escape velocity and can leave Earth's gravitational field forever.

As we have mentioned, the inverse-square form of gravity accounts for Kepler's first and second laws, as Newton was able to prove. The proof is simple with the aid of fairly elementary modern mathematics, although Newton himself had to invent calculus almost as he went along. While the derivation itself is beyond the scope of our discussion here, we can comment on a few of the results. Consider two bodies in orbit around each other, one with mass M_1 and the other with mass M_2. Kepler's second law merely requires that the force of gravity must act along the line connecting the centers of the two bodies; such a force is called a *central force*. Kepler's first law narrows the possibilities; it requires that the force either obey the inverse-square law or else increase linearly with distance. Newton realized that gravity must decrease with distance; hence Kepler's first law, combined with this observational fact, pins down the form of the force law to an inverse square. Possession of the correct formulation of universal gravitation then made it possible for Newton to derive the full form of Kepler's third law:

$$G(M_1 + M_2)P^2 = 4\pi^2 r^3, \tag{3.5}$$

where r is the length of the semimajor axis of the ellipse describing the orbit and P is the period of the orbit. In this equation, we may use any set of consistent units and are not restricted to years and astronomical units. Notice that the factor which relates the period and the semimajor axis involves the *sum* of the masses of the two bodies. In the solar system, the mass of the Sun is completely dominant, and so the sum is essentially equal to the mass of the Sun alone. As a consequence, the relationship $P_{yrs}^2 = R_{AU}^3$ holds for all the planets, for all practical purposes; if this had not been true, Kepler might never have discovered this law in the first place. In contrast, if we were to study a binary star system, the masses of the two stars might well be comparable. For such a system, we could measure the period of their mutual orbit, and if we could resolve them telescopically, we might be able to find the length of the semimajor axis of the orbit. Without additional information, however, we could at best determine the sum of the two masses, not each individual mass. Fortunately, in cosmology we generally want to know the *total* mass of large systems and do not need to determine the masses of specific components of the systems. Kepler's third law, as modified by Newton, thus enables us to measure the masses not only of stars, but of galaxies and clusters of galaxies. Kepler's law provides a way to weigh the universe.

The three laws of motion and the law of gravitation are the fundamental relationships that make up Newtonian dynamics. At last, after more than 17 centuries of fumbling, humanity could comprehend the motion of the heavens. The Sun is the center of attraction of the solar system; the planets and comets orbit it. The motions may be described by a succinct, precise set of laws. The impact of this discovery upon European cosmology cannot be understated. Just as the rediscovery of Aristotle's writings paved the way for the Renaissance, the elucidation of the laws of motion was a factor in the shift in thought known today as the Enlightenment, and the industrial revolution which accompanied it.

Once these laws were disseminated among the intellectual elite of Europe, humanity's understanding of the cosmos increased rapidly. The new mechanics were quickly applied to many observations and experiments. For example, Newton's friend Halley conjectured that the bright comets observed in the years 1531, 1607, and 1682, all of which shared some similarities, might actually be the same object. Halley worked out an orbit for what is now called Halley's Comet, and predicted it would reappear in 1758. He did not live to see his prediction validated. The return of Halley's Comet was first spotted by an amateur astronomer on Christmas night of 1758. It has returned faithfully, approximately every 76 years, ever since. It is fitting that the man who played such a role in the publication of Newton's masterwork should be immortalized by his own application of Newton's laws.

Newton's laws are extraordinarily simple in form, but, unfortunately, it is difficult to compute the consequences of the laws for gravitating systems of more than two objects. Mathematicians have shown that it is impossible to find *exact* solutions for the mutual orbits of three or more bodies. However, approximate solutions may be found with pencil and paper, provided that one body is much more massive than the others. In that case, orbits are determined, in the main, by the two-body equations, with small corrections. A planet's orbit is very nearly determined by considering just the planet and the Sun. The gravitational influence of other planets produces small perturbations on that orbit. Thus for the solar system it is possible to apply *perturbation theory* to compute orbits. Newton had known only the five planets familiar to the ancients, but he suspected that the mutual attraction of Jupiter and Saturn might be detectable, and he even asked John Flamsteed, the astronomer royal, whether the two showed any anomalies in their orbits; none were observed at the time.

In 1781, more than half a century after the death of Newton, the German-English astronomer William Herschel discovered Uranus by direct observation, using a telescope he designed and built himself. Uranus is, in principle, visible to the unaided eye, but barely so; Herschel's knowledge of the sky enabled him to spot a dim star where no star should have been. Astronomers duly recorded observations of Uranus and computed its fundamental orbit. By 1845, the data were sufficiently precise to

Figure 3.9 Edmund Halley (1656–1742), the English astronomer who persuaded Newton to publish the *Principia*. His application of Newton's laws to cometary orbits allowed him to predict the return in 1758 of the comet that now bears his name. Courtesy of Yerkes Observatory.

show that the orbit of Uranus could not be explained by perturbations from the known planets. This small anomaly soon led to a great triumph of Newtonian mechanics: the prediction of an unseen planet. John Adams in England and Urbain Leverrier in France simultaneously predicted that a new planet must lie beyond Uranus, and both gave a location for that planet. At first the observers paid insufficient heed to these predictions, especially in England; Adams' work was ignored by George Airy, the astronomer royal at the time. Finally, in 1846 an astronomer at the Berlin Observatory discovered the new planet on his first attempt, within 1° of its predicted location. In keeping with the practice of naming planets for Graeco-Roman deities, the new planet was christened Neptune, for the god of the sea.

Compared with what was possible with the Ptolemaic tables, the power of Newton's mechanics was intoxicating. The whole universe, for all time and space, seemed within the grasp of humanity's understanding. The Newtonian universe was infinite in extent and populated evenly with stars similar to the Sun. Each star had its own mass and a specific instantaneous velocity. Given the mass of every planet and star in the universe, and their velocities and positions at one instant in time, Newton's equations are fully deterministic, predicting both the future and the past evolution. The gravitational law provides a force, the second law determines the acceleration, the acceleration determines the velocity, and the velocity determines the new position. The practical difficulties of actually computing the evolution of the universe are not so important. The watershed was the transformation of the universe from something intrinsically mysterious and unknowable to something deterministic and calculable.

In modern times the availability of tremendous computing power reduces somewhat the practical problem inherent in Newton's laws. The equations of gravitation among large numbers of bodies are routinely solved with great accuracy on computers; we shall later see how such "N-body" simulations are helping to investigate the formation of galaxies in the universe. However, it turns out to be impossible to predict orbits for all times with arbitrary accuracy. Self-gravitating systems are known to be *chaotic*; eventually, very small errors in our knowledge of the current orbits become large errors in our projections of future orbits. For the solar system, with relatively few bodies and one very dominant mass, these errors grow only over billions of simulated years, so for the needs of determining the orbits of, for example, space probes, we may solve Newton's equations to any desired precision. Nevertheless, the chaotic behavior of gravity shows that we can still find surprises in Newtonian mechanics.

Just as Aristotle's cosmology fitted the attitudes of the Middle Ages, so did Newton's cosmology suit the prevailing philosophy of the Enlightenment. The universe was like a grand clockwork, the stars and planets turning to the pull of gravity like the bearings of a finely balanced watch. It was a confident age, when knowledge of both science and technology increased rapidly. The Industrial Revolution was stirring, and Europe was well established in its colonial adventurism. Among the educated, affluent classes of both Europe and North America, a popular theology was deism, which was heavily influenced both by Newtonian cosmology and by the growing precision in technology. Deism views the universe as a kind of majestic machine, created by a master machinist and set into eternal motion; it is natural law that reveals the divine. Newton's clockwork universe ticked along for almost two centuries before new complications arose, which once again changed our views of the universe.

Newton was well aware of the majesty of his accomplishments, yet he was also aware of their limitations. A particular difficulty, for which others criticized him, was the appearance that gravity exerted its influence instantaneously at a distance, with neither an intermediary nor obvious causal contact. Newton conceded that he could not find the cause of gravity, but it was, for the moment, enough to elucidate its effects. For insights into cause, the world would have to wait for Einstein.

> I do not know what I may appear to the world; but to myself I seem
> to have been only like a boy playing on the seashore, and diverting
> myself in now and then finding a smoother pebble or a prettier shell
> than ordinary, whilst the great ocean of truth lay all undiscovered
> before me.
>
> —ISAAC NEWTON

The Geologists

If cosmology was the grandest science, based upon the stars, geology began as one of the humblest, a purely practical exercise in locating exploitable minerals, planning roadbeds and canals, and the like. Perhaps the old Aristotelian notion of the Earth as debased persisted; in any case, the study of the Earth for its own sake did not begin to become established as a science until the eighteenth century. However, just as the contemplation of the heavens above produced an understanding of the vastness of space, contemplation of the rocks at our feet was to produce an awareness of the vastness of time.

Geologists noticed very early on that many rock formations were stratified, apparently built layer upon layer by some process. In 1669, Nicolaus Steno, a Danish-Italian naturalist, published the suggestion that older rocks were below and newer ones were on top, an idea that seems perfectly obvious today; but at the time, all rocks were thought to be the same age. It slowly became clear that rocks were marked by the history of the Earth, and in the late eighteenth century, the British geologist James Hutton proposed the theory of *uniformitarianism*, the assertion that the same geological processes that we observe today, such as wind, water, and volcanism, also operated in the past. Geologically, the past can be explained by an understanding of the present. Uniformitarianism did not gain immediate favor, as it conflicted with the prevailing beliefs in Europe at the time. Nevertheless, the very thickness of the layers of rock could not be ignored, and in the early nineteenth century geologists began to entertain the idea that the Earth might possibly be very old. Within some of the layers fossils were found, strange traces of creatures which matched no known living animals. At first many Europeans believed that these animals were still alive somewhere else, but as more and more of the world became known in Europe, this became increasingly untenable.

The theory that found favor during much of the early nineteenth century, especially in France, was *catastrophism*, the belief that the Earth had experienced numerous and frequent upheavals in the past, each catastrophe wiping out the animals of that geological layer. The extinct animals were replaced, either by a separate creation or else by colonization by animals from other regions. But there were always doubts. Whatever the explanation for these mysterious imprints, the geologist William Smith demonstrated during the end of the eighteenth century and the beginning of the nineteenth that a given type of rock layer was uniquely associated with a particular set of fossils. The strata could be arranged in relative order by examination of the fossils they contained. This was not inconsistent with catastrophism, but the depth of the layers implied such a large number of catastrophes as to be troublesome. Moreover, there were resemblances in animals from one layer to another. By the 1830s, geologists had realized that the strata showed a progression of complexity of the fossils they bore; the oldest rocks contained no detectable fossils. Next came layers which held only invertebrates, and finally came newer layers which successively were dominated by fishes, reptiles, and finally mammals and birds.

One of the earliest proponents of a transformation theory in biology was the French scientist Jean-Baptiste Lamarck. Lamarck held that animals changed over time, gradually and in response to environmental conditions, a view which put him into direct conflict with catastrophism. Lamarck's explanation for such changes was that acquired traits could be passed from parent to offspring; the famous and standard example is the giraffe, stretching its neck to reach higher leaves and passing the elongated neck to its offspring. We know now that such acquired traits cannot be inherited, but during the whole of the nineteenth century, the mechanisms of inheritance were entirely mysterious. Gregor Mendel's work in establishing the discrete and predictable nature of inheritance was carried out from 1862 to 1865 but was ignored until after the turn of the century. Lamarck, like many of his time, believed in Aristotle's *Scala Naturae*, the great Ladder of Life, a hierarchical arrangement of creatures in order of increasing perfection, with the pinnacle of life, humans, at the top. The driving force for evolutionary change, in Lamarck's view, was not survival, but an urge to climb the ladder toward greater complexity. Although Lamarck's theories have been completely discredited, it should be noted that it was the first consistent proposal that animals change over time, that species are not fixed and perfectly suited for their niches.

Meanwhile, in Great Britain, Charles Lyell published between 1830 and 1833 his *Principles of Geology*, a book which placed uniformitarianism on a firm foundation. Lyell's work is often considered the beginning of modern geology, and it clearly showed that the Earth was ancient, although no one knew at the time how old.

Darwin and Wallace

Charles Darwin (1809–1882) was the scion of a wealthy and influential family. His paternal grandfather Erasmus was a major figure in the elite circles of the day, and his maternal grandfather was Josiah Wedgwood, founder of the famous china and pottery company. Charles, however, was a woolgatherer, indifferent toward his studies, and most enthusiastic about wandering the countryside collecting specimens of interesting animals and plants. When Charles received an offer in 1831 from Captain Robert Fitz-Roy to travel as naturalist aboard the H.M.S. *Beagle* on a five year voyage of exploration, he eagerly accepted. Darwin's adventures were to write the final chapter of the Copernican revolution, by removing humankind from its assumed splendor as a special kind of creature.

One of the books that Darwin took along with him was Lyell's first volume on geology. As he observed both the variety and the similarities of animals all around the world, Darwin came to see the evolution of species as itself a form of uniformitarianism; the same processes occurred throughout the history of the Earth, leading to slow and gradual changes in the animals that occupied it. It was the competition for survival, and the survival of offspring, that drove these changes. In any generation, those best suited to their environments left more offspring, of which in turn the best adapted reproduced most successfully. The natural variations in individuals was the raw material of change. Some of the offspring of the giraffe's ancestors had necks a little longer than others; by their ability to reach higher leaves, they gained a better diet, and produced more offspring, of whom the longest-necked survived best. There was still no understanding of the biology of heredity, but Darwin was thoroughly familiar with artificial selection, especially pigeon breeding, by which the breeder chooses for breeding stock those young that display some desired characteristics. Of course, nature could work in a similar manner. Natural selection, operating upon the inherent variability of a population, could, over the eons of time provided by the new understanding of geology, produce the great array of species from a few ancestors.

By the time he returned from his journey in 1836, his ideas had already crystallized, but Darwin had little courage to confront their implications. He feared dissent and the disapproval of his family, and thus he delayed publication for nearly 20 years, until a fateful letter arrived from a young man who, like Darwin many years before, had set off to see the world. The young man was Alfred Russel Wallace (1823–1913), and unlike Darwin, he had grown up and lived in poverty and hardship. Wallace had also traveled the tropics, and he had reached the identical conclusions as Darwin. In fact, it was Wallace who first broached the topic to the wider world, while Darwin dallied and procrastinated, endlessly reworking his notes. Wallace published a small paper in 1855, proposing that species came into existence from earlier species. It was only then that Darwin was persuaded to publish. Darwin's friends made arrangements for papers by both men to be read at the same scientific meeting in 1858. (Wallace, the working-class outsider, seems to have been deeply grateful at the opportunity to be heard by the learned men of British science, a profession generally reserved for members of the upper class at the time.) The credit for the theory of natural selection rightly belongs to both Darwin and Wallace, but it was Darwin who published, in 1859, the landmark book *The Origin of Species*, and Darwin whose name became associated in popular parlance with evolution.

The *Origin of Species* was described by Darwin himself as "one long argument." The book marshals many facts and shows how simply they fit the hypothesis of natural selection, but at the time, no mechanism was known that could account for the process of gradual change in species. It was not until the beginning of the twentieth century that the work carried out by Gregor Mendel in his monastery's garden was rediscovered and replicated, leading to an understanding of heredity. The elucidation of the biochemistry of inheritance began only in 1944, when it was established that the unit of heredity, the gene, was composed of the molecule deoxyribonucleic acid, or DNA. Although much remains to be learned about genetics, the broad outlines are now well understood. Moreover, evolution itself, in response to environmental pressures, has been directly observed, on small scales. The examples with the most ramifications for human societies are the development of drug resistance in bacteria, and pesticide resistance in many insects. A dose of antibiotics may kill most, but not all, of a population of bacteria. Those that survive carry a trait that enables them to resist destruction by the drug. The resistant bacteria are able to reproduce extravagantly in the ecological space cleared by the deaths of their erstwhile competitors, and the frequency of the gene which endows the bacteria with resistance increases. Now, after fifty years of routine treatment with antibiotics, many common bacteria are resistant to such drugs as penicillin, and pharmaceutical companies must engage in a constant search for new substances that are effective, at least until new resistant strains arise.

If all creatures gradually changed through a process of natural selection, then the Earth could not be young. The process of evolution requires a great deal of time. Yet the geologists had already reached this conclusion; their evidence pointed to a slowly evolving Earth, and antiquity of the Earth provided time over which biological evolution could occur. The obvious changes that had taken place in the crust of the Earth itself made the idea of change of species tenable. The controversy made an accurate estimate of the age of the Earth one of the most important problems of the late nineteenth and early twentieth centuries.

Ironically, one of the first scientific challenges to face the new theories came from physics. William Thomson (Lord Kelvin) and Hermann Helmholtz independently computed the age of the Sun to be only approximately 100 million years, a figure which contradicted the evidence from geology, and which did not provide sufficient time for evolution to occur. With perhaps the unfortunate tendency of some physicists toward arrogance, Kelvin declared that the geologists were wrong, since the Earth cannot be older than the Sun, although he did hedge by commenting, parenthetically, that he had considered only the known laws of physics, and new phenomena could alter his result. Kelvin's mathematics were correct, but his assumptions were wrong; he had assumed that the Sun shone by means of the release of its gravitational energy as it contracted, in which case it would, indeed, be young. In fact, the Sun, as well as all other stars, is powered by nuclear reactions, a physical process unknown in Kelvin's time. The physics of the atomic nucleus was developed during the first 30 years of this century, and the notion occurred to several scientists, including Arthur Eddington and George Gamow, that nuclear reactions might play a role in stars. The fusion of protons was proposed as early as the 1920s as an energy source for the Sun, but nuclear physics was barely understood at that time, and the details were not correctly worked out. It was not until 1938, after further progress in the theory of quantum mechanics, that Hans Bethe elucidated some of the reactions by which the stars shine.

The discovery of nuclear physics not only removed the apparent age problem of the Sun, but also provided a means for directly measuring the age of the Earth. By the 1920s, the technology for **radioactive dating** had become established. For the dating of rocks, one of the most useful isotopes is uranium 238 (^{238}U), which decays to lead 206 (^{206}Pb) at a known rate. By comparing the ratio of ^{238}U to ^{206}Pb, it is possible to determine the time since the rock solidified into its present form.

The technique works best for igneous rocks, those brought up from the depths of the Earth's mantle by volcanic action, since such rocks contain the greatest quantities of radioactive elements, but the principle can be applied to anything in which radioactive nuclei are present. The results indicate that the oldest rocks on Earth are approximately 3.9 billion years old. Material from meteorites, whose surfaces were never molten and are essentially unchanged from their origin, show ages of approximately 4.5 billion years. The oldest of the moon rocks returned by the Apollo astronauts show about the same age as the meteorites. Various other dating techniques, as well as theoretical computations of the age of the Sun, agree quite well that the solar system is 4.6 billion years old. The most ancient established fossils, of bacteria, have been found to be 3.5 billion years old, very nearly the same age as the oldest surviving rocks.

The age of the universe itself is much more poorly known and is still a matter of some controversy. We shall study this question in more detail later, but it is certain that our Galaxy is at least as old as the solar system. The best estimates of the age of the Galaxy come from determinations of the age of globular clusters, which are thought to be the oldest objects in the Milky Way, and from white dwarfs. These methods give an age for the Milky Way of between 10 and 20 billion years.

Taking Down the Ladder

From Copernicus onward, the understanding has grown that the universe is not static, not "perfect" and immutable, but dynamic and ever-changing. The Earth, assumed through most of human history to be stationary and central, is a small chunk of rock in orbit about a middling star. Throughout the history of the Earth, its life has always been, and always will be, dominated by bacteria. Bacteria are found in every environment that can support life and are by far the most common organisms. For more than 2 billion years, bacteria were the *only* life forms on Earth. Eukaryotes, cells with true nuclei, first appear in the fossil record scarcely a billion and a half years ago. Multicellular organisms have existed for only 750 million years. The genus *Homo*, to which modern humans belong, arose on the plains of Africa some 2 million years ago, while anatomically modern humans go back at most a mere 200,000 years. In contrast, the dinosaurs were the dominant vertebrates for over 100 million years. Most of Earth's history took place without the presence of humans. If we disappeared, the Earth and its major life form, the bacteria, along with whatever other organisms might exist at the time, would continue unperturbed.

In the medieval Ptolemaic cosmology, the physical construction of the cosmos was hierarchical. At the center was Hell, the basest and lowest possible state. At the other extreme, outside the sphere of the stars, was the realm of the spirits. Humanity lay in the middle, on the surface of the Earth. Similarly, the great Ladder of Life placed humans at the peak of the Earthly species, but falling short of the perfection of heavenly beings. What Copernicus did to the heavenly spheres, Darwin did to the Ladder of Life. Humans sit at no pinnacle, either at a physical center of the cosmos, or at the peak of biological perfection.

The Earth is but one small planet orbiting around one common star; should it be distinguished as the only place in which life occurs? A straightforward adoption of the Copernican principle would argue that we are not alone. Life may not be common in the universe, and it may not exist anywhere else in the Milky Way. But if planets formed around one ordinary, unexceptional star, by a process which, to the extent that it is understood, does not require unusual conditions, then planetary systems must be abundant, especially around stars that lack binary partners. It is true that life is fairly sensitive, placing demands upon the conditions it requires, at least for the carbon-based life with which we are familiar. Life, as we understand it, requires reasonable stability of star and planet, the presence of a good solvent such as liquid water, and protection from disruptive radiation from the star, so that the weak chemical bonds that hold together the complex molecules of life are not broken. But under the right conditions, the great antiquity of life on Earth indicates that it develops readily. Of the unknown billions of stars in the uncounted billions of galaxies, it is difficult to argue that there cannot be other planets that support life. Whether intelligent life would exist on such planets, we cannot, as yet, say. The development of intelligent life, or, at least, life forms that are capable of asking questions about the universe in which they live, does not even seem to have been inevitable on Earth.

For many such cosmological questions, we have no definite answers. But we have come far from the geocentric, anthropocentric world of Aristotle. With the realization of our true place in the universe, humankind has been forced to accept humility. In exchange we have found that the universe of which we are a part is far larger, grander, and more fascinating than could have been imagined even a century ago.

KEY TERMS

Newton's first law
force
mass
Newton's third law
weight

law of inertia
acceleration
law of conservation of
  momentum
law of universal
  gravitation
radioactive dating

uniform motion
Newton's second law
vector
gravitational constant

Review Questions

1. When fully loaded with fuel, a certain aircraft has a mass 1.25 times greater than its mass when carrying minimal fuel. The acceleration of the aircraft for takeoff must be the same in both cases, since the length of the runway available, and the takeoff speed required, are both fixed. How much more "thrust" (force), relatively, must the engine exert to accelerate the aircraft for takeoff when it is fully loaded?

2. Airplanes, especially smaller ones, often "crab," that is, fly at an angle relative to the desired direction of travel. What conditions might make this necessary?

3. Why is circular motion not "natural"? Why does the velocity of an object in circular motion change even though its speed is constant?

4. You are an astronaut floating in space, while holding an object with a mass which is 1/100th of your mass. You throw this object in some direction. What happens to you? Is there a difference between how Newton's laws work in space and how they work on the Earth?

5. In a certain science fiction story written for youngsters, an accident causes an untethered astronaut to float away from his spaceship. Fortunately, he manages to return safely to the ship by making swimming motions with his arms. What is wrong with this? What is the difference between swimming in water and "swimming" in space?

6. Aristotle says "To keep your automobile moving down the highway requires a steady force, hence you must keep your foot on the accelerator pedal." What would Newton say in rebuttal?

7. Zorlo has a mass that is 1.5 times that of the Earth, and a radius 1.25 times greater. How large is the acceleration due to gravity at the surface of Zorlo compared to the acceleration at the surface of the Earth? (Hint: you do not need to know the value of the gravitational constant for this problem.)

8. Which of Kepler's laws enables modern cosmologists to compute the mass of a distant cluster of galaxies? How might such a measurement be performed?

9. What is the principle of uniformitarianism? Do you think it can be applied to the universe as a whole as well as to the Earth?

10. When Lord Kelvin computed the age of the Sun, what critical assumption did he make? What was his result, and what did he conclude about the age of the Earth? What later discovery showed Kelvin's result to be incorrect, and why?

11. Astronomers sometimes try to estimate the number of planets on which technological beings might live. Part of the process is to estimate (1) the fraction of planets capable of supporting life on which life actually appears, (2) the fraction of those planets with life where some form of life achieves intelligence, and (3) the fraction of those planets where intelligent life develops the technological capabilities necessary to send radio signals. What do you think these fractions might be? Discuss your choices.

Notes

1. Gottfried Leibniz, in Germany, simultaneously and independently invented the concepts, and for years, a nationalistic dispute raged between England and Germany over the credit for this important work. Today both men are given credit as the developers of calculus.

2. See Appendix C for definitions of systems of units.

PART II
Background

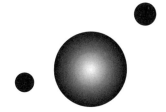

CHAPTER 4

Lighting the Worlds

Humanity contemplates the cosmos from the restricted vantage point of a small planet near the edge of an out-of-the-way galaxy. How it is that we can learn anything at all about so vast a thing as the universe? The universe is more than its contents alone; the physical laws that govern the interactions between objects tell us the properties of the cosmos. From the smallest elementary particles to the largest galaxy cluster, the rules of the universe leave their mark. By understanding how nature works in our own backyard, we can uncover the workings of the cosmos itself. In only 300 years, science has made great progress in elucidating these rules, and may be close to an understanding of how the universe operates at its most fundamental level. Science builds upon what is known; since cosmology deals with the overall principles of the universe, it draws upon knowledge from many fields. In this chapter, we will provide a brief, and highly selective, overview of a few topics relevant to our later studies: basic properties of matter, the fundamental forces of physics, and some properties of light. From there we shall begin our detailed exploration of the grandest science.

The Nature of Matter

Humans have long searched for the fundamental basis of matter. Throughout the Middle Ages, Europeans generally held the Aristotelian view that Earthly matter consisted of four elements: earth, air, fire, and water. The heavenly bodies were made of the celestial ether, an ill-defined, perfect, and immutable substance. The modern view of matter began to take shape when chemistry developed into a science in the eighteenth century, distinguishing itself from the mystical pursuit of alchemy. Antoine Lavoisier (who was guillotined by overzealous French revolutionaries) and Joseph Priestly showed that chemical behaviors could be attributed to certain substances into which most chemicals could be broken down. These substances, which took many forms, were themselves chemically irreducible; they are the **elements**. By 1810, it was accepted that each element corresponded to a unique type of particle, an **atom**, a theory first developed in its modern form by the British chemist John Dalton. The atom is the smallest subdivision of matter that retains fixed chemical properties. The enormous variety of chemicals is created by links between atoms, their *chemical bonds*, which hold them together. Combinations of two or more atoms are called *compounds*; the behavior of a compound is, in general, nothing like the behavior of any of the elements that make it up, but depends in a fairly complicated way on the elements present and how they are bonded.

In 1869, Dmitry Mendeleev and Lothar Meyer, working independently, arranged the known elements into a table according to their atomic weights. Remarkably, the elements were found to show regularities in their behavior that repeated themselves nearly uniformly along a column of the table. These regularities were so predictable that Mendeleev was able to shift at least one element, indium, whose atomic weight had been incorrectly determined. He left spaces for undiscovered elements, predicting not only their atomic weights, but also their general chemical properties. When the first "missing" element, gallium, was discovered in 1875, it created great excitement, for it made clear that there was a unifying principle to chemistry that could soon be understood. After the development of atomic theory early in the twentieth century, chemists realized that another characteristic of atoms, the **atomic number**, was the key to chemistry. When the elements are arranged in order of atomic number, the regularities along the columns of the modern *Periodic Table* are nearly exact.

For astronomy, the most important elements are the first two in the periodic table, hydrogen and helium. Hydrogen makes up about 75%, and helium approximately 24%, of all the matter in the universe. The rest of the elements, while far less abundant, play an obviously important role: the Earth and everything on it, including

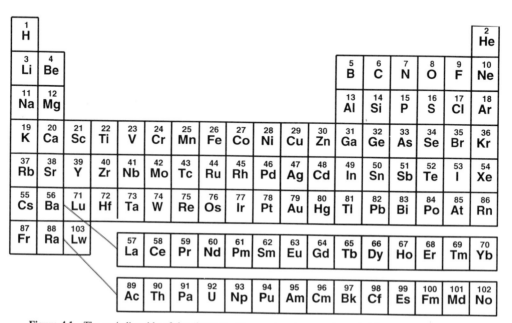

Figure 4.1 The periodic table of the elements. The number above each symbol is the atomic number, the number of protons in the nucleus. The two lightest elements, hydrogen (H) and helium (He), are the most abundant in the universe.

us, are made of these elements. We now have an explanation for the formation of the natural elements and their relative abundances, one of the great successes of modern astronomy and cosmology. As we shall see, the cosmic abundances of the elements also severely constrain the possible models of the universe. The atoms themselves can tell us something about the history of the cosmos.

The first discoveries of **elementary particles** soon clarified the Periodic Table. Atoms, while they do indeed represent the smallest particle of a particular element, are not indivisible. The **electron**, discovered by J. J. Thomson in 1897, was the first elementary particle found. Thomson initially measured only the ratio of the charge of the electron to its mass and made what amounted to a leap of faith that he had discovered a new, subatomic particle; his insight was later confirmed by more exacting experiments. Electrical charge per se had been known since the ancients observed that amber (Greek *elektron*) could, when rubbed with fur, attract small bits of straw. In the eighteenth century, Benjamin Franklin studied electricity and proposed that it was of two varieties, which he dubbed "positive" and "negative." The discovery of the electron made it clear that charge could be associated with individual particles; the electron carries one unit of negative charge. The charge, if any, controls the electrical behavior of the particle. If a particle has a charge, it is either positive or negative; if there is no charge, the particle is neutral. Like charges repel one another, while opposite charges attract. *Currents*, such as those that power electrical devices, consist of moving charged particles.

Even after the discovery of the electron, a great deal of confusion over the structure of the atom persisted for quite a long time. Atoms were known to be electrically neutral, but the atomic weight and number were not understood. The "plum pudding" model proposed by J. J. Thomson envisioned a structure with electrons embedded like raisins ("plums" in certain baking contexts) in a "cake" of positive charge. This model was accepted for a while, but it never really worked well. Between 1909 and 1911, Ernest Rutherford set out to test it, by shooting *alpha particles* (now known to be helium nuclei) at an extremely thin gold foil. Most of the alpha particles passed through the foil with only slight deflections, but there were a few which were deflected by large angles, in some cases nearly reversing their directions. Although

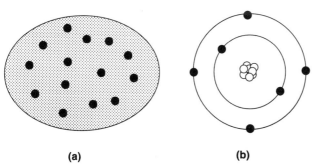

(a)   **(b)**

Figure 4.2 (*a*) The "plum pudding" model of the atom. Individual electrons are embedded in a general "cake" of positive charge. (*b*) The "solar system" nuclear model. Individual electrons orbit a compact, positively charged nucleus.

very few such extreme scattering events occurred, the result completely contradicted the predictions of the "plum pudding" model; at the time, no known model of the atom could explain it. Rutherford later said, "It was quite the most incredible event that has ever happened to me in my life. It was almost as incredible as if you had fired a 15-inch shell at a piece of tissue paper and it came back and hit you."

In 1911, Rutherford developed a new model for the atom; it consisted of a tightly packed **nucleus** surrounded by orbiting electrons. This theory explained his data perfectly. Most of the bombarding alpha particles, which are positively charged, passed far from the nucleus, and were scarcely affected, especially since the cloud of negatively charged electrons partially cancelled the positive charge from the nucleus. But a very few alpha particles happened to penetrate the electron cloud and pass very close to the nucleus, and for these particles, the repulsive force was very large. Rutherford was even able to predict mathematically the probability of such large deflections, based upon his model; the predictions fitted the data extremely well. Further work eventually led Rutherford to the realization that the atom with the smallest atomic number, hydrogen, had a nucleus consisting of a single particle. The new particle was christened the **proton**.

Almost immediately, theorists began to work out the details of this new model of the atom. The first, and most obvious, model was based on an analogy with the solar system. After all, the electrostatic force between charges obeys an inverse-square law, just like gravity, and the nucleus is much more massive than the electrons, just as the Sun is far more massive than any planet. Unfortunately, the analogy broke down. By the time of Rutherford's discoveries, it was known that accelerated charged particles radiate away some of their energy in the form of light. The original "solar system" model of Rutherford and Arnold Sommerfeld was thus untenable; the orbiting electrons should lose energy and spiral into the nucleus, which would have had most unfortunate consequences for chemistry!

Clearly, a better model for the atom was required. Niels Bohr, working on the hydrogen atom, hit upon a solution in 1913. Bohr's work was a major contribution to the nascent quantum theory, as it showed that electron orbitals were *quantized*; they could not be arbitrary. Unlike planets, electrons could occupy only discrete orbits of fixed energy. As long as the electron occupied a permitted orbital, it did not radiate, but if it jumped from one orbital to another, it emitted or absorbed a single **quantum**,

Figure 4.3 Rutherford scattering by an atomic nucleus. Most α-particle (energetic helium nuclei) shot toward gold foil are barely affected, but those passing very close to a nucleus undergo large deflections. No such extreme deflections would be observed if the atom's positive charge were smoothly distributed; hence the large deflections provide evidence for compact nuclei within atoms.

an indivisible unit, of light. It was already known that light could be characterized, under certain circumstances, as discrete particles called **photons**; according to the Bohr model, each transition involved only that photon whose energy equalled the difference in the electron energy levels. The new theory explained, with elegance and simplicity, many experimental results for light emission from hot gases.

By the mid-1920s, scientists had developed most of the modern picture of the atom. The atomic number is simply the number of protons in the atom, and this atomic number uniquely determines which element the atom represents. The protons reside in the nucleus, while electrons orbit far away, on the scale of the atom. The swarm of electrons is arranged in *shells* of increasing energy levels. Shells are further subdivided into *orbitals* of slightly differing energies. Only two electrons may occupy each orbital, but the total number of orbitals differs for each shell. The innermost shell, of lowest energy, has just one orbital and can contain only two electrons, whereas outer shells can hold more electrons, always in multiples of two. The chemical properties of an atom are determined by the number and arrangement of its electrons. Chemical reactions involve interactions between the outer electrons of one atom and those of others. Electrically neutral atoms rarely posses a set of fully occupied shells. Only the *inert gases* (sometimes called the noble gases)—helium, neon, argon, krypton, xenon, and radon—possess complete sets of filled orbitals; consequently, they participate in almost no chemical reactions. The strict regularity in the occupation of electron shells accounts for the patterns in the Periodic Table; atoms with similar numbers of unpaired electrons have similar chemical properties.

An atom which literally gains or loses an electron, thereby acquiring a net electrical charge, is called an **ion**. One way that an atom may become ionized is by temperature; a sufficiently large heating can provide enough energy to an outer electron to liberate it from the nucleus. Since the temperature required for this to happen varies with different elements, the ionization state of a distant cloud of cosmic gas can provide clues to the temperature of the gas. In the early universe, most of the hydrogen was ionized; since one of the characteristics of ionized gas is that it is more opaque to light than is neutral gas, this phenomenon places a fundamental limit on how far out, and hence back in time, that we can see.

Two ions of opposite charge that approach closely can be electrically attracted to each other and thus can adhere to form a chemical compound; such a bond is said to be *ionic*. The most familiar example of a compound held together by an ionic bond is sodium chloride, ordinary table salt. Most atoms cling fairly tightly to their electrons, however, and engage in chemical reactions in an attempt to complete their partially filled electron shells. The most common type of chemical bond is the *covalent* bond, in which the atoms share electrons. Only the electrons are affected by any type of chemical reaction.

Nuclear Physics

During the 1920s, physicists concentrated on atomic theory, arriving, with the help of the new quantum theory, at the model just described. Not much attention was paid to the nucleus. The only elementary particles known were the electron and the proton; it was thus assumed that electrons were present in the nucleus, as well as in the surrounding shells, although all nuclei still had a net positive charge. This model of the nucleus was probably the best that could have been devised at the time; besides, it was aesthetically pleasing to most scientists to think that the universe consisted of two particles of opposite charge. Unfortunately, this simple picture met the fate of many others: new discoveries which contradicted it. In 1932 a new elementary particle was discovered, the **neutron**. The neutron is somewhat more massive than the proton, and, as its name indicates, it has no net electric charge. It was quickly realized that the neutron was the "missing" particle of atomic theory; it was the true nuclear partner of the proton.

Neutrons and protons together make up the nucleus of atoms, and are collectively known as **nucleons**. The electrical charge of the nucleus establishes the atom's electron structure; hence the atomic number determines the type of element. Two atoms with the same number of protons but different numbers of neutrons are **isotopes** of the same element. Atoms are denoted by a symbolism of the form $_{p}^{n+p}Z$, where Z stands for the one-or-two-letter symbol for the element, p indicates the number of protons (often omitted, since that is always the same for a given element), and $n + p$ is the total number of neutrons and protons. For example, the isotope of carbon which contains 6 protons (making it carbon) and 6 neutrons, for a total of 12 nucleons, is symbolized as $_{6}^{12}C$. Isotopes occur in different *abundances*; for most elements, one isotope dominates, while the others are relatively rare.

In retrospect, of course, there had long been clues that the universe was not so simple as to consist of only two kinds of particle. In 1896, Henri Becquerel discovered that a crystal of a uranium compound resting atop a sealed, unexposed package of photographic film left an image. Becquerel had discovered **radioactivity**, the first known **nuclear reaction**. Chemistry occurs only among the electrons in the cloud surrounding the nucleus; the nucleus itself is never affected by any chemical reaction. Nuclear reactions, on the other hand, directly involve the nucleons. The nuclei of radioactive isotopes are unstable and emit radiation of some form, which may transform the nucleus. This radiation is of three types: *alpha particles*, which consist of two protons and two neutrons, *beta particles*, which are electrons, and *gamma rays*, which are essentially light rays of very high energy. When a nucleus emits an alpha particle, it *transmutes* into another element, that which has two fewer protons; it also drops in neutron number by two. For example, $_{92}^{234}U$ (uranium 234) is an alpha emitter; the result is a nucleus of $_{90}^{230}Th$ (thorium 230). Emission of a beta particle also causes transmutation, because when a beta particle (an electron) is emitted, a neutron is converted into a proton; therefore, the atom becomes an isotope of another element, that which has one additional proton and one less neutron. Beta decay causes $_{82}^{210}Pb$ (lead 210) to transmute to $_{83}^{210}Bi$ (bismuth 210). Emission of a gamma ray, on the other hand, does not change the elemental identity of the atom. Gamma rays may be emitted either on their own or in conjunction with alpha or beta particles.

The decay of any particular radioactive nucleus is completely unpredictable; however, if a sample of many nuclei is prepared, after a certain time interval, called the **half-life**, half the members of the original sample will have decayed. In another half-life interval, another half will decay, leaving only one-fourth as many as were initially present, and so forth. Radioactive decay provides an excellent means of dating specimens. For example, a sample of uranium decays into lead at a known rate; therefore, comparing the ratio of the amount of lead to the remaining quantity of uranium provides an accurate estimate of the time elapsed since the original sample of uranium accumulated. In the crust of the Earth, radioactive elements are most abundant in igneous rocks—those formed by volcanic eruptions. The oldest volcanic rocks on Earth are found in remote regions such as Antarctica, Greenland, and parts of Canada. They provide a lower limit to the age of the Earth, showing that the planet is at least 3.9 billion years old. Similar principles can be applied to determine the age of the galaxy, but since there are more potential sources of error and observational difficulties in this case, radioactive dating is less reliable than for the Earth; nevertheless, estimates from radioactive decay are at least consistent with other evidence.

A nuclear reaction even more extreme than radioactivity was identified shortly after the discovery of the neutron. In 1934, Enrico Fermi was attempting to create heavy elements by bombarding uranium with neutrons. He thought he had succeeded, but his interpretation of his data was criticized by Walter Noddack, who suggested that the uranium had, instead, actually split apart. This possibility was taken up by Otto Hahn, Fritz Strassmann, Otto Frisch, and Lise Meitner, who worked on the problem for 5 years. Frisch and Meitner developed the theory of atomic fission, while the others searched for experimental evidence. In 1939, Hahn and Strassmann succeeded in demonstrating that $^{235}_{92}$U would, upon absorbing a neutron, undergo **fission**, the spontaneous splitting of the nucleus into two much lighter nuclei. Only a few very heavy isotopes participate in fission reactions; the most important natural fissionable isotope is $^{235}_{92}$U. It can split in a number of ways, with the most probable outcome yielding barium, krypton, three neutrons, and a great deal of energy.

Uranium, with 92 protons, is the heaviest naturally occurring element. All of its isotopes are radioactive, but their half-lifes are mostly quite long, up to several billion years; hence uranium is fairly abundant in the Earth's crust. The heavier elements, called *transuranic elements*, are much more unstable and occur only under very special conditions, which typically must be engineered by humans. Of the transuranic elements, the most important is plutonium. Plutonium is very readily fissionable and liberates a great deal of energy when it splits. If it does not fission, it decays by emitting alpha particles. Plutonium is the basis for fission weapons, whereas most nuclear reactors for the production of electrical power use uranium (some use plutonium). Plutonium can be created from uranium 238 by neutron bombardment in a nuclear reactor.

Fission is one kind of nuclear reaction that involves the heaviest atoms. Some of the lightest elements will take part in another kind of reaction, **fusion**, in which two nuclei combine or fuse into a heavier element, with the liberation of various particles, as well as *much* energy. Fusion reactions occur only at extremely high temperatures and densities, as the two nuclei must be forced very close together before fusion will occur. Nuclear fusion is an especially important process in the present universe, as it is the source of energy in the stars. The Sun, and other stars like it, shines by converting 4 hydrogen nuclei (protons) to 1 nucleus of $^4_2$He deep within its core. Humans have been clever enough to learn how to initiate uncontrolled fusion

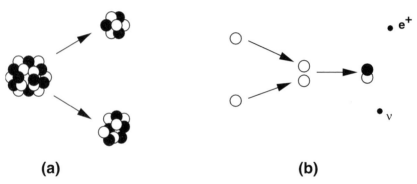

Figure 4.4 (*a*) Nuclear fission occurs when a heavy nucleus splits into two lighter nuclei. (*b*) Nuclear fusion occurs when two light nuclei fuse to form a heavier nucleus. In the figure, two protons fuse to become a deuterium nucleus, plus a positron and a neutrino. Both types of nuclear reaction can release a great deal of energy.

reactions, in bombs, by first raising the temperature and densities to the necessary levels via a fission explosion. We have not yet learned to control fusion reactions as a power source, however, primarily because such high temperatures and densities are extraordinarily difficult to create and maintain on Earth. In the Sun and other stars, fusion occurs in the innermost core, where the tremendous pressure due to the weight of the overlying layers confines the nuclei, and creates the high temperatures and densities required.

Nuclear theory opened up grand new vistas in physics. In the nineteenth century, the only forces known were the electromagnetic force and gravity. It quickly became apparent that neither of these could possibly have anything to do with nuclear reactions. Beta decay was particularly enigmatic, once it was realized that no electrons are present in the nucleus. If that was the case, where did the electron, which was generally ejected from the nucleus with a high energy, come from? In 1934, Enrico Fermi developed a theory of beta decay that introduced the idea that a neutron could be converted into a proton and an electron. In order to satisfy the conservation of momentum, Fermi postulated the existence of a new particle, the **neutrino**. (The neutrino had been suggested previously by Wolfgang Pauli, but Fermi first developed the mathematical theory.) "Neutrino" means "little neutral one"; Fermi gave it such a name because, in addition to being electrically neutral, it had, at most, a very tiny mass. Later theories assumed that the neutrino was strictly massless, but recently, evidence has grown that it does have a small mass; as we shall see, this question has significant cosmological ramifications.

The World of Modern Physics

At the beginning of the twentieth century, many of the era's leading scientists had refused even to accept that atoms existed; less than four decades later, the structure of the nucleus was nearly established. Even the stars themselves had yielded their deepest secret, the source of their energy. The nucleus itself soon seemed to represent only one aspect of the interactions among particles; after the Second World War, particle physics began to emerge as one of the most active subfields of physics. Although the proton, neutron, electron, and possibly the neutrino are the most important of the elementary particles in the present universe, there are many others. As physicists studied *cosmic rays*, which are high-energy rays impinging upon the Earth from space, they found new kinds of particles. With the construction of accelerators, even more particles were discovered. Many of these less-familiar particles are unstable; after some very short half-lifes, they decay into other particle species. As more and more "elementary" particles were found, physicists realized that there had to be a classification scheme to make sense of them.

Quantum mechanics is the system of physical laws that governs the behavior of the elementary particles and of nuclei and atoms. It is a formal, mathematical system that developed from the work of many of the greatest scientists of the twentieth century, such as Niels Bohr, Max Planck, Albert Einstein, Werner Heisenberg, and Erwin Schrödinger. Quantum mechanics made it possible to sort out the confusing extravagance of particles and to understand their behaviors. The salient feature of any quantum property is that it is quantized; it cannot occur in arbitrary amounts but only in multiples of a certain inherent value. Electric charge is an example of a quantum property; any particular particle, such as an electron, always has the same, specific quantum of electric charge. Particles possess many quantum properties and may be classified in various ways, depending upon the problem at hand; for now we shall be concerned with only one important property. According to modern particle physics, there are two fundamental classes of particle, with the division based upon the *spin* of the particle. The spin of a particle is similar to the spin of a macroscopic object, such as a baseball, but with the important difference that it is quantized. Remarkably, the spin of a particle may take *either* integer *or* half-integer values; that is, values of 0, 1, 2, and so on, *or* 1/2, 3/2, 5/2, and so forth, are permitted, but nothing else. A particle with integer spin is a **boson**, while one with half-integer spin is a **fermion**. The photon has a spin of 1; it is a boson. The electron has a spin of 1/2; it is a fermion.

Spin has a direction as well a magnitude, and this is also limited to discrete quantum amounts. The number of possible orientations depends on the spin. For example, the spin of an electron may be "up" or "down," relative to any particular direction the experimenter might choose. By convention, an "up" spin is positive, 1/2 for an electron; while a "down" spin is negative, $-1/2$ for the electron.

There is a fundamental division of labor between bosons and fermions. The primary duty of bosons is to carry force and energy. Fermions, on the other hand, make up matter. In addition to their different "jobs," bosons and fermions have drastically different properties. The most important of these is related to their "sociability," in a loose manner of speaking. Bosons are content with one another's company, and arbitrary numbers of them can crowd arbitrarily close together. Fermions, in marked contrast, obey the Pauli **exclusion principle**, a property worked out by the Austrian physicist Wolfgang Pauli. The exclusion principle is a limitation upon the **quantum state** of fermions, where a "state" consists of a description of everything that quantum mechanics permits us to know. The state of a particle might include its energy, its spin, whether the spin is up or down, and so forth. According to the Pauli exclusion principle, *fermions of the same species which can interact with one another may not* *simultaneously occupy the same quantum state.* The exclusion principle explains why only two electrons are permitted to occupy each orbital around an atom; one has spin up, the other spin down, but otherwise their states are the same. The exclusion principle also demands that fermions cannot crowd together, since interacting fermions must have distinct quantum states. No matter how bizarre it might seem, the exclusion principle controls much of the behavior of matter at the scale of atoms, nuclei, and particles. In astronomy, it is of fundamental importance in the structure of white dwarf stars and neutron stars.

The first quantum theory was nonrelativistic, but soon quantum mechanics and special relativity were combined into relativistic quantum mechanics. (Quantum theory has yet to be combined with general relativity theory.) This new theory contained a remarkable prediction: the existence of **antimatter**. Every particle has a partner called an *antiparticle.* The antiparticle is, in some respects, the "mirror image" of the particle, as it has the identical mass; an antiparticle differs from its partner by possessing the opposite sign of electrical charge, as well as opposite sign of some other quantum properties. (A neutral particle has a neutral antiparticle.) Only the antiparticle of the electron has its own name; it is called the **positron**. When a particle collides with its antiparticle, both are converted to pure energy, in the form of gamma rays. A few particles, most importantly the photon, are their own antiparticles. The universe today appears to be composed entirely of matter, although early on both matter and antimatter were present in great abundance.

No nineteenth-century scientist would even have dreamed of the menagerie of particles that were known by 1940. Although right now nature might seem to be excessively complicated, there is an underlying simplicity, which is partially understood. The myriad particles interact with one another in various ways, but all of the known interactions can be explained as due to one of only four fundamental forces of nature. These four forces are the **strong interaction**, which holds nucleons together in the nucleus; the **weak interaction**, which mediates nuclear reactions such as fission and beta decay; the **electromagnetic force**, and **gravity**. According to modern theories of particle physics, these four fundamental forces arise due to the exchange of "carrier" bosons called, for rather obscure historical reasons, *gauge bosons.* The binding between particles by a given force is thus much like the tie between two people playing catch by tossing a ball back and forth between them: the ball carries momentum and energy from one player to the other. The electromagnetic force is particularly well understood; its gauge boson is the particle of light, the familiar photon. Just as the ballplayers can toss a lighter ball farther, the range of a fundamental force is determined by the mass of its gauge boson. The photon is massless; therefore the range of the electromagnetic force is unlimited. Gravity is also carried by a massless boson, the *graviton,* which has so far eluded detection. The weak interaction is mediated by a massive particle; hence its range is limited. The strong interaction has an unusual behavior: as the distance grows, the nuclear force increases. Within the nucleus, the strong interaction has a massive carrier known as the *pion.* At higher energies, in the strange world of **quarks**, the carrier boson is a massless particle known as the *gluon.* Quarks are the fermions from which the nucleons, as well as some other particles, are constructed; gluons hold quarks together to form the nucleons and their kin.

Of all the forces, the strong interaction is, as the name implies, the strongest; it exceeds the electromagnetic force by about a factor of 100. It would have to be so, or it could not accomplish the job of holding the positively charged nucleus together in the face of powerful electromagnetic repulsion. The weak nuclear force is much weaker than the electromagnetic, by a factor of 10^{11}. But even that huge difference is dwarfed by the full range between the strong interaction and gravity: about 10^{41}. The two nuclear forces, or interactions, operate only over very small distance scales of around 10^{-15} cm, comparable to the size of a nucleus. The electromagnetic and gravitational forces, in contrast, are *long range*; both diminish as the inverse square of the distance between two particles. We shall not return to the nuclear forces until we study the early universe; here we will describe the other two.

The electromagnetic force is the force that exists between charged particles; it is ultimately responsible for many of the everyday forces we experience. It directly holds ions together in ionic bonds, by the attraction of positive and negative charges. It also causes molecules to stick to one another because molecules almost always have some distribution of charge, even if they are neutral overall. It is the adherence of molecules, through the relatively weak electromagnetic forces between them, which holds together almost all everyday objects, including our bodies. Glues work by causing various molecules to link together. The floor does not collapse under your weight because its molecules are electrostatically bound to one another. Friction is nothing but the very weak attraction of the surface molecules on one object to the surface molecules on the other object. The electromagnetic force is also responsible for the generation and transmission of electromagnetic radiation, that is, light.

The last of the four fundamental forces is gravity. Although gravity is an incredibly weak force compared to the others, it nevertheless is the most important for the universe as a whole. This is because the nuclear forces are short range, and the electromagnetic force is almost always *shielded*, or reduced, because most things in nature are, overall, electrically neutral. If this were not so, there would be enormous forces, as unshielded negative and positive charges attract one another very strongly, much more strongly than their gravitational attraction. Huge currents would result as charges were pulled to one another, until approximate neutrality would quickly prevail. Although mass plays the role of "gravitational charge," it is of only one type; there is no possibility for shielding by a charge of opposite type. Therefore, over scales larger than approximately 10^{-6} cm, the typical distance between molecules, gravity dominates. It is gravity which shapes the universe we observe, and through most of this book, we will be concerned with the gravitational force.

A convenient mathematical way of representing a force is by means of a **field**. A field is a function that describes the strength of the force at any point in space. Thus we may speak of the *gravitational field*, a representation of the force of gravity at all points. Similarly, we speak of the *electromagnetic field*, which can itself be broken into an electric field and a magnetic field. More generally, the term *field* can describe any physical entity that has an extension in space, such as the distribution of temperatures in a solid.

Allied with the concept of force are **work** and **energy**. In physics, the quantity *work* is defined very precisely, as the exertion of a force to produce a displacement.

Although both force and displacement are vectors (displacement is distance plus direction), work has only a magnitude, not a direction. A quantity that is fully described by its magnitude is called a *scalar*; thus, force is a vector, but work is a scalar. Since both the quantities that enter into computing the work are vectors, but the result is a scalar, you might guess that there is a special way of combining vectors to obtain scalars. In fact, there are a number of ways to obtain a scalar output from vector input; work is computed by one very useful method. We have already mentioned that a vector can be resolved into *components* along any two mutually perpendicular directions; all that is necessary is to construct a right triangle, with the "legs" along the two component directions, and the original vector the hypotenuse. Computing the work is essentially a matter of finding what part of the force is along the direction of the displacement; we then multiply this component of the force with the distance. If the force and the displacement are perpendicular, there is no work at all!

A quantity related to work is *energy*. Energy can be defined as the *capacity to do work*. If you do work against gravity to lift a ball to a certain height, the ball acquires **potential energy**. If it fell, it could strike the fin of a turbine and turn it, causing work to be done. The motion of the ball was the direct cause of the turning of the turbine, and therefore energy must be associated with motion; it is called **kinetic energy**. As the ball fell, it lost potential energy and gained kinetic energy. Its kinetic energy was then (mostly) converted into work when it hit the turbine. The recognition of the intimate connection between work and energy, by Sir James Joule, was a great step forward in the understanding of **thermodynamics**, the science of energy in general, and heat in particular.

Heat is another, very important, form of energy. Heat is related to the aggregate energy of the random motions of the individual molecules which make up an object,

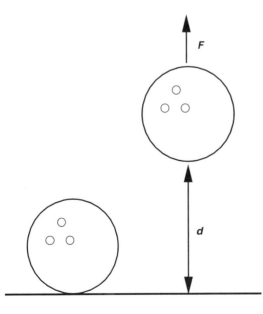

Figure 4.5 Lifting a bowling ball requires the expenditure of a quantity of work equal to the force applied times the distance the ball is lifted, $W = Fd$.

in contrast to what we specifically call kinetic energy, which is a consequence of the bulk motion of an object as a whole. Heat is used every day to produce work, by, for example, expanding the gas in the cylinders of an automobile engine. Although heat and temperature are related, they are not the same thing. **Temperature** is a function of the mean random speeds of molecules, whereas heat depends on such quantities as the individual random kinetic energies of the molecules, and the density of the substance. Thus it is possible for an object to have a very high temperature but relatively little heat energy. The corona of the Sun is a tenuous halo of ionized gas that surrounds the Sun and is visible during solar eclipses; it has a temperature of millions of degrees but is so thin that its heat content is not extreme. As a general rule, however, higher temperature is associated with greater amounts of heat energy.

Other examples of energy include *chemical energy*, the energy required to create or break a chemical bond, and *energy of deformation*, the energy required to transform an object, such as the fender of your car, into a deformed state. One of the most important laws of physics is the law of **conservation of energy**, which states that *energy is neither created nor destroyed, but is only converted from one form into another.* In classical physics, there is a companion to this law, the law of **conservation of matter**, which similarly states that *matter is neither created nor destroyed.* When we study special relativity we shall learn that mass and energy are equivalent, through Einstein's famous equation $E = mc^2$; mass itself is just another form of energy. Special relativity shows that the separate laws of matter and energy conservation must be superseded by a new principle, that of the conservation of matter plus energy. Both the law of conservation of matter and the law of conservation of energy can be considered to be individually valid to a high degree under ordinary conditions. However, in cosmology we shall often encounter circumstances that are far from "ordinary," so we must keep this grander principle in mind.

The conservation of energy is also known as the first law of thermodynamics. Since there is a first, there must also be a second. The second law of thermodynamics is one of the most significant laws of physics, and one of the least understood. Many equivalent statements of the second law exist; for now let us give the version presented by the German physicist Rudolf Clausius in the middle of the nineteenth century: *No cyclic process exists whose sole effect is to transfer heat from a cooler to a warmer body.* There are many devices, such as refrigerators and heat pumps, which transfer heat from cooler to warmer bodies; but this process is always accompanied by the exhaust of waste heat. The second law denies the possibility of perpetual-motion machines. Some energy is always dissipated into waste heat in any real, macroscopic process, and for this reason no machine, no matter how clever or carefully designed, can run forever without an input of energy. No perpetual-motion machine has ever been built. Every one that has been claimed has been found wanting upon close examination. Some have been outright frauds; others were so carefully balanced that they could operate for a very long time but not indefinitely. A notorious device of recent years whose inventor claimed it to produce *more* energy than it consumed was shown to be nothing but a simple power converter, and a very inefficient one at that. (A machine of this kind would also violate the first law of thermodynamics.) The second law has always triumphed, no matter how ingeniously humans have tried to circumvent it.

More modern versions of the second law connect this inevitable dissipation of energy to an increase in **entropy**. Precise, mathematical definitions of entropy exist, but loosely, entropy is a measure of the *disorder* of a system; the higher the entropy, the greater the disorder. The second law can be restated as *in any process, the overall entropy increases, or at best remains the same.* Since an ordered system has a greater potential to do work, an increase in entropy is accompanied by a reduction in available energy. The second law does not deny the possibility of the *existence* of order, however. Order can always be created locally by the expenditure of energy. As an example, biological systems—living creatures—represent highly ordered states, perhaps the most highly ordered in our region of the universe. Nevertheless, their mere presence is not a refutation of the second law; on the contrary, modern research on the theory of ordered systems indicates that dissipation is *required* for complex, ordered states to arise naturally. But there is a price for order, and that price is the conversion of available energy into waste heat whose capacity for useful work is greatly diminished. Biological entities obtain their energy ultimately from the Sun, or in a few species from geothermal energy. Like any other macroscopic process, life results in an overall increase in the entropy of the universe. Neither are artificial processes immune to the second law; energy must be expended to support manufacturing and transportation, with the inevitable consequence that entropy increases, and the Earth's supply of utilizable energy is reduced.

It may seem that the second law exists only to frustrate human attempts to get something for nothing. A consistent system of thermodynamics could be developed without it; yet it is always confirmed, not only in experiments, but in the realities of engineering and everyday life. The second law seems intimately related, in ways which we cannot yet understand, to the earliest moments of the universe, as well as to its ultimate fate. The second law appears to determine the *arrow of time*, the relentless march of time in one direction only. The second law of thermodynamics may be one of the deepest, most fundamental rules of the universe.

The third law of thermodynamics essentially completes the foundation of the system. (A fourth law, called the zeroth law, provides a statement that thermal equilibrium is possible.) The third law is a consequence of the observation that cooling to very low temperatures is difficult, and becomes more difficult as the temperature is lowered. The lowest possible temperature is **absolute zero**. The third law of thermodynamics states that *absolute zero can never be attained, but only approached arbitrarily closely.* The Kelvin temperature scale widely used in science sets its zero point at absolute zero, with its unit size equal to a Celsius (centigrade) degree. Absolute zero (0 K) corresponds to -273.15 C.

Waves

The material world of particles and atoms seems concrete and familiar; yet just as important to cosmology is the incorporeal world of **waves**. A wave is a disturbance in some quantity that propagates in a regular way. We are all familiar with water waves, ripples in a body of water that move across the surface, leaving it undisturbed after they pass. Waves carry energy with them as they travel, as anyone who has ever stood in the surf should realize. Waves are characterized by maxima called *crests* and minima called *troughs*. The maximum displacement from the undisturbed position is the **amplitude** of the wave. The number of crests that pass an observer in a specified unit of time is the **frequency** of the wave, while the distance from one crest to the next is the **wavelength** of the wave. For a pure or *monochromatic* wave, the wavelength is a well-defined constant. In general, however, an arbitrary wave is a *superposition* of many pure waves, and the wavelength is not so easy to define. The distribution of frequencies in a superposed wave is called its **spectrum**. The energy carried by a wave is related to its frequency; generally, the higher the frequency, the greater the energy transmitted.

Important types of wave include sound waves, which are oscillations in the pressure of a gas or liquid, and water waves, which are displacements of parcels of water. (Sound waves can travel in water, also, but they differ from what is defined as a water wave.) Sound and water waves are examples of waves which require a *medium* for their propagation. The particles of the medium move very little, whereas the wave can move, and transmit energy, over great distances. Water waves easily travel across the Pacific Ocean, while similar waves in the Earth's atmosphere can circle the globe. For astronomy, the most important type of waves are electromagnetic waves, that is, light. Light differs from the other waves described here in that it does not require a medium for its transmission, but otherwise its properties are similar to those of any wave.

Waves have several unique behaviors. Waves can undergo *reflection*, partially or completely, when they strike a surface; that is, part or all of the wave train turns back and travels in the opposite direction. When a wave passes from one type of

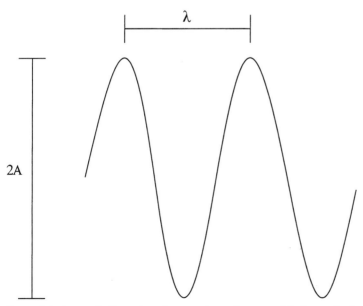

Figure 4.6 Schematic illustration of a monochromatic wave. *A* is the amplitude, while λ is the wavelength. A pure monochromatic wave would extend in both directions forever.

medium to another, *refraction*, a change in the wave's speed, and thus its direction of motion, occurs. An example of refraction in water waves is the bending of the waves as they move from deeper water into a shallow inlet. When a wave passes through an opening that is comparable in size to its wavelength, it undergoes *diffraction*, the bending around the obstacle. Diffraction of sound waves enables you to hear voices through an open door even when the speaker is not aligned with the doorway.

Waves can *interfere* with one another. When two waves pass through one another, two crests or two troughs may meet and reinforce each other, creating *constructive interference* and resulting in a greater displacement at that point than is present in either individual wave. If a crest and a trough meet, the result is *destructive interference*, in which the net displacement is reduced, sometimes even exactly cancelled. If the waves are *linear*, the **interference** may be computed at each point simply by summing the two amplitudes (positive for crests, negative for troughs) at that point. For *nonlinear* waves this simple law does not hold, but interference still occurs.

One important interference effect occurs when two waves superpose to form a pattern of alternating light and dark bands, called *interference fringes*. An example is provided by monochromatic light passing through two closely spaced slits and projecting onto a screen. The wave crests from the two slits alternately reinforce and cancel one other, creating a characteristic pattern. The appearance of these interference fringes is a definite indication of the wave nature of light.

One of the most important consequences of wave properties, from the point of view of astronomy, is the **Doppler effect**. The Doppler effect is familiar to everyone when it affects sound waves. As a train approaches a grade crossing, the driver waiting in his car hears the pitch of the whistle rise. After the train passes, the pitch drops. If the train and the driver were at rest with respect to one another, the sound waves

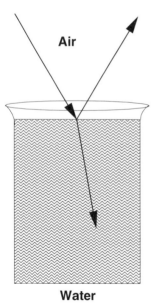

Water

Figure 4.7 The wave phenomena of reflection and refraction occur at the boundary between two media through which the wave propagates with different speeds. When a wave encounters such a boundary, usually it is partially reflected and partially refracted (deflected from traveling along a straight line). The ratio of reflected to refracted energy depends upon the properties of the interface. The angle of refraction depends upon the change of the wave's speed as it enters the new medium, whereas the angle of reflection is equal to the angle of incidence.

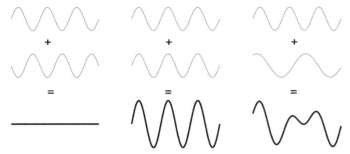

Figure 4.8 Wave interference. Two waves that pass through one another reinforce or cancel, partially or wholly. The net wave (*bold line*) is the sum of the amplitudes of the interfering waves at any point in space. In the first column, two waves of equal wavelength and amplitude, but precisely out of phase, cancel exactly. In the second column, two waves of equal wavelength and amplitude, and in the same phase, add coherently. The third column depicts a general case with two waves of different wavelengths. Interference is difficult to visualize, but it can sometimes be directly observed in water waves.

from the whistle of the train would move outward in a roughly spherical pattern; the constant wavelength between successive crests would determine the fundamental pitch of the whistle. When the train is approaching, however, it is moving in the same direction as are the wave crests that reach the driver; thus each successive emitted crest follows the previous one at a shorter interval than if the train and driver were mutually at rest. Conversely, as the train recedes, it is moving opposite to the direction of motion of the wave crests reaching the driver's ears, so successive waves arriving at the driver's position are spaced at longer intervals than they would be if the train and driver were at rest.

We can illustrate this phenomenon more concretely. Suppose you decided to learn to play tennis, but you could not find a human practice partner patient enough to put up with your attempts to bat the ball around. You might then use a device similar to a miniature cannon which shoots tennis balls at a constant rate, as seen by you when you stand at rest near the rear of the court and watch the balls fly past. If you ran toward the cannon, the interval between the balls you would encounter would be shortened, because each successive ball would now have less distance to cover before meeting you. Conversely, if you ran away from the cannon, each ball would have to make up the extra distance caused by your recession before it could reach you, and thus the interval between balls would, as seen by you, increase. The individual tennis balls could correspond to the crests of a wave, the cannon to any kind of source. In this situation, the receiver, you, is moving, but there is still a Doppler effect, so clearly it cannot depend upon whether the source or the observer is moving. This example also demonstrates that the Doppler effect can occur for *any* kind of periodic phenomenon.

The Doppler effect is, therefore, a consequence of the *relative motion* between the source and the observer. The effect depends only upon the nature of waves and upon the motion of the source relative to the receiver, and thus this phenomenon affects light waves in exactly the same way as sound waves. If the source is approaching the observer the light waves "bunch up" in the observer's frame of reference, and are shifted toward higher frequencies; this is a **blueshift**. If the source is receding from

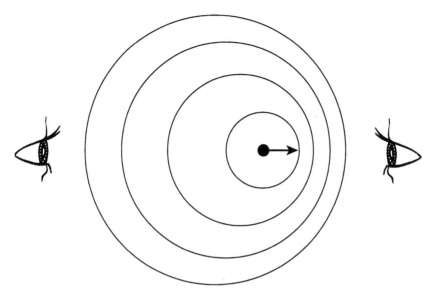

Figure 4.9   The Doppler effect. When the source is approaching the observer, the wave crests "bunch up," resulting in a shorter observed wavelength; this is a blueshift. When the source is receding from the observer, successive wave crests are "stretched," giving a longer observed wavelength; this is a redshift. When viewed from the transverse direction (i.e., perpendicular to the direction of motion), the wavelength is unchanged.

the observer, the light waves shift to lower frequencies, resulting in a **redshift**. The formula for the Doppler shift of light is, for relative speeds v much less than the speed of light c:

$$z = \frac{\lambda_{\text{rec}} - \lambda_{\text{em}}}{\lambda_{\text{em}}} = \frac{v}{c}, \tag{4.1}$$

where z is the standard notation for redshift, λ_{em} is the wavelength in the frame of the emitter, and λ_{rec} is the wavelength in the frame of the receiver. A negative value of z indicates a blueshift; positive z gives a redshift.

On Earth, the Doppler effect has found numerous applications, including the radar "speed guns" by which a highway patrolman may measure the speed of approach of your automobile. (In a radar system, a transmitter emits radio waves that reflect from a target and return to their source.) Astronomy depends particularly heavily upon the Doppler effect. For stars, nebulae, and nearby galaxies, the Doppler shift can tell us how fast the object is moving toward or away from the Earth. The major shortcoming of this technique is that it cannot give us the absolute velocity but only its radial component. If an object is moving transversely to the Earth, then it is neither approaching toward nor receding from us, and there is no Doppler shift. (To be more precise, there is a transverse Doppler shift that is a consequence of time dilation at relativistic speeds, a topic which will be covered in chapter 7. But almost all objects move at small speeds compared to c, and the relativistic transverse Doppler shift is insignificant.) Consequently, we cannot detect the transverse component of the velocity of most objects by means of a Doppler shift. Even with this limitation on our knowledge, however, considerable useful information can be determined. For example, if we measure, from different regions of an object, both a redshift and a blueshift, we can conclude that the object is rotating, and we can measure its rotational speed.

The Nature of Light

In astronomy, the most important wave is *light*. We spend our lives immersed in light, but how many have wondered what it is? The nature of light was argued and debated for centuries, but only over the past 300 years, since the development of experimental science, has any significant progress occurred. Newton performed some of the most important early experiments; it was Newton who showed that white light was a combination of all colors. Newton also studied a specific kind of interference fringes, now called *Newton's rings*, which occur when a glass plate with a very slight curvature is placed over a flat plate, and the whole assemblage is illuminated from beneath. Oddly, although interference fringes are an unmistakable signature of a wave, the rings were not recognized as such. Newton believed quite firmly that light was corpuscular, consisting of a stream of particles, since he could not accept that a wave could account for the apparently straight and narrow propagation of a beam of light. Other scientists of the time, most prominently Christian Huygens, were equally convinced that light was some kind of wave, but this faction held almost equal disregard for what experimental evidence then existed. The matter seemed finally resolved in 1803, when Thomas Young passed light through two very narrow slits in a solid plate and obtained interference fringes. A century later, however, Einstein revived the corpuscular theory of light, but in a form which Newton would not have recognized and which he probably would have disliked. In the modern view, light can show both particle and wave natures, though never at the same time. We shall often have need only of light's manifestation as a wave; but occasionally the particle nature of light will be important, especially when we study the early universe. For now, let us concentrate on the wave properties of light, with some allusions to its particle manifestation.

Visible light is a specific type of *electromagnetic wave*. Electromagnetic waves are traveling disturbances in the electromagnetic field. Unlike other kinds of wave, they do not require a medium for their propagation, although this important fact was not understood until early in the twentieth century. All electromagnetic waves are of the same nature, differing only by their wavelength. The full range of such waves is called the **electromagnetic spectrum**. For the convenience of humans, the electromagnetic spectrum is divided into *bands*, or groups of frequencies. At low frequencies, we call the waves *radio waves*. Progressing to higher frequencies, we have *microwaves, infrared radiation, visible light, ultraviolet radiation, X-rays*, and finally, at the shortest frequencies, *gamma radiation*. The difference in names is due to the separate discoveries of different portions of the electromagnetic spectrum before it was realized that all these waves were of the same kind. The division into bands is also quite arbitrary and has no particular physical significance. The most obvious subdivision is visible light, defined as that band which the human eye can

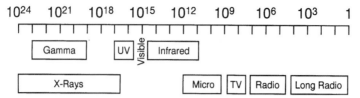

Figure 4.10   The electromagnetic spectrum. The scale shows the frequency of the waves in cycles per second. The high-energy, short-wavelength end on the left is the regime of gamma rays and X-rays. The low-energy, long-wavelength end on the right is the radio region. The division into bands is conventional and somewhat arbitrary. Visible light, that portion of the spectrum to which our eyes respond, is located between the ultraviolet and infrared.

detect. However, even here there is some ambiguity, as different people can see slightly different ranges; in fact, some people who have had the lenses of their eyes removed can see into the ultraviolet. (The lens absorbs ultraviolet rays and prevents them from striking the retina, which might be damaged by the higher-energy light. Exposure to ultraviolet light has been implicated in the development of cataracts.) It is common to employ the word "light" as a generic term for all the electromagnetic waves, and we shall do so unless there is some need to distinguish one band from another. Remember that visible light is not fundamentally different from any other part of the electromagnetic spectrum.

The relationship between wavelength and frequency for an electromagnetic wave traveling in a vacuum is very simple:

$$\lambda\nu = c, \tag{4.2}$$

where λ is the conventional symbol for wavelength, and ν is the symbol for frequency. In this formula, c is a constant of proportionality between the two quantities. It has units of speed and turns out to be the speed of motion of the wave in the vacuum; it is called the **speed of light**. *All* electromagnetic waves travel at this same speed in a vacuum. In a medium, however, a group of electromagnetic waves initially traveling together will traverse the medium at different speeds, always less than c; this phenomenon is called *dispersion*. When white light, which consists of a superposition of all the wavelengths in the visible band, is passed through a prism, the different wavelengths travel through the glass with slightly different speeds, causing them to refract differently at each of the two surfaces they cross. As a result, the prism breaks white light into its monochromatic components. In the field of *spectroscopy*, the analysis of spectra, the superposition of all wavelengths is called the *continuum*.

The speed of light is enormous in comparison to almost any other speed we can imagine, but it is finite, and that has important implications for cosmology. When we look at a star, we see that star not as it is "now," but as it was when the light departed from it. Looking into space is equivalent to looking back in time. The distance light travels in a year is called a **light year**. In MKS units, the speed of light c is 2.998×10^8 m/sec; hence a light year is 9.5×10^{12} km, or about 6×10^{12} miles. Notice that the light year is a unit of *distance*, not of time.

As mentioned, light acts not only like a wave, but sometimes like a particle; the photon is the particle of light. Although the photon is massless, it (and other massless particles) still transports energy. Each photon carries an amount of energy given by the formula

$$E_\nu = h\nu, \tag{4.3}$$

where h is a constant called **Planck's constant**, and ν is the frequency of the corresponding wave. A single photon carries one quantum of energy; hence it is associated only with a single wavelength and frequency, that is, a monochromatic wave. According to the laws of quantum mechanics, light will reveal either its wave or its particle nature in a given experiment, but never both at once.

Radiation is the general term for the emission of energy from an object, often in the form of a wave. The word is frequently applied to the emitted wave or particle itself, as in the expression "ultraviolet radiation." Nuclear radiation, which we have already discussed, may consist of particles, specifically helium nuclei (alpha) or electrons (beta); only gamma rays are actually photons. Charged particles radiate electromagnetic waves when some of the particles' energy is converted into photons. One example is a transmitting antenna, which converts current (moving charge) into electromagnetic radiation. In this case, some of the kinetic energy of the charged particles is transformed into electromagnetic energy.

One of the most important sources of electromagnetic radiation in nature is **thermal radiation**. In any substance with a temperature greater than absolute zero, the constituent particles (atoms or molecules) vibrate, jiggle, and possibly rotate. Energy levels are associated with those overall motions; macroscopically, the collective energy of these random motions is what we experience as heat. A portion of this heat energy is converted into photons and radiated away.

The spectrum of thermal radiation from an arbitrary object can be quite complex, depending upon such variables as the composition of the object, its shape, how much external energy it is capable of absorbing, and so forth. The only general rule is that the hotter the object, the higher the energy of the photons it emits. We all know this from everyday life. An iron for pressing clothing emits no visible light but glows brightly when photographed with film sensitive to the infrared. The coils of an electric stove set on "high" glow with red light; red is the lowest-energy visible light. The stove also emits a great deal of infrared radiation, which cooks the food, but it is hot enough that some of its emission is in the visible. Hotter objects, such as a very hot poker, emit more and more in the visible until they emit all visible wavelengths and thus appear white. Still hotter objects acquire a bluish color, as their emission shifts into the higher-energy visible blue and beyond to the ultraviolet.

There is one extremely important special case in which thermal radiation is easily predictable. This is the thermal emission from a *perfect absorber*, called a **blackbody**. By definition, a perfect absorber is also a perfect emitter. Radiation from such an object is called **blackbody radiation**. Of course, a perfect blackbody is an idealization, but close approximations abound, even on Earth. One excellent approximation is *cavity radiation*. As an example, imagine a kiln for firing pottery. As the kiln heats, its interior fills with thermal radiation emitted by the walls. Since the temperature of the walls surrounding the cavity is the same, the emission and absorption of energy within the cavity must come into balance, regardless of the nature of the kiln walls. This is the key characteristic of blackbody radiation: it represents a state of **equilibrium**, or balance, in the photons.

If we drill a small hole in one wall of the cavity and sample some of the radiation within, we will find that the shape of its spectrum does not depend upon the configuration or composition of the walls, but *only* upon the temperature. The spectrum rises to a maximum intensity at a certain wavelength, then falls back down toward zero emission. Moreover, the wavelength (or, equivalently, the corresponding frequency) at which the peak of the spectrum occurs is uniquely correlated with the temperature; from just this single datum, we can determine the temperature of the radiation. Specifically, the peak wavelength of the blackbody spectrum is inversely proportional to the temperature of the emitter. The formula relating these two quantities is called the *Wien displacement law*, and is given approximately by

$$\lambda_{pk} \approx \frac{0.29}{T} \text{ cm,} \qquad (4.4)$$

where the temperature T is in degrees Kelvin. The surface of the Sun is a close approximation to a blackbody. Its surface temperature is about 6000 K, and its spectrum peaks in the visible range.

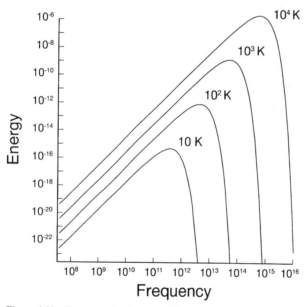

Figure 4.11 Representative spectra of blackbody radiation for different temperatures. The shape of each curve is the same; only the magnitudes and positions differ. The peak frequency depends only upon the temperature of the blackbody emitting the radiation. Reference to the electromagnetic spectrum in Fig. 4.10 shows that blackbodies emit significant visible light only when their temperatures are greater than about 1000 K.

For astronomy, the most important property of blackbody radiation is that the shape of the spectrum is always the same; it is simply shifted to shorter wavelengths (i.e., higher frequencies) at higher temperature. It is also shifted to higher energies. As the temperature of a blackbody is increased, the total amount of energy per unit volume goes up as the fourth power of the temperature,

$$\text{energy density} \propto T^4. \tag{4.5}$$

If the blackbody radiates into space, the amount of energy radiated per unit area will also be proportional to T^4. Stars are approximately blackbodies, so a hot star with a surface temperature twice that of the Sun would be radiating 16 times as much energy per unit surface area. If that hot star had the same surface area as the Sun, its total energy emission would correspondingly be 16 times as great. Because such hot stars are typically considerably *larger* than the Sun, their total energy output is usually much greater than the Sun's.

Blackbody radiation is especially important to cosmology. As we continue our study, we shall discover that the universe is filled with such radiation, with an associated temperature of a mere 2.7 K. The presence of this background radiation indicates that the universe was once much hotter, providing direct evidence for what is known as the big bang. How this background radiation originated, and what it can tell us about the early universe, is a story to be developed in later chapters.

At the end of the nineteenth century, the explanation for the spectrum of blackbody radiation had stymied every great physicist who had worked on it. The peaking and overturning at high energies could not be predicted by the classical laws of thermodynamics. The lower-energy portions ascending up to the peak could be explained by the physics of the era, but the resulting classical formula continued to rise indefinitely, predicting infinite energy at the shortest wavelengths! Since this was obviously impossible, it was called the *ultraviolet catastrophe*.

In 1900, the German physicist Max Planck presented a formula which fitted the data nearly perfectly. Planck had set out to find a theoretical explanation for blackbody radiation. He tried many possibilities, but the breakthrough came when he made the assumption that radiation could be emitted and absorbed only in discrete units, the *quanta* (singular *quantum*) which we have previously discussed. The explanation of blackbody radiation was the very first hint of quantum mechanics, a theory which did not develop fully for another 25 years. The earlier classical formula worked reasonably well for low energies, where the quantum nature of the light was not very important, but failed at high energies, where only quantum effects could explain the data.

Advances in the new quantum theory quickly led to another triumph in the understanding of light. When light from a radiating sample of tenuous gas is analyzed, it will be found to consist of bright, narrow lines; such a spectrum is called an **emission spectrum**. Because it consists of distinct spectral lines, this type of radiation is often called **line radiation**. Niels Bohr's work on the quantization of electron orbitals in atoms provided the explanation for this form of electromagnetic radiation; it originates from the *quantum transitions* of the orbiting electrons. Each electron bound to a nucleus must have a well-defined energy, specified by the orbital it occupies. Under certain circumstances, the electron may drop into an orbital of lower energy, emitting a photon in the process. If we define E_i as the energy of the initial orbital, and E_f as the energy of the final orbital after the transition, then the frequency of this photon is obtained from equation (4.3) above:

$$E_f - E_i = h\nu. \tag{4.6}$$

The orbitals cannot have arbitrary energies, and therefore this difference is always some discrete amount, which depends upon the transition. Obviously, the lower-energy state must be available or the transition cannot occur; recall that only two electrons may occupy each orbital. Since the exact electron configuration is specific to each particular element, the transitions permitted to the electrons depend upon which type of nucleus they orbit. Moreover, because of various rules of quantum mechanics, it may turn out that for a given atom, some transitions are much more probable than others, and some are almost forbidden, even if those orbitals are available. Each atom thus emits a unique spectrum of frequencies that is so characteristic that every element may be identified from its spectrum alone.

The inverse process also occurs; an electron may absorb a photon and be boosted from a lower-energy orbital into a higher-energy orbital. Because this can occur only if a photon of exactly the right energy happens to be available, it is also highly specific and characteristic. The element absorbs exactly those frequencies which it would emit in the opposite process. If white light strikes a collection of atoms in the gaseous state, each atom will absorb photons of precise frequencies, and when the light that has passed through the gas is analyzed with a prism, the "missing" frequencies

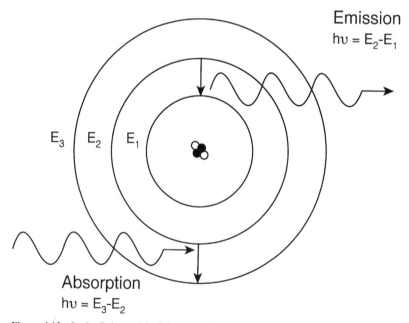

Figure 4.12 In the Bohr model of the atom, electrons surround the central nucleus in specific orbitals that correspond to particular energy levels. When an electron makes a transition from a higher energy level to a lower, a photon is emitted with precisely the energy corresponding to the difference. An electron can jump from a lower to a higher energy level if it absorbs a photon with exactly the required energy.

will be unique to the specific element. Such a spectrum is called an **absorption spectrum**.

The spectrum of the Sun shows a forest of absorption features called *Fraunhofer lines*, after their discoverer. Atoms in the relatively cool outer layers of the Sun absorb some of the photons generated from deeper, hotter layers. However, this was unknown when the lines were first resolved. (The corresponding emission lines were found two decades after Fraunhofer's death; they are more difficult to detect.) The realization came to Gustav Kirchoff and Robert Bunsen, over the period between 1855 and 1863, that these lines could be identified with laboratory spectra of Earthly gases. This stunning discovery made possible astronomical spectroscopy and paved the way for modern cosmology. Until then, most astronomers had believed that it would never be possible to determine the chemical composition of heavenly objects. But by the 1920s, most of the time of all large telescopes was devoted to spectroscopy, as is still the case today. The demonstration that the Sun, and the more distant stars, were made of the same elements as were found on Earth was a powerful vindication of the Copernican principle.

Emission, absorption, and thermal radiation can coexist in the same spectrum. When an astronomer photographs the spectrum of an object such as a distant galaxy, she will find lines superimposed on a continuum. Most of the continuum is thermal radiation from the object, while the line radiation consists of discrete, resolvable transitions that are specific to the particular elements present in the object. For example, a portion of the atoms in a cloud of interstellar gas might be directly excited by the light from a bright star embedded in the cloud. The spectrum of such a cloud would show emission lines, which by their frequencies and strengths would reveal the kinds and abundances of elements present. The background due to the thermal radiation of the cloud as a whole would provide an estimate of its temperature.

How Brightly They Shine

Almost all the information we can gather about the universe and its contents comes from the photons we detect. Astronomy is an *observational* science, as opposed to an experimental science; we cannot arrange controlled experiments to study the universe as a whole but can only observe it. It is worthwhile, therefore, to review briefly a few of the basic quantities that arise when measuring the light from the sky.

Stars, and other astronomical objects, give off light. The total amount of electromagnetic energy emitted per unit time by a source is called the **luminosity**, generally symbolized as L, and in astronomy often expressed in terms of the **solar luminosity** $L_\odot$. When an object such as a star shines, light travels outward from all points on its surface. If we consider photons which are traveling through the nearly-empty space around the star, we can ignore absorption or other losses of energy.

The luminosity is never directly measured. At the Earth, we intercept only a portion of the total radiation emitted by an object; only that small fraction of this energy that intercepts a detector can be measured. Most people are aware that the brightness of a source goes down with distance. This can be made mathematically exact by imagining a spherical surface surrounding the star, at some distance R from it. From energy conservation, the total amount of luminous energy crossing such a sphere is the same as that which was emitted at the surface of the star. Since the surface area of a sphere increases as the radius squared, the energy per unit time crossing such surfaces at greater and greater distances from the star must decrease as the inverse of the square of the distance from the star. This argument is succinctly expressed mathematically as

$$\text{brightness} = \frac{L}{4\pi R^2}. \tag{4.7}$$

Brightness is also called the energy **flux**, and it specifies the amount of energy per unit area per unit time. The distinction between luminosity L and energy flux is that the former refers to the *total* energy emitted per unit of time over the entire source, whereas the latter designates the energy received per unit time and per unit area at a detector located anywhere outside the source.

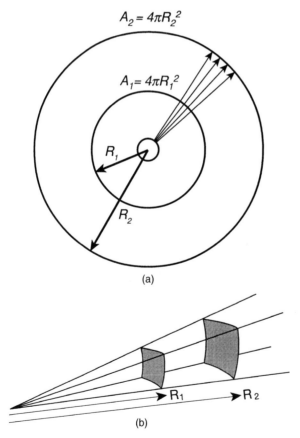

$A_2 = 4\pi R_2{}^2$

$A_1 = 4\pi R_1{}^2$

R_1

R_2

(a)

R_1 R_2

(b)

Figure 4.13 (*a*) Because the surface area of a sphere increases as the square of its radius, (*b*) the power per unit area (flux) of light emitted by a point source diminishes as the light travels farther into space.

The relationship between distance and brightness (flux) is very important to astronomy, and especially to cosmology. Astronomers can measure the energy flux, but unless we know the distance to the star we cannot compute its luminosity. Conversely, if we know, e.g., from a study of nearby stars, that a certain type of star has a characteristic luminosity, we can compute the distance to any star of that type by measuring the flux of radiation we receive from it. Such a measurement of distance is often called the **luminosity distance** in order to emphasize how it was obtained and to indicate the errors that might occur.

Of course, the assumption of no loss (or gain) of energy between us and the source is a major restriction. Space is indeed quite empty, but between us and even nearby stars, there is a lot of it, so that even a very small density of matter means that the light which reaches us may have been diminished by more than just the geometrical factor in equation (4.7). Any decrease of flux due to interactions with matter is called **extinction**. The amount of extinction varies in complicated ways with such factors as the quantity and type of intervening matter and the frequency of the light. Because so many unknown factors play a role, constructing a model for the dimming of light by extinction is often not all that easy, but if we are to be able to use luminosity distances with any degree of confidence, it is necessary to account for extinction. This adds considerably to the difficulty of measuring cosmic distances, but the luminosity distance is nearly our only possibility for gauging the farthest reaches of the universe, so we have no choice but to do our best with it.

In addition to the luminosity, another quantity of great interest in astronomy is the *mass* of an object. We cannot construct gigantic balances to measure the mass of a star or galaxy directly, so how can it be determined? We know that the mass of an object determines the gravitational force it exerts on other objects. For the Sun, we can easily measure its gravitational influence to very high precision, and from this we learn that the Sun has a mass of about 2×10^{30} kg. The mass of the Sun, in whatever stated units, is called one **solar mass**, denoted by $M_\odot$. This mass is so large that it may be difficult to comprehend; for comparison, the mass of a typical human is roughly $3 \times 10^{-29} M_\odot$. This is close to the ratio of the mass of a human to that of a single proton.

We cannot readily determine the mass of an isolated, distant star, but many, perhaps as many as half, of all known stars are members of binary systems, a system of two stars that mutually orbit one another. From Kepler's laws, we can determine the sum of the masses of the two objects; if we can also find the force between them, we can calculate the masses of the individual objects. By such means, astronomers have found that stars range in mass from approximately 20% of the mass of the Sun to over 50 $M_\odot$. In addition to binary systems, stars also occur in large groups. It is difficult to apply Kepler's and Newton's laws to such clusters because the equations of Newtonian gravitation can be exactly solved only for two bodies. All we can obtain from star clusters are statistical properties, but, again using Kepler's laws, those statistics can provide a good estimate of the cluster's aggregate mass. If we can also observe how many, and what types, of stars are present, we can estimate the mass in luminous objects (usually mostly stars) of the cluster, from our knowledge of the properties of other stars of the same types. We will return to this topic in detail in later chapters.

Where Are We?

The solar system consists of the Sun, a smallish star resident in the suburbs of an average galaxy, and all the lesser objects that are gravitationally bound to it. The Sun dominates its system completely; the second-largest object, Jupiter, has only 0.096% the mass and 2% the diameter of the Sun. There are nine planets and innumerable smaller bodies. We can construct a scale model of the restricted solar system, consisting only of the Sun and the nine planets, to make it easier to grasp the scale of the system. We are going to need a great deal of room to accomplish this, since the solar system is very large and very empty. As a start, suppose that the Sun were the size of an orange. The Earth would then be about the diameter of a small BB pellet (1 mm) at a distance of 11 m from the orange. The moon is 0.25 mm in diameter, and located about an inch (2.5 cm) from the Earth. Jupiter is about 1 cm in diameter and resides 60 m, over half the length of a football field, from the orange. Tiny Pluto is only 0.2 mm in diameter, and its mean distance from the orange is 430 m, about four football fields. Yet even these staggering distances are just down the street, compared to separations in interstellar space. The nearest star to the Sun, at a distance of 4.3 lt-yr, is Alpha Centauri, a star (more precisely, a stellar system) visible only in the Southern Hemisphere. On our scale model, Alpha Centauri is about 3000 km from the orange. Interstellar distances are really too large to be comprehended by human intuition, yet they are still small compared to the scale even of the galaxy. It is only through the symbolism of mathematics that we are able to understand the nature of the cosmos.

As far as we can tell, almost all stars occur within **galaxies**. Galaxies are great clusters of stars, gas, and dust, which make up the fundamental population of the universe. Galaxies are divided into three major categories. **Spiral galaxies** are great disks of stars, with grand patterns of spiral arms threaded through them like the fins of a pinwheel. The spirals themselves cannot be rigid objects, or they would have long since wound themselves up to a much greater degree than we observe; they are thought to consist of density waves which drift through the stars and gas like ripples on a pond. The spiral arms are delineated by their overabundance of bright, young stars and glowing gas clouds and may be the major location for star formation. Spirals have a range of masses, from a few billion to several hundred billion stars.

The other major category of galaxy is the **elliptical galaxies**. As their name implies, these galaxies are ellipsoidal, that is, shaped roughly like a football or a flattened basketball. Some are, or appear to be, nearly spherical, especially the largest ones. Ellipticals cover an enormous range, from the dwarfs, with as few as a million stars, up to the giants which contain thousands of billions of stars. In contrast to spirals, ellipticals seem to contain scant gas or dust, and show little evidence of recent star formation.

The third category is something of a catch-all for any galaxy that does not fit into the previous two: the **irregular galaxies**. Irregular galaxies show no particular structure, although many might be distorted by their interactions with other galaxies. Some irregulars, especially the *dwarf irregulars*, might be prevented from pulling themselves into a spiral shape by the gravitational dominance of large galaxies which they orbit. Others may simply show no structure, or even a tendency toward structure, at all. How galaxies formed, why they take the shapes they do, and why so few types are observed, is one of the major outstanding puzzles of cosmology and astronomy.

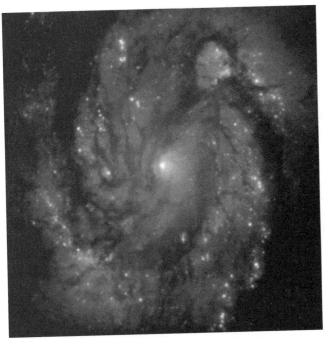

Figure 4.14   A spiral galaxy, M100, as observed with the *Hubble Space Telescope*. This galaxy is located in the Virgo Cluster of galaxies, at a distance of about 17 Mpc. (J. Trauger, JPL; STScI/NASA.)

Figure 4.15   The giant elliptical galaxy M87, located at the heart of the Virgo Cluster. An elliptical galaxy is a spheroidal mass of stars, showing no overall rotation or structures such as spiral arms. (Image Copyright AURA Inc./ NOAD/NSF.)

Figure 4.16   The Large Magellanic Cloud is an example of an irregular galaxy. This small galaxy is a satellite of the Milky Way and can be seen from the Southern Hemisphere of Earth. Copyright © Anglo-Australian Observatory/Royal Observatory Edinburgh, 1984.

The galaxy in which the Sun and its solar system are located is called the Milky Way galaxy, or just the Galaxy. Although we cannot, of course, observe it from the outside, the distribution of stars in our skies immediately shows that the Milky Way consists of a flat disk. We cannot see its center in visible light because thick clouds of obscuring dust intervene between us and the core, but we know that the center of the Milky Way lies in the constellation Sagittarius and is one of the brightest radio sources in the Galaxy. Our inability to see our own Galaxy from the exterior inhibits detailed understanding of its structure. We can, however, estimate it to contain approximately 100 billion stars. The Sun is about 30,000 lt-yr from the center, roughly two-thirds of the way to the visible edge of the galaxy. (Galaxies have no strict cutoff, but at some point become faint enough to define a boundary.) The solar system completes one revolution around the Galactic center in 200 million years.

Galaxies show a strong tendency to bunch into **galaxy clusters**. Our own Local Group is a modest cluster of perhaps a few dozen galaxies, dominated by the two large spirals called the Milky Way and the Andromeda Galaxy. This is a typical configuration for such loose clusters, and, like others observed, ours is asymmetrical; the Andromeda Galaxy is about twice as massive as the Milky Way. Other galaxy clusters are much richer and denser, containing anywhere from a few hundred, up to thousands, of large galaxies and unknown numbers of small, dim galaxies. Whereas the dominant galaxies of loose clusters are generally spirals, rich clusters contain a mixture of galaxy types. Most ellipticals reside in fairly dense clusters, and giant ellipticals are often found at the very center of a large cluster. The spatial scale of galaxy clusters also varies, from the 2 million light years of the Local Group, to the 6 million light years of a cluster such as the Virgo Cluster.

Figure 4.17   The center of the great galaxy cluster in the constellation Virgo. This irregular cluster contains approximately 2500 galaxies. (Image Copyright AURA Inc./NAOD/NSF.)

Galaxy clusters are gravitationally bound; that is, the galaxies orbit one another. Beyond this scale is the suggestion of even larger structures, called **superclusters**. The largest superclusters seem to be too large to be fully gravitationally bound; their origin is a mystery. Perhaps the galaxies are merely particles in some great overarching structure of the universe. How big are the largest structures, and how could they have originated? These are some of the most important questions in cosmology.

We do not know how many galaxies inhabit the universe. Beyond the reaches of the Milky Way itself, nearly every object we see is a galaxy. There are at least as many galaxies in the observable universe as there are stars in the Milky Way. Galaxies may be the glowing tracers of the mass of the universe, the visible spots in a great flow of matter; or they may contain all the matter. Galaxies formed very early in the history of the universe; but are galaxies fundamental, or did they condense from the larger structures we observe? What creates and maintains spiral patterns? Do ellipticals result from the merger of spirals, or are they of different origin? Galaxies are the starry messengers which tell us of the origin and structure of the universe itself, if only we could understand their stories.

KEY TERMS

elements	atom	atomic number
elementary particle	electron	nucleus

proton	quantum	photon
ion	neutron	nucleon
isotope	radioactivity	nuclear reaction
half-life	fission	fusion
neutrino	quantum mechanics	boson
fermion	exclusion principle	quantum state
antimatter	positron	strong interaction
weak interaction	electromagnetic force	gravity
quark	field	work
energy	potential energy	kinetic energy
thermodynamics	heat	temperature
conservation of energy	conservation of matter	entropy
absolute zero	wave	amplitude
frequency	wavelength	spectrum
interference	Doppler effect	blueshift
redshift	electromagnetic	speed of light
light year	spectrum	radiation
thermal radiation	Planck's constant	blackbody radiation
equilibrium	blackbody	line radiation
absorption spectrum	emission spectrum	solar luminosity
flux	luminosity	extinction
solar mass	luminosity distance	spiral galaxy
elliptical galaxy	galaxy	galaxy cluster
supercluster	irregular galaxy	

Review Questions

1. What is an isotope, and how is it related to an element? Why does atomic number determine the chemical properties of an element?

2. ^{238}U (uranium) decays to ^{206}Pb (lead) with a half-life of 4.5 billion years. If the ratio of ^{238}U/^{206}Pb in a meteor is equal to 1/3, how old is the sample? If the meteor originally contained some ^{206}Pb from a source other than radioactive decay, how would that affect your age estimate?

3. Describe and distinguish nuclear fission and fusion. Which types of element are involved in each of these processes?

4. Describe two differences between a boson and a fermion. To which family do the electron, the proton, and the photon belong?

5. Name the four fundamental forces of nature. Which are the strongest? Which one creates most everyday forces? Which one dominates the universe at large scales? Why does only one force dominate at large scales?

6. The wavelength of a particular hue of green light is 5.0×10^{-7} m. What is the frequency of this light? What is the energy of a photon of this light? (Values of some important constants of physics and astronomy are given in Appendix A.)

7. A man, charged with going through a red light, comes before a traffic court. He argues that the Doppler shift made the light appear green to him. If red has a wavelength of

7000 angstroms (Å; 1 Å is equal to 10^{-8} cm) and green has a wavelength of 5500 Å, then, from the Doppler shift formula, what was his speed as a fraction of the speed of light c?

8. The diameter of a telescope's mirror determines how much light it can gather. The amount of energy collected over the area of the mirror from the light of a particular star can be measured with sensitive instruments. How does the inverse square law then tell you the *total* energy given off by that star? At the Earth's orbit the light of the Sun is distributed evenly over a sphere with a radius equal to that of the Earth's orbit (about 10^{11} m). The telescope has a 1 m radius (2 m diameter). What fraction of the Sun's light can the telescope capture?

9. What is the unique characteristic of blackbody radiation?

10. How does the surface temperature of a reddish star compare with the surface temperature of a bluish-white star? Does the diameter of the star matter when determining the temperature? What is the ratio of the peak wavelength emitted by star 1 to the peak wavelength emitted by star 2 if the surface temperature of star 1 is twice that of star 2?

11. Explain the significance of luminosity distance. What sort of errors can occur in the measurement of this quantity? How can astronomers correct for these complications?

12. The Andromeda Galaxy is about 2 million light years away from us. What distance would that correspond to in the scale model discussed in this chapter, in which the Sun is the size of an orange?

CHAPTER 5

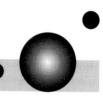

The Lives of the Stars

> ...the glorious sun
> Stays in his course, and plays the alchemist
>
> —SHAKESPEARE, *King John*, III, I

The stars change very little during the course of a human lifetime. Indeed, they have hardly changed in appearance over the length of recorded human history. In pre-Newtonian cosmologies, the stars were eternally affixed to a single, unchanging celestial sphere. Even after the age of Newton, they were the "fixed stars," whose distribution coincided with absolute space. The occasional appearance of a supernova indicated that perhaps the heavens were not immutable, but it was only in this century that these rare events were associated with the deaths of stars. As for stellar birth, astronomy textbooks dated as recently as the 1950s speculate that there might perhaps be places where we could observe a new star being formed, as if such an event would be quite rare.

We know now that stars are not eternal; they come into existence, go through a "life cycle," and die. Through observations, and through the careful construction of detailed models based on an understanding of the laws of physics, astronomers have learned a good deal about the lives of the stars. The type of existence a given star has, and the method of its death, depend upon the *mass* of the star, and to a lesser extent upon its *chemical composition*. Less massive stars, such as the Sun, burn their fuel slowly, live long, and when they exhaust their fuel stores, they flicker out as slowly cooling white dwarfs. More massive stars live fast and die young, and end their existences in some spectacular cataclysm, leaving behind a compact and enormously dense cinder called a neutron star. The most massive stars have the most violent ends; they may blow themselves to nothingness in a supernova, or, if a core is left behind, they may collapse until they cut themselves off from the rest of the universe.

In comparison to the grand galaxies that fill the huge volume of the universe, individual stars might seem insignificant. Certainly, one smallish star is of great importance to life on one tiny planet, but what roles might stars play in the cosmos at large? Most obviously, the stars make it possible for us to be aware that anything else exists. If all matter other than the Sun were dark, we would not even know, at least directly, of our own Galaxy, much less of the billions of other galaxies which inhabit the universe. Some light is emitted from very hot gas near the centers of galaxies, but most of the visible light in the universe, and much of the energy in other bands, originates directly or indirectly from stars. The populations of stars make galaxies visible, but more than that, they enable us to measure the masses and compositions of the galaxies. Certain kinds of bright stars provide a means to gauge the distances of galaxies; furthermore, when a massive star collapses, the resulting explosion is so brilliant that it can be seen across enormous distances, providing a means to measure the expanse that the light has crossed. Humbler, lower-mass stars have an equally important role to play in our cosmological investigations. Such stars can have ages comparable to the age of the universe itself. A star is a much simpler object, and much more amenable to observation, than is the universe as a whole, so that stars provide us with the best estimate for the age of the cosmos. Finally, stars play an active role in the evolving cosmos; their nuclear furnaces are the sole source of all the elements beyond lithium. As arguably the most important denizens of the universe, the stars are of great significance in the study of cosmology.

A Star Is Born

Between the stars lies interstellar space. By Earthly standards, it is a very good vacuum, with an average density of about one atom of hydrogen per cubic centimeter. Nevertheless, within this space are enormous clouds of gas, consisting mostly of hydrogen, with a lesser quantity of helium. Some clouds also contain considerable *cosmic dust*, which consists primarily of tiny specks of minerals and soot, sometimes coated with various ices. The matter between stars is collectively known as the **interstellar medium**. Clouds of interstellar material, although very tenuous, are so large that their masses can be quite significant, up to thousands of solar masses. They are called **nebulae** (singular *nebula*), from the Latin word for "cloud." One important effect of the nebulae is their influence on starlight. They absorb some of the photons, and *scatter*, that is, send in all directions, others. Looking at stars through interstellar clouds is much like trying to see the headlights of vehicles through a fog. How much light is transmitted depends upon the thickness of the fog, as well as on the types of particles which make it up. Some nebulae, mainly those that are very cold and contain much dust, are almost completely opaque. Other nebulae contain bright stars embedded within them and glow themselves due to their reemission of the energy they absorb from the star. In any case, the presence of these obscuring clouds of gas and dust complicates our measurements of the luminosity distances to stars that lie behind them. The resultant dimming of the light makes the stars seem farther than they really are.

    But obscuration is far from the only role these great clouds play in the universe; their most important function is to be the birthplace of stars. A star is born when a clump of gas contracts under its own gravity. Today such a statement may seem obvious, but it was a daring hypothesis when it was first put forward late in the

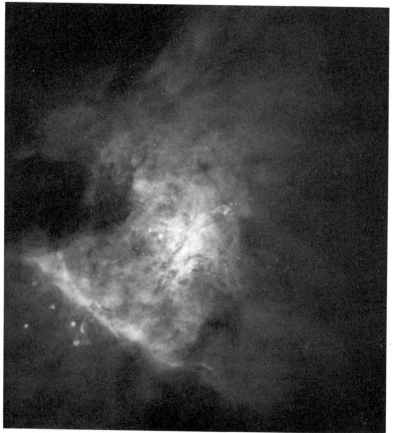

Figure 5.1 *Hubble Space Telescope* image of the Orion nebula, a region of active star formation, visible in the "sword" of the constellation Orion. Glowing gases, heated to emission by embedded hot young stars, are seen interspersed with dark, dusty clouds. (C. R. O'Dell, Rice University; STScI/NASA.)

eighteenth century. The philosopher Immanuel Kant, among others, had arrived at inklings of this model, but it was first introduced in a developed form by the French mathematician and physicist Pierre Simon de Laplace. Laplace proposed that a rotating cloud of gas, as it pulled itself together, would flatten into a disk. The central portion of the disk would gather itself into a ball to form a star, while the outlying regions serenely coalesced into planets. The disk hypothesis neatly explained why the planets orbit the Sun in such an orderly manner; all of them, or at least all known to Laplace, obediently lie nearly in a single plane. During his lifetime, the mathematical tools barely existed to study his proposal carefully; indeed, Laplace himself invented many analytical techniques for working with gravitating systems. It should not be surprising, then, that the details of his model did not quite work out. Nevertheless, Laplace's insight is still a useful conceptualization today, and it gives a good qualitative description of the process that creates new stars.

Star formation is still understood only in outline; the details remain elusive, and a subject of active research. The basic ideas are simple, however. The Galaxy is filled today with nebulae, as must have been even more true in the past. The most likely stellar nurseries are gigantic *molecular clouds*, huge aggregations of cold gas. These clouds contain a significant number of molecules, rather than just solitary atoms. Hydrogen forms a molecule consisting of two atoms chemically bound together, whenever it can. In the near vacuum of interstellar space, a lone hydrogen atom has little opportunity to encounter another one, and most of the hydrogen is atomic. A molecular cloud, however, has a sufficient density that about half its gas takes the molecular form. In the present-day Galaxy, these clouds generally also contain many other kinds of relatively simple molecules, such as carbon monoxide, water, formaldehyde, ethanol, and ammonia. Dust grains are also abundant in the molecular clouds. Dust grains are very efficient at radiating away energy, which is an important reason that these clouds are likely progenitors of stars. The dust cools the cloud and helps to shield the molecules from high-energy photons, leaving the clouds with temperatures ranging from 10 K to 100 K. Dust is also particularly opaque to most visible light, scattering it away so that it never reaches our telescopes; thus clouds containing dust grains are dark, and the dustier the cloud, the blacker it appears through a telescope.

Gravity will, as always, try to pull dense regions into collapse; the cloud resists this through gas pressure. If sound waves can cross the condensing region faster than it contracts, then the sound waves, which are waves of pressure, will be able to restore a balance between gravity and the pressure of the gas. The speed of sound decreases with temperature; therefore, the colder the gas, the greater its chance of collapsing before pressure can build up. Under the right conditions, the cloud, or a portion of it, will be unable to maintain itself and will begin to contract. The gravitational instability that initiates the contraction will probably also cause the condensation to fragment into many smaller clumps; thus most stars likely formed as members of clusters. There are many known clusters of young stars, such as the famous Pleiades, and most stars that are known to be young reside in groups. Many older stars, however, travel solo through space, or perhaps in the company of one or a few other stars. An isolated star, such as the Sun, probably escaped in its youth from its nursery mates as a result of gravitational interactions among the young stars, and with external objects.

As each would-be star collapses due to its self-gravity, the gas retains its spin, or angular momentum, and forms a disk much like that imagined by Laplace. The collapse compresses the gas, causing it to heat up. Some of the rotation of the gas is carried away, possibly by magnetic fields threading the cloud, allowing further collapse and compression at the center. Eventually, most of the matter accumulates at the center, while the rest remains in an encircling disk. The central sphere, now a protostar, continues to contract and heat. As its temperature rises, more and more of its hydrogen ionizes, that is, the atom loses its single electron. Free electrons scatter and absorb photons very effectively, so the more electrons that are liberated, the more opaque the protostar becomes. If photons cannot escape from the gas, their energy is trapped within the protostar, causing the temperature to rise even further. At last, the temperature within the core of the protostar rises to a sufficient level to ignite nuclear fusion; the energy generated from this process finally provides the newborn star with the pressure required to prevent further collapse.

Meanwhile, the protostellar disk is undergoing changes of its own. The heat and pressure from the particles and photons streaming from the surface of the infant star blow the lightest and most volatile elements, such as hydrogen and helium, away from a region immediately surrounding the star. Hence the clumps, or *planetesimals*, that are able to form near the star are likely to be rocky, composed of mostly nonvolatile substances. Farther away, planetesimals form with large quantities of lighter matter, such as ices. As the young stellar system develops, these planetesimals collide and clump together; the largest clumps sweep up the smaller particles they encounter, becoming larger *planetoids* in the process. Planetoids in the outer, cooler part of the disk can attract and retain hydrogen and helium, becoming gas-giant planets. A new planetary system has formed.

To create a star, the core temperature in the collapsing protostar must become high enough to ignite nuclear reactions. It is likely that some globules of gas that begin to contract are too small for this ever to occur. The planet Jupiter is an example of such a failed star; if it had had approximately 80 times the mass, it would have become a small, dim star rather than a giant planet. The minimum mass required for star formation is probably in the neighborhood of 0.08 $M_\odot$.[1] Condensed objects below the minimum mass for stellar ignition have been dubbed **brown dwarfs**. Brown dwarfs might still radiate heat as they slowly contract, converting some of their gravitational energy into infrared radiation, and, in principle, this infrared signal could be detected. However, they would be extremely dim even in the infrared, making them very difficult to see.

The lower mass limit cited for the creation of a star is based on a calculation of the mass required to produce a sufficiently high temperature at the core of the object. But just how many such failed stars exist? They might be scattered throughout the Galaxy and could, if they exist in large numbers, make a significant contribution to the total mass of the Galaxy. Humble objects are the most abundant in nature, and this is a simple fact, not an anthropomorphism: small stars should be easiest to form. Massive stars are rare, while the majority of stars are smaller than the Sun. At the low end of the mass range, the stars are cool, glowing only with a faint red light; these stars are called **red dwarfs**. The fraction of stars that form with a given mass seems to be mostly determined by the mechanisms of star formation. When we consider all such fractions for all masses, we obtain a function called the *initial mass function* (IMF). Unfortunately, the IMF is only partially known. The observed IMF seems to show a fairly sharp cutoff at around 0.2 $M_\odot$, considerably greater than the theoretical minimum mass for a star. It may be that some factor other than mass alone places a lower limit upon the IMF. It may also be that we are simply not observing the near-minimum stars because they are so dim.

New technology has made it possible to look for brown dwarfs, and the dimmest of the red dwarfs. One of the best candidates for a brown dwarf was found late in 1995; it is the tiny companion to a star known as Gliese 229. Deeper searches have found evidence for very dim red dwarfs, although not in the numbers expected. Work continues, but for now the abundance of brown dwarfs, and the faintest of the red dwarfs, remains uncertain.

We have given only the barest sketch of the formation of stars; there are sure to be many variations on this theme. Many effects are still poorly understood, such as the

Figure 5.2 *Hubble Space Telescope* image of the faint brown dwarf companion of the cool red star Gliese 229. The brown dwarf is located 19 lt-yr from Earth in the constellation Lepus. Estimated to be 20–50 times the mass of Jupiter, it is too massive and hot to be a planet but too small and cool for a star. (T. Nakajima and S. Kulkarni, Caltech; S. Durrance and D. Golimowski, JHU; STScI/NASA.)

role which might be played by magnetic fields. Moreover, our qualitative description might seem to apply only to solitary stars; yet many, perhaps half or more, of all stars are members of binary systems, two stars that orbit one another. Some stellar systems of three or even four stars exist nearby. For example, Mizar, a star in the handle of the Big Dipper (Ursa Major), is a doublet; those with good eyes can easily make out the companion, Mizar B, on a dark night. Many such *optical doubles* are coincidental, the two stars at vastly different distances, but Mizar A and B are, in fact, a pair; they compose a *visual binary*. It turns out that *both* Mizar and Mizar B are themselves double stars, making the system a quadruple star! Yet the broad outline we have sketched surely still applies to such stars. Whether each star within a system might have an associated disk, at least near the time of its birth, is uncertain. The disks might be disrupted if the companion is too close, or they might survive but be unable to produce planets, or there may be planets around the members of some binary systems. It is difficult enough to understand thoroughly the formation of one star and its disk; multiple-star systems are another step upward in complexity.

Astronomers cannot even be certain that a planetary system, or even a protoplanetary disk, forms around all single stars. Theory indicates that it should, although the subsequent formation of planets may well not be inevitable. Even if they form, they may not survive; the "wind" of high-energy particles and the intense radiation from a very massive, bright star might well sweep away *all* of its disk, not just carve out a small region depleted in light elements. On the other hand, there is direct evidence that the solar system is not unique in the Milky Way, much less in the universe. One nearby, young star, Beta Pictoris, has yielded up photographs of a disk of dust. Unfortunately, no planets could possibly be directly resolved at the distance of the disk, even if they might be present or forming. Hence all we can conclude with certainty from Beta Pictoris is that it provides a wonderful example of a star which *does* possess a dusty disk during the early stages of its life. Disks are difficult to see even for close stars; the glare from Beta Pictoris ordinarily overwhelms the weak emissions, most of which are in the infrared, from its disk, and the star's image must be artificially covered for the disk to become visible.

Late in 1995, the first extrasolar planet orbiting an ordinary star was found around 51 Pegasus, a star similar to the Sun. Discovered by astronomers at the Geneva Observatory and confirmed by observations at the Lick Observatory, the planet has at least half the mass of Jupiter. This planet cannot be seen directly even with powerful telescopes; its presence was inferred from the wobble in the star's motion produced by the planet's gravity. The Lick Observatory team has since detected several additional planets using this technique; their findings suggest that planetary systems may be relatively common. All of the planets discovered so far have been large, comparable in mass to Jupiter; a planet as small as the Earth would be extremely difficult to detect even with improving technology, although eventually it may become possible. In any case, there is now unequivocal proof that some other stars do have planetary companions, though as yet no firm evidence exists of extrasolar planets suitable for the formation of life as we know it. Most astronomers were always confident that other planetary systems would be found, although they were still quite excited by these discoveries; at last, there is more than one such system to study, so that theories of planet formation may begin to be tested.

Holding Its Own

What, then, is a star? All stars are huge balls of gas, mostly hydrogen, held together by gravity. Throughout the life of a star, two opposing forces determine its structure: gravity and pressure. Gravity works to pull the gas toward the center of the star, and, as the gas is compressed, its pressure rises until a balance is reached. This state of balance between two competing forces is known as **hydrostatic equilibrium**, and it holds for most of the lifetime of a star.

To understand stars, we must understand how they generate and radiate the energy that offsets the omnipresent pull of gravity. This much was long understood, but at the beginning of the twentieth century, the mystery was the mechanism of energy generation. One possibility, ordinary chemical reactions, is certainly insufficient to keep the stars shining for very long. If the Sun were made entirely of coal, and some source of oxygen allowed the coal to burn, a star's entire life would last only a few hundred thousand years. But geologists had plenty of evidence that the Earth was much older than this. In the late nineteenth century, the physicists Hermann Helmholtz and William Thomson (Lord Kelvin) independently suggested an alternative power source: gravity itself. Energy is released when a body is dropped in a gravitational field; for example, water falling over a water wheel performs work. Perhaps, Kelvin and Helmholtz conjectured, the balance provided by hydrostatic equilibrium was not quite perfect. Perhaps the star continued to contract under its own gravity at a very slow rate. As it did so, its gas would be compressed and heated, so that some of its gravitational energy would be converted into heat and light.

The belief that gravity powered the stars held sway for many years, although there were hints that this answer was not right. Calculations indicated that gravity could keep the Sun shining for many millions of years, but mounting terrestrial evidence suggested

that the age of the Earth was in the *billions* of years. The discovery of radioactivity, near the end of the nineteenth century, provided a possible means out. Here was a previously unknown energy source, clearly neither chemical nor gravitational. As more and more came to be known about the atom, physicists realized that the nuclei of atoms could be broken apart or fused together, and that in many cases this would release energy, possibly in enormous amounts. Einstein's famous formula $E = mc^2$, which in essence states that energy and mass are equivalent, shows just how much energy nuclear reactions can release. Multiplying the mass of the Sun by the speed of light squared, c^2, and dividing by the solar luminosity, the energy radiated per second, shows that the Sun could potentially shine for

$$\frac{M_\odot c^2}{L_\odot} = 14{,}000{,}000{,}000{,}000 \text{ years.}$$

Thus only a small percentage of the total mass of the Sun need be converted to energy to enable it to shine for tens of billions of years. Nuclear reactions could easily provide more than enough time for the Earth and its inhabitants to form and evolve. By 1938, enough was known about atomic physics for physicists to work out one sequence of fusion reactions by which the stars shine.

The stars we see in the sky, and our own Sun, are furnaces burning nuclear fuel. The heat generated by those nuclear reactions provides the gas pressure to keep the star from collapsing under its own weight. How can a gas accomplish such a Herculean task? The gases inside the Sun consist of the nuclei of atoms, and the electrons that have been stripped away from them, all moving about at high speed, and colliding with each other. Is there any way to make sense of this chaos? Fortunately, there is. The number of particles is so huge, and the way they interact sufficiently simple, that the behavior of the gas as a whole can be described in an averaged, statistical way.

Any particular particle will have some mass m, and will be moving at some velocity v until it collides with some other particle and changes its velocity. Velocity and particle mass can be combined to yield an energy due to motion, that is, a *kinetic* energy. The Newtonian formula for this energy is $E_k = \frac{1}{2}mv^2$; this holds for any particle as long as Newtonian physics is valid, which is mostly true even in the interior of the Sun. Because the collection of particles is constantly interacting, the gas comes into an equilibrium characterized by some average particle kinetic energy. The quantity we call temperature is defined by this average energy per particle. The higher the average kinetic energy of the particles, the higher the temperature. Moreover, two gases which have the same temperature but different particle masses must differ in their average particle velocities; the gas with the lower-mass particles would necessarily have a higher average velocity.

If a gas at temperature T is confined within a rigid box, the particles will collide not only with each other, but also with the walls of the box. Since each collision changes the velocity of the incident particle, a force must be exerted upon the particle; but by Newton's third law, the particle must also exert a force upon the wall. Multitudes of such collisions by the constituent particles of the gas can be averaged to yield a macroscopic force per unit area upon the wall, resulting in a *gas pressure*. Working through the units, we find that force per unit area has the same dimensions as energy per unit volume; indeed, pressure can also be characterized by the average

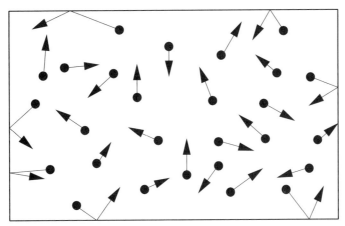

Figure 5.3 Gas molecules within a box move about with different velocities. The higher the temperature, the higher the average speed of the molecules. Collisions with the walls of the box exert a force per unit area, or pressure, on those walls.

energy in a gas per unit volume. But we have just argued that for a gas in equilibrium, the temperature specifies the average kinetic energy per particle. A higher average kinetic energy ($mv^2/2$) should result in a larger change in momentum (mv) when gas particles strike their surroundings. (Recall the momentum form of Newton's second law from chapter 3.) We conclude that temperature and pressure ought to be related. For many gases, including stellar gas, the pressure and temperature are related by the **ideal gas law**

$$P = nk_BT. \tag{5.1}$$

This equation states that pressure P equals the number of gas particles per unit volume n, multiplied by the average energy, k_BT, of those particles. The quantity k_B is a number called the *Boltzmann constant*, and its role is, essentially, to convert the units of temperature into energy units. The temperature T is measured in kelvins, where 0 K corresponds to absolute zero. Classically, absolute zero is the lowest temperature possible, corresponding to no particle motion at all. (Quantum mechanics does not permit the complete absence of motion, but absolute zero still represents the minimum allowed temperature.) The higher the temperature, the more rapid the motions, the larger the kinetic energies, and the greater the pressure, as indicated by the ideal gas law.

The ideal gas law shows that the temperature and pressure are proportional for a fixed amount of gas held to a fixed volume. Of course, a star is not a rigid container; its gas has some freedom to expand or contract, and such changes in volume affect the pressure and temperature through the number density n. However, under most circumstances during a star's life in equilibrium, the changes in volume are quite small relative to the size of the star.

A star must be supported by the pressure of its gas. The deeper into a star, the greater the weight of the overlying layers, and the higher the pressure and temperature. Imagine a thin shell of gas located at a radius r from the center of the star. If the star is to remain stable, the net inward force on any such shell must equal the net

force outward. The net force inward is the force of gravity at that location, plus the inward-directed force due to the pressure from the gas lying beyond r. The only available outward force is provided by pressure from the gas beneath the shell. Setting the outward force equal to the total inward force leads to the conclusion that pressure must increase deeper into the star. A similar argument shows why water pressure must increase at greater depths in the ocean; the higher pressure supports the overlying layers.

In a careful treatment of stellar structure we would consider each infinitesimal shell of gas, calculating the pressure needed to provide support down through the star. Such calculations show that the larger the total mass of the star, the greater the central pressure. We are also interested in the temperature structure of a star, since this determines nuclear reaction rates, the interactions of photons with the star's ionized gas, and so forth. This demands that we find some relationship between pressure and temperature in the star. For normal stars, the ideal gas law provides this relationship. We might be concerned that this simple law would fail for the extreme conditions in the deep interior of a star, but real gases actually obey the law to an excellent degree even at very high temperatures. Applying the ideal gas law tells us that the more massive the star, and hence the higher the central pressure, the hotter it must be at its core.

The mass is not the only quantity important to a star, although mass plays the major role. It is slightly less straightforward to visualize, but the elemental composition also affects a star's structure, through the average mass per gas particle. If the particles have greater mass, then there will be fewer of them for a given stellar mass. This means the number density n that appears in the ideal gas law will be lower, on average, requiring higher temperatures to produce the same central pressure. Most stars have similar compositions, consisting of approximately 75% hydrogen and 25% helium, by mass, with other elements present in small quantities. However, the compositional variations from star to star, even though relatively small, can produce subtle differences. A careful analysis of the equations of stellar structure shows that the nature of a star is almost entirely controlled by its mass and its composition.

Up to this point we have concerned ourselves with the implications of hydrostatic equilibrium. If this were the whole story, the star would just sit indefinitely, without changing. But stars radiate heat and light into space; were it not for nuclear reactions, which replenish the lost energy, the stars would cool and go out of equilibrium. The high temperature in the stellar core is just what is needed to make those reactions go.

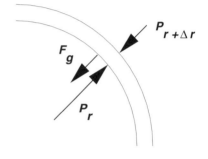

Figure 5.4 Forces on a spherical shell of gas in a star. The gravitational attraction of the mass interior to the shell creates a downward force on the shell. The gas pressure above the shell (pushing down) is less than the gas pressure below (pushing up); this difference balances gravity.

Twinkle, Twinkle, Little Star

Why is the temperature so important in nuclear fusion? The core of a star is composed predominantly of free protons and electrons whipping around at very high speeds. Under conditions even close to what we Earthlings might regard as "ordinary," two protons repel one another, since they both have positive electrical charge. The closer the protons approach, the more strongly they repel one another, because the electrostatic force, like gravity, follows an inverse-square law. This mutual repulsion creates something called the *Coulomb barrier*, and ordinarily keeps the protons apart. In an atomic nucleus, however, protons manage to stick together despite the electrostatic repulsion. Why is this? The protons are bound together by another force, called the strong interaction, and this force is much stronger than the electrostatic force. But the strong force is very short range. The trick, then, is to force the protons sufficiently close together that the nuclear interaction can take effect.

The higher the temperature, the closer the protons can approach each other. Thus, the temperature determines the rate at which nuclear reactions take place. In the core of the star, the protons have very high energies (temperature), and are forced extremely close together. Under such conditions, occasionally a purely quantum effect, called *tunneling*, can occur; the protons pass through the Coulomb barrier and merge. The product of this *proton-proton process* is not two protons stuck together, but is a *deuteron*, the nucleus of the **deuterium**, or heavy hydrogen, atom; the deuteron consists of one proton and one neutron. In this process, a proton is converted to a neutron, and a positron and a neutrino are ejected from the new nucleus. (Technically, this reaction also involves the *weak interaction*; this can be seen from the presence of a neutrino. The weak force is distinct from the strong force. For the present purposes, however, the general picture we have developed here is adequate.) The positron immediately annihilates with an electron, releasing energy; the neutrino also carries away some energy. Neutrinos interact so little with ordinary matter that the energy they carry is essentially lost immediately from the star. An amount of energy equal to the *binding energy* of the deuterium nucleus is liberated. The binding energy is the amount of energy required to break apart the nucleus; hence when such a nucleus *forms*, the same amount of energy must be released.

Deuterium readily fuses with another proton to form ^{3}He, releasing a high-energy photon (γ) in the process. The new ^{3}He nucleus quickly reacts with another to create ^{4}He, a very stable nucleus; two protons are also produced, which may then reenter the cycle. The net result is the fusion of four protons into one nucleus of ^{4}He, the creation of two neutrinos, and the liberation of the binding energy of the helium nucleus. The total mass-energy released by fusing hydrogen to helium is about 0.7% of the rest mass-energy of the reactants.

Schematically, we can write the reactions involved in the proton-proton process as

$$p^+ + p^+ \rightarrow D + e^+ + v,$$

$$D + p^+ \rightarrow {}^3\text{He} + \gamma,$$

$$^3\text{He} + {}^3\text{He} \rightarrow {}^4\text{He} + p^+ + p^+.$$

There are additional reactions routes that convert the ^{3}He into ^{4}He, but the final result is largely the same.

There is another process, the *CNO cycle*, in which ^{12}C (carbon-12) goes through reactions with protons, passing through ^{13}N, ^{14}N, ^{15}N (nitrogen 13, 14, and 15) and ^{15}O (oxygen-15) before the last step, in which ^{15}N fuses with a proton and emits an alpha particle, that is, a nucleus of ^{4}He, thereby reverting to ^{12}C. Although it is much more complicated than the proton-proton process, the net result of the CNO cycle is the combination of four protons to create one nucleus of ^{4}He, along with the emission of two positrons and two neutrinos. The ^{12}C reemerges at the end unchanged; it thus functions as a *catalyst*, a substance which participates in and assists a reaction, but itself is unaffected overall. The rate at which the CNO cycle proceeds is temperature sensitive, and it is rare in stars like the Sun; it is important only for stars more massive, and thus hotter, than the Sun. The CNO cycle also, obviously, depends upon the presence of carbon atoms. We shall eventually learn that only hydrogen and helium were created in significant quantities in the early universe, before the first stars formed; all other elements were built up by fusion processes within stars. The CNO cycle thus depends upon earlier generations of stars that made carbon; it was not available to the first stellar generation.

The Sun, to give a specific example, has a luminosity of 3.9×10^{26} J/sec. To find out how much hydrogen burning is required to maintain this amount of energy production, divide the luminosity by $0.007c^2$ for the fraction of rest mass released; this yields 6×10^{11} kg of hydrogen per second, or about 600 million metric tons. The energy thereby released slowly makes its way in the form of photons to the outer layers of the star. The way is difficult, for the hot inner layers are opaque, and photons are constantly scattered, absorbed, and reemitted. In the Sun, a star of average density, a photon generated by the core takes hundreds of thousands of years to work its way to the transparent outer layers; from there it can at last stream into space. The light falling on us today was generated by nuclear reactions in the Sun's core that occurred before our species walked the Earth.

The fusion processes in the cores of stars create new, heavier elements. Only hydrogen and helium, and a small fraction of the light element lithium, are **primordial elements**, in existence since the beginning of the universe. All other elements are manufactured in stars. The newly created elements return to interstellar space when the star sheds most of its gas at the end of its life. There the enriched gas may join other clouds of gas to bring forth later-generation stars, such as our own Sun. The oxygen we breathe, the carbon and nitrogen and sulfur and phosphorus that make up much of our bodies, the iron and aluminum and the silicon upon which our industries and economies are based, indeed, almost all of the matter on Earth, and in our own bodies, was created within ancient, massive stars that lived and died before the Sun was born.

The details of the nuclear processes are not as important as is the realization that they provide the energy to maintain the star in hydrostatic equilibrium. We can make some further progress in understanding stars without any knowledge of nuclear reaction rates. The mass of a star is the most important factor in establishing its core temperature. Temperature, in turn, determines the rate at which nuclear reactions proceed in the star's core. The energy released in the core must work its way through the star to be released at its surface, thus ultimately determining the star's luminosity. Therefore, there must be a relationship between the mass of a star and its luminosity. The ingredient needed to complete that relationship is an approximate relationship between temperature, luminosity, and stellar radius. This in turn depends upon the rate at which energy can be transported through the star. A very simple physical argument, which assumes that photons diffuse through the dense gas deep within the star till they reach the thin outer layers, yields an approximate relationship between luminosity and mass of

$$L \propto M^3. \tag{5.4}$$

For example, a star of 10 solar masses, 10 $M_\odot$, would have a luminosity not 10 times, but 1000 times that of the Sun. The luminosity, and hence the fuel consumption, of a star thus goes up quite rapidly with its mass.

So far, we have discussed only theory; what about observations? There is a straightforward relation between the luminosity of a star and its *surface* temperature, T_s. Stars are nearly blackbodies, so the energy per unit area they radiate is proportional to the fourth power of the surface temperature, T_s^4. Again, the total luminosity will be the energy per unit area times the total surface area (proportional to R^2); hence $L \propto R^2 T_s^4$. Luminosity cannot be observed directly, but surface temperature is relatively easy to measure. It is only necessary to observe the continuum spectrum of the star, determining where that spectrum peaks; the blackbody relationship then gives the surface temperature. (If that is not sufficiently accurate, known corrections can be applied to make a better model of the radiation of the star, and an improved value for the temperature can then be computed.) The color of a star is related to its surface temperature. The redder the star, the cooler its surface. Bluer stars are hotter. If the distance to the star can be measured by independent means, then the observed flux can be converted into total luminosity. Another approach is to study a group of stars at the same distance, such as a star cluster. In either case, it is possible to measure the luminosity and the temperature for a number of stars and make a plot. The plot should reflect the underlying stellar physics we have described with our simple physical stellar models.

And indeed it does. If the surface temperature of the star is plotted as a function of its luminosity, we obtain a graph called the *Hertzsprung-Russell diagram* (generally shortened to the "H-R diagram"). The points are not scattered about but fall into very narrow and well-defined curves. Most stars lie on the **main sequence**. Along the middle portion of the main sequence, the luminosity is related to the mass by $L \propto M^{3.1}$, very close to the value obtained by a simple physical argument. Thus, the observations indicate that the processes occurring in main sequence stars are controlled primarily by the conditions required for hydrostatic equilibrium, the balance between gravity and the pressure supplied by the heat from nuclear reactions. For as long as the fusion of hydrogen to helium dominates, the star resides, usually quietly, on the main sequence of the H-R diagram.

If we wished to develop a realistic stellar model, we would have to write down the *differential* equation of hydrostatic equilibrium. Then we would be obliged to include rate equations for the nuclear reactions in the core, and we would be required to solve the difficult equations of radiative transfer. In fact, a realistic model of a star

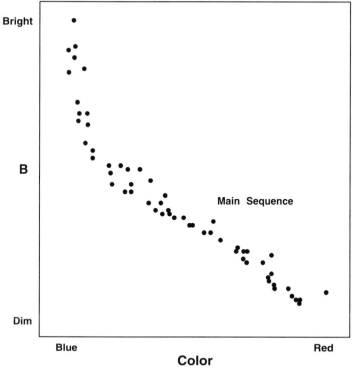

Figure 5.5 Composite Hertzsprung-Russell diagram for some of the stars of the Pleiades, a young stellar cluster. The vertical axis is the logarithm of the brightness of the star; for stars at a fixed distance the observed brightness will be proportional to the luminosity. The color is a measure of the temperature of the star. The stars do not appear randomly, but lie along a curve called the main sequence. Stars on the main sequence fuse hydrogen in their cores.

is sufficiently complicated that it is necessary to solve the resulting equations on a computer using numerical techniques. Although not all of the phenomena are perfectly understood, especially those having to do with the transport of energy within the star, stellar models are still good enough to reproduce the main sequence to a high degree of accuracy.

Do you find it believable that stars can be understood quite reliably? In fact, stars are probably among the best-understood structures in the universe. They are very important to cosmology because their lives are uniform and predictable. And their ages, and their deaths, have significant ramifications for the universe and its contents.

Stellar Ages

Astrophysicists like to joke that "we understand every star except the Sun." The problem with the Sun, of course, is that we have an overwhelming amount of data on every detail of its existence, including its every magnetic outburst and minor shudder. We cannot forecast, or sometimes even explain, the day-to-day workings of the Sun. Even so, we do understand the fundamentals of the construction of the stars and the grand outlines of their lives. We can exploit this knowledge to determine the ages of stars, by comparing observations to the predictions of our models.

The stringent physical constraints that govern the evolution of a star result in a predictable life history. Stars wage a constant battle against their tendency to collapse under their own weight. The nuclear reactions deep within the interior provide the energy that is radiated away by the star; as long as the lost energy is constantly replenished by fusion, the temperature at the core can be maintained high enough to fight the inexorable pull of gravity. For nearly all of a star's life, the most important nuclear reaction is the fusion of hydrogen into helium. But such things must end, and every star eventually runs out of hydrogen fuel. Although we, who depend upon the stability of one main-sequence star, must be thankful that the time available is long, the events which happen *after* the star leaves the main sequence are of the greatest interest for cosmology. How quickly this point is reached depends upon the mass of the star. Very massive stars burn their candles at all ends, blazing gloriously for a few million years before exhausting their supplies. Less massive stars, such as the Sun, burn their fuel more frugally and exist in a stable state for many billions of years. This fact can be used to set a limit on an important cosmological measurement: the age of the universe.

When the hydrogen in the core of a star is used up, the core contracts and the star changes its structure. As it contracts, the temperature in the core rises sufficiently to initiate fusion of helium into carbon. At this point the star becomes a **red giant**. The subsequent evolutionary phases of the star, its old age and death, are interesting in their own right, but for now, let us concentrate on this critical event when the star ceases to burn hydrogen in its core. This is when the star "leaves" the main sequence, increasing in luminosity, but also expanding in size so that its surface temperature drops. The star can now be found above and to the right of the main-sequence line in the H-R diagram.

The great majority of a star's lifetime is spent on the main sequence as a hydrogen-burning star. For all practical purposes, we can define the age of a star to be its main-sequence lifetime. What sort of lifetimes would we expect? Recall that main-sequence stars have a luminosity-to-mass dependency of roughly $L \sim M^3$. The lifetime of a star must be determined by the amount of fuel available to it, divided by the rate at which it consumes that fuel. Hence the stellar lifetime is proportional to its mass (fuel) divided by its luminosity (burn rate), that is, $t_* \sim M/L$. Together, these relationships imply that stellar lifetimes decrease with the square of increasing mass, $t_* \sim M^{-2}$. This is just a rough calculation, but it indicates that massive stars live much shorter lives in comparison to low-massed stars.

Studies of stellar ages have determined that the stars of the Milky Way Galaxy fall into two broad categories, called **Population I** and **Population II**. Population II (or just Pop II) stars are very old, probably nearly as old as the galaxy itself, whereas Population I (Pop I) stars are much younger, and continue to form today. The Sun is a member of Population I. The major difference between Population I and II stars, other than their ages, is their *composition*. Old stars have far fewer **metals**, which to astronomers means any element heavier than helium, whether chemically a "metal" or not. This is consistent with the formation of heavy elements within stars; the early generations of stars must have formed from gas which had little metal content, since there were no earlier stars to create the metals. Population I stars formed from the debris of older, massive stars that exhausted themselves quickly. There is some speculation that an even older group of stars, Population III, might exist, but there is currently no definite evidence for their presence.

While most Population I stars are found in the spiral disk of the Milky Way, most Population II stars are found in the Galactic bulge toward the center of the Galaxy, or in the halo surrounding the disk of the Galaxy. In the halo Pop II stars are often found in **globular clusters**, huge, nearly spherical clusters of about 100,000 stars each. The globular clusters orbit the Galactic center within a roughly spherical volume. Globulars are found not only around the Milky Way but are also seen around every other galaxy close enough for objects of their size to be resolved. Globular clusters are thought to be the oldest objects in the Galaxy; thus the age of their most ancient stars provides a lower limit to the age of the Galaxy, and hence of the universe itself.

Consider such a cluster of stars, whether a globular cluster or a younger "open" cluster. The cluster members formed at about the same time, from the same great nebula of gas. Thus the stars should all have about the same initial composition. Stars of all masses were created, in a distribution given by the IMF. As time goes by, the most massive stars use up their hydrogen and evolve off the main sequence. Slowly, the main sequence disappears, starting at the high-mass end and moving toward the low. Of course, we can observe only a "snapshot" of a star cluster, at one particular time in its evolution. If we plot an H-R diagram of all the stars that are members of the cluster, we will find many dots spread along the main sequence, with an abrupt cutoff corresponding to those stars that are on the verge of ending their main-sequence lives. (Stars more massive than this have already left the main sequence.) The cutoff point represents a specific mass, the **turnoff mass**. If you know the lifetime of stars of that mass, then you know the age of the cluster.

There are, of course, many uncertainties in age determinations. Variations in composition, mass loss by stars during the stable stages of their lives, and the effects of turbulent mixing in stellar layers, are examples of potential sources of error. Much of the uncertainty lies with unknown stellar compositions; there are also difficulties in determining precise main-sequence turnoff points, and matching those points with theoretical models. Even so, experts in stellar ages have reached a consensus. The oldest globular clusters in the Milky Way galaxy and its neighbors have been determined to be at least 12 to 15 billion years old. Most of the uncertainties tend to extend the "error bars" out to *greater* ages, so if anything, the clusters may be older than the estimates. At the present time, it appears to be quite difficult to find a reasonable combination of error and uncertainty that would produce stellar ages in the oldest globular clusters of less than about 12 billion years.

White Dwarfs to Black Holes

Stellar ages are determined from main-sequence evolution. But what happens to a star when it leaves the main sequence? If a star like the Sun began its life entirely composed of hydrogen and gradually converted all this hydrogen to helium, it could

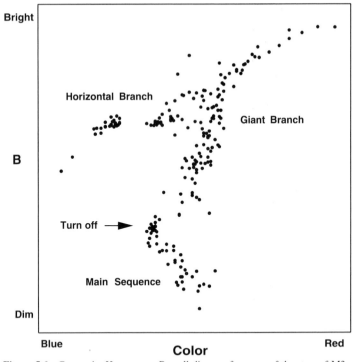

Figure 5.6 Composite Hertzsprung-Russell diagram for some of the stars of M3, an old globular cluster, plotted in the units of brightness versus color. Only the stars in the lower center of the plot lie along the main sequence. Stars that have left the main sequence lie above, on the horizontal branch, and to the right, on the giant branch. (Compare with Fig. 5.5) The point at which the stars leave the main sequence is the "turnoff" point. A comparison of the observed main-sequence turnoff with the predictions of stellar theory gives the age of the cluster.

shine for 100 billion years. However, nuclear reactions take place only where the temperature and density are high enough, and in most stars this is only deep within its core. At best, a typical star can convert no more than about 10% of its hydrogen to helium. When hydrogen can no longer be fused, the pressure in the core drops, allowing gravity again to compress the gas, and hence to raise the central temperature. The increase in temperature causes the outer regions of the star to expand, and the expansion cools the surface layers. The star balloons to enormous size, creating a **red giant**. When the Sun reaches the red giant phase, perhaps 5 billion years into the future, its surface will extend to near the orbit of Mars. The intense radiation falling upon Earth will destroy any life that might remain, and the planet itself will spiral into the bloated Sun, vaporizing in its hotter inner layers. The Sun will reclaim its innermost planets.

The increase in the core temperature is important because the next nuclear fuel to be burned, helium, does not fuse at the lower temperatures present during the Sun's main-sequence life. When the core becomes hot enough, helium will begin to fuse into carbon in the deep interior, while hydrogen continues to burn in a relatively thin shell surrounding the core. The new fusion reaction is called the *triple-alpha* process because three nuclei of ^{4}He fuse to ^{12}C, with the unstable nucleus ^{8}Be (beryllium-8) created as an intermediate product. This new energy source stops the gravitational contraction and stabilizes the star, allowing it to continue to shine, once again in equilibrium. The star exits the main sequence for the *horizontal branch*, the region on the H-R diagram occupied by stable helium-burning stars.

But again, this stage can last only so long, for the helium is consumed even more quickly than the hydrogen before it. What happens next depends upon the mass of the star. For stars of modest mass, such as the Sun, the end is quiet. The heavier the nucleus, in general, the higher the temperature and density required to force it to participate in fusion reactions. Stars up to about 6 times the mass of the Sun are not sufficiently massive for gravity to be able to raise the core temperature to a high enough level for further fusion reactions to occur. Once the usable helium fuel has been converted to carbon in such stars, nuclear reactions cease, and the core once again contracts under its own gravity. This continues until the matter in the core becomes so compact that electrons cannot be squeezed together further. This state is called **electron degeneracy**, and it is a quantum-mechanical consequence of the Pauli exclusion principle. Electrons are fermions, and thus, by the exclusion principle, no two can occupy the same quantum state. In electron degeneracy, all low-energy quantum states are occupied, forcing many electrons into high-energy states. It would take considerable energy to squeeze the electrons even closer together, so the electrons provide a new source of pressure that does not depend on temperature. This is quite different from the ideal gas law; most significantly, it means that the star can now resist gravity with no further generation of heat. It is somewhat analogous to the intermolecular electrostatic forces that give a crystal such as quartz or diamond its great rigidity. As the core settles down to its degenerate state, nuclear burning can continue in the surrounding stellar envelope. This eventually causes the star to eject its swollen outer layers; if we happen to see the expanding shell of gas, it might take the form of a lovely object called, for historical reasons, a *planetary nebula*. Finally, only the degenerate core is left behind, in a remnant known as a **white dwarf**. A white dwarf star no longer burns nuclear fuel and shines only because it takes many millions of years for light to percolate out to the surface from deep within its core. Eventually the star will cool, and the white dwarf fades away. This is the eventual fate of our Sun.

White dwarfs have sufficiently low luminosity that the only ones we can observe directly are in our solar neighborhood. The bright star Sirius is actually a binary; the tiny companion, invisible without a good telescope, is a white dwarf, the first discovered. All white dwarfs are very small and very dense. (Imagine packing the mass of the Sun into a volume the size of the Earth.) This immediately tells us that their gravitational fields are relatively strong. The chemical composition of white dwarfs probably varies somewhat, but observations are consistent with the theory that they should consist predominantly of carbon, with some oxygen, the final products of helium burning. The unusual state of the matter in a white dwarf has some interesting consequences. For one thing, the greater the mass of a degenerate white dwarf, the *smaller* its radius. For another, as the white dwarf cools, it can actually *crystallize*; its nuclei, long separated from their electrons, behave much more like a solid than like the gaseous plasma of which the star was previously composed.

Since a white dwarf is no longer generating energy, it cools at a rate determined mainly by only a few quantities: its surface temperature and area, which control the rate at which energy is radiated into space, and the length of time required for a photon to work its way from the interior to the surface. White dwarfs have extremely high surface temperatures, as much as tens of thousands of degrees, but not a lot of surface area, so overall they radiate rather slowly. Moreover, they are so dense that it takes a very long time for photons to diffuse outward. As photons slowly trickle to the surface of the white dwarf and stream away, the star loses energy and cools; with time, a white dwarf will shift its color from blue-white to yellow to red, and then will finally cease to emit in the visible at all. White dwarfs cool so slowly, however, that the universe is probably still too young for a significant number to have disappeared from visibility.

You might have realized by now that if we could compute the rate of cooling of a white dwarf, we could deduce the time elapsed since it formed. In principle, this is possible; in practice, there are many difficulties. Since we cannot fetch any samples of white dwarf matter, nor can we recreate it in the laboratory, we must rely upon theory to construct models of the characteristics of the material of which the dwarfs are composed, then attempt to compare the predictions of the model with observations of real white dwarfs. Another unfortunate limitation is that our sample of white dwarfs is small. They are dim and tiny—the brightest have a luminosity of approximately $0.1\ L_\odot$—and we can see only those in our Galactic neighborhood, even with the best of modern telescopes. Most of those we can find are the binary companions of normal stars. Nevertheless, many efforts have been made to estimate the age of the oldest white dwarfs in the Galaxy, since such a datum would obviously set a lower limit to the age of the Galaxy itself. The best estimates obtain an age of approximately 12 billion years for the most ancient white dwarfs, consistent with the ages of the globular clusters.

Occasionally, a white dwarf can revive, if it has a companion. When two stars orbit, their gravitational fields overlap, since gravity's range is infinite. Each of the stars is surrounded by a region, called the *Roche lobe*, within which its gravity dominates that of its partner. The Roche lobes of the members of a binary touch at a point known as the *Lagrange point*; this is where the gravitational tug of each star is equal in magnitude. In a typical binary system, each star is much smaller than its Roche lobe. If the separation between the two is large, both stars will spend their lives well within the confines of their Roche lobes. However, as stars age and leave the main sequence, they swell up to giant size. In a close binary system, when one of the stars reaches the red giant phase, it can overflow its Roche lobe, and some of the distended star's outer layers can be transferred onto the smaller companion.

What happens to this transferred gas? Since the members of a binary star system are in mutual orbit around one another, any gas flowing from one star must partake of this rotation. Thus we encounter a situation reminiscent of our argument about the formation of a protostellar disk; as gas flows from one star to the other, it falls inward along an orbital trajectory. If the star toward which the gas is falling is small enough, and this will certainly be true for a white dwarf, the inflowing gas stream misses the star's surface and goes into orbit. In this case, the inflowing gas creates an **accretion disk**. Dissipation of the angular momentum of the gas through turbulence in the disk means that a parcel of gas cannot orbit its new primary at a fixed radius. Rather, it spirals toward the star. The fall of the gas in the gravitational field releases energy; the gas in the disk is compressed and heated. The accretion disk may emit high-energy radiation, even X-rays, which can be detected from Earth. Eventually, the gas crashes onto the surface of the star, emitting a burst of energetic radiation.

If the accreting star happens to be a white dwarf, the transfer of mass can have some other interesting consequences. A white dwarf cannot incorporate the new material in a smooth manner, as would a normal star, since a dwarf's pressure support comes not from ordinary gas pressure, but from degeneracy pressure. Whereas ordinary pressure can adjust itself with temperature and density, degeneracy pressure is *independent* of temperature. The infalling gas is thus compressed and heated as it strikes the unyielding surface of the white dwarf. When enough gas has piled up, it can reach the 10^7 K required for the initiation of hydrogen fusion. The white dwarf suddenly flares in brightness and becomes a **nova**. After this thermonuclear explosion from its surface, it once again fades away. Often, the cycle is repeated, when enough gas again accumulates to reach the ignition point. At their peak brightness, novae seem to have fairly uniform luminosities, which means it might be possible to use them to determine distances. Unfortunately, they are not perfectly standard; work continues to determine whether novae can help to calibrate the cosmic distance scale.

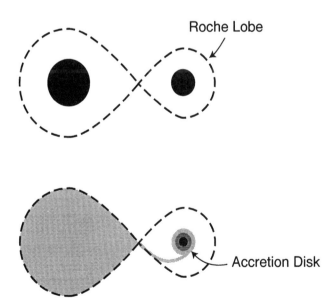

Figure 5.7 (*a*) The Roche lobe surrounding binary stars is the curve that marks the region within which a particle is gravitationally bound to one star or the other. Such a curve would be spherical around a single star; here it is distorted by the tidal forces of the binary. The crossing point in the figure "eight" is the point at which the two stars' gravitational attractions are equal but oppositely directed. (*b*) If one star fills up its half of the Roche lobe, it can overflow and transfer gas to the other star, possibly creating an accretion disk around its companion.

The fate of lower-mass stars is to cease fusion after using their available supply of helium, then to eject their outer layers quietly, with the cores left to cool slowly as white dwarfs. Stars more massive than a few solar masses experience more phases at the ends of their lives, going through one nuclear fuel after another to battle the crush of gravity. After the star's helium is exhausted, the core contracts and heats again, and the outer layers expand, sending the star up the *asymptotic giant branch* of the H-R diagram. In very massive stars, carbon may first ignite; for sufficiently massive stars, increasingly heavy elements are subsequently burned, fusing all the way to iron. The star becomes a gigantic cosmic onion, consisting of concentric shells in which increasingly heavier elements are fused. The final fusion product is iron. Unlike the lighter elements, iron demands an *input* of energy to be forced to fuse into heavier elements. Iron cannot provide the energy the star needs so desperately to support its weight, since further fusion would actually consume the star's precious energy; once the available matter has been burned to iron, the star is out of usable fuel. Iron is the end of the nuclear road.

A star with an iron core must seek an equilibrium with gravity that does not require further expenditure of energy. Smaller stars could find their final equilibrium in electron degeneracy. However, for any object with a mass greater than 1.4 times the solar mass, the pressure from even electron degeneracy is not sufficient to support the star against its own weight. In 1930, Subramaynyan Chandrasekhar realized that in order to provide the incredible pressures required to maintain more massive stars, the electrons supplying that support would have to move at greater than the speed of light, which was known from the special theory of relativity to be an impossibility. Thus special relativity demands an upper mass limit, today called the **Chandrasekhar limit**. If the dying star fails to eject enough of its matter to allow its collapsing core to drop below this limit, the electrons cannot supply the necessary pressure. But if electron degeneracy pressure falls short, the star does not just slowly contract. It collapses catastrophically, sending a shock wave into its outer layers and blowing them off in a single cataclysmic explosion called a **supernova**. A supernova is so bright that for a brief interval of a few weeks, it may outshine the galaxy in which it occurs, a blazing beacon visible across enormous distances. The explosion is so powerful that most of the star's matter is blasted away and dispersed into space.

This may seem like a cruel fate for a star, to end in a huge explosion, but the supernova plays a vital role in the history of the cosmos. It is in supernovae that the heavier elements, forged near the center of the star during its last stages of existence, can find their way into space, and thence into later stars and planets. Indeed, so much energy is liberated in the blast that elements heavier than iron can be created, even though, as remarked above, these reactions consume energy rather than releasing it. The gold and silver with which we ornament ourselves, and which we hoard and covet, came into abrupt being in the final moments of the life of a massive star. Cobalt and nickel and zinc, and the uranium of our nuclear-power plants and of our weapons—all these heavy elements are the ashes of massive stars.

A supernova which arises from the collapse of a massive star is designated by astronomers as Type II. This name suggests that there must be another type of supernova, the Type I supernova. The Type I supernova originates from the explosion of a white dwarf in a binary system. If the companion star overflows its Roche lobe, gas accretes around and onto the white dwarf. The accretion of matter can lead to a nova. A nova outburst probably blasts away much of the gas accumulating on the surface of the dwarf, but not all of it. After each cycle of nova activity, the mass of the white dwarf increases slightly; eventually, it may acquire more gas than it can support. It is constrained by the Chandrasekhar limit throughout its existence. Should its mass rise above that limit for any reason, including mass transfer from a binary companion, electron degeneracy pressure cannot continue to support it. The star collapses violently. The sudden increase in temperature detonates the carbon; because of the degeneracy of the matter, *all* the matter in the white dwarf fuses almost simultaneously. The resulting explosion rivals the death of the supermassive star in its brilliance.

Type I supernovae have a property that is of particular interest to the cosmologist. In a Type I supernova, the progenitor was very near to the Chandrasekhar limit, or else it would not have collapsed in the first place. Therefore, Type I supernovae tend to peak in energy output at very similar luminosities. If all Type I supernovae had exactly the same luminosity, they would make excellent distance indicators, since supernovae are visible for such huge distances. They can be seen easily as far away as the Virgo cluster, approximately 50 million light years distant. Alas, the progenitors are not completely identical, and they do not necessarily explode at precisely the Chandrasekhar limit, so there is some variability. Astronomers continue to study how uniform the Type I supernovae really are.

There is much more variability in the light output of a Type II supernova than for a Type I. Still, the progenitors of Type II supernovae are all rather similar; all have followed life histories that depend mostly upon the mass of the star, and all have arrived, just before exploding, at the stage at which their cores are presumably quite near the Chandrasekhar limit. The variability in their maximum luminosity is small enough that astronomers can compute the approximate distance of a Type II supernova just by comparing its apparent brightness to the expected absolute brightness. Unfortunately, while such a distance estimate is adequate for many purposes, it still contains much too much intrinsic error to be of use for high-precision cosmological measurements. Type II supernovae have not yet been sufficiently characterized that their absolute brightness can be securely related to their behavior, but research into this subject is advancing.

Although a great deal of the star is blown out into interstellar space by a Type II supernova, some fraction is probably left behind in a core remnant. If the mass of the remnant still exceeds the Chandrasekhar limit, what can the star do? It cannot settle down as a white dwarf star; so what remains? As the star collapses to greater and greater compaction, the electrons are squeezed into the atomic nuclei themselves, where they are forced to merge into the protons, forming neutrons. The neutrons, which are much more massive than electrons, can themselves exert a degeneracy pressure known as **neutron degeneracy** pressure. The entire star is compressed essentially to the density of an atomic nucleus, but composed only of neutrons. This massive neutron nucleus is known as a **neutron star.** A neutron star is astonishingly compact; if an object with the mass of the Sun were to collapse completely to a neutron star, its radius would be only about 10 km, roughly the size of a typical large city on Earth. The neutron star is a remarkable object. Its existence was predicted as early as 1934 by Fritz Zwicky and Walter Baade, although their suggestion was ignored for decades; a neutron star seemed too bizarre to consider.

This attitude changed in 1967 when the first **pulsar** was detected. A pulsar emits highly regular, energetic bursts of electromagnetic radiation, generally as radio waves. The pulses from the first pulsar were so regular that the discoverers, Jocelyn Bell and Anthony Hewish at Cambridge University, first thought they had received signals from another civilization! No familiar astronomical process was known at the time that could produce electromagnetic bursts of such sharpness and regularity, at such a rapid rate. Ordinary oscillations would be inadequate to explain the signal. As more and more pulsars were observed, however, the mystery slowly yielded. Thomas Gold first suggested that pulsars might be associated with the exotic neutron star. Subsequent observations have borne this idea out; no other mechanism is remotely plausible to explain the properties of pulsars. But a neutron star is no longer generating energy from fusion; how might it send pulses into space?

Suppose a "hot spot" is present on the surface of a rotating star. The light emissions from such a spot would sweep through space like the beacon from a rotating lighthouse lamp. Just as a sailor sees the beam from the lighthouse only when it points at him, so we see the radiation from the pulsar at intervals equal to the rotation period of the star. But what kind of star could rotate once per second? If it were a white dwarf, about the size of the Earth, such rapid rotation would tear it apart. A neutron star, on the other hand, would have only about the diameter of a typical city and could easily rotate at such speed without breaking up. The case was clinched in 1968 by the discovery of a pulsar in the center of the famous Crab Nebula, an untidy blob of gas in the constellation Taurus. The Crab Nebula is well identified with a supernova observed in A.D. 1054 by Chinese astronomers; it is the shocked, disordered remnant of the outer layers of the star. Calculations by J. R. Oppenheimer and others had shown by 1939 that a likely outcome of supernova explosions was a neutron star. The association between a pulsar in the Crab Nebula and the known supernova that had occurred there made the identification of pulsars with neutron stars all the more certain. The Crab pulsar emits approximately 30 pulses per second, one of the most rapid rates of any pulsar. It pulses in optical wavelengths as well as radio, also unusual. Since the date of the supernova is known, we know that the Crab pulsar is young. Fast pulse rates and high energy output are associated with recently formed pulsars. As they age, they lose rotational energy, and the period of their pulsations increases.

Although astronomers have learned much about neutron stars, their structure is still somewhat mysterious. The matter in a neutron star is compressed into an even stranger state than that of a white dwarf. A white dwarf is somewhat like a very dense solid; unusual, but not mind-boggling. A neutron star is much weirder, more like a huge atomic nucleus than it is like anything familiar. The interior of the neutron star is probably in a fluid state, meaning that the neutrons move around freely. They move so freely, in fact, that the interior is said to be a *superfluid*, a fluid in which no friction is present. The fluid of degenerate neutrons is surrounded by a thin crust of fairly normal matter, consisting of crystalline iron nuclei, free electrons, and free neutrons. The tiny radius, for the large mass, of a neutron star implies an enormous, almost incomprehensible gravitational field near its surface. Occasionally, the intense gravitational field causes a defect in the crystalline structure of the crust to crack, resulting in a *starquake* as the crust readjusts. The starquake causes a *glitch* in the pulsar, a small, but very sudden, drop in the period of its pulsation. These starquakes provide valuable information into the nature of neutron star matter.

If a neutron star is such a dense, exotic object, how could it be set into such rapid spinning? It is a consequence of an important law of physics, the **conservation of angular momentum**. Angular momentum is a measure of the resistance of a body to changes in its rotation and is defined as

$$L = I\omega, \tag{5.5}$$

where ω is the rotation rate of the body, in angle per unit time, while I is a quantity called the *moment of inertia*. The moment of inertia describes the matter distribution of the object; the farther the mass from the axis of rotation, the greater the moment of inertia. Conversely, the more concentrated the matter near the axis of rotation, the smaller the moment of inertia. The law of angular momentum states that if no outside torque, or twisting, acts upon the body, its angular momentum does not change. Therefore, if the moment of inertia changes, the rate of rotation must change in such a way that the angular momentum remains the same.

Perhaps the most familiar illustration of the conservation of angular momentum is the figure skater executing a spin. The skater usually begins with arms outstretched, spinning at a certain rate. As he draws his arms toward his body, his moment of inertia decreases; to conserve the total angular momentum, there must be an increase in his rate of spin. Occasionally the skater even crouches, pulling all parts of his body close to the axis of rotation to increase his rate of spin even further. As he unfolds, his moment of inertia increases and his spin decreases, until he is spinning slowly enough to stop easily by exerting a small torque with the skate blade.

Even a falling cat must obey the law of angular momentum as it rights itself. High-speed photography of cats dropped from safe distances clearly shows them to twist their front legs in one direction, while their hind legs twist oppositely. The cat is still able to turn its body to land feet downward, but at each motion, its angular momentum must be conserved as it falls.

It is no accident that the moment of inertia is reminiscent of the inertial mass; its role for rotational motions is analogous to the function of inertial mass for linear motions. Since the moment of inertia of a star depends upon its mass distribution, the gravitational collapse will change the moment of inertia drastically. The radius of the core of the dying star can shrink abruptly by a factor of perhaps 1000 or more. For a sphere, the moment of inertia is given by the formula

$$I_s = \frac{2}{5}MR^2. \tag{5.6}$$

Thus, if little mass is lost, the newly formed neutron star must spin approximately a million times faster than its precursor. Typical pulsars rotate with periods of 1 second to approximately one-fourth of a second. For the idealized example of a neutron star executing one rotation per second, conservation of angular momentum would imply that the precursor rotated about once per month, which is comparable to the rotation rate of the Sun.

Along with rotation, the neutron star must possess a hot spot in order to emit the beamed radiation. A lighthouse mirror may turn, but unless the lamp is lit, there will be no beam. How does the hot spot generate such radiation? As far as we know, all stars possess magnetic fields; the field is tightly coupled with the ionized gas of which the star consists. When the core of a massive star collapses, its magnetic field is pulled along with it, greatly concentrating the field and producing huge magnetic forces. Most astronomical objects, including the Sun, have overall magnetic fields that look somewhat similar to, though are stronger than, the field of a bar magnet; there is a north and a south pole, and the field lines run continuously from one pole to the other. The collapse probably does not change the basic configuration of this field, although it does greatly amplify it, much as the spin rate is increased by the tremendous decrease in radius. A neutron star is thus like a bar magnet, of extreme strength, in space. Associated with the magnetic field should be a strong electric field, which rips charged particles from the crust of the star. These particles are trapped in the magnetic field and forced to accelerate to high, perhaps relativistic, velocities.

The details of how a fraction of the rotational energy of the star is converted into narrow pulses, as opposed to more diffuse radiation from around the star, are not very well understood, but some general statements can be made. The photons that we receive as pulses are likely emitted from the regions around the magnetic poles of the neutron star. The magnetic poles would not, in general, be aligned with the rotation axis of the neutron star. (This is hardly unusual; the magnetic axis of the Earth is misaligned with its rotation axis.) If the magnetic and rotation axes were coincident, we would receive a constant beam of radiation, and that only if our line of sight happened to look along the axis. However, if the emission comes from the magnetic poles, and these poles do not line up with the rotational poles, then the rotation will carry the beam around, sweeping it into our line of sight once per rotation. Only if the searchlight is pointed at an angle to its rotation axis can the lighthouse send a beam around the cape.

Many neutron stars seem to be solitary. This is not surprising, as we might expect that the violence of a supernova explosion would tear apart a binary, liberating, or perhaps even destroying, any companion the progenitor might have had. But some of the most interesting neutron stars are not alone. For example, two known pulsars apparently have planets. It seems unlikely that primordial planets would survive a supernova, so it may be that a stellar companion was obliterated in the blast, then recondensed into a disk and assembled itself into one or more planets around the neutron star. Such a system, if this is what happens, must be very bizarre—a former star reincarnated as a planet, orbiting the corpse of its erstwhile companion. Other neutron stars are members of normal binary systems. The dynamics of such a system are quite similar to that of a white dwarf binary, with some interesting twists due to the presence of the neutron star. The accretion disk around a neutron star would be much hotter and more energetic than around a white dwarf. Gas which piled up on the surface of a neutron star would find the crust to be even more unyielding than that of a white dwarf, and repeated episodes of sudden thermonuclear fusion might occur. Such a model explains the *X-ray bursters*, sources which emit spurts of X-rays at irregular intervals. Most such bursters are located near the center of the Galaxy or deep within dense clusters, environments where the density of stars is fairly high, and thus where a significant population of neutron stars could be expected to have formed.

Figure 5.8 A pulsar has a hot spot that is carried around by the pulsar's spin. The location of the hot spot corresponds to the star's magnetic axis. Because the magnetic axis is not aligned with the rotation axis, as the neutron star rotates the hot spot beams radiation into space like a searchlight, producing the observed pulses of radiation.

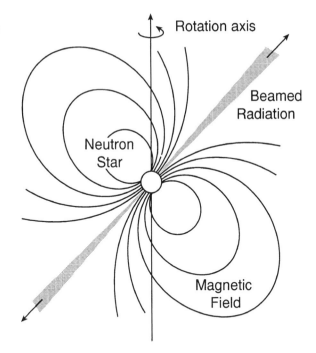

An even more bizarre effect can occur in the vicinity of a neutron star. Gas spiraling toward the rapidly rotating neutron star could be flung out at relativistic velocities in two *jets* collimated along the axis of rotation. The enigmatic object SS 433, a star system located approximately 16,000 lt-yr from Earth, might be an example of such a system. The spectrum of SS 433 reveals a mixture of approaching and receding gas with unusually large Doppler shifts; the spectra also show a smaller, regular shifting with a period of 164 days. The Doppler shifts indicate gas speeds up to one-fourth of the speed of light. The best explanation for this object is that it is a binary, one of whose members is a star that has overflowed its Roche lobe; the invisible companion is probably a neutron star. The strong emission lines emanate from a pair of relativistic jets, one directed toward our line of sight (approaching) and the other away from it (receding). The regular shifting occurs because of *precession*, a wobbling of the axis of rotation due to gravitational torques upon the neutron star. SS 433 exhibits on a small scale behavior similar to that seen in the cores of a fraction of galaxies, and especially in the cosmological objects known as *quasars*. Accretion around a neutron star, or around even more massive objects at the centers of galaxies, may be a common phenomenon throughout the universe.

Interactions with other stars can affect neutron stars in other ways. Although most pulsars have periods of a few tenths of a second, in the 1980s a new class was discovered which spin with a mind-boggling period of a few *thousandths* of a second. These *millisecond pulsars* are thought to be the product of mass transfer from a companion. The accreting matter would be rotating, so as it struck the surface of the neutron star, it could add angular momentum to it, thus increasing the neutron star's rotation rate. Neutron stars are so small and so rigid that they spin with amazing periods, but even so, the millisecond pulsars are probably near the breakup limit. The pulsar might later be ejected from its binary system by an encounter with another object; of 40 or so millisecond pulsars in globular clusters, about 10 are known to be members of binaries, while another 30 or so seem to be lone travelers.

Even stranger is the binary pulsar. If it is unlikely that a binary system could survive one supernova, it seems nearly impossible for it to survive *two*. Yet, at last count, three binary neutron star systems have been discovered. Perhaps both progenitor stars lost quite a lot of mass prior to exploding, so their supernovae were not excessively violent. Alternatively, perhaps the two pulsars did not form together. A solitary pulsar might have interacted with an existing binary, displacing the normal star. Whatever the formation mechanism, the binary pulsar is a fascinating and important object because it provided the first firm, albeit indirect, evidence for the existence of gravitational radiation, waves in the fabric of spacetime itself.

The white dwarf is supported by electron degeneracy pressure and has an upper limit on its mass. The neutron star is supported by neutron degeneracy pressure, and it too has an upper limit to the mass that can be so supported. Astrophysicists are not entirely certain what that upper bound is; the physical state of matter at these extreme densities and pressures is not as well understood as we would like. However, the limit almost certainly lies between 2 and 3 times the mass of the Sun. If an imploding stellar remnant finds itself with more than this mass, this unfortunate star cannot halt its collapse as a neutron star. Modern physics knows of no force sufficient to prevail against gravity; the star collapses to a *black hole*. The black hole has properties so strange that we cannot appreciate them until we have made a more careful study of the structure of the universe.

KEY TERMS

interstellar medium
red dwarf
deuterium
red giant
metal
electron degeneracy
nova
neutron degeneracy
conservation of angular
    momentum

nebula
hydrostatic equilibrium
primordial element
Population I
globular cluster
white dwarf
Chandrasekhar limit
neutron star

brown dwarf
ideal gas law
main sequence
Population II
turnoff mass
accretion disk
supernova
pulsar

Review Questions

1. What objects in the Galaxy are the most likely stellar nurseries? What properties make them good locations for star formation?
2. Distinguish between brown dwarfs and red dwarfs. Are brown dwarfs common? What would be the significance of a huge number of brown dwarf stars?

3. What is hydrostatic equilibrium, and why is it important to the existence of stars?

4. What are the main physical characteristics that control the life of a star?

5. What is the main sequence? What is the main-sequence turnoff point in a cluster of stars, and how can that be used to obtain an estimate of the cluster's age?

6. What happens in a nova? How is this different from a supernova? Some science fiction stories have had plots in which the Sun threatens to become a nova or supernova. Is this a possible scenario?

7. Why is there an upper limit to the mass that can be supported by electron degeneracy pressure?

8. Explain why a Type I supernova makes a better indicator of distance than a Type II supernova. Why does a supernova make a good distance indicator for cosmology compared to ordinary stars?

9. Describe three ways in which the study of stars can provide important cosmological information.

10. Suppose a stellar core with a radius of 30,000 km rotates once, that is, 2π radians, every 5.2×10^7 s (about 25 days). Let the mass of the core itself be M_c. The star undergoes a supernova and the core collapses to a neutron star. Assume that no mass is lost from the core (an unrealistic assumption, but adequate for this example), but the radius decreases to 30 km. Assume that both the progenitor and the neutron star are approximately spherical. What is the new rotation rate of the star?

Note

1. The symbol $M_\odot$ represents 1 solar mass.

PART III
Relativity

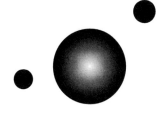

CHAPTER 6

Infinite Space and Absolute Time

Creating the Universe

What is the "universe"? If the universe is "everything," can there be anything outside it? Where do we fit into the universe? How was the universe created? What will be its eventual fate? With all the complexity that we see immediately around us, how can we hope to understand something so intricate on even larger scales?

Such questions have been asked for as long as we have any traces of human thoughts. The answers that have been imagined have been profound, or philosophical, or fanciful, or stern, but until the development of modern science, the answers offered had less to do with the way the universe was than with the way humans imagined it might be. The universe remained mysterious and ineffable. This slowly began to change with the ancient Greeks, who saw a universe built on geometry, a universe that was just as beautiful in its mathematical harmony as any mythological cosmology. With the development of Newtonian mechanics, the universe began to seem comprehensible to the human mind. The advent of science meant the development of models of the universe that could be compared with and tested against observation. These models incorporate the inferred natural laws that give coherence to our observations and enable us to predict previously unobserved phenomena. We may comprehend no reason that the real universe must obey any laws at all, particularly those of human construct, but we can say with confidence that our rules describe *something* about the real universe.

The universe that is accessible to science is the *physical* universe: the universe of material objects, of energy, of space, and of time. This universe contains all that is physical, including all things that are observed, anything that affects or influences other observables, all that is affected by physical things, and hence everything that is subject to experiment and scientific proof or disproof. Atoms, particles, energy, forces, the laws of nature, even space and time, are physical. Everything composed of matter, or subject to the laws of nature, must also be physical and hence part of the universe. Anything that is not part of the universe cannot, by definition, have any physical properties. This definition keeps our cosmological considerations meaningful and consistent.

It might seem obvious to regard matter, energy, physical laws, forces, and the like, as physical things, but the inclusion of space and time in this list requires further justification. You may feel little certainty that you know precisely what time and space are, but they do not seem physical. Space is usually imagined as some sort of immutable, great arena in which existence occurs, while time is simply the absolute demarcation of occurrences as they happen. Expressed rather flippantly, space is that which keeps everything from happening at the same place, and time is what prevents everything from happening at once. Beyond that, few have much further conception of the properties of space and time.

Time as a physical quantity seems especially troubling to some, since it appears to be at odds with much human experience. The rate of the passage of time can seem to vary depending upon one's mood; a pleasant day may fly past, while an unpleasant hour may seem to last forever. Time might even seem to be a human construction. Yet this clearly cannot be true. Human perceptions of a quantity are distinct from that quantity. The human brain is capable of keeping track of short time intervals with impressive accuracy, but if distracted or bored, it can be easily fooled. This is just as true of space as of time. Many well-known illusions depend upon tricking the systems in the brain that estimate distance intervals or relative sizes; yet space often seems more concrete than time. Moreover, the conceit of time as a human construction smacks of anthropocentrism. There is clear evidence that the universe has changed, that it has a history; but most of this history, not only of the universe, but even of the Earth, has passed without the presence of humans. Thus time must have existed before humans came into being. Furthermore, the universe is very much larger than the sphere of human influence, yet periodic physical processes clearly occur in all parts of the universe, so time must exist where there are no humans. Time and space play a role in the laws of nature independent of humans. The issue that has faced scientists is *how* time and space enter into the construction of the universe.

Philosophers have debated through the centuries whether or not space and time can be said to exist in their own right, or whether they are only relations between physical things, where "things" can make the sole claim to existence. As we shall see in later chapters, however, the modern theories of special and general relativity make it quite clear that space and time *are* physical; they can influence matter and energy, and, in turn, be affected by matter and energy. They are active participants in the history of the universe. Indeed, it is possible to construct models of the universe that contain space and time alone, yet still change and evolve.

The inclusion of space and time as physical components of the universe has certain consequences. Any model of the universe must include and explain space and time along with every other physical phenomenon. It is not permissible to invoke a preexisting space and time in which to construct the universe. For example, it is not meaningful to ask, "what happened before the universe existed?" or "what is outside of the universe?" because both of these questions assume the existence of attributes ("before" and "outside") which must posit space and time as properties distinct from the universe itself. Yet time did not exist before the universe, and space does not exist outside it. The big bang did not happen "somewhere." The universe is not expanding into "space," nor even into "spacetime." Do not think of the universe as embedded in something "larger."

The confusion over the physical nature of space and time carries over into one of the thorniest cosmological questions: the creation of the universe. When humans ponder the creation of the universe, generally the question they ask is "Why is there *something* rather than *nothing*?" Why is there a universe at all? In asking that question, you might very well imagine the state of "nothingness" as a great emptiness, extending in all directions and lasting an exceedingly long time. The flaw in this image is that time and space are physical entities, so empty space moving forward in time already describes "something." How, then, were space and time created? Since we cannot help but imagine an act of creation, or, for that matter, any action, in terms of space and time, how can we contemplate some unknown metastate in which this ultimate act of creation occurred?

This issue is sufficiently disturbing to some cosmologists that they attempt to sidestep it by extending the history of the universe into an indefinite, infinite past. If there is no point at which $t = 0$, the reasoning goes, there is no need for creation. However, the question of existence is not answered by supposing that the universe is infinitely old. Time is physical, and an "infinite time" would be just another physical attribute of the universe. Whether or not the universe has infinite extent in time is a question not much different from the superficially less disturbing issue of whether or not the universe is spatially infinite or finite. An infinitely old universe is not *nothing*, so it must have been created; it was simply created with time that extended infinitely, in the same way that the universe may have been created with infinite spatial extent.

Some relativists and cosmologists, among them Stephen Hawking of Cambridge University, have pointed out that in general relativity, finite space and finite time can form a completely self-contained, finite *spacetime* with no boundary or edge at all. The point we call $t = 0$ only appears to be a boundary in time because of the way in which we have divided spacetime into "space" and "time." Such a hyperspherical universe can be contemplated with the help of an analogy to the Earth. On the Earth, the North Pole is the limit to how far it is possible to travel in the direction we call "north," but it is nevertheless just a point on a continuous, boundaryless globe. Similarly, the point $t = 0$ in a spherical big bang model of the universe represents merely an arbitrary demarcation in "time." Without boundaries (spatial or temporal), there is no need to imagine the universe to be contained within some meta-universe.

Both the infinitely old universe and the hyperspherical universe attempt to avoid the question of creation by eliminating $t = 0$ as a special point in time. There may be any number of reasons to prefer a universe of infinite or finite time, infinite or finite space; there are certainly detectable differences among these types of models. But the presence (or absence) of a $t = 0$ point in time provides no answer to the mystery of creation, nor does it have implications for the existence of a creator, beyond those provided by the mere fact of existence. There is little, if anything, that can be said about the metaphysical creation of the universe. Since our observations are of physical attributes, and science deals with physical things, the issue of creation, which must necessarily be metaphysical, cannot be addressed. The universe might be here because of the action of some creator, or maybe it "just so happened." At present, it is not possible to ask this question in a way that is scientifically testable.

In scientific cosmology, we confine our attention to well-posed questions, those we might hope to answer. For example, we can ask, what is the universe like right now? How did it arrive at this state? Traditionally, we would answer such questions with a description of the observed universe, and a statement of the laws of physics, laws which we believe describe the time history of the universe. If we trace the evolution of the universe backward in time, we can ask whether or not there was a point $t = 0$. If there was, our exploration must eventually arrive at the question of *initial conditions*, the description of how things were at the earliest possible moment that we can consider. The science of cosmology aims to describe those initial conditions and to answer the question of how the universe evolved from them.

There are many possible sets of initial conditions, and we must adopt criteria for what we shall hold as "good" initial conditions. As an example, suppose we were to assert that the universe was created at 7:20 this morning. In such a case, everything we know must have been created from nothing at that moment, including the stores of memories in our minds, light arriving from distant stars at the Earth, fossil bones in the ground, and history books with words describing a past. This is clearly a very complicated set of initial conditions. Moreover, such a model cannot be disproved, because any condition one might propose as a test could simply be lumped into the initial state that was created at 7:20. This lack of testability means that such a model fails as a scientific theory. If we compare the initial conditions in the "7:20" model with the big bang initial conditions, we find that in the big bang model, the universe began in a much simpler state. There was a certain amount of energy and matter, certain physical laws, and certain fundamental constants. The complexity of the universe we observe existed as a potentiality, and developed naturally in the subsequent evolution.

In formulating our cosmological models, we would like to be able to describe the initial state of the universe in as few terms as possible. In science we generally adhere to the principle of Occam's razor; in the absence of compelling evidence to the contrary, the simplest of competing explanations is preferred. The big bang universe has the virtue of relative simplicity of its initial conditions. As our understanding advances and theory approaches ever nearer to $t = 0$, the initial conditions of the big bang seem to become even simpler.

Even with the comparatively simple set of initial conditions afforded by the big bang model, there are interesting and challenging questions to consider. For example, the fundamental constants of nature, such as the gravitational constant G, the speed of light c, and Planck's constant h, are held by current physical theories to be constant in space and time, and hence part of the initial conditions. The particular values of the fundamental constants, along with the basic laws of physics, determine what is possible in the universe. If any of these conditions were changed, even slightly, then the universe that would result might be quite different from the one we observe. What if nuclear reactions were not possible at the densities and temperatures prevailing in the cores of gravitationally bound conglomerations of gas? Would there still be stars? What if chemical constants were sufficiently altered that carbon could not form the long chains found in organic molecules? In either such hypothetical situation or in many others, life, as we understand it, might not develop.

We do not know why the fundamental constants have the values they do, or whether it is possible that they could have other values. But we can imagine that all things were possible, and, out of all possible universes, ours is special by dint of our presence in it. The fact that our existence carries implications for the nature of the universe is known as the **anthropic principle.** Its most basic form, the *weak anthropic principle*, states that the conditions we observe in the universe must be compatible with our own existence. The weak anthropic principle sifts out all possible universe models that do not admit the possibility of the development of life. The cosmologist Fred Hoyle is said to have invoked the weak anthropic principle to predict the existence of an excited state of the carbon atom, because such a state allows the triple-alpha nuclear reaction to create carbon in stars.[1] Since Earthly life depends on the existence of carbon atoms, we can infer that the necessary excited state must exist, a conclusion that was confirmed by laboratory experiments.

Many find the anthropic principle appealing because it appears to give a special role to our existence in the universe, but in fact it says nothing inherently more profound about life per se than it says about atoms, or stars, or galaxies. In the example above, the mere existence of carbon is sufficient; the carbon has no compunction to form a basis for life. In a universe with different physical conditions, carbon may still have been able to form by some other means or else life might be based on another atom. By itself, the weak anthropic principle is not even really a testable scientific hypothesis; it is merely a restatement of the requirement that our models be consistent with observation.

A more stringent, and controversial, form of the anthropic principle, the *strong anthropic principle*, states that the initial conditions occurred *because* we are here; that is, our presence here and now somehow affected the initial conditions such that we could eventually arise. Thus according to the strong anthropic principle, the conditions necessarily existed so that we can exist; the "purpose" of the universe is to create life. The strong anthropic principle does not explicate how this backward influence might have been exerted but does seem to require forethought on the part of the universe. This takes it beyond the bounds of science and into *teleology*, the attribution of intent to the universe as a whole.[2]

Some people are drawn to the strong anthropic principle because it asserts a meaning to the universe, and that meaning is us. To the student of history, however, this is very familiar. As we have seen, most myths included a central role for humans. In the absence of any scientific basis for the strong anthropic principle, we again enter the realm of mythology. It is sometimes argued that even though we do not have any basis for the strong anthropic principle, the fact that we are here, and the apparent specialness of the universe, must be telling us something. The weakness in this position is that we have no grounds for concluding that this universe is really so special. We have but one example; we have no way of knowing what might be possible, or what the alternatives might mean.

As an illustrative example, consider what might have happened to you if your father had been killed in a war before you were conceived. If that had occurred, the "you" that exists here and now, would not exist to ask such a question. Hence your very existence necessarily (and tautologically) implies that your father lived at least long enough for you to be conceived. But it does not imply that the purpose of your father was to produce you, and hence the war's outcome was preordained. In this case it was a matter of chance. Things happened as they happened. Each of us is here by a happy accident of conditions.

It may be that the universe must contain life. It may be that the initial conditions had to be what they were. But it is equally conceivable that it "just so happened." For the moment we have no scientific basis for any conclusions. The "whys" of creation remain a mystery. But describing the subsequent unfolding of that creation will prove challenge enough.

The Cosmological Principle

The nature of time and space have always been at the heart of humanity's cosmological musings. Early anthropocentric cosmologies placed humans at the center of the All, creating in the process a very special attribute of space: a center. Similarly, creation stories tended to place specific restrictions on time, such that the history of the universe coincided more or less exactly with that of humanity. The geocentric universe of Aristotle was more physical than the earlier anthropomorphic mythologies, but it still placed Earth, the home of humankind, at the spatial center. Aristarchus, and later Copernicus, moved the center of the universe from the Earth to the Sun, the first significant loss of status for humankind. The Sun-centered, or heliocentric, view is correct for our solar system; the Sun *is* at the center of motion of the planets. But what about the universe as a whole? Is there a center to the universe?

The center of the universe, if it exists, must be a special place, if for no other reason than that it is unique. But the universe is, in virtually every model since Newton, a very large place. What are the chances that the solar system would occupy such a special location? Essentially zero, of course. Observations have progressively demonstrated that the Earth is not the center of the solar system, that the Sun is not the center of the Milky Way Galaxy, and that our Galaxy is not the biggest we can see, nor is it even at the center of its modest cluster of galaxies. While we cannot decisively prove that we do not lie near some center of the universe, the history of human cosmological thought suggests that a certain humility is in order. The principle that the Earth or the solar system does not occupy any special place in the universe is usually called the **Copernican principle**. This principle does not claim that no center exists; only that we are not located there.

Even if we accept that we are not at the center of the universe, might there yet be a center somewhere? Since we cannot see all of the universe, we are unable to answer this question from direct knowledge. Instead, we must bring to bear certain concepts that will aid us in understanding the overall structure of the cosmos. Two very important such concepts are *isotropy* and *homogeneity*.

Isotropy is the property of uniformity in all directions. No single direction is special or distinct from any other. As an example, imagine that you are standing in the middle of a forest, with identical trees and level terrain as far as you can see, regardless of where you look; nothing enables you to pick out any particular direction. Such a forest is isotropic. Suppose that you walk a while in some random direction

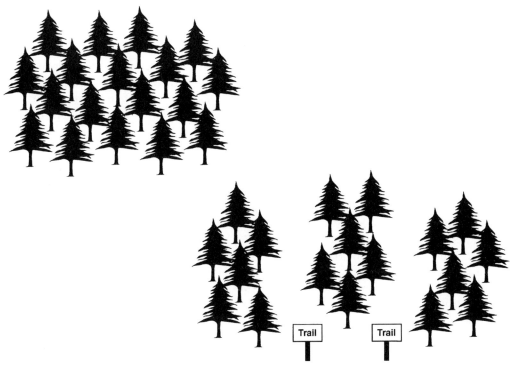

Figure 6.1 (*a*) An isotropic and homogeneous forest and (*b*) a homogeneous but anisotropic forest. The isotropic and homogeneous forest looks the same at all points and in all directions. The homogeneous anisotropic forest has a preferred direction, that selected by the trails, but on sufficiently large scales looks more or less the same everywhere.

and come upon a trail. The forest is no longer isotropic; the trail selects a preferred direction.

The surface of an unmarked sphere provides another example. Such a sphere is isotropic: all directions are equivalent. Contrast this with the surface of a cylinder. On a cylinder there is a long direction parallel to the axis, and a short direction around the axis. The cylinder is not isotropic.

To a certain extent, we can test the universe for isotropy. We need merely make observations in all spatial directions and determine whether there is any systematic trend, or dependence upon direction, for any measurable property. We can certainly define special directions in space, such as the directions toward the Sun or the Galactic center, but these are strictly local properties, rather than universal attributes. To test for isotropy, or anisotropy, of the universe as a whole we must examine the largest scales, such as the overall distribution of all observable galaxies or the distribution of quasars throughout the sky. As far as we can tell from the observations, the universe is indeed isotropic at the largest scales. Such measurements are prone to various observational errors, however, not the least of which is the fact that the most distant galaxies are the most difficult to see. Thus we cannot unequivocally declare from these indications that our universe is isotropic, although isotropy remains the most viable, as well as the simplest, interpretation of the data.

The strongest evidence for the large-scale isotropy of the universe is the cosmic background radiation. This background radiation consists of microwave energy that is present at every point in the sky and has the spectral distribution of a blackbody, at a temperature of 2.735 K. The best explanation for this radiation is that it is the "afterglow" of the big bang. It is observed to have very nearly equal strength and temperature in all directions, after we account for the motion of the Earth. The uniformity of this cosmic relict constitutes an important testimony for the isotropy of the universe.

The second concept which will aid us in our quest for the universe is **homogeneity**, the property of similarity of all locations. Something that is homogeneous is the same, on average, everywhere. The surface of an unmarked sphere is homogeneous: every point is the same as every other point. The surface of a cube is not homogeneous: the edge points are different from the points on the cube faces. A dense forest can seem quite homogeneous when you are lost in it; you can walk for many miles without detecting a noticeable change.

If the universe is homogeneous, then all points throughout all space are more or less equivalent, and everywhere the same physical laws are obeyed. It is impossible to prove that the universe is homogeneous. We cannot visit, or even see, all possible points in the universe. But from what we can see, it looks fairly homogeneous. Distant stars and galaxies resemble nearby stars and galaxies. The same elements we find on Earth are present in the farthest quasars. Despite our inability to examine all of space, we can infer that the universe is probably homogeneous by noting that it appears to be isotropic on the largest scales. If the universe seems to be isotropic, then either it really is the same everywhere or else we live at a unique point where the universe gives the appearance of isotropy.

It is possible for a universe to be isotropic but not homogeneous; however, this occurs only in the special, and rather contrived, case that some central point exists, and isotropy holds only at that single point. An example of an isotropic, but inhomogeneous, situation would be the pinnacle of the only hill in a huge forest. All around you, the scenery would look the same in every direction. But the observed isotropy holds only at the top of the hill. Once you left the peak, there would always be a special direction: upward to the summit of the hill. If we apply the Copernican principle to state that we are not in a special location, then the universe must look more or less isotropic to all observers and must, therefore, be homogeneous. Thus, isotropy plus the Copernican principle implies homogeneity.

It is important to understand that while isotropy implies homogeneity, the converse is not true. Any universe that is isotropic and has no special point must be homogeneous; whereas the universe could be homogeneous, but not isotropic. Remember that isotropy demands that there be no preferred direction, whereas homogeneity merely requires that the universe have the same appearance everywhere. Consider again the trail through the forest. Such a forest cannot be isotropic, since the trail clearly defines a special direction. If there is only one trail, then it would also delineate a set of special locations, so this forest would not be homogeneous, either. But suppose that there is a network of trails running north and south, cut through the forest every kilometer. This forest would be homogeneous on large scales, but not isotropic. Geometrical figures provide other examples: the surface of an infinitely long, uniform cylinder is homogeneous, but not isotropic, because there are distinguishable directions, along the axis and around it. A spherical surface, on the other hand, is both homogeneous and isotropic.

Figure 6.2 shows sections of two-dimensional "universes." Imagine that these are simply representative sections (ignore the borders), and that they actually go on forever. Figure 6.2a is both homogeneous and isotropic. It is a bland, uniform gray; every point is the same, and every direction looks the same. Figures 6.2b and 6.2c are homogeneous on a sufficiently large scale. While there are variations (e.g., the black and white squares of the chessboard), these same patterns appear everywhere. These figures do have a sense of direction, however. On the chessboard it is possible to proceed along the squares or in a diagonal direction; these are quite distinct. Hence Figures 6.2b and 6.2c, while homogeneous, are not isotropic. Finally, Figure 6.2d is a random assortment of shapes, neither isotropic nor homogeneous.

The concepts of isotropy and homogeneity of the universe are combined into one overall principle, the **cosmological principle**, which asserts that all points and directions in the universe are more or less equivalent, and thus that the universe is both homogeneous and isotropic. Given the cosmological principle, we conclude that *there is no center of the universe*. All points in space are basically equivalent, and there is no single point that is central, or in any other way special.

The adoption of the cosmological principle completes the process begun by Copernicus. Not only is the Earth not the center of the universe, there is no center at all. It is not surprising that the cosmological principle came rather late in the history of humanity's thinking. When you look up at the night sky, the universe appears to be

(a)   (b)

(c)   (d)

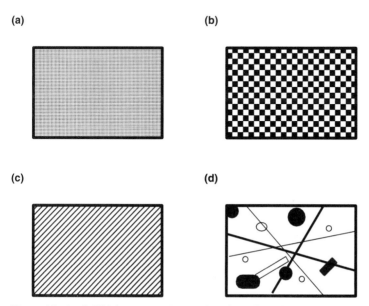

Figure 6.2 (a–d) Which representative samples of two-dimensional universes are homogeneous or isotopic, or both?

anything but isotropic. The stars are not distributed evenly, but are concentrated in a broad band, the Milky Way, which stretches across the sky, delineating a direction. The uniformity of brightness in this band led astronomers at first to conclude, incorrectly, that the Sun was near the center of a great disk. Only later was it realized that appearances are deceiving, and we are actually near the edge of the disk of the Galaxy. Astronomers formerly assumed that the Milky Way constituted the bulk of the universe, but improvements in telescope technology laid that fallacy to rest. In our universe we see galaxies, and clusters of galaxies, and clusters of clusters of galaxies, with galaxies of different sizes and shapes, for as far as we can detect their light. The Milky Way is nothing unique, after all. On what scale does the universe become truly homogeneous? Even now that question is as yet unanswered, but it does appear that on the largest scales, those most suitable for cosmology, the universe is isotropic, and, by implication, homogeneous.

The cosmological principle goes far beyond a simple assertion that the universe has the same *appearance* everywhere, to include all physical properties. Only by an appeal to the cosmological principle can we posit that the same laws of physics discovered on Earth also apply to distant galaxies, and that all objects, no matter how far from us, are composed of the same fundamental substances as we find on Earth and in its vicinity. This is clearly a sweeping generalization that might seem to reach beyond our capabilities; but without something like the cosmological principle, how could we ever hope to understand anything about our universe? It can be argued that the most important aspects of the cosmological principle relate to the uniformity of physical laws. We might easily imagine a universe that was not isotropic in its distribution of matter, even at very large scales; such models have been advanced, even quite recently. But we could not hope to understand a universe in which physical laws varied willy-nilly from one region to another. If the same spectrum originated from different elements, or indeed if the elements themselves had different properties elsewhere, we could say precious little about distant galaxies or quasars. The cosmological principle is an assumption about the nature of the universe. Like all scientific postulates, it is unprovable. It is, however, *disprovable*; its continued justification depends on the coherence and success of the models which utilize it.

If there are no special directions or locations to space, what about time? The cosmological principle asserts that the universe is homogeneous, but such a universe need not be static or unchanging; it requires only that at a given time, all points must appear the same. In the analogy of the forest, we might begin with an empty clearcut onto which Douglas-fir seedlings are planted uniformly. The trees grow at roughly equal rates, and at any given time the forest looks the same, but it still changes with time. What if we had not a Douglas-fir monoculture, but an old-growth forest; would that be considered homogeneous? It depends upon the scale we wish to consider. Such a mixed forest would not be homogeneous in the same way as a stand of Douglas fir, but if a helicopter dropped you into the middle of it, you would have difficulty distinguishing one location from another over a fairly large scale. One patch of forest would have about the same number of redwood trees, fir, spruce, pine, and so on, as would any other patch.

There is a more restrictive principle which holds that there is no special point in time, as well as in space. This is known as the **perfect cosmological principle**, and it states that the universe is the same at every point in space and at every point in *time*. Continuing the analogy of the forest, you might imagine an old-growth forest in which new trees grow at exactly the rate needed to replace those that die; the age of the forest would be as undeterminable as a location within it. Any universe that obeys the perfect cosmological principle must appear to be the same, on the average, everywhere and for all times. Such a principle is extremely restrictive. As we shall see, the perfect cosmological principle goes too far, and has been disproved. Observations indicate that the universe *does* have many special points in time, and does evolve with time.

Cosmological models are intimately linked to the philosophy behind the physical laws held to govern the universe. The cosmological principle is one possible paradigm. Before the advent of the modern model, other physical theories informed other cosmologies. The interdependence of cosmology and physical philosophy is sufficiently great that the failure of one could bring down the other as well. From the age of the Greek philosophers until the present, cosmology and physics have advanced, or declined, hand in hand.

The Aristotelian universe is an example that is clearly neither homogeneous nor isotropic in space, not only in its appearance, but also in its physical laws. According to Aristotelian physics, Earthly objects moved through space linearly, toward the location that was appropriate to their percentages of earth, fire, water, and air, while celestial motions were perfect circles executed forever. The Aristotelian cosmology was in accord with Aristotelian physics. Special points and directions were inherent to the model. Space was defined only in terms of the objects it contained; Aristotle could not conceive of the vacuum of space, and he stuffed his model with tangible physical entities. Not only the Earth and heavenly bodies were physical, but also the spheres that bore the planets and stars on their daily travels had real physical existences. Aristotle would have denied any possibility of travel to the Moon, for the traveler would be unable to continue with linear motion in the celestial realm and would probably smash into the Moon's crystalline sphere as well. On the other hand, the Aristotelian model was, more or less, unchanging in time. Aristotle's concept of time seems to have been rather ill-defined, but it functioned as a marker of occurrences. Even here, however, the inhomogeneity in space played a role; change occurred only on Earth, not in the heavens.

Newtonian physics, in contrast, makes no special distinctions in space or in time. Newton's laws of motion contain no preferred directions, nor does location have any inherent effect upon mechanics. Newtonian physics depends implicitly upon the existence of an absolute space and time to which motion is referred. Whether an acceleration is present can be determined by measuring the change in velocity with respect to markers laid down in absolute space and time. The markers themselves, which might consist of the background of fixed stars, or any other appropriate standard, are merely convenient references that have no intrinsic significance of their own. Space and time have an independent existence, regardless of how we choose to measure them.

Newton's cosmology reflected his mechanics. The universe consisted of stars scattered about uniformly everywhere in space; the stars either lived forever or died and were recreated. This grand machine was set into motion at some specific point in time, but throughout its existence, the universe looked the same for all locations and all times. Newtonian physics was everywhere valid, and a knowledge of the initial conditions would, in principle, enable a perfect computer to calculate the entire destiny of the universe.

Just as the claustrophobic and rather judgmental Aristotelian-medieval universe troubled some thinkers of its time, so did Newton's aggressively deterministic cosmos create doubts among many philosophers of the Enlightenment. Not only did it seem to preclude any free will on the part of humans, but it made some strangely rigid assumptions. One difficulty was that Newton's law of gravity required a force to act instantaneously across empty space; what conducts that force? Absolute space and time, which affected everything but which were affected by nothing, were also particularly repugnant to some scientists of the day. Moreover, it was recognized even then, and by Newton himself, that the Newtonian universe depended on a very delicate balance; since gravity is strictly attractive, its force would inexorably pull lumps of matter together. The only way to prevent the Newtonian clockwork from collapsing onto itself was to assume an infinite, perfectly uniform distribution of matter. Despite these background rumblings, however, Newtonian mechanics was an indisputable success, and the weaknesses of the corresponding cosmological model were swept under the rug for two centuries. After all, it had no compelling competitors at the time.

The modern viewpoint that has arisen during the twentieth century flows from and around the cosmological principle; to understand modern cosmology, we must explore its relation to the form of modern physical theory. This journey will take us from grand galaxies to the elementary particles, but underlying all of it will be the meaning of space and time. Let us begin, then, by contemplating how we can quantify the relationships among space, time, and our observations of the universe.

Taking Measurements

The scientific revolution introduced the importance of measurements into our conceptions of the universe. Pure thought alone cannot reveal the nature of the universe any more than it can manufacture gold. Careful measurement is fundamental to the attainment of scientific knowledge through scientific *observations*. We must measure physical properties in a repeatable manner that is unaffected by the observer or by the instrument. We all may feel intuitively that spatial and temporal relationships exist between objects and events, but vague impressions are of little use to science. In order to form precise conclusions, these relationships must be described objectively, but this demands that we describe the process of measuring. We must learn to distinguish between those things that are physically significant and those that are relative to how they are measured.

The most obvious datum is position. Any object in the universe has a location in space at each instant of time; these points in space and time are labeled with **coordinates**. The customary notation for coordinate locations is (x, y, z, t), where x, y, and z represent the spatial quantities, and t represents time. (If we simplify matters by working with only one spatial dimension, we shall refer to its coordinate as x.) Coordinates are merely convenient labels, not physical attributes of space or time, so the symbols and units chosen are arbitrary. The coordinates of a point have no intrinsic significance; their only importance lies in the relationships between two sets of coordinate values, such as relative locations. We measure space by means of a standard, which we shall generically call a "ruler," regardless of what it actually might be. We measure time by means of "clocks," where a "clock" could be any standard periodic physical process and need not literally refer to a wristwatch. Distance and time intervals have physical significance, but whether we measure a distance interval in inches, yards, or meters is not important. A measurement in yards is merely an expression relative to an arbitrary standard, but the distance itself represents a real, reproducible, quantifiable measurement.

Since measurements of separations in space and time are among the most important, we will concentrate initially on understanding the meaning of these quantities. As a specific example, if we wish to know the distance between two points in space, we may begin by laying down coordinate lines that run at right angles to each other (i.e., a grid) in x and y. Next we assign spatial coordinate locations to each point, say, (x_1, y_1) for point 1, and (x_2, y_2) for point 2; then we determine the difference between those coordinates, $\Delta x = x_2 - x_1$ and $\Delta y = y_2 - y_1$.[3] The Pythagorean theorem enables us to find the desired distance by summing the squares of the two sides to obtain the square of the distance between points 1 and 2; specifically:

$$s^2 = \Delta x^2 + \Delta y^2, \qquad (6.1)$$

where we have used the symbol s to indicate distance. The Pythagorean theorem has an obvious generalization to three dimensions—think of a cube rather than a square—but for the present illustration, the familiar two-dimensional version is sufficient.

Quite often what we really want to know is how long it took to travel a specified distance. The quantity that describes the change of position with time is **velocity**. Velocity is a vector, meaning that it has a *direction*, as well as a magnitude, associated

Figure 6.3 The Pythagorean rule in a two-dimensional Euclidean space. The distance between two points (x_1, y_1) and (x_2, y_2) is given by the square root of the sum of the squares of the individual coordinate separations, Δx and Δy.

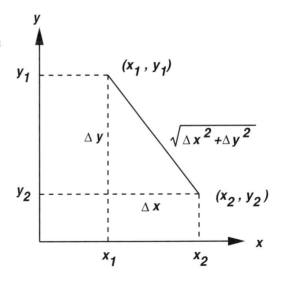

with it; the magnitude of the velocity is the **speed**. For example, if you travel 20 km in half an hour, you can say that your *speed* is 40 km per hour. If you further state that you traveled from east to west, then your *velocity* is specified as 40 km per hour toward the west. If you drive on a winding road at a constant speed of 40 km per hour, your velocity changes with each turn of the wheel. The distinction between speed and velocity can be very important and should be kept firmly in mind. In general, velocity is defined as a *derivative*, that is, the rate of change, of the three-dimensional position vector. However, by confining our attention to motion along one specific direction (x) we can write this as $v = \Delta x/\Delta t$, simplifying the mathematics without much loss of qualitative content. Velocity, then, is the change in space position, divided by the accompanying change in time.

A change of velocity with time is an **acceleration** and is written $\Delta v/\Delta t$. Since acceleration is defined in terms of velocity, it too carries directional information. An object may have both an acceleration and a velocity, and they need not be in the same direction at all. If you jump up from the ground, your velocity is initially in the upward direction, but the gravitational acceleration is directed down toward the Earth, which is why you eventually reverse your motion and fall back to the surface. Orbits provide another example. The velocity of an orbiting planet is nearly perpendicular to the line between the planet and the Sun, whereas the gravitational acceleration is directed from the planet toward the Sun. In the case of purely circular motion of any kind, the velocity and the acceleration are exactly perpendicular to one another. Moreover, since acceleration is the change in velocity, there may exist an acceleration *even if the speed never changes*. Riding along that winding road at a very steady speed of 40 km per hour, you will nevertheless feel an acceleration at each curve.

It may seem that velocity and acceleration are very similar quantities, both mere descriptions of how something changes with respect to time, but there are important physical distinctions between the two. Newton's second law, $F = ma$, tells us that a force is required to produce an acceleration. By Newton's first law, in the absence of a force, a body in a state of uniform motion will continue in that same state indefinitely. Stated simply, uniform motion means constant velocity. (Rest is a special case for which the velocity is zero.) Uniform motion is the natural state and will last indefinitely; only a force can cause an acceleration. This means that there is some attribute of a body, its **inertia**, which causes it to "resist" changes in its velocity. How do we quantify inertia? We do so through a property we call **mass**. Inertial mass is *defined* by Newton's second law; we measure mass by applying a known force and observing the resulting acceleration. We refer to *unaccelerated* motion as **inertial motion**. Thus any uniform motion is an inertial motion.

It is important to grasp that there is a real, physical difference between inertial motion and accelerated motion. But, you may ask, if our units are arbitrary, how may we determine whether a motion is accelerated or inertial? Suppose we had a measuring device that changed its length scale or a clock whose mechanism made it run at different rates. With this ruler and clock, it would seem that velocity is continually changing. How could we distinguish a measurement made with such odd measuring devices, from true accelerated motion? There are two ways to answer this. First, we can note that mathematically, the properties of true accelerated motion are expressed by Newton's second law, and this relationship does not depend upon our measuring units. Second, we can appeal to experience: the difference between inertial motion and accelerated motion is *palpable*. Acceleration requires a force, and this has consequences for objects such as the human body. Acceleration is often measured in units of g, the gravitational acceleration at the surface of the Earth. Consequently, accelerations are sometimes loosely called "g-forces." Pilots of fighter planes must wear special "g-suits" because they may experience very high accelerations in a tight turn or dive. (Remember that circular motion, even if only part of the circle is traversed, or if the radius of the circle is changing, means that there is an acceleration and hence a force.) Such large forces can cause humans to pass out, or if strong enough, can even be fatal.

To clarify the distinction between acceleration and distorted units of measurement, we must introduce the concept of a **frame of reference**. The frame of reference is a system of coordinates attached to an observer whose viewpoint we are considering. Suppose you define the origin of an (x, y, z) coordinate system coincident with your navel, and time coincident with the watch on your wrist. These specifications define a frame of reference. With respect to the coordinates attached to your body, you are always at rest, since your coordinates move with you; thus this frame is defined to be your *rest frame*. How is your rest frame related to other frames, such as the frame defined by the distant "fixed stars"? In particular, what is your state of motion relative to the fixed stars? Are you at rest, moving with a constant velocity, or undergoing acceleration? This has implications for your frame of reference.

Suppose you and your coordinate system are in deep space; you are unaccelerated. What does this imply? Your accelerometer reads zero, and you feel no forces acting on your body. You float along at a constant velocity, in inertial motion. In such a case, you reside within a very special reference frame, called an **inertial reference frame**. An inertial reference frame is *any frame in which a free particle executes uniform motion,* that is, moves at a constant velocity, as specified by Newton's first law. In an inertial frame, a particle set into motion at constant velocity would continue in such uniform motion indefinitely.

Conversely, a noninertial reference frame is one that is not inertial, but what does this mean? Most obviously, it implies that forces are acting upon all objects within the frame; but how do forces affect the reference frame itself? A familiar example might help to clarify the issues here. Imagine that you are riding the "Rotor" at an amusement park. The Rotor, and similar rides, consists of a tube whose inner walls are covered with a rough material, such as burlap. The riders stand against the walls of the cylinder, and it begins to rotate. When the Rotor reaches a certain angular speed, the floor drops several feet, and the riders adhere to the walls. It is the friction between clothing and the burlap that prevents the riders from sliding down, but the force that presses them against the burlap comes from the acceleration they experience due to the circular motion. What happens to the motion of free particles within such a reference frame? Suppose that while you are riding the Rotor, you decide to play catch with your friend who is directly across from you; you toss a ball toward your friend. What happens? The ball curves off to the side! But suppose another friend watches from above, outside the Rotor, as you throw the ball. Your overhead friend insists later that the ball, once released from your hand, flew in a straight line, as

we may have difficulty in visualizing a true inertial frame, but from our definitions, it is clear that the presence of gravity creates a noninertial frame. Any object moving near the Earth will be affected by a gravitational acceleration. Like the centrifugal and Coriolis forces, we are aware of gravity and can account for it when we write our equations of motion for objects moving near the surface of the Earth. Gravity also acts only vertically, so that horizontal motions and forces are unaffected by it. (In fact, the direction of gravity defines the "vertical.") If a reference frame is moving at constant velocity in a gravitational field, then motions occur within that frame of reference *exactly* as they would at rest, with the effects of gravity included.

An airplane can approximate such a frame of reference if the air is smooth. Suppose you are riding in a jet at cruising speed and altitude, with no atmospheric turbulence in your path. The flight attendant hands you a can of soft drink and a cup; you pour the soda into the cup exactly as you would if you were sitting at rest in your kitchen at home. The cup and the stream of liquid share the same constant horizontal velocity, so no effect of that velocity can be detected within your frame. If the airplane accelerates, however, either by speeding up or by changing its direction, you are likely to spill the soda as you attempt to pour it. This is similar to the arguments used by Galileo and others to demonstrate that motion is not always detectable from within the moving frame of reference.

Since gravity is ubiquitous throughout the universe, what *would* be a truly inertial reference frame? One example would be a spaceship traveling at constant velocity in deep space, where gravity is, locally, negligibly small. Another example of an inertial frame is one which is *freely falling* in a gravitational field. What is special about freefall in a gravitational field? Recall that mass appears both in Newton's second law and in Newton's law of universal gravitation. We have even written these two masses with the same symbol m, but they are really two distinct concepts. In the second law, mass is a measure of the inertia, or the resistance to acceleration. In the law of gravity, mass is a measure of "gravitational charge," analogous to the role of electric charge in the theory of electromagnetism. If the "gravitational mass" is the same as the inertial mass, then we may combine these two equations and cancel the mass of the "test" object. For any object falling in the gravitational field of Earth, we may write

$$g = \frac{G M_{\text{Earth}}}{R_{\text{Earth}}^2}, \tag{6.2}$$

which does *not* depend on the mass of the falling body. This is the mathematical expression of the experimental result that all objects fall at the same rate in a grav-itational field. In the absence of any nongravitational force such as air resistance, a feather and a cannonball dropped from the same height at the same time will hit the ground together. Apollo astronaut David Scott performed exactly such an experiment on the airless Moon during the *Apollo 15* mission in 1971, dropping a feather and a hammer at the same instant. Both fell with the same acceleration and struck the surface simultaneously.

The independence of gravitational acceleration from the inertial mass is also the solution to a famous "trick" question of physics. A hunter aims at a monkey who is holding the branch of a tree. Just as the hunter fires, the monkey lets go of the branch in an attempt to evade the bullet. Does the bullet hit him? The answer is "yes," if the hunter's aim is accurate, because both the bullet and the monkey fall in the vertical direction at exactly the same rate, despite their large difference in mass. (Anyone who shoots targets is aware that bullets certainly do fall while in flight, just like any other object.) Because in freefall everything falls together at exactly the same rate, gravity is effectively "cancelled out"; all motions relative to the freefalling frame will be consistent with Newton's Laws. In the theory of general relativity, we shall find that this seemingly innocent and obvious equivalence of inertial and gravitational mass will have some amazing and profound consequences.

The Relativity of Space and Time

> Either this man is dead, or my watch has stopped.
>
> —Groucho Marx

Given that inertial frames of reference exist, why are such frames important? When making measurements using a coordinate system, it must be possible to distinguish those things that are physically significant from those which are related only to the specific reference system by which they are measured. We have suggested such a distinction by emphasizing that acceleration has physical consequences, independent of the coordinate system used to measure it. As a more trivial example, if someone tells you to time your heartbeat but hands you a defective watch, you should not conclude that your heart is malfunctioning.

We can clarify matters even further by means of some definitions, beginning with an **inertial observer**. An inertial observer is simply an observer whose rest frame is inertial. Next comes the concept of **invariance**. A quantity is said to be *invariant* if all inertial observers would obtain the same result from a measurement of this quantity. On the other hand, a quantity is said to be *relative* if different inertial observers obtain different results from their measurements. **Relativity**, which is a general term and does not apply only to Einstein's theory, tells us how to relate observations made in one inertial frame of reference to observations in another such frame.

As a first example, let us consider the frame of a train moving at constant velocity. One of the passengers drops a ball onto the aisle. Another passenger who observes the fall of the ball will see exactly what she would see if the ball were dropped on the surface of the Earth; the ball lands at the feet of the person who dropped it. Suppose that another observer, who is at rest with respect to the Earth, watches the same ball as the train goes by. The Earth-based observer measures the path of the ball, relative to his own frame, to be a parabola, since the ball shares the horizontal velocity of the train. Both observers agree that the ball accelerated downward due to the force of gravity. Both agree on the magnitude of that force, on the mass of the ball, on the value of the acceleration, and on the length of time required for the ball to fall. Both observers can apply Newton's laws of motion to compute the theoretical path of the ball. However, they disagree on the velocity of the ball, the path it took while falling, and its final position. These differences are all attributable to the motion of the train. The quantities acceleration, mass, force, and time interval are invariant. The observers disagree on the coordinates because they are using different coordinate frames; they also obtain different results for the position and the net velocity of the ball at any given time. Quantities such as coordinates, position, and velocity are relative.

Inherent in this example is the assumption that space and time are absolute. All observers agree on space and time intervals; that is, 1 s of time and 1 m of distance are the same in all inertial frames. Further, all inertial frames are equivalent. There is no absolute motion per se, no single "correct" inertial frame that is "better" than any other. Since all inertial frames are equally valid, we need only find the procedure for relating measurements in one frame to the measurements in another, thereby accounting for the relative quantities. The equations that relate measurements made in one Newtonian inertial frame to those made in another are called, collectively, the *Galilean transformation*. They are very simple and intuitive; basically, the equations of the Galilean transformation simply adjust the observed velocities by the relative velocity between the two frames. In our example above, if the train is moving with speed v_{train} toward the west, as measured by the observer on the ground, and the

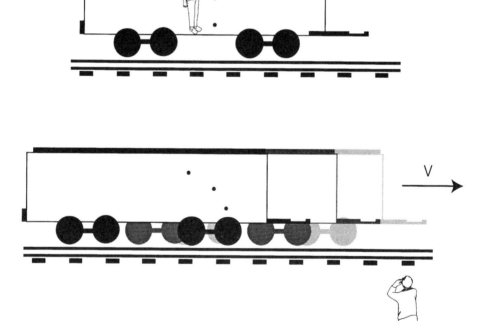

Figure 6.6 Path of a dropped ball, as seen (*a*) in the frame of a train moving at constant velocity and (*b*) in the frame at rest with respect to the Earth. *Relativity* describes how to relate measurements made in one inertial frame to those made in another.

passenger throws the ball down the aisle toward the west with horizontal speed v_{ball}, as measured by an observer on the train, then the horizontal velocity of the ball, as measured by the observer on the ground, is

$$v_{ground} = v_{train} + v_{ball} \qquad (6.3)$$

toward the west. On the other hand, if the thrower faces the back of the train and tosses the ball toward the east with horizontal speed v_{ball}, then the horizontal velocity of the ball measured by the ground-based observer is

$$v_{ground} = v_{train} - v_{ball} \qquad (6.4)$$

toward the west.

Since physical laws are intended to describe some objective properties of the universe, we can see that they must be invariant under the transformation from one inertial observer to another; otherwise they would depend upon the coordinate system used to make measurements, and coordinate systems, as we have emphasized, are purely arbitrary. **Galilean relativity** is the formal statement that Newton's laws of motion are invariant under the Galilean transformation. That is, Newton's laws work equally well, and in the same manner, in all inertial reference frames, if those frames are related by the Galilean transformation. If Galilean relativity gives the correct relationship between inertial frames, then Newton's laws provide an accurate description of the laws of mechanics, since they do not change their form under a Galilean transformation. Note too that since Newton's laws operate precisely the same in all inertial reference frames, no experiment can distinguish one such frame from another; this implies that you can never tell if you are "really moving" or "really at rest," as long as your motion is unaccelerated.

A Fly in the Ointment

Up to this point we have dealt exclusively with Newton's laws of mechanics, which are invariant under the Galilean transformation. What about other laws of physics? Do they obey the principle of invariance under Galilean relativity? From what we have learned about mechanics so far, it seems entirely reasonable to expect that all laws of physics should be Galilean invariant. Why should any inertial reference frame be better than any other? Certainly not any reference frame tied to the Earth, for one specific example; this would violate the Copernican principle. Perhaps there could be some absolute cosmic frame of rest, although the introduction of any special frame of reference would tend to vitiate the spirit of relativity.

During the middle part of the nineteenth century, scientists were fairly certain that all of physics must be invariant under the Galilean transformation. However, the laws of physics were still being uncovered; one area of especially active research at the time was electricity and magnetism. In the 1860s, the British physicist James Clerk Maxwell (1831–1879) developed a theory of electricity and magnetism which showed that these two forces were actually manifestations of one "electromagnetic" force. A consequence of Maxwell's equations was that fluctuating, time-varying electromagnetic fields traveled through space at the speed of light. It soon became clear that this electromagnetic radiation *was* light itself. Maxwell's equations, which describe the evolution of electric and magnetic fields, depend specifically upon a velocity: the velocity of light. Yet as we have seen, velocity is a quantity that is *relative* under Galilean transformations, so Maxwell's equations are also not invariant, but are relative, under Galilean relativity.

When Maxwell's equations were developed, their lack of Galilean invariance was not immediately troubling to most physicists. Waves in matter, such as elastic waves or sound waves, require a medium in which to propagate; the speeds that describe these waves, such as the speed of sound, are specified with respect to the medium through which the wave travels. The net velocity of the wave, as seen by an observer not moving with the medium, is the vector sum of the velocity of propagation, plus the velocity of the medium, if it is moving. Since all the waves familiar in the middle nineteenth century were of this nature, the reasoning of the day concluded that light too traveled through a medium, called the **luminiferous ether** or just the *ether*. This "ether" has nothing to do with the volatile chemical substance of the same name, nor is it the same as the celestial "ether" of Aristotle; the luminiferous ether had no other reason for its existence than to provide the expected medium for the propagation of light.[5] It had no particular tangible properties of its own; it was massless and invisible. This seems rather peculiar: why should the universe contain this strange substance with such a specialized function? After all, air does not exist solely to carry sound waves. But so strong was the mechanical picture of waves in the minds of nineteenth-century scientists that no other alternative was seriously entertained.

It was thus assumed that Maxwell's equations were valid only in the frame of the ether. Many physicists of the time even concluded that the rest frame of the ether could be identified with the Newtonian absolute space. But if the ether has a frame, then that must be some kind of *preferred* frame of reference, which presumably fills all space. As such, it must be possible to detect the ether through its special frame of reference; in particular, a carefully designed experiment should be able to measure the motion of the Earth through the luminiferous ether. Once the ether was observed, it was believed, the theory of electromagnetism would be complete; together with Newton's laws, the description of the fundamental properties of nature would also then be finished. In 1887, two American physicists, Albert Michelson and Edward W. Morley, set out to measure the motion of the Earth with respect to this frame.

Michelson and Morley set up an experiment in which a beam splitter broke a beam of light into two. One half of the original beam was sent in one direction, struck a mirror, and was reflected back to another, angled mirror. The other half of the beam traveled precisely the same distance perpendicular to the first direction, where it was also reflected and returned. The experiment asked whether the transit time was equal for the two perpendicular round trips. If light behaved like a mechanical wave, the experimental setup would be analogous to two swimmers in a river, one traveling across the current and back, and the other swimming the same distance downstream and then returning upstream. The swimmer who had only to cross the current twice would complete the trip faster than the swimmer who had to battle the current on the way back. The difference in swimming time could be used to derive the flow speed of the river.

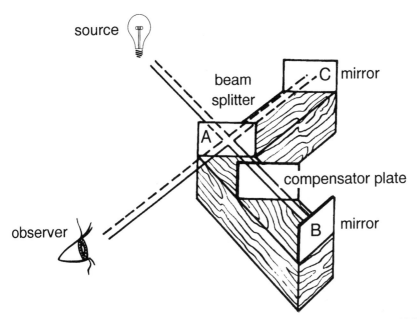

Figure 6.7 Schematic illustration of a Michelson and Morley interferometer experiment. Light from a source travels to a half-glazed mirror that splits the light beam into two, sending light down both arms of the apparatus. If the round-trip time along AB differed from that of AC due to differences in the speed of light with direction, the observer would see interference fringes when the beams recombined. The entire apparatus was floated in a pool of mercury, allowing it to turn freely so that all directions could be tested.

Michelson and Morley measured the transit time for the light by recombining the light beams upon their arrival, thus superposing the two light waves. If the light beams had different round trip times, they would be out of phase when recombined. Adding light waves with different phases results in alternating constructive and destructive interference, producing a pattern of light and dark known as *interference fringes*.[6] A device to observe such fringes is called an **interferometer**; this particular experimental setup is known as a Michelson-Morley interferometer. The apparatus was constructed so that it could be rotated, turning one arm and then the other toward the direction of motion of the Earth. Since the speed of light plus ether was expected to differ for the two arms in a predictable way as the device was rotated, the change in the interference fringes would provide the difference in the travel time of the light along the two paths, and hence the velocity of the Earth with respect to the ether.

That, at least, was the idea. To their great surprise, however, *no* difference in light travel times was observed. Michelson and Morley repeated their experiment numerous times and at different times of the year. In the end, they determined that the velocity of light was the same to less than 5 km/s per second in the two mutually perpendicular directions. Upon taking their experimental limitations into account, the outcome of their experiment was the declaration that the speed of light was equal in both directions. While this might seem at first glance to be an experimental failure, their "null" result was one of the most important experimental observations of the late nineteenth century.

The Michelson-Morley experiment left physicists in some disarray for nearly 20 years. Except for a few small difficulties such as the minor confusion about the ether, physics had seemed to be more or less wrapped up.[7] Yet this seemingly small inconsistency led directly to the development of a new and startling theory, and a new way of looking at space and time that will form the foundation for our modern cosmological theories. We turn now, to Einstein's theory of relativity.

Einstein

By the last quarter of the nineteenth century, the universe had expanded dramatically, both in space and in time. The distance from Earth to the Sun had been accurately measured about a century before, in 1769, finally setting the absolute scale of the solar system; the solar system alone was discovered to be much larger than the size of the entire universe in the Aristotelian cosmos. Evidence was mounting that Earth was several billions of years old. Change and evolution throughout the universe was becoming an accepted paradigm. Newtonian mechanics seemed to give humanity a glimpse of the architecture of the universe itself. Physicists felt they had every reason to feel proud, perhaps even a little smug. And yet, a few pieces of the electromagnetic theory still could not be made to fit.

Albert Einstein was only 8 years old when the crucial experiment performed by Albert Michelson and E. W. Morley demonstrated that Newtonian mechanics could not accommodate the behavior of light. The resolution of this conundrum had to wait until Einstein reached adulthood and created one of the most daring theories in the history of science, as radical in its own way as the Copernican revolution and yet relentlessly logical and beautiful.

Albert Einstein (1879–1955) was born in Ulm, Germany, the son of a less-than-successful businessman. An unspectacular, although not untalented, student, he left Germany in his teens and traveled through Italy. Eventually he settled in Switzerland, where he attended the Swiss Federal Institute of Technology, finally obtaining his doctorate in 1900. Unable to find employment as a scientist, he accepted a position as a patent examiner with the Swiss Patent Office in Bern. He later reminisced nostalgically about his days as a patent clerk. He enjoyed the work of evaluating patent applications, and his life as a scientific "outsider" seems, if anything, to have stimulated his creativity. In 1905 he published at least three epochal papers. One was a work on Brownian motion, the jiggling of tiny particles due to the many impacts of molecules of air or water upon them. Another was his explanation of the photoelectric effect, a mysterious phenomenon that occurs when light strikes the surface of a metal. This paper employed and elaborated upon the quantum theory of radiation developed a few years previously by Max Planck to explain the blackbody spectrum. Einstein's grand hypothesis was that light itself was quantized; we now refer to a quantum of light as a *photon*. The explanation of the photoelectric effect was one of the earliest applications of quantum mechanics, and eventually won Einstein the Nobel Prize in physics. The third paper, *Zur Elektrodynamik bewegter Körper* (*On the electrodynamics of moving bodies*), published in the German scientific journal *Annalen der Physik*, laid out the special theory of relativity.

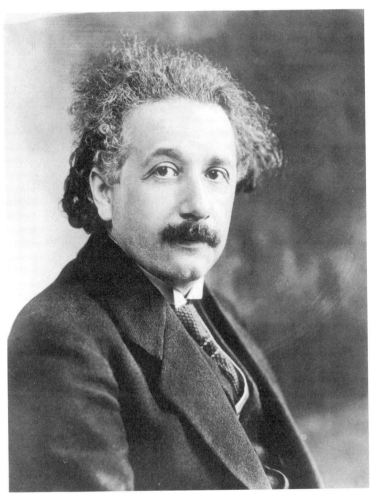

Figure 6.8   Albert Einstein (1879–1955). Best known for formulating the special and general theories of relativity, Einstein made many fundamental contributions to the development of quantum mechanics as well. Courtesy of Yerkes Observatory.

The special theory of relativity wrought dramatic changes in our view of the universe. No longer could we imagine the absolute, pristine space and time of Newton. Space and time were not the stage upon which the drama of dynamics unfolded; they became actors in the play. The special theory showed that the electromagnetics of Clerk Maxwell was more accurate than the mechanics of Newton. It does not denigrate Newton's great achievements in the least to discover that his physics was not quite right; he could not have arrived at the correct formulation even with his intimidating genius, as the necessary understanding of electromagnetics was lacking in his day. Newton's mechanics is an approximation, valid only in the limit of speeds that are very small relative to the speed of light. Since essentially all Earthly motions occur at such speeds, certainly for all macroscopic objects, Newton's theory seemed completely adequate.   The need for the special theory of relativity was not perceived until a contradiction was discovered with what seemed, at first glance, to be a completely separate arena of physics.

After his triumph with mechanics, Einstein turned to gravitation. This proved a tougher nut to crack, and occupied Einstein for the next 10 years. By then, he had become a member of the scientific establishment, securing prestigious positions at universities in Prague, Zurich, and finally Berlin. Although he arrived quickly at the physical foundations of what became the general theory of relativity, the mathematical representation of the ideas was far from obvious, and Einstein reached many dead ends. Finally, around the time of the First World War, his friend Marcel Grossman introduced him to a branch of mathematics known as Riemannian geometry. Einstein found his answer there; the equations of general relativity were published late in 1916. Almost immediately, they were applied to cosmology, first by Einstein himself, in 1917, and later by scientists such as Alexander Friedmann, Willem de Sitter, and Georges Lemaître.

It is unfortunate that both special and general relativity have acquired such an intimidating reputation. The special theory requires no more than algebra for a basic understanding of its workings, although details of its application demand somewhat higher mathematics. The general theory is, of course, more complex and cannot be fully understood without higher mathematics; the fundamental ideas, however, are not intrinsically difficult. The real impediment to the understanding of both theories is not the mathematics, but the new way of thinking they demand. Our intuitions often mislead us in our attempts to understand even Newtonian mechanics. The theories of relativity require a mental flexibility that the complacent of mind may not be willing to attempt. Yet a little effort can provide a basic understanding of these great ideas that have so significantly shaped physics in the twentieth century.

KEY TERMS

anthropic principle	Copernican principle	isotropy
homogeneity	cosmological principle	perfect cosmological
coordinates	velocity	principle
acceleration	inertia	speed
inertial motion	frame of reference	mass
inertial force	inertial observer	inertial reference frame
relativity	Galilean relativity	invariance
interferometer		luminiferous ether

Review Questions

1. Describe the weak and strong anthropic principles. What philosophical assertion does each make about the universe? What do you think about them?
2. Why is it not a scientifically valid question to ask what happened before the universe came into existence?

Figure 6.9   (Exercise 6.3)

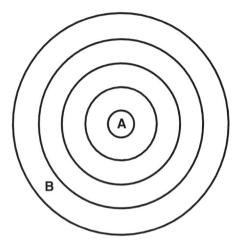

3. Flatlanders live in a two-dimensional universe. Suppose such a universe were described by the figure, such that all matter is confined to the indicated rings. Does this universe appear isotropic to an observer at point A? To an observer at B? Is this universe homogeneous for either observer? What would the observer at A conclude if she or he applied the Copernican principle? Explain your answer. Draw an example of a universe that is homogeneous but nowhere isotropic.

4. Is the cosmological principle consistent with the existence of a center or an edge to the universe? Explain.

5. Explain the distinction between the cosmological principle and the perfect cosmological principle.

6. We have mentioned that galaxies are grouped into clusters. How can the existence of such clusters be consistent with a homogeneous universe?

7. Explain the distinction between invariant and relative quantities.

8. An airplane is traveling at 300 mph toward the west. A rambunctious child seated in front of you throws a ball toward the tail of the aircraft, that is, toward the east, at 6 mph. According to Galilean relativity, what is the velocity of the ball relative to an observer in the airplane? Relative to an observer at rest on the surface of the Earth?

9. You wake to find yourself in an airplane with all its windows covered. Is there any experiment you can perform to determine whether you are flying with a uniform velocity, or at rest on the runway? (Ignore external effects such as engine noises, which could be simulated as a diabolical plot to trick you.) If the airplane changed its velocity, could an experiment show this? If so, give an example of an experiment you might perform that could detect an acceleration of the airplane.

10. Why did the appearance of the speed of light in Maxwell's equations create a problem for Galilean relativity theory?

Notes

1. Chapter 5 discusses this stage of stellar evolution.

2. If the intent of the universe is to create life, then it has done so in a very inefficient manner. For example, Aristotle's cosmos would satisfy the strong anthropic principle and would give a much greater amount of life per cubic centimeter to boot!

3. The symbol Δ is the standard mathematical notation for the concept of *change* in a quantity. Thus the expression Δx indicates "the change in the spatial coordinate x"; the Δ and the x are inseparable in this context.

4. Although the Coriolis force affects all motion on the Earth, it is, despite what you may have heard, nearly impossible to observe in an ordinary bathtub or bucket. On those scales the force is much too small to see, for practical purposes. Water draining from a tub swirls mainly because of effects that are far more significant than the Coriolis force at that scale; you would be as likely to see water draining from a kitchen sink in a clockwise sense in the Northern Hemisphere, as in a counterclockwise sense. However, the Coriolis force *does* significantly affect the trajectories of long-distance artillery shells, as some British naval gunners found to their embarrassment in a conflict in the Southern Hemisphere; they used Northern Hemisphere tables of the Coriolis force for their targeting corrections, but this force acts in the opposite direction in the Southern Hemisphere. Long-range shells fired with the incorrect aiming missed their targets by miles.

5. A similar ether had also been proposed to account for the transmission of gravity over distance.

6. Wave properties such as interference are described in chapter 4.

7. Another "minor" problem was explaining the blackbody spectrum. The resolution of that problem led to the creation of quantum mechanics.

CHAPTER 7

The Special Theory of Relativity

Einstein's Relativity

Imagine that you are the captain of a starship traveling at 99.99% of the speed of light. An asteroid is on a collision course with one of your world's space outposts and would surely destroy it if it hit. The asteroid is at a distance of 3,000,000 km, as measured by the sensors on the space station. You are flying toward the asteroid; at the instant you pass the space station, your ship fires its laser cannon at the asteroid. What would you observe? What would the officer on duty at the station see? When will your laser light beam reach its target? Both you and the watchman at the station want to know how quickly the threat will be eliminated, but the station's officer is especially concerned, of course.

How would these questions be answered in Newton's universe? In Newtonian cosmology, space and time are absolute, and the same for all observers. The ship and the laser beam are traveling in the same direction; therefore, Galilean relativity tells you that to obtain the net speed of your laser cannon beam, you should simply add the speed of the ship to the speed of light, which is about 300,000 km/s in the vacuum. Dividing the distance to the asteroid by that net speed, you would compute that the laser beam will hit the asteroid 5.0025 s after firing. Is this the correct answer? What would happen if the asteroid were *behind* you, so that the ship and the laser beam were traveling in *opposite* directions as you fired the cannon? In that case, you must subtract your velocity from that of the laser beam. Does that mean that the watchman would see the laser beam crawl through space at 0.01% of the usual velocity of light? What if you were traveling at exactly the speed of light while moving away from the direction in which you fire the beam? Would the light then have zero velocity in the frame of the space station? Can we even define a light beam with zero velocity?

If light were analogous to sound waves, we could use Galilean relativity to find the correct answers. Sound waves are waves of pressure moving through a fluid, such as the air. Because they are waves in a medium (e.g., the air), they move at a specific velocity (the speed of sound) relative to the medium. Wind carries sound along with it, and the total speed of the sound relative to the ground is the speed of the waves relative to the air, plus the speed of the wind, taking directions of motion into account. If we regard light as moving through some medium, historically called the luminiferous ether, then the light waves will always move at the speed of light, *relative to the ether*. We are now in a position to deduce what will happen in the spaceship problem posed above; we need only know how fast the spaceship is moving with respect to the ether, just as we might compute how rapidly sound waves would travel if emitted by a loudspeaker mounted atop a moving vehicle.

Prior to 1887, nearly every scientist in the world would have proceeded in this manner. After 1887, the universe no longer seemed so simple. This was the year of the Michelson-Morley experiment, one of the crucial experiments that once in a great while turns our science upside down. The most careful measurements that Michelson and Morley could possibly make found, to well within experimental error, *no* difference in the speed of light, regardless of whether the light was moving parallel or perpendicular to the motion of the Earth. Yet Earth's orbital speed is large enough that, if such an effect existed, Michelson and Morley would have detected it easily. What did this mean?

Newton's laws of motion are invariant under Galilean transformations. There is no absolute frame of rest, and all inertial frames are equivalent. The Galilean transformations provide the way to link observations made in one inertial frame with how things would look in another inertial frame. However, the equations of electromagnetism, the Maxwell equations, are *not* invariant under the Galilean transformation. The speed of light c enters into Maxwell's equations in a fundamental way, yet speed is not a quantity that is Galilean invariant. Maxwell himself believed that there should exist some special frame of reference in which his equations were correct as written; this would correspond to the frame in which the ether is at rest. Measurements in any other frame would be related to the ether's rest frame by taking into account the relative motion between the ether and that other frame.

But Michelson and Morley were unable to detect any evidence for motion with respect to this purported ether. In the absence of an "ether" to establish a frame for the speed of light, physicists were left with two unpalatable alternatives. The first possibility was that Maxwell's equations were incorrect, or perhaps that the physics of light was simply not the same in all inertial frames. The other alternative was that the Galilean transformation is invalid; but this would imply that something was amiss with Newton's mechanics. Yet Newtonian mechanics works so well for computing orbits; how could it possibly be wrong? On the other hand, the Maxwell equations were just as successful at explaining electromagnetism as the Newtonian equations were at explaining mechanics. How can we reconcile the invariance of one set of physical laws with the noninvariance of another?

One of the first attempts to account for the null result of the Michelson-Morley experiment was made in 1889 by George F. FitzGerald, who suggested that objects moving through the ether at velocity v were physically contracted in length according to

$$L(v) = L_0\sqrt{1 - (v^2/c^2)}. \tag{7.1}$$

That is, a moving object would literally shrink by this amount in the direction of its motion through the ether. Such a contraction of the arm of the Michelson-Morley interferometer, in the direction parallel to the motion of the Earth, would shorten the travel distance for the light moving in that direction by precisely the amount needed to compensate for the change in the light propagation speed. Thus, the round trip time would be equal for both arms of the apparatus, and no interference fringes would be seen. There was no fundamental theory to explain why objects would so contract; this was simply an ad hoc suggestion that reconciled the null result of the Michelson-Morley experiment with the existence of an ether. A hypothesis was put forward based on the recognition that intermolecular forces are electromagnetic in nature, so perhaps the very structure of matter was affected by motion through the ether. Yet this hypothesis seems very strange. How would an object be compressed? What if a living creature were to travel at a speed, relative to the ether, that was very close to that of light; would it be squeezed to death if v became very close to c?

Many scientists rejected the FitzGerald contraction, clinging instead to a more conservative interpretation. They struggled to explain the null result of the Michelson-Morley experiment as a consequence of "ether drag." If moving bodies dragged the ether along with them, then near the surface of the Earth, no relative motion of Earth and ether could be detected. There was even, apparently, some experimental evidence for this; it had been known since the 1830s that the speed of light propagating through a moving fluid was different from its speed in a fluid at rest. When light travels through a medium, its speed is always less than its speed *in vacuo* and depends upon the properties of the medium. For light traveling through a fluid such as water, some of the velocity of the fluid seemed to be imparted to the light, a phenomenon attributed to a partial entrainment, or "dragging," of the ether by the fluid. This explanation preserved the old mechanical view of light, but at the expense of attributing to the ether even more strange properties, such as some kind of viscosity. If the Earth were dragging the ether, should it not lose energy, slow down in its orbit, and eventually fall into the Sun? This certainly had not occurred, nor was there any evidence for a systematic shrinking of the Earth's orbit. Moreover, the ether-drag hypothesis predicted an effect on starlight as it entered the ether surrounding, and dragged by, the Earth; but no such effect was observed.

A bolder proposal was put forward by the Austrian physicist Ernst Mach. No motion relative to the ether was observed because there *was no* ether. An elegant experiment had been carried out to test the existence of the ether. The ether was not found; hence the ether theory was disproved. Accepting Mach's point still required the development of a new theory to replace the discredited ether theory.

Several scientists took up the idea of the FitzGerald contraction, in particular Hendrik Lorentz, and later Henri Poincaré and Joseph Larmor. They demonstrated that the Maxwell equations were invariant under a new kind of transformation law which makes use of the FitzGerald contraction. The new transformation law, now known as the **Lorentz transformation**, contains a dilation, or slowing, of time, in addition to the length contraction proposed by FitzGerald. Thus the Maxwell equations are invariant when using the Lorentz transformation, whereas Newton's equations are invariant under the Galilean transformation. But neither of these transformations is arbitrary; they derive from fundamental ideas about the nature of space and time, so both cannot be correct. The Lorentz transformation, with its dilation of time and contraction of space, stands in direct opposition to something that was still regarded as more fundamental than Maxwell's equations, namely, Newton's absolute time and

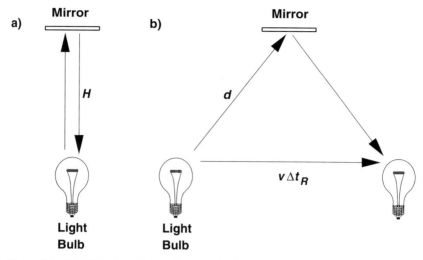

Figure 7.2 Relativistic time dilation. The path of a flash of light traveling from a bulb, up to a mirror, and back, as seen in (*a*) the rest frame of the train and (*b*) the rest frame of someone in the ground frame watching the train go by. In the ground frame the train moves to the right a distance $v\Delta t_R$ during the round trip of the light. The resulting light path is longer for the observer in the ground frame, but since the speed of light is the same in all frames, the time intervals in the two frames must be different.

Just how much slower is each tick of the moving clock? The robber remembers his geometry and uses the Pythagorean theorem to compute the distance d traveled by the light in his frame of reference:

$$d^2 = H^2 + \left(\frac{1}{2}v\Delta t_R\right)^2. \tag{7.2}$$

Recall that in the robber's frame, $\Delta t_R = 2d/c$, and in the passenger's frame $\Delta t_P = 2H/c$, so we can eliminate d and H to obtain a quadratic equation

$$\frac{1}{4}(c\Delta t_R)^2 = \frac{1}{4}(c\Delta t_P)^2 + \frac{1}{4}v^2(\Delta t_R)^2. \tag{7.3}$$

Working through the algebra leads us to the result

$$\Delta t_R = \frac{\Delta t_P}{[1 - (v^2/c^2)]^{1/2}}. \tag{7.4}$$

Since we have assumed that v is less than c, $1 - (v^2/c^2)$ is less than 1, and therefore Δt_R is greater than Δt_P. This is the mathematical expression of what we have stated above, that the light travel time measured by the robber is larger than the light travel time measured by the passenger.

We have stated that this bulb-and-mirror apparatus could be used to construct a clock; one complete transit could be regarded as one "tick." Thus, one tick of our bulb-and-mirror clock will be longer for a moving clock than the same tick will be for a clock at rest. Does this result follow just because we have constructed some unusual clock with mirrors and light beams? What if we used an "ordinary" clock? But what is an ordinary clock? Suppose we used an atomic clock to measure the time interval between the departure and the return of the flash. Would that make a difference? A clock is just a physical process with a regular periodic behavior. The details of the clock's construction are irrelevant; no matter how we choose to measure the time interval, we shall always find that an observer moving with respect to the clock will observe that the interval for one tick is longer than it is for one tick in the clock's own rest frame. Put more succinctly, "moving clocks run slow."

In equation (7.4) we found that the relationship between the time intervals in the two frames contained the factor

$$\frac{1}{[1 - (v^2/c^2)]^{1/2}} \equiv \Gamma. \tag{7.5}$$

Does this look familiar? It appeared in equation (7.1), the Lorentz contraction. Perhaps we are now beginning to understand its significance. It does not tell us anything about a physical contraction of moving matter, but rather describes the way in which space and time are related for observers who are moving with respect to one another. It is often called the **boost factor** between two inertial frames.

As an example, let us suppose that the train has a boost factor of $\Gamma = 2$, relative to the frame of the robber; this corresponds to a velocity of about $0.87c$. The passenger's clock measures an interval of 30 s between some sequence of two events occurring on the train, such as the entrance of another passenger into the car and his exit on his way to the dining car. According to the robber, the interval between these two events is 60 s.

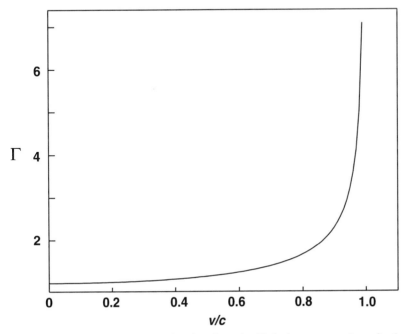

Figure 7.3 The boost factor Γ as a function of speed, with the latter expressed as a fraction of the speed of light. The boost factor is 1 when $v = 0$ and becomes infinite when v is equal to c.

Perhaps you are now thinking that the passenger would experience a strange world inside a train traveling close to the speed of light. Clocks run slow, people move in slow motion. But that cannot be correct. There is no such thing as an absolute frame of rest; all inertial frames must be equivalent. The passenger must observe that his own inertial frame is perfectly normal.

To emphasize this point, let us invert the situation. Suppose the robber has the bulb-and-mirror apparatus, and the passenger measures the interval between the flash and its return to the source. What do you suppose the passenger will measure? Would you expect that the robber's clock would run slow, as observed by the passenger? That is exactly what happens. By assumption, the train is an inertial frame, so with respect to the train, it is the passenger who is at rest, and the robber who is moving with velocity $-v$. The clock that is *moving* with respect to some rest frame always runs slow relative to that rest frame. This consequence is required by postulate 1 above; both observers must obtain the same result when performing such an experiment. If the passenger were to observe that the robber's clock ran faster, then when the two of them compared their results, they would be able to agree that it was the train that was "really" moving. But there is no absolute motion in special relativity; no inertial frame is preferred, and therefore no inertial frame is "really" moving. Any two inertial frames are equivalent, and both must measure the same relative speed between them. This is the **principle of reciprocity**.

Perhaps the easiest way to avoid confusion is to remember that any clock at rest with respect to a given inertial observer will run "normally," that is, at the fastest rate, as measured by that observer; all clocks moving with respect to an observer run slow. No matter where you go in the universe, or how fast you travel, the watch on your wrist will continue to run at the usual rate, according to you. We call the time measured by a clock that is at rest with respect to a specific inertial observer the **proper time**. Proper time is a very important concept, one to which we will later return.

Already we have found one astonishing result: time intervals are not invariant. What about space? We can anticipate that if moving clocks run slow, then moving rulers must be short if the speed of light is to remain constant for all observers, since speed is simply a distance divided by a time interval. Furthermore, we can guess that the boost factor must specify the contraction. But it is not difficult to see how this effect arises.

We measure time by the ticking of clocks, that is, by counting the number of cycles of some repetitive phenomenon. In our example above, we used a pulse of light bouncing back and forth between mirrors set a fixed distance apart. How do we measure separations in space? We do so by comparing the length of something to that of a fixed standard, a "ruler," at some specific time. More explicitly, if we wish to measure the length of an object, we hold our ruler against it such that one end of the object lines up with one fiducial mark on our ruler, and *at the same time* the other end of the object lines up with another fiducial mark. Then we count the number of marks between the two ends. Another method of measuring a length, such as the length of a jogging trail, is to travel from one end of the object to the other at a constant, known speed, and measure how long the transit takes.

For example, let us return to the train and train robber, and consider two telephone poles beside the tracks. The robber wishes to measure the distance between the poles, and to do so he will make use of the train, which is traveling at known velocity v. The robber simply measures the time interval required for the front edge of the train to pass from the first pole to the second, Δt_R; he thus determines that the distance between the poles must be $\Delta x_R = v \Delta t_R$. Now suppose the passenger on the train decides to measure the distance between the same two telephone poles, which are moving with respect to him at speed v. Both the passenger and the robber agree on the relative speed v, as they must if inertial frames are to be equivalent. The passenger uses a similar timing technique of noting when the first, and then the second, pole passes the edge of his window; he measures a time interval Δt_P between the passage of the first pole and the second. Thus the distance between the poles in the train frame is $\Delta x_P = v \Delta t_P$. You can see clearly from this example that space measurements are tied in with time measurements, which should not be surprising since time and space intervals are related by speed or velocity. We have already solved for Δt_R in terms of Δt_P; hence we can obtain

$$\frac{\Delta x_P}{\Delta x_R} = \frac{\Delta t_P}{\Delta t_R} = \left[1 - (v^2/c^2)\right]^{1/2} \tag{7.6}$$

or

$$\Delta x_P = \Delta x_R \sqrt{1 - (v^2/c^2)}, \tag{7.7}$$

which is exactly the Lorentz-FitzGerald contraction, equation (7.1) Thus we have demonstrated that the passenger measures the distance between the moving poles to be *shorter* than the distance measured by the robber, in whose frame the poles are at rest. The distance measured by the passenger is specified by the self-same factor that was first proposed as an ad hoc explanation for some unexpected experimental results. Now it appears naturally and elegantly from the two fundamental postulates of special relativity.

Reciprocity applies to length measurements, just as to time measurements. If the robber measures the length of one of the cars of the train, he will find this length to be shorter than that measured by the passenger, who is located within the rest frame of the car. Specifically, if the boost factor of the train is again 2, and the length of the car 10 m as measured by the passenger, the robber will observe its length to be 5 m. The result always depends on who is doing the measurement and the relative velocity of the object that is being measured.

Similar to the proper time, we can define the **proper length** to be the length of an object as measured in its own rest frame. The proper length of an object is always the *largest* possible. The meter stick you hold in your hand always has the expected length of 1 m. Meter sticks rushing past you at any velocity are shorter than 1 m, as measured by you.

So far we have spoken only about measurements of a distance that is parallel to the direction of motion. The Lorentz contraction occurs only along that direction, never in the direction perpendicular to the motion. To understand this, first recall that

FitzGerald's proposal to explain the Michelson-Morley experiment invoked contraction only along the direction of the motion of the apparatus. It is easy to see that the situation would not have been clarified if both arms of the apparatus contracted by the same amount. As another illustration, let us apply the principle of reciprocity, and consider an argument in terms of the relativistic train. Suppose the train is 4 m tall and is traveling at such a speed that its boost factor is 2; it approaches a tunnel that has a 5 m clearance. If there were a Lorentz contraction in the vertical direction, the robber, who is at rest with respect to the tunnel, would measure the height of the train as 2 m, and would thus conclude that the train will easily fit into the tunnel. By reciprocity, however, the engineer must observe that the tunnel has only a 2.5 m clearance, while the train is still 4 m tall in its own rest frame; the train will be wrecked. But there cannot be a wreck in one frame and not in the other. Hence there can be no contraction along directions that are not in relative motion. Only that component of the relative velocity that is along the direction of the relative motion has any effect upon the Lorentz transformation.

Special relativity and the Lorentz transformation lead to a quite unexpected view of the nature of time and space intervals. Why did we not notice the need for special relativity until near the end of the nineteenth century? Why did Newtonian mechanics and Galilean relativity seem to work so well for 200 years? Let us compute the boost factor for one of the fastest macroscopic motions that you might personally experience on the Earth, the flight of a supersonic airplane such as the Concorde. The Concorde has a length of approximately 60 m and flies, at top speed, at about twice the speed of sound. For this velocity, the boost factor is $\Gamma = 1.000000000002$, which makes the Concorde's Lorentz-contracted length approximately 10^{-8} cm less than its rest length. This difference is about the diameter of an atom. For Earthly motions and speeds, the stuff of our everyday experiences, the effects due to the Lorentz transformation cannot be detected.[4] When v/c is very small, Γ is very close to one, and the Lorentz transformation reduces to the familiar Galilean transformation.

The Meaning of the Lorentz Transformation

Let us pause for a moment and reconsider these conclusions. Equations (7.4) and (7.7) together give the Lorentz transformation, the formulae for relating measurements in one inertial frame to those in another inertial frame. Length contraction and time dilation demonstrate that length and time intervals are different for observers moving with different velocities. This may seem disturbing; we are accustomed to thinking of space and time in Newtonian terms, as absolute, unvarying, and universal. Now they seem to depend on how fast an observer is moving, relative to some other frame. But what *are* space and time? How do we measure them? We have discussed briefly how distances can be measured. Since distance is defined to be the spatial separation between two points at some *simultaneous point in time*, it is clear that at a very basic level, measurements of space and time are connected. We measure time in terms of the number of occurrences of a regular physical process, such as the swinging of a pendulum. Today we may replace the pendulum with the vibration of a quartz crystal or the oscillation of a cesium atom, but the concept is the same. These are all physical processes, and we find that we define our concepts of space and time intervals in comparison with standard physical processes.

Since our measurement of time is so closely tied to the behavior of physical systems, all means of measuring time within a given frame must give consistent results. Readers of science fiction occasionally encounter characters who are aware that time is running at some strange rate because they see their own clocks running slowly or, if it is a time-travel story, even running backward. But this is clearly absurd. If time slowed down (or stopped or reversed, whatever that means), then all physical processes would behave the same way, including the pulse of your heart, the speed of your thoughts, the swing of the pendulum, or anything else by which you might determine the passage of time. The rate at which physical processes occur gives us our measure of time, and if all those rates changed together, an observer could not notice it. Try to imagine a way of measuring time that does not involve some periodic physical process!

Modern physics has shown that physical processes depend on the interaction of fundamental forces. The most important forces for everyday events are gravity and electromagnetism. These long-range forces result from the exchange of massless particles: the graviton for gravity, and the photon for electromagnetism. Massless particles move at the speed of light; hence gravitational and electromagnetic forces propagate through space at the speed we call the speed of light. Light, per se, is not particularly important here. We could just as well call it the speed of gravity, but few would know what we meant. For physical processes, the exchange of the particles that produce forces has ultimate importance. From this, you may see how it was correct for Einstein to put the constancy of the speed of light ahead of the invariance of individual time and space measurements. Moreover, the crucial distinguishing factor of special relativity is not so much the speed of light, as it is the existence of a finite speed of propagation of forces. Any such finite speed limit would result in a transformation law like the Lorentz transformation; conversely, we may regard the Galilean transformation as that which would hold in a universe in which forces propagated with infinite speed. Thus special relativity does not stand in isolation but is linked in a very profound way with the laws of nature.

The finite speed of propagation of the gravitational force solves one further nagging problem with Newtonian mechanics. Newton himself was somewhat perturbed by the instantaneous action at a distance that was implied by his gravitation law. Maxwell's equations described the transmission of electromagnetic force, but at a finite propagation speed. What happens to electromagnetism if we let the speed of light go to infinity? If c became infinite, the term v^2/c^2 would always be zero for any finite velocity, and, as we have stated, the Lorentz transformation reduces to the Galilean transformation. This demonstrates explicitly that the Galilean transform is appropriate for instantaneous force transmission, such as is implied by Newton's law of gravity, whereas the Lorentz transformation is appropriate for finite speeds of force propagation. Framed in this way, perhaps Newton himself would agree that his gravity law needs modification so that it will behave like Maxwell's equations. However,

Einstein's special relativity alone does not accomplish this task. It is the general theory of relativity that reformulates the theory of gravity to include, among other features, a finite speed of propagation.

More Transforms

We began this chapter by contemplating what we might see from a spaceship if we shot a laser beam into space. Although we are now in a position to describe what happens in that situation, let us return instead to the simpler case of the train traveling down the tracks at speed v. The train passenger throws a ball down the aisle toward the front of the car. The speed of the ball in the frame of the train is v_{ball}. The Galilean transformation would tell us that the speed of the ball in the robber's frame is given by the sum of its speed in the passenger's frame, plus the speed of the train, $v + v_{ball}$. (If the passenger threw the ball toward the back of the car, the Galilean transformation would give us a speed in the robber's frame of the train speed minus the ball speed.) But we have just found that the Galilean transformation is not correct. We also know from the postulate of relativity that the speed of light must be the same in all frames; hence we need a formula that yields, schematically, $\alpha(v+c) = c$, where α is a factor yet to be determined. From the Lorentz transformation, it is possible to work out exactly how velocities must add. We shall omit the details and merely present the *relativistic* velocity-addition formula

$$v_{tot} = \frac{v_{ball} + v}{1 + v_{ball}v/c^2},\tag{7.8}$$

where v is the relative velocity of train and robber, v_{ball} is the velocity of the ball in the train frame, and v_{tot} is the velocity of the ball as seen by the robber. This equation has the desired property that if we replace the ball with a light beam, thus setting $v_{ball} = c$, we find that the robber also measures $v_{tot} = c$. Hence, as required, the speed of light is the same in both inertial frames. Notice that this law also says that we cannot hope to achieve superluminal (faster than light) speeds with a hypercannon mounted on a relativistic spaceship. No matter how close two (sublight) speeds are to the speed of light, they can never add to a speed greater than that of light.

The relativistic addition formula also explains the apparent experimental observation of "ether dragging" by a moving fluid. Nineteenth-century scientists had observed that light seemed to travel faster when it was propagating within a moving fluid than when it traveled within a fluid at rest; this was often cited as experimental support for the dragging of ether before Einstein developed special relativity. In view of our present knowledge, we can see that what was truly observed was the relativistic addition law, although this was not realized until fully 2 years after the publication of Einstein's paper on special relativity. Light propagates through a fluid at a speed less than c; the exact speed is a function of the index of refraction of the substance. If the fluid, in turn, is moving with respect to the experimenter, the observed speed of the light in the frame of the experimenter is given by equation (7.8). It could be said that special relativity was confirmed before it was even conceived, but the correct interpretation could not be seen until Einstein was able to break through the prevailing patterns of thought.

Although the *speed* of light is the same in all inertial frames, this does not mean that all inertial observers will see light in quite the same way. Time dilation and length contraction play roles in the propagation of light waves if the source and receiver are in relative motion. The classical Doppler effect occurs with relative motion because the crests of the light waves bunch together (for relative approach) or stretch out (for relative recession). Relativity adds to this effect a correction: the frequency of the light, which is an inverse time interval, is less at the source than at the receiver, due to time dilation. The relativistic Doppler formula is given by

$$z + 1 = \sqrt{\frac{1 + (v/c)}{1 - (v/c)}}. \tag{7.9}$$

A relativistic Doppler effect also occurs in the direction perpendicular to the relative motion, exclusively as a consequence of time dilation. The frequency of a light wave is like a clock; the electromagnetic fields oscillate some number of times per second. If you observe a moving source, you will see its clock running slow; hence the frequency of the light must be reduced, even if the light is coming to you along a direction perpendicular to the motion of the emitter.

The transverse Doppler effect is generally very small; it is practically unobservable for most motions. However, a few astronomical objects do exhibit relativistic motions, with detectable transverse Doppler shifts. One such object is the remarkable SS 433, a star in the constellation Aquila, about 16,000 lt-yr from Earth. In radio images, the star evinces jets of gas extending from the main source. SS 433 is a binary system consisting of a normal star and a compact object, most likely a neutron star; its location within an ancient supernova remnant supports this scenario. Gas from the normal star is drawn into an accretion disk around the neutron star, and some of it is squirted at relativistic velocities in two oppositely directed jets, along the axis of the neutron star. The jets wobble, and when the beam is directly perpendicular to our line of sight, a purely relativistic, transverse Doppler shift can be observed in the spectrum, corresponding to gas speeds of about one-fourth the speed of light. No matter how extreme the behavior, somewhere in the observable universe there is usually some object which demonstrates it.

Even though no two velocities can add to give a speed in excess of light *in vacuo*, there is still a great difference between, for example, $0.9c$ and $0.99c$. Suppose our starship has on board a particle accelerator that can eject a particle beam of matter at $0.9c$, as measured in its own rest frame. If the spaceship wishes to project a particle beam at a speed of $0.99c$, as measured by an observer on the space station, at what speed must the spaceship travel, relative to the space station? The velocity-addition equation (7.8) tells us that the spaceship must have a relative velocity of a little under $0.9c$ in order for the beam to reach $0.99c$. In other words, $0.99c$ is approximately twice as fast as $0.9c$! If we compute the boost factors, we find $\Gamma = 2.3$ for $0.9c$, while $\Gamma = 7.1$ for $0.99c$, a very large difference. In terms of velocity increase, it is approximately as difficult to go from $0.9c$ to $0.99c$ as it is to go from zero velocity to $0.9c$.

The strange velocity addition rule of special relativity hints at another important consequence besides the intermingling of space and time, time dilation, and the Lorentz contraction. It leads us to what is perhaps the most famous equation in history, $E = mc^2$. But what does this renowned equation *mean*, and how does it fit into relativity theory? First we must specify what we mean by "energy." We have previously defined energy as "the capacity to do work." In the Newtonian universe, energy is not created or destroyed, but only transformed from one form to another. Similarly, there is a separate conservation law for matter; matter is neither created nor destroyed. One of the most important forms of energy is kinetic energy, or the energy of motion. In Newtonian mechanics, it can be shown that the kinetic energy of a particle is given by

$$E_k = \frac{1}{2}mv^2, \tag{7.10}$$

where m is the mass of the particle, and v is its speed. The Newtonian kinetic energy of a particle at rest is zero. The Einsteinian equation is the relativistic generalization of this concept of kinetic energy. The equation is more correctly written as

$$E = \Gamma m_0 c^2, \tag{7.11}$$

where Γ is our new acquaintance, the boost factor, and m_0 is the *rest mass* of the particle, that is, its mass as measured in its own rest frame. Notice that, since the boost factor is 1 for a particle at rest, this definition of energy does not vanish for $v = 0$. Thus we find that in relativity, there is a **rest energy**, given by $m_0 c^2$, associated with every massive particle. As the speed of the particle increases, its energy also increases. For *small* velocities, it is possible to show that the relativistic energy equation reduces to

$$E = m_0 c^2 + \frac{1}{2}m_0 v^2 + \text{additional terms}, \tag{7.12}$$

where if v is very much less than c, the additional terms are very, very small, and we recover the Newtonian law, with the *addition* of the new concept of the rest energy. At the other extreme, as the speed increases and begins to approach that of light, the relativistic energy becomes very large, much larger than the simple Newtonian rule would predict; it is arbitrarily large for speeds arbitrarily close to the speed of light.

As an example, consider how much energy would be required to accelerate 1 kg of matter to $.87c$, for which $\Gamma = 2$. In order to compute the relativistic *kinetic* energy, we must subtract from the total energy, as given by equation (7.11), the rest energy specified by $m_0 c^2$. Carrying out this procedure, taking care to keep our units consistent, we obtain a result of 9×10^{16} joules (J) of kinetic energy. In units that might be more familiar, this is 3×10^{10} kilowatt-hours, or about 20 megatons TNT equivalent. You would need all the energy released by a very large thermonuclear bomb in order to accelerate just 1 kg of matter to a speed close to c. This is a serious limitation on our ability to boost anything, even elementary particles such as protons, to speeds approaching that of light. At accelerator laboratories around the world, scientists do just that, accelerating protons to speeds near light speed. It is no coincidence that very large power lines can be seen going onto the grounds of these accelerators!

The rest energy is an interesting concept. Does it represent some irreducible amount of mass that is always conserved, or does it mean that energy and mass are truly equivalent and can be transformed into one another? When Einstein wrote down his famous equation, there was no experimental evidence to decide that issue, but he chose to interpret his equation boldly, asserting that the equals sign meant just that, equality. Mass can be converted into energy, and energy into mass. In fact, the rest energy can be interpreted as the energy due to the inertial mass; in this view, inertial mass is itself just another form of energy. Subsequent events have proven that this assertion is correct.

If even a tiny fraction of the rest energy of a particle is converted to another form of energy by some means, the yield can be enormous. For most people, nuclear weapons most dramatically illustrate the principles of special relativity. Nuclear reactions, both fission and fusion, are modestly efficient means of extracting rest mass energy, converting about 1% of the rest mass involved into other forms of energy. Ironically, nuclear reactions are relatively ineffective at extracting rest energy; they just happen to be the best mechanisms available on Earth. After all, 1% of the rest energy of even a few kilograms of fuel is an enormous amount of energy. Although weapons may be the most familiar application of nuclear reactions, fusion in particular is of utmost importance to humans; fusion reactions occurring at the core of the Sun are ultimately responsible for the existence of life on Earth. The Sun has so much mass that 1% efficiency provides power for tens of billions of years at its current luminosity, so this small efficiency is adequate for our needs. The most efficient process possible is matter-antimatter annihilation, in which a particle and its antiparticle are both converted *completely* into energy.[5]

Sometimes it is believed that equation (7.11) applies only to such exotic reactions as matter-antimatter annihilation or nuclear reactions. In fact, *any* release of energy, including such mundane ones as chemical processes, results in a change of mass. In the ordinary nonrelativistic world in which we live, this change in mass is unmeasurable; yet it occurs. Conversely, energy can be converted into matter, although this process is somewhat more exotic than the other. In particle accelerators, high-speed particles are slammed together. At their large boost factors, they have energy to spare to produce particles and antiparticles; that is, matter is created from energy. So just as we can no longer think of "space" and "time" individually, we must not think of "energy" and "matter" as distinct quantities. Mass and energy are revealed to be two aspects of the same entity, "mass-energy"; and it is "mass-energy" that is conserved.

Examples of relativistic speeds occurring naturally near Earth are not easy to find. One "everyday" relativistic motion is the flight of muons through the atmosphere. Muons, a kind of heavy particle with an extremely short half-life, are created high in the upper atmosphere when cosmic rays, which are actually high-energy particles from the Sun and other sources, collide with atoms. In the collision, some of the kinetic energy of the impinging particle is converted into matter, such as muons; this is in itself a stunning illustration of the equivalence of mass and energy, and hence of the special theory of relativity. In its own rest frame, the muon decays with a mean life expectancy of only 2 microseconds (μs). Atmospheric events produce muons traveling at speeds such as $0.99995c$, corresponding to a boost factor of about 100. Even at this speed, without relativistic effects a typical muon would travel only 600 m over its brief existence, and practically none would ever reach the surface of the Earth. Yet the surface is constantly bombarded by relativistic muons. In the frame of the Earth, the muon exists 200 μs, in which time it travels 60 km! The arriving muons also have a relativistic mass 100 times greater than their rest mass, as measured in the frame of the Earth.

Spacetime

We have discovered that time intervals, space separations, and simultaneity (whether two events occur at the same time), are not absolute, but depend upon the frame of reference of the observer. Suddenly, the orderly Newtonian universe has been replaced by a much more unruly one, where space and time are relative to the observer and where a measurement of time depends on space, and vice versa. Most of us grow up thinking of space and time as absolute, distinguishable quantities; now we find that they somehow intermingle. We can no longer think of "space" and "time" as separate entities; rather than three space dimensions and one time dimension, our new view of the universe is a four-dimensional **spacetime**.

The idea of spacetime was developed by Hermann Minkowski in 1908 as a way to unify the mixing of time and space, as given by the Lorentz contraction and time dilation, into one four-dimensional structure. Unfortunately, it is difficult to think in four dimensions, and impossible to visualize any four-dimensional object. However, a very useful device for representing occurrences in this spacetime is the **spacetime diagram**. Usually we simplify matters by plotting only one space dimension, since we cannot show more than two of them anyway. We cannot draw, or even model on a tabletop, four mutually perpendicular axes, but one space dimension is generally adequate for the purpose of understanding a physical process. The remaining axis is labeled as time. Together, the time and space coordinates represent some inertial reference frame. The position x and time t of an object can be plotted on such a graph, tracing out a continuous curve on a spacetime diagram. Any point on such a curve is an event, and the curve itself is called a **world line**. (Strictly speaking, the world line describes only the path of a point particle; any larger object is a collection of points, each moving along a world line, so the object as a whole traces out a *world tube*. The distinction is not very important here.) On such a diagram, a straight line corresponds to an object moving with a constant velocity, that is, an inertial observer; the slope of the line is proportional to the speed of the observer. If the world line curves—that is, if its slope changes—then the velocity is changing, and the curve corresponds to a noninertial observer.

When drawing spacetime diagrams, it is customary to calibrate the time variable as ct, so that both axes are labeled with the same units. An observer at rest with respect to these coordinates traces a world line that is vertical, that is, it remains at a constant space position. A light beam follows a world line given by $x = ct$; with our units convention, such a path is thus a straight line at an angle of 45°. A second massive observer, moving inertially with respect to the observer at rest in the frame, has a world line that is also straight but always makes an angle greater than 45° with respect to the space axis, so that the velocity is less than c. Conversely, if the slope of the line connecting two events is less than 45°, then those events could be connected only by moving faster than light.

Figure 7.4 A spacetime diagram showing three world lines. A point on a world line is called an event. In the inertial frame of (x, t), any straight world line corresponds to inertial motion. A curved world line corresponds to accelerated motion.

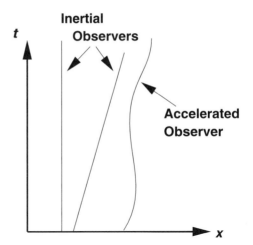

If spacetime were like our usual (x, y) space coordinates, we would be on familiar ground. Given two points in (x, y) coordinates, we could use the Pythagorean theorem to obtain the distance between them. Recall that the square of the distance along the hypotenuse of a right triangle is equal to the sum of the squares of the other two sides:

$$(\Delta r)^2 = (\Delta x)^2 + (\Delta y)^2. \tag{7.13}$$

This Pythagorean formula defines the distance between two points on a flat surface, in terms of perpendicular coordinates x and y.

But spacetime is not like ordinary space. Given two events which lie close to one another on a world line, how can we define a "distance" between them? And what would that distance mean? Each of the two events occurs at points labeled by appropriate (x, t) coordinates. We define the distance, or **spacetime interval**, between them by

$$\Delta s = \sqrt{(c\Delta t)^2 - (\Delta x)^2}. \tag{7.14}$$

The factor of c ensures that we are subtracting two quantities with the same units. This distance measures separations in what is now known as **Minkowski spacetime**, in honor of Hermann Minkowski, who first formulated its properties. The principal difference between the spacetime distance formula and the Pythagorean formula is the negative sign between the time interval and the space interval. Time may be the fourth dimension, but it is not simply like the three space dimensions. The properties of this spacetime geometry are different from the ordinary flat *Euclidean space* to which we are accustomed.

The spacetime interval derives its fundamental significance because it is *invariant* under the Lorentz transformation. All observers can measure both the space and time intervals separating two events. We have learned that different observers will obtain different results for the individual space and time intervals, with the measurements related for inertial observers by the Lorentz transformation. However, when any observer combines his own values for the space and time intervals between two events into a *spacetime* interval, as specified by equation (7.14), *all* observers will obtain the

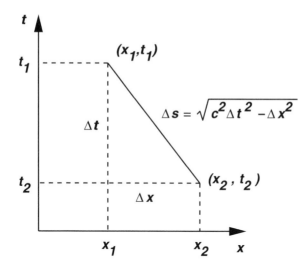

Figure 7.5 The rule for the spacetime interval between events located at (x_1, t_1) and (x_2, t_2). It differs from the Pythagorean rule (cf. Fig. 6.3) by the presence of the negative sign between the space interval and the time interval. The spacetime interval is invariant under the Lorentz transformation.

same result. This means that no matter what coordinates you use, or what your velocity is with respect to the (x, t) coordinates, you will obtain the same result for the value of the spacetime interval. If you are so inclined, you may convince yourself of this fact by applying equations (7.4) and (7.7). It is comforting that after we have thoroughly mixed up space and time, we finally have something that is invariant.

Can we provide a physical interpretation for the spacetime interval? Suppose an inertial observer measures this interval using a clock at rest with respect to herself. In her own rest frame, she is always at the same location, so Δx is zero. All that remains is the time interval; hence the spacetime interval corresponds to the *proper time*, which we have already encountered. Since it is invariant, *any* inertial observer can measure it, and each will obtain the same value, even if different observers may disagree about the separate space and time intervals. In simplified terms, everyone will agree about how much time elapses on a given observer's watch as that observer travels from one event to another, although different observers may disagree on how that proper time interval compares to the time elapsed on their own watches.

Suppose we are interested in the spacetime interval between two events along some arbitrary *curved* world line. In that case, we cannot use equation (7.14) directly. In order to compute this quantity, we must consider a large number of pairs of events close together along the curve, calculate each spacetime interval, and add them up.[6] The spacetime interval between two events depends on the path between them, which at first might seem quite remarkable. With some reflection, however, this ought not seem so odd; after all, the distance from your home to the nearest grocery store depends on the path you choose to get there. Since the spacetime interval equals the proper time along that world line, we see that the elapsed proper time kept by two clocks that travel on different paths through spacetime from one event to another, can be different.

What is the spacetime interval between two nearby events on the world line of a light ray? Since the light ray always travels at speed c, for it, $\Delta x = c\Delta t$, so the spacetime interval is zero. But the sum of any number of zeros is still zero; therefore, the spacetime interval between *any two events* on the light beam's world line is *always* zero, for *any* inertial observer! This result shows us immediately that the spacetime interval does not behave like the Pythagorean distance of Euclidean space, for the Euclidean distance is always positive for any two distinct points and is zero if and only if the points coincide. In contrast, the *square* of the spacetime interval can be positive, negative, or zero, and we can use this fact to divide all spacetime intervals into three classes. If the spacetime interval, and its square, are positive, the interval is said to be **timelike** (i.e., the time interval $c\Delta t$ is greater than the space interval Δx). Timelike world lines describe paths that can be traversed by massive, physical particles, all of which must travel at less than the speed of light. If the square is negative, the interval is **spacelike**. If two events are separated by a spacelike interval, they cannot be connected either by a light ray or by the world line of a particle traveling at a speed less than the speed of light.[7] What is the spacetime interval itself, in the case that its square is a negative number? The square roots of negative numbers make up the set called the *imaginary* numbers. The imaginary numbers are probably less familiar than the real numbers, but they represent a perfectly valid mathematical concept, useful in many tangible fields such as engineering. However, we shall have no need to make explicit use of imaginary numbers in this text.

If the spacetime interval is zero, it is said to be **null** or **lightlike**. Any particle that travels on a null world line must have zero mass, that is, the particle is *massless*. The converse is also true; a massless particle must always travel at the speed of light. Photons can be created or destroyed, but they cannot slow to less than c, the speed of light in a vacuum. The speed of light itself may differ in different media; for instance, the speed of light in water is less than its speed in a vacuum. This occurs because the atoms or molecules of the substance absorb photons and then, after some interval of time, reemit them. The photons themselves always travel at speed c between interactions.

Up to now we have talked mostly about an observer who is at rest with respect to the inertial frame (x, t). What about an observer who is moving with respect to this inertial reference frame? Can we deduce how the space and time axes of a moving observer can be drawn onto our spacetime diagram? First, construct an inertial set of coordinates (x, t) and draw the spacetime diagram. Suppose we label the moving observer's own coordinates by (x', ct'). The time axis is easy: it must lie along the observer's world line because an observer is always at rest in his own coordinate frame. The position of the space line is less obvious. Our first impulse would be to draw it perpendicular to the ct' axis. This is appropriate for an ordinary spatial coordinate system such as (x, y), but remember, spacetime is different. The one thing we must preserve is the constancy of the speed of light in all frames. In our (x, ct) frame, the world line of a light beam runs halfway between the x and t axes such that $\Delta t = \Delta x/c$. The same must be true for the (x', ct') frame. Hence we must draw the x' axis so that the light beam makes an equal angle with both the ct' and the x' axes, as shown in Figure 7.6.

It may look strange to have perpendicular axes that do not make a 90° angle with respect to one another, but Minkowski spacetime is not like the familiar flat-space Euclidean geometry. The canting of the (x', ct') axes occurs because Minkowski spacetime can be represented only approximately on a flat sheet of paper. As an analogy, consider a drawing of a three-dimensional cube on two-dimensional paper. Even though the edges of a cube always form an angle of 90° where they meet in the three-dimensional space in which the cube exists, the lines in the drawing meet at different angles. A drawing of a three-dimensional cube on a two-dimensional sheet of paper can only approximate the actual shape of the cube.

We can notice some interesting features from Figure 7.6. Recall that an observer says that two events occur at the same place if their spatial location is the same. For the observer at rest with respect to the (x, ct) frame, two events that occur at the same place will lie along a vertical line; that is, Δx between those events will be zero, although they will be separated by a time interval. The moving observer (the "primed" frame) will not see such events occur at the same point in space, but will see a $\Delta x'$ as well as a $\Delta t'$, in general. These statements are not remarkable; they hold in Galilean relativity as well. In four-dimensional spacetime, we must extend this concept to include time. Two events are simultaneous if they occur at the same time but at different places; that is, simultaneous events have zero Δt. The stationary observer will say that simultaneous events lie along a horizontal line, such as the x axis. But the moving observer will *not* observe that those events are simultaneous. In the frame of the moving observer, simultaneous events must lie along lines parallel to the x' axis, which is slanted with respect to the x axis. Similarly, for events simultaneous in the primed frame, $\Delta t'$, but *not* Δt, is zero. Simultaneity is relative.

Is there anything about which all observers agree? The spacetime interval is an invariant, and any function, such as the square, of an invariant is also invariant; therefore, all observers will agree that a given interval is either timelike, spacelike, or null. Let us draw a spacetime diagram to illustrate this. First, draw the possible world lines of a photon that passes through some particular event, labeled A. The photon can travel toward the left or right but must always move with speed c, so its possible

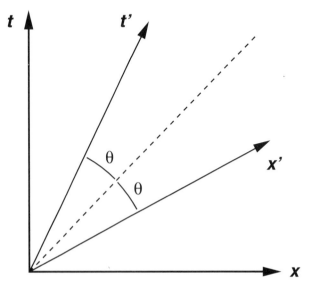

Figure 7.6 Relationship between the spacetime coordinates (x', t') of an inertial observer moving with respect to the spacetime diagram of another inertial observer (x, t). The angles θ that the t' and x' axes make with respect to the light cone are equal.

world lines are two lines at angles of 45° to the horizontal. If we were to plot more spatial dimensions, the possible world lines of the photon would lie on a cone, so this dividing surface is called the **light cone**, or sometimes the *null cone*.

The surface of the light cone divides spacetime into distinct regions for all observers. The region inside the cone corresponds to all events that are separated by timelike intervals from event A. The region outside the cone contains all those events separated from A by spacelike intervals. The half-cone below A is called the *past light cone*. The past light cone and the timelike region within it make up the **past** of event A. Similarly, the half-cone above A is called the *future light cone* of A, and this half-cone and the timelike region it encompasses compose the **future** of A. Events outside the light cone of A are in the **elsewhere** of A. Given two events, B within the past light cone of A, and C within its future light cone, it can be shown that *all* observers will agree that B is in the past of A and C is in its future, although they may not agree about where and when within the past and future, respectively, these events occur. Thus all observers agree that B occurs before C. On the other hand, for an event D in the elsewhere of A, observers may disagree on the order of events A and D; some may see that A occurs first, while others may observe that D happens first, and still others may regard the two events as simultaneous.

But if different observers do not agree on the ordering of events that are spacelike separated, are we in danger of losing the idea of cause and effect? Can we find a frame in which the lights go on before the switch is thrown? Most of us believe intuitively that a cause must always precede its effect. This has been formalized into the **principle of causality**, and it is one of the guiding principles of physics. It cannot be proven from any physical laws; it is, in some respects, one of the axioms of physics. But without it, we cannot make sense of the universe. Science is based on the belief that we *can* understand the universe; its success at this endeavor is ample demonstration of the power of its axioms.

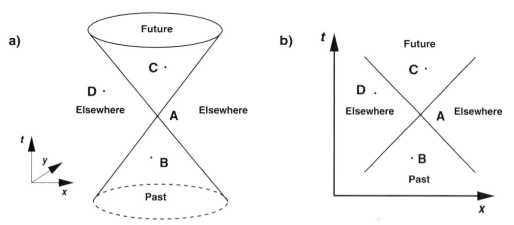

Figure 7.7   The light (null) cone seen (*a*) with two spatial and one time dimension, and (*b*) as a simple (*x, t*) plot. A light cone can be constructed at each event in a spacetime. The light cone divides spacetime into those events in the past (e.g., event B), in the future (e.g., event C), and elsewhere (e.g., event D) from A.

It is the principle of causality, ultimately, which asserts that no physical particle can travel faster than the speed of light. We already have ample evidence that there is something special about the speed of light. For one thing, the boost factor becomes infinite at that speed, and its reciprocal is zero; time dilation becomes infinite, while objects are Lorentz contracted to zero length. If we attempt a speed greater than that of light, then the boost factor becomes imaginary. Moreover, Einstein's famous equation, in its more precise form (eq. [7.11]), shows that the "relativistic mass" of a particle becomes infinite at the speed of light unless the rest mass of the particle is zero; this justifies our statement that photons are massless, since we know that photons have finite energy. (It also demonstrates, although it does not prove, the converse, that massless particles move at the speed of light.) These strange phenomena suggest that the speed of light sets the upper limit to speed in the observable universe. Yet it is causality that *requires* this to be so. The ordering of events is guaranteed only for *timelike* or *lightlike* spacetime separations. If information could travel faster than the speed of light, that is, along spacelike world lines, then the principle of causality could be violated; an effect could precede its cause in some frames. Events separated by a spacelike interval *cannot* be causally connected to one another. Therefore, no information, or any physical particle that could carry it, can travel faster than the speed of light.

The light cone affects even the nonrelativistic world; when you look up at night to see the first star of the evening, you are looking back along the past light cone to the event when the photons that form the image of the star on your retina left the surface of the star. Looking to a distant star is equivalent to looking backward in time. In fact, every image is a picture along the past light cone, for photons bring us the information by which we see. For the things of this Earth, the time delay is generally of no significance. But when we seek to study the contents of the universe, those objects which are at farther and farther distances are seen as they were at greater and greater times in the past. We can never form a picture of the universe as it is "now." In some respects, though, this is a benefit, for it means that as we look out through space we can see the history of the universe laid out before us.

The Twin Paradox

Our study of spacetime diagrams will make it much simpler to understand one of the most famous paradoxes of special relativity, the so-called "Twin Paradox." Andy and Betty are fraternal twins. Betty is chosen to go on the first mission to Alpha Centauri. She rides in a spaceship at nearly the speed of light, visits Alpha Centauri, then returns at nearly the speed of light. Andy stays on Earth and waits for her return. While Betty was traveling at relativistic speeds, her clocks, including her life processes, ran slow relative to Andy's frame; therefore, upon her return, she is younger than Andy. This seems like a straightforward solution, until we consider Betty's point of view. To Betty, Earth receded at nearly the speed of light and Alpha Centauri approached. Subsequently, Earth returned at nearly the speed of light. In her frame, her clock was always running normally. Since it was Andy who receded and approached at nearly the speed of light, *Andy* should be younger. Who is right?

The resolution lies with the realization that Betty started from rest, turned around at Alpha Centauri, and returned to a state of rest on Earth. Consequently, she must have accelerated at least 4 times on her trip. The accelerations mean that she did *not* remain in *one* inertial reference frame, whereas Andy did remain in his inertial frame. It is only *inertial* frames that can be equivalent, so there is no paradox here; the accelerated twin moved from inertial frame to inertial frame before returning to Andy's frame, which he never left. Betty, the traveling twin, is younger.

If Andy had decided, after Betty's departure, to join her, he would have had to hop onto the next mission in a spaceship fast enough to catch up to her. In this case, when he reached her ship, we would find that Andy would now be younger than Betty. If Andy leaves a given inertial reference frame, hence experiencing an acceleration, and then rejoins the same inertial frame, which necessarily requires another acceleration, Andy will find that his clocks show less elapsed time than do the clocks of those who never left the original inertial frame.

We can illustrate the Twin Paradox on a spacetime diagram. First we must recognize that there are *three* inertial reference frames relevant to this problem. The first is the frame of the stay-at-home twin, Andy. (We will ignore the motions of the Earth, since they are so tiny compared to the speed of light. We will also assume that Alpha Centari is at rest with respect to Earth, although this is not essential.) The second is the inertial reference frame of Betty while traveling to Alpha Centari, and the third is her inertial reference frame during her return voyage. Figure 7.8 illustrates the round trip in all three of the inertial frames. Betty departs for Alpha Centauri, traveling at constant, very high velocity, at event 1. She turns around at event 2, reversing her direction and returning at the same high, constant velocity. She arrives home at event 3. Notice that in all three reference frames, Andy's world line is straight, indicating that he remains inertial for the entire trip. In contrast, Betty's world line is not straight in all three diagrams, but changes direction at event 2.

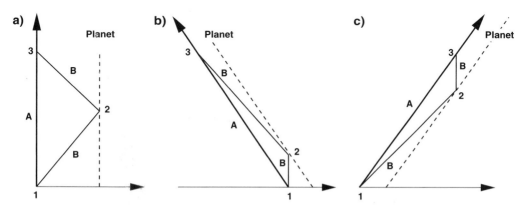

Figure 7.8 Three spacetime diagrams describing the twin paradox. Each diagram shows the point of view of a different inertial frame. (*a*) The twin paradox in the frame of the stay-at-home twin A. (*b*) The twin paradox in the inertial frame of B during her outward voyage. (*c*) The twin paradox in the frame of B during her return voyage. The essential asymmetry between A and B is that B does not remain in *one* inertial frame for the whole journey.

The elapsed proper time along any world line is obtained by summing the space-time intervals along that world line. Since the paths through spacetime differ, we should not be surprised that the proper times recorded by clocks traveling along different world lines between the same two events (Betty's leaving and returning) will be different. The *maximum* amount of proper time between *any* two events is that recorded by a clock that follows the straight line through spacetime between those two events; that is, the clock which remains in one inertial, constant-velocity frame. This means that in Minkowski spacetime, the *longest* time between any two events is a straight line! The fact that the straight line is a maximum, rather than a minimum, is another consequence of the negative sign in the spacetime interval, and another way in which relativity can confuse us. Euclidean space and Minkowskian spacetime have different properties.

If the maximum proper time is obtained by the inertial clock moving along the straight line in spacetime, which clock would show the minimum time? If you wish to record zero time between two events, there is only one way to do it: follow the light beam. A beam of light sent out into space and bounced back to Earth would follow a noninertial, yet still lightlike world line, and the spacetime interval along any lightlike path, accelerated or not, is always zero. Objects with nonzero mass, such as Betty, cannot travel at the speed of light, but Betty can minimize her proper time between two events by traveling as close as she can to the light cone. In all three frames of the Twin Paradox, only the world line of Andy is a straight line through spacetime. Betty travels from event 1 to event 3 by a noninertial route close to the light cone. Hence her clock reads less elapsed proper time than does Andy's, and the faster Betty travels, the smaller her elapsed proper time. There is no "Twin Paradox" once we understand this.

You might hear it said that the Twin Paradox requires general relativity for its resolution, but this is completely incorrect. It is a common misconception that special relativity cannot accommodate acceleration, that general relativity is required to deal with it. As we shall see in the next chapter, general relativity is a theory of gravitation and has nothing particular to say about acceleration per se that cannot be treated within special relativity. There is a formula for the relativistic acceleration, as well as a relativistic generalization of Newton's second law. We have not shown them because they are rather advanced, not often used, and not essential to our story here. But they exist. It is more difficult to carry out the calculation of proper time when acceleration occurs, almost always requiring calculus and some rather messy manipulations. But if we know the accelerations, we can do it. In our discussion above we simply assumed that Betty's intervals of acceleration during her journey were small, and we ignored the small corrections that would be obtained by integrating carefully along those curved paths in her world line. Regardless, we find that if Andy stays home, Betty is unequivocally the younger twin upon her return.

One More Paradox

Let us test our new-found understanding of special relativity with another example, illustrated in Figure 7.9. A succession of outposts is established in space. Each has a master clock, and all the space stations, and their clocks, are at rest with respect to one another. By a clever arrangement, we switch on all the clocks at the same time; that is, the events consisting of the starting of the clocks are simultaneous in the frame of the clocks. Thus the clocks will be synchronized. A spaceship approaches the line of clocks at close to the speed of light. When the spaceship passes station 1, its on-board clock and the master clock of the first station, clock 1, read the same time, say, 03:00, by a previous arrangement. The watchman on station 1 notes that the clock on the spaceship is running slow relative to his clock. By reciprocity, an officer on the spaceship will find that clock 1 is running slow. The spaceship passes the line of space stations, reaching Station 5 at 03:10, according to the shipboard clock. What does the master clock of station 5 read at this instant?

Begin by considering what the officer on the spaceship sees. To him, all five clocks aboard the space stations are moving and hence are running slow. Therefore, since the station clocks are all synchronized, clock 5 must read the same as clock 1, and as they are running slow, clock 5 must read a time less than 03:10 when the ship passes.

That sounds perfectly reasonable until we consider the point of view of the observers on the space stations. They see the spaceship rushing past them, and observe its on-board clock running slow in their frame. Since clock 5 reads the same as clock 1, clock 5 must read a time later than 03:10 as the ship passes. This too sounds fine, until we realize that it is exactly the opposite conclusion to that which we reached by reasoning from the point of view of the officer on the spaceship. What is wrong?

What is wrong is that not all the statements made about the spaceship's point of view can be correct. We asserted in both arguments that the clocks were synchronized,

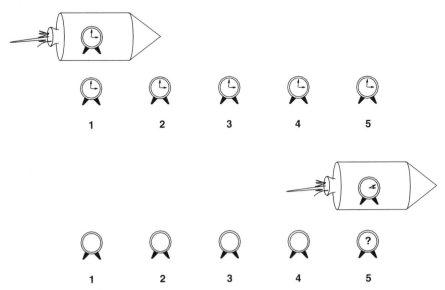

Figure 7.9   A relativity paradox. When the rocket ship passes the first clock, all the ground-based clocks are synchronized in the ground frame. The rocket ship's clock is identical with clock 1 as it passes. What does clock 5 read when the rocket ship passes it at a later time? Is it ahead of or behind the rocket's clock?

but that is a statement about simultaneity, which is relative. The station clocks are synchronized in their own rest frame, *not* in the frame of the ship. The five spacetime events that correspond to the moment when the station clocks read 03:00 do not occur simultaneously in the frame of the ship. In the ship's frame, clock 5 reads 03:00 at an earlier time than clock 4 reads that time, which in turn reads 03:00 earlier than does clock 3, and so on.. As the spaceship passes each clock, it sees that clock running slow, but they were never synchronized in its frame. In the ship's frame, clock 5 was ahead of clock 1, so when the spaceship passes this last station, clock 5 is still ahead of the spaceship's clock, and therefore it reads a time later than 03:10. This is the careful explanation. The second argument above reached the correct conclusion because it referred only to observations made in the frame in which the station clocks were, in fact, synchronized. The first argument reached an incorrect conclusion because it attempted to make use of a synchronicity that did not exist in the frame where it was employed.

Some Misconceptions

It is sometimes believed that clocks seem to run slow due to the fact that the light from a receding clock takes time to catch up with the observer, so that time dilation is just an effect of the travel time of light. But by that reasoning, should a clock not run fast as it approaches? It is true that what we actually "see," in terms of light striking our eyes, will be influenced by light-propagation effects, but this is a separate issue. It is important to realize that relativistic effects such as time dilation and Lorentz contraction do *not* arise because of any failure to take into account the finite travel time of the light. All observers know the speed of light and are able to make use of this information in computing what the clocks in other frames are reading. If the spaceship officer from the previous example continued to watch clock 1, which might be a radio signal, while passing the other space stations, he would see clock 1 running slow as it receded. But this has nothing to do with the fact that the light must travel farther and farther to reach the ship. The shipboard observer has a clock and knows how fast clock 1 is moving relative to him, and he can take into account the travel time of light when he computes the reading of clock 1. The clock still runs slow relative to the shipboard clock. Similarly, when we synchronized the clocks in the space stations' inertial reference frame, we implicitly assumed that the speed of light was accommodated in our startup apparatus.

It is also widely believed that it might be possible to find a way to travel faster than the speed of light if only we would try hard enough, that scientists who say it cannot be done are fogies who cannot imagine new technologies. There are many reasons why it might be said that something cannot be done. One reason is simple ignorance. An example of this is the famous editorial of 1920 in *The New York Times*, which asserted that Robert Goddard's rockets could not possibly operate in a vacuum, since, according to the writer, "Professor Goddard ... does not know the relation of action to reaction, and of the need to have something better than a vacuum against which to react—to say that would be absurd." Using Newton's laws of motion, you should be able to spot the flaw in this argument. Of course, the "reaction" is against the *rocket*, or the jet airplane, for that matter.[8]

Another reason that something cannot be done is inadequate technology. For example, until the 1940s, many people believed that the speed of sound could not be exceeded. They based this belief on the fear that the large stresses induced by supersonic travel could not be withstood by any material. It is correct that supersonic speeds create severe stresses and heating; supersonic aircraft must be specially designed and are built with unusual materials. (The fastest acknowledged airplane in the world, the SR-71 Blackbird, has a hull of titanium, an exceptionally strong metal.) But successful designs were created, and now it is even possible for the everyday wealthy and famous, not just military fighter pilots, to break the "sound barrier" in supersonic commercial jetliners. It was always recognized, however, that the sound barrier was an engineering problem, not a fundamental limit. The speed of light is intrinsically different; it is not a mere technological challenge. It is a *fundamental* part of the way the universe works.

The special theory of relativity has been tested repeatedly. We have accelerated elementary particles to near the speed of light in large accelerators, and the relativistic effects we have discussed have been observed and measured. Special relativity has been subjected to some of the most exacting experiments ever performed, and it has in every case been found to give an accurate description of the observations. Denying relativity would also deny the validity of Maxwell's equations, and they are amply confirmed every time you turn on a radio. The absolute limit of the speed of light is as much a part of special relativity as is the Lorentz contraction. Denying this barrier would repudiate one of the most successful theories of modern physics.

The theory of relativity has profoundly altered the way in which we view the universe. It has merged concepts previously thought to be unrelated; space and time become spacetime. Matter and energy are united into mass-energy. In special relativity, electromagnetism becomes consistent with mechanics. Special relativity also shows that electric and magnetic fields are essentially the same phenomenon. There is no ether, and hence no special frame at rest. The first postulate of relativity constrains all theories of physics, since all natural laws must be the same in any inertial frame. On the level of philosophy, the special theory of relativity eliminates the last vestige of Earth, or humanity, as a privileged observer, since it denies the existence of *any* preferred inertial frame. If special relativity seems to defy common sense and intuition, it simply means that the universe is more than our limited human awareness perceives it to be.

KEY TERMS

Lorentz transformation	Lorentz contraction	relativity principle
event	simultaneity	time dilation
boost factor	principle of reciprocity	proper time
proper length	rest energy	spacetime
spacetime diagram	world line	spacetime interval
Minkowski spacetime	timelike	spacelike
null	lightlike	light cone
past	future	elsewhere
principle of causality		

Review Questions:

1. How does the transition from Newtonian to relativistic mechanics illustrate one or more of the five criteria for a scientific theory?

2. A man watches a football game from the window of a train moving with constant velocity parallel to the football field. Consider the situation both from Galilean and Lorentzian points of view. Does the observer in the train measure the football field to be longer or shorter than the usual 100 yards? Is the length of the period as measured by the man in the train longer or shorter than the usual 15 minutes?

3. If it is impossible to exceed the speed of light, why is it nevertheless possible to get to Alpha Centari, a distance of 4 lt-yr away, in less than 4 years of time, as measured by a space traveler?

4. A relativistic train approaches you at a speed of 0.9c. What is its boost factor? If the train has a length of 50 m in its rest frame, what is its length in your frame? If you were the dispatcher of the relativistic train line, you would have to keep track of the clocks on the train as well as your own clock. If your station clock records an interval of 10,000 s between two given events, how many seconds does the conductor's on-board clock measure between the same two events?

5. In what way can time be called the fourth dimension? How does it differ from the other three?

6. Two spaceships (A and B) approach your spaceship at 9/10 the speed of light (0.9c) from opposite directions. They send out radio messages. What is the speed you measure for the radio waves from A and B? What is the speed of the radio waves from A as measured by B? What speed does B measure for your motion? How fast does spaceship B observe spaceship A to be moving?

7. Consider the situation as depicted in question 6. Whose clock (his, yours, or A's) does B observe to run slowest? Whose clock does A observe to run slowest? If all three ships have the same length when measured in their own rest frames, which ship does B observe to be shortest? If spaceship A sends out a pulse of red light, will B see it to be blueshifted or redshifted?

8. Draw a spacetime diagram. Put two events on it, labeled A and B. Indicate their separations in space and time, that is, their Δx and Δt. Draw the world line of a stationary observer. Draw the world line of an observer moving at constant nonzero velocity. Draw the world line of an accelerating observer. Draw another spacetime diagram and include a moving inertial observer. Draw the space and time axes that correspond to the moving observer's "at-rest" frame and label these axes x' and t'.

9. In the accompanying spacetime diagram, which pairs of events may be causally connected? Which cannot be causally connected?

10. Using $E = mc^2$, we found that 1 kg of mass has the energy equivalent of 20 Megatons. Estimate how many Megatons of energy would be required to accelerate a spaceship with a mass of 1 million kg to a speed of 0.99c. What does this suggest to you regarding the practicality of space travel at relativistic speeds?

11. Explain why we are unaware of the effects of special relativity in our everyday lives.

12. Describe two quantities considered invariant in Newtonian physics that are relative in special relativity. Describe two new quantities that are now known to be invariant.

13. (More challenging.) Suppose that a train robber decides to stop a train inside a tunnel. The proper length of the train is 60 m, while the proper length of the tunnel is 50 m.

Figure 7.10   (Exercise 7.9)

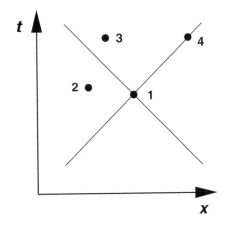

The train is traveling at 4/5 the speed of light. According to proper lengths, the train would not fit inside the tunnel, but the robber plans to use relativity to his advantage. The length of the moving train in the rest frame of the tunnel, and of the robber, is 36 m. The robber computes this and decides to trap the train inside the tunnel, since, in his frame, the train should fit. From the point of view of the train's engineer, however, the *tunnel* is only 30 m long, just half the length of the train. The engineer knows that his 60 m train will not fit completely into the tunnel. The robber thinks that the train will fit, whereas the engineer is sure it will not. But either the train will fit, or it will not; it cannot do both. Who is correct? *Hints.* Consider the following events: the locomotive enters the tunnel. The locomotive reaches the end of the tunnel. The caboose enters the tunnel. The caboose reaches the end of the tunnel. Which events are necessarily causally connected? Which are not? Draw a spacetime diagram, and label these four events. In which order do these four events occur in the robber's frame? In the train's frame?

Notes

1. The word "ether" survives to this day in colloquial use. References to radio and TV signals moving through the ether are still common, and the "Ethernet" links computers.

2. The term "thought experiment" can lead to confusion among those unfamiliar with the concept. A thought experiment refers to the deductive process of predicting the outcome of a specific experiment using the general principles of (in this case) special relativity. While these experiments may be difficult to carry out in practice, there is nothing in principle that prevents doing so. In fact, many such relativity experiments have been performed.

3. Note that the constancy of the speed of light is the crucial assumption in special relativity. If Galilean relativity were applicable, the velocity of the light along the trajectory *d* would be the vector sum of its vertical and horizontal velocity components, and its speed would be the magnitude of whatever vector was thus obtained. But the second relativity postulate requires that the speed of light be the same for all observers.

4. Atomic clocks, the only timepieces capable of measuring extremely tiny time differences, have been placed in airplanes and flown around the world. The cumulative effect due to time dilation, relative to an identical stay-at-home atomic clock, has been measured and found to agree with the predictions of special and general relativity.

5. For good reason, the writers of the television show *Star Trek* chose matter-antimatter engines to power their starship. With 100% conversion of mass to energy, they are the most efficient engines possible. One practical problem with such engines is obtaining fuel. As far as we know, the universe is composed almost entirely of ordinary matter.

6. If you have studied calculus, you will remember that the distance along a curve is found by integrating the differential arc lengths from the beginning to the endpoint. The spacetime interval corresponds to the arc length in spacetime.

7. There has been some speculation about whether or not particles could exist that must always travel *faster* than the speed of light. Such particles are called *tachyons*, and if they existed, they would have very strange properties indeed.

8. Jet airplanes do depend upon the presence of air for steering; a rocket ship cannot "bank" against a vacuum, despite what you might have seen in movies. But both jet aircraft and rockets obtain their propulsion from the reaction of their gas exhausts against them.

CHAPTER 8

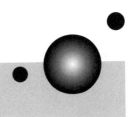

The General Theory of Relativity

The Need for a General Theory

Einstein's conviction that the universe obeyed the relativity principle led him not only to revise Newton's mathematical equations of mechanics, but even more drastically, to discard the concept of absolute space and time. It proved unnecessary to alter Maxwell's equations of electromagnetism, since they already obeyed the correct invariance law; it was their lack of Galilean invariance that had motivated Einstein in the first place. Special relativity thus brought mechanics and electromagnetics into full consistency.

But what about Newton's other great contribution, his law of universal gravitation? Special relativity seems incapable of dealing with gravity. The special theory describes the relationship between measurements in inertial frames. In our discussion so far, we have been tacitly assuming that our laboratories on Earth constitute inertial frames. But this can be true only in approximation; we constantly experience a gravitational force, yet an object experiencing a force is accelerated and hence cannot reside in an inertial frame. Moreover, there is no way to shield ourselves from Earth's gravity, or of any other massive object. Any object with mass will produce a gravitational force, in accordance with Newton's law. The universe is filled with masses, and the gravitational force extends indefinitely, so in principle, there is no point in the universe where gravity's influence does not reach; gravity is the dominant force in the universe. Newton's law of gravitation also requires that the magnitude of the force depend on the square of the distance between the masses. Yet we have just learned that distance is not absolute but relative; how, then, can we accommodate an inverse-square law? Which distance should we use? Does the force vary according to the frame of reference? Finally, Newton's law implies that gravitational force is felt instantaneously at a distance. But we now know that this cannot be, for nothing can propagate faster than the speed of light. Somehow time, and a finite propagation speed, must be incorporated into the gravitational force law.

In the light of all these considerations, how can we incorporate gravity into the theory of relativity? We need a more *general* theory which will accommodate all frames, both inertial and noninertial, and which can describe the effects of gravity. This broader theory is Einstein's general theory of relativity. And just as special relativity had surprising, even astonishing, consequences, so we will find remarkable consequences of general relativity that will forever change our view of the cosmos.

Some examples can help us begin our investigation of general relativity. In this space age, most of us have seen films taken aboard the *Space Shuttle* and its predecessors, enabling us to visualize experiences in which gravity seems to be absent. In particular, consider the *Space Shuttle* approaching a malfunctioning satellite 400 km above the Earth. A spacewalking astronaut attempts to snare the satellite, but the slightest touch sets it spinning. Finally, the 75 kg astronaut catches the satellite, which on the surface of the Earth would have many times the weight of the astronaut; yet in orbit, he handles it as if it were made of foam. Back inside the crew quarters, the crew members float about the compartment as if they were filled with helium. The television commentator says that this all occurs because of the weightlessness of outer space. But why is "outer space" weightless? Is there "no gravity" in outer space? No, although this is a common misconception.[1] Newton's law of universal gravitation states that the gravitational force is inversely proportional to the square of the distance from the center of one body to the center of the other. The radius of the Earth is approximately 6500 km. Since 400 km is only about 6% of this number, the force of gravity at the altitude of the Space Shuttle still has fully 88.7% of its value at the surface of the Earth. If Earth's gravity is still present, why are the astronauts weightless in space?

The answer lies in the realization that the concept of an inertial frame, and the effects of gravity, are intimately linked. Think back to the "Rotor" ride at the amusement park. As the cylinder spins faster and faster, your body is pressed against the wall of the tube. You feel very heavy. When you reach some dizzying angular speed, the floor of the Rotor suddenly drops away, and you are left hanging on the wall. Somehow the gravity you are experiencing has changed. A similar, more serious, device is the centrifuge used in astronaut training. It consists of a small car attached by a metal arm to a central hub. A motor in the hub drives the arm and the car in a circular rotation. If you could ride in this car, your body would feel as if it were very, very heavy. Your weight would seem to increase as the car spun faster around the central hub. Soon you would find that you could scarcely raise your arm at all. What forces are acting here? Newton's first law requires that a force act in order for circular motion to occur; without such a force, an object will move in a straight line. That is why the car must be attached to the hub by a heavy metal arm; otherwise it would fly off in a straight line, with unhappy consequences for you. The "real" force you are experiencing is called *centripetal force*; it is the force that acts toward the central hub and causes your car to execute circular motion. In your rotating frame of reference, however, you experience *centrifugal force*, which acts away from the center. Physicists often refer to an inertial force such as centrifugal as a *fictitious force*, because it is an artifact of the rotation of your frame of reference. But how can a "fictitious" force make you dizzy, or seem to increase your weight unbearably, or pin you to the side of a metal cylinder?

All these phenomena are connected to frames of reference. Fictitious, or inertial, forces occur when an observer is in an accelerated, or noninertial, frame of reference.

Nonaccelerated, inertial frames do not experience these forces. We have learned that the special theory of relativity relates observations made in inertial frames to one another, and because inertial frames are special, we call it the special theory. But what makes inertial frames special? You know they are special by experience. When you change from one inertial frame to another, you feel an acceleration that has real, palpable physical consequences. (It is not the fall from a high building that would kill you, it is the sudden deceleration at the pavement!) We are not asking here how to treat acceleration mathematically or to account for its effects. Special relativity is perfectly capable of dealing with acceleration per se. However, special relativity presupposes the existence of inertial frames. It accepts Newton's first law as valid and defines an inertial frame as one in which that law holds; that is, any free particle executes strictly uniform motion. We now need to know what determines an inertial frame of reference in the first place, and what creates the accelerations we feel when we are not in an inertial frame.

The Equivalence Principle

The first clue in our development of general relativity can be found in our contemplation of the weightlessness of astronauts. We may think of the phenomenon of weightlessness as some sort of "antigravity" effect, but what it really represents is a good inertial frame. When in orbit, the space shuttle is falling around the Earth in a state of **free fall**. The shuttle, the astronauts, their equipment, and the target satellite, are all falling together. As was demonstrated by Galileo, all objects, regardless of their mass, fall at the same rate in a gravitational field. When a body is freely falling it is weightless, and hence in the state of free fall it *feels* as though gravity has been cancelled. This simple idea will be developed into one of the fundamental principles of the general theory of relativity. Like the special theory, the general theory is derived from only a few simple, powerful postulates, and this first basic principle, which we will discuss in some detail, is called the *equivalence principle*.

From Newton, we learned that a *force* exists whenever a body is accelerated, and that the constant of proportionality between the force and the acceleration is the *mass* of the body. Symbolically, this is the famous equation $F = ma$. The m in this equation is the inertial mass, and it measures the resistance of an object to being accelerated. Even in orbit, in a state of weightlessness, it is still necessary to contend with an object's inertia; it remains difficult to push around massive objects. Newton also gave us the law of universal gravitation, which tells us that the gravitational force on a body is proportional to its mass. Experiment has shown that the inertial and gravitational masses are equivalent. But *why* should these two masses be the same? After all, the electric force is dependent on electrical charge, a quantity which is unrelated to the inertial mass of the charged object. Why should the gravitational force not depend on some special "gravitational charge," which we might call the gravitational mass m_g, rather than on the inertial mass? There is no *a priori* reason why gravitational force should have any connection whatsoever with inertial forces. Yet experiment has clearly shown that the acceleration due to the force of gravity acting upon some body is *independent* of the mass of that body; this could be true only if inertial and gravitational mass were equal. In the absence of any other forces, such as air resistance, *all bodies fall with the same acceleration* under the influence of gravity.

In practical terms, this means that gravitational and inertial forces produce effects which are indistinguishable. While standing on the floor of your kitchen, you drop something, and it falls. This does not surprise you. Now imagine that you are traveling in a spaceship far from any source of gravity, between the distant stars. The spaceship's main rocket engine is engaged, and the spaceship is accelerating at one *g*, the same acceleration as due to gravity on the surface of the Earth. If you drop something, it will fall against the direction of the acceleration of the spacecraft. Is that what you would have expected? Sitting in a chair in your living room can be just like sitting in a chair in a rocket ship whose engine is operating.

Now suppose your spacecraft assumes an orbit around some planet. You shut off the rocket engine. Remember that an *orbit* is a state of perpetual free fall around another body. No power is required to maintain it, provided that no energy is lost due to some deceleration, such as friction from the tenuous outer edge of the planet's atmosphere. The Moon orbits the Earth because it is accelerated toward the Earth, in accordance with Newton's law, but since the Earth is curved and finite and the Moon has some tangential motion, the Moon never approaches the Earth's surface. Thus the Moon is constantly falling. In an orbiting, freely falling frame, the inertial forces such as "centrifugal force" exactly cancel the gravitational force. *This* explains why

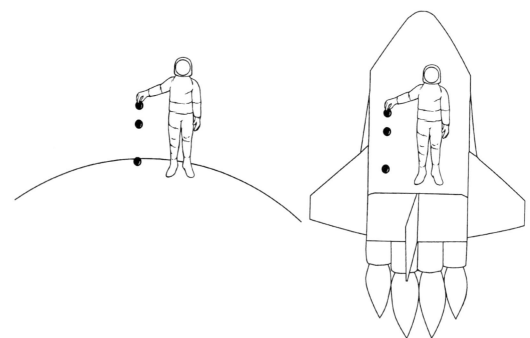

Figure 8.1 The weak equivalence principle states that gravity is indistinguishable from any other form of acceleration. The experience of a person in a spaceship accelerating with one *g* is the same as that of a person standing on a planet with gravitational acceleration *g*.

the astronauts aboard the Shuttle are weightless while in orbit; it is *not* due to any "lack" of gravity in space. Gravity is "cancelled" by free fall. This also explains the effect of the centrifugal forces felt in the Rotor ride or in the centrifuge. Those forces pull on you as if you were on a planet with a much stronger gravitational field. You cannot distinguish the inertial force from a gravitational force.

The assertion that gravity is completely indistinguishable from (or equivalent to) any other acceleration is called the **weak** (or Newtonian) **equivalence principle**. It is not necessary to go into orbit to find examples of the weak equivalence principle at work. For instance, suppose you enter an elevator to travel upward. The elevator starts and accelerates toward the top of the building, but you feel yourself pushed down against the floor. The relevant force is the push of the floor upward on you, by which the elevator compels you to share its acceleration. In your frame, however, it is you that are pushing down on the floor, and this force is indistinguishable from a gravitational force toward the floor. You feel "heavier" than normal. When you return to ground level via the elevator, the elevator begins to accelerate downward. You feel lighter on your feet and experience a fluttery feeling in your stomach, as though your viscera are floating. Can you explain why you feel these sensations? As the elevator starts downward, its acceleration is toward the ground. Therefore, the elevator floor is falling away from your feet and pushes up on you with *less* force than when the elevator is stopped. Thus, you press down on the floor with less force than you would if the elevator were at rest, and you experience this as a gravitational force that is less than the usual force due to the Earth. Notice that these effects occur only when the elevator is accelerating. After it has reached its constant operating speed, your weight will feel completely normal.

What if the cable were to break? You and the elevator would go into free fall down the shaft. The elevator and all its contents, including you, would then fall with acceleration g. Therefore, the floor of the elevator would exert *no* force upon you, nor would you exert a reaction force upon it. You would become weightless, *not* because gravity has suddenly been switched off, but because in your "elevator" frame, your apparent weight is the force you exert against the floor, and in free fall that force is zero.[2]

It is desirable to expose astronauts to free fall during their training, before they go into space. Rather than load them into an elevator and cut the cable, NASA flies them in an airplane. For a portion of the flight, the airplane follows a parabolic trajectory that mimics the path of a free-falling object, such as that of a body thrown into the air. For a short but significant interval of time, the astronauts experience free fall and weightlessness.[3] You may have had a similar experience if you have ever been aboard an airplane that encountered a severe stretch of turbulence and suddenly lost altitude. When the dinner trays and coffee cups start floating around the cabin, that is a good clue that the plane is momentarily in free fall.

You might find it surprising that an acceleration *upward*, "against" gravity, makes you feel heavier. If gravity and acceleration are indistinguishable, then should not a downward acceleration feel like gravity? This counterintuitive effect stems from the fact that we all spend our lives in a noninertial frame, one tied to the surface of the Earth, and consequently we have difficulty in visualizing inertial frames. We are able to employ Newton's laws in our noninertial frame only by explicitly including the effects of gravity. When we drop a ball, we claim that it accelerates "downward"; that is, we are adopting the point of view that our Earth-based frame is inertial, in the sense of Newton's *second* law. We have now learned that our home frame is not an inertial frame and would not be so even if it were to stop rotating, because it sits upon a large, gravitating mass. When we drop a ball, it occupies an inertial frame (temporarily, until it collides with the surface of the local gravity source) and is actually *not* accelerating. Thus it is *we* who are accelerated; and if a ball falling downward is not accelerated, then we must be accelerated *upward*. Hence what we call gravity is equivalent to an upward acceleration, as seen from the freely falling frame. This is a subtle and perhaps difficult point, but important to a full understanding of the meaning of inertial frames in general relativity.

But what if you are sitting at rest in your living room? You will agree, no doubt, that you constantly experience gravity. But if you are sitting motionless, then how can you be accelerated? And if you are in free fall, are you not accelerated, with acceleration g, toward the center of the Earth? How can we reconcile our usual view of gravity as an acceleration with the claim that freely falling observers are unaccelerated? Perhaps the Rotor will again help to clarify these issues. To your friend watching you from an inertial frame, you are most certainly accelerated, else you would not be executing circular motion. Your friend watching from overhead says that you are experiencing a centripetal force, which is provided by the wall of the metal cylinder. In your own frame, however, you are motionless; there is no centripetal force (toward the center), but rather a centrifugal force (away from the center), which exactly balances the reaction force from the wall. In your frame you are not moving, so you claim that you are unaccelerated. There is just some force ("gravity") pinning you to the wall of the Rotor. Gravity is just like the "fictitious forces," for example, centrifugal and Coriolis forces, which we have previously identified as artifacts of a noninertial frame of reference. A freely falling observer is truly unaccelerated; it is you who are accelerating relative to the inertial frame. Yet just as we may still apply Newton's second law within a rotating frame, such as the Rotor or the Earth, provided that we introduce the "fictitious forces" to account for our noninertial motion, so may we introduce the "fictitious force" we call gravity, and continue to make use of Newton's second law for the conditions prevailing within our noninertial frame.

It may sound as though we have just arrived at the conclusion that gravity does not exist! However, this is not the case. What we have found is that gravity has no separate existence but is related to the concepts of inertial frame and acceleration, which in turn are fundamentally tied to the nature of space and time. General relativity incorporates all these separate ideas into a unified picture.

Our discussion of the equivalence principle has so far been somewhat theoretical. Newton *assumed* the equivalence of gravitational and inertial mass, based on somewhat sketchy evidence and his intuitive sense of aesthetics. Yet this assertion can be tested. The earliest such experiments were performed long before the equivalence principle was formulated. Galileo discovered that bodies fall at a rate independent of their inertial mass, which motivated Newton to set the two forms of mass equal in the first place. Newton himself carried out experiments on pendula to test this hypothesis. He found no change in the period of pendula whose bobs were made of different substances, but were otherwise identical; of course, his experimental errors were large. The first highly accurate experiment to test the equivalence principle was performed in 1889 by Baron Roland von Eötvös. Eötvös constructed a device called a *torsion balance*. He suspended two bodies, of nearly equal mass but different composition, from a beam which hung from a very fine wire precisely at its center. If the magnitudes of the Coriolis force (from Earth's rotation) and the gravitational force had differed between the bodies due to their differing composition, Eötvös would have detected a twisting of the wire. None was seen, and Eötvös was able to conclude that inertial and gravitational mass were equal, to approximately one part in 10^9. In the twentieth century, Robert Dicke and others have pushed the limit of such an experiment to the level of one part in 10^{11}, but the Baron's results were sufficient to convince many, including Einstein, that inertial and gravitational mass are equivalent.

We have so far confined our discussion to mechanics, the physics of motion. But the mechanical equivalence of inertial frames with free-falling frames hints at something deeper, namely the **strong** (or Einstein) **equivalence principle**. This principle is similar to the first postulate of special relativity in its sweeping generality. The strong equivalence principle states that *all* inertial and freely falling frames are completely equivalent, and there is *no* experiment that can distinguish them. Where the weak equivalence principle addresses only mechanics, the strong equivalence principle speaks of *all* of physics. In particular, it makes the relativity principle, and thus special relativity, applicable in freely falling laboratories. The strong equivalence principle is fundamental to general relativity; henceforth the expression "equivalence principle" shall refer only to the Einsteinian principle.

The strong equivalence principle has profound consequences, stemming from the fact that it is genuinely impossible to distinguish an inertial frame from a free-falling frame. A few thought experiments will make this clear. Imagine a beam of light shining from one side of an elevator to the opposite side, such as from a flashlight in your hand. If this elevator were in deep space, in a good inertial frame, far from any gravitational field, then an observer inside the elevator would see the beam trace a straight line across the elevator. Now consider the same situation in an elevator that is freely falling in a gravitational field. By the strong equivalence principle, we must observe exactly the same result as before: the beam passes straight across

Figure 8.2 Torsion balance, such as used by Baron Eötvös to test the equivalence principle. The spheres have identical mass, but are made of different substances; Eötvös used wood and platinum. The spheres experience a gravitational force and an inertial force, the Coriolis force due to the rotating earth. If the ratio of the inertial to gravitational mass in the two spheres were not the same, there would be a net twist on the central wire.

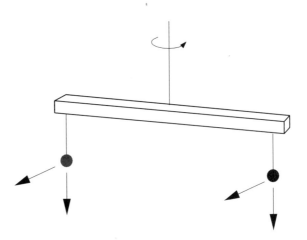

the elevator. But for the elevator in a gravitational field, this result tells us that the light must be falling along with all the other contents of the elevator. Otherwise, the elevator would fall some distance while the light was traversing it, and the beam would appear to the elevator-based observer to bend upward! This implies immediately that the gravitational field must have forced the light to travel on a curved path, relative to the distant fixed stars.

Thus the strong equivalence principle leads us to the conclusion that light falls in a gravitational field. What does this mean for the theory of relativity? In the Minkowski spacetime of special relativity, light always travels on straight, lightlike lines through spacetime. Now we find that the equivalence principle demands that these trajectories must curve in the presence of gravity. Hence the presence of gravitating matter affects spacetime itself, changing inertial world lines from the straight lines of special relativity to curved lines. The effects of gravity, then, can be incorporated into the theory of relativity by allowing spacetime to curve. This is the fundamental basis of general relativity.

The elevator thought experiment can lead us to still more interesting results. Suppose a rocket ship in deep space accelerates forward. A bulb at the nose of the rocket emits a beam of light that is observed by a receiver on the rocket's floor. Because of

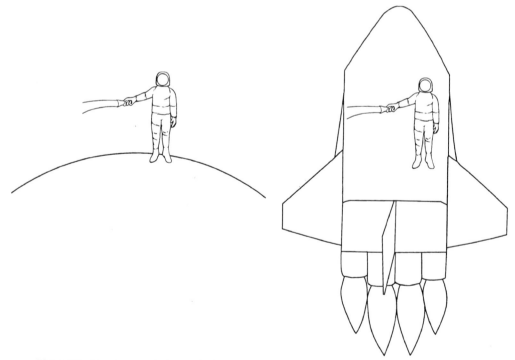

Figure 8.3 The strong equivalence principle asserts that all laws of physics are identical in any inertial frame, including the laws governing the behaviour of light. A person in a spaceship accelerating with one g will see the light beam from the flashlight curve downward. A person standing on a planet with gravitational acceleration g will see exactly the same curve in the light beam.

the acceleration, the receiver's velocity will increase between the time of the emission of the light and its reception. This results in a relative motion between the receiver and the bulb at the moment of light emission, producing a blueshift in the light. According to the equivalence principle, the ship's forward acceleration is indistinguishable from a gravitational field directed toward its floor. The scenario is completely equivalent to an observer on the surface of the Earth receiving light from a bulb at a higher elevation. We have again shown something quite remarkable using only the strong equivalence principle: light traveling downward in a gravitational field, that is, toward the source of the field, gains energy. Conversely, by a very similar argument, light traveling upward, "against" a gravitational field, is redshifted, meaning that it loses energy. The loss of energy experienced by a light wave as it climbs against a gravitational field is called the **gravitational redshift**. Similarly, if light falls down toward an observer deep within a gravitational field, that observer will see a gravitational blueshift.

Next we return to the centrifuge for another thought experiment. Imagine that you sit at the central hub, but have placed a clock on the end of the arm. You start the centrifuge and watch the clock spin around. As the centrifuge speeds up, the clock whirls faster and faster and experiences, in its own frame, greater and greater centrifugal acceleration. Since the clock is moving at a high rate of speed relative to your (nearly) inertial frame, by special relativity, you will see the clock run slow. However, the clock is not in an inertial frame; it is accelerated. By the equivalence principle, this is the same as sitting in a gravitational field, such as on the surface of a planet. From the equivalence principle, we thus conclude that a clock in a gravitational field runs slow relative to an observer in an outside inertial frame; we have discovered gravitational time dilation. We can similarly use the equivalence principle to show that there is a purely gravitational length contraction as well.

Not only the clock, but all physical processes, run slow in the presence of gravity, so we cannot detect the effect by looking at our own clock. We must compare the ticking of clocks at two *different* points in a gravitational field. There is no reciprocity; given two observers with clocks, one deep in a gravitational field and the other high in orbit where the field is weaker, then the clock at the lower point (stronger gravity) runs slower than the orbiting clock (weaker gravity), according to *both* observers. They disagree only in their judgment as to whose clock is running at the "right" speed, since both think their own is correct. The observer deep in the gravitational field observes that the clock of the observer in orbit is running fast. Today's atomic clocks are sufficiently accurate that they can detect gravitational time dilation effects even here on Earth. The atomic clock kept by the National Institute for Standards and Technology at Boulder, Colorado, at an altitude of approximately 1600 m, runs faster than the similar atomic clock near sea level, close to Washington, D.C. The effect is small but detectable and must be taken into account for high-precision measurements.

Two Viewpoints on the Nature of Space

The equivalence principle has led us to some interesting, perhaps even startling, conclusions about physics in a gravitational field. The equivalence between free fall and inertial motion, and between gravity and acceleration, suggests that somehow

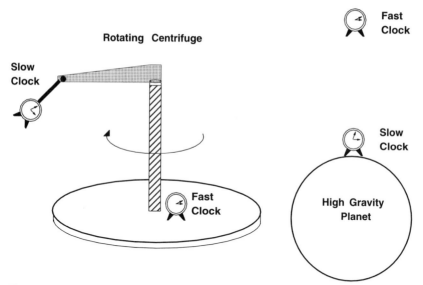

Figure 8.4 Time dilation from the equivalence principle. From special relativity we know that the moving clock on the centrifuge will run slower than the clock at rest. The clock on the centrifuge is not in an inertial frame; it experiences a centrifugal force. We conclude that inertial forces make clocks run slow. Gravity is equal to acceleration by the equivalence principle, so a clock experiencing gravitational acceleration will run slow compared to an unaccelerated clock.

gravitating mass *defines* inertial motion and acceleration, a significant departure from Newton's concepts. Let us examine two contrasting viewpoints regarding space and time.

In Newton's universe, acceleration is defined with respect to a space and time that exist absolutely, in their own right, independent of the existence of matter. They are simply laid down throughout the universe, and all action takes place with reference to them. They are unaltered and unaffected by any of the universe's events. Even special relativity, which blends space and time into spacetime, does not question the existence of an absolute spacetime. Accelerations occur with respect to absolute space or spacetime. In the twin paradox, one twin feels an acceleration, the other does not, and this situation is not symmetric.

Newton's contemporary (and rival), Gottfried Leibniz, first disputed this viewpoint, arguing that space cannot exist apart from matter. Admittedly, it is difficult to imagine what absolute space would be in the absence of matter, or how we would measure distances without some sort of yardstick. Furthermore, absolute space postulates the existence of something that affects everything, but is itself affected by nothing. Although philosophers quarreled over this for two centuries, scientists overwhelmingly adopted Newton's viewpoint. Newton's mechanics *worked*, after all.

The first scientist to systematize the alternative worldview and place it into the context of physics, rather than metaphysics, was the German physicist and philosopher Ernst Mach. The statement that inertial frames are established only by the distribution of matter in the universe has come to be known as **Mach's principle**. Mach insisted that absolute space made no sense. To him, space, and hence inertial frames, are meaningful only in relation to the distribution of matter in the universe. Where there is nothing, one cannot define motion, much less acceleration.

To illustrate this, consider an everyday example. Suspend a bucket of water from a rope, and give it a spin. Most of us have spun a bucket of water at one time or another, and we know that when the water starts to spin with the bucket, its surface curves, rising upward toward the rim of the bucket. Newton himself performed this experiment. The curving of the water is due to the centrifugal force, that is, to an inertial force resulting from the water's noninertial (rotating) frame. Now imagine a universe that is empty except for Newton's bucket of water, and repeat the experiment. If the universe contains nothing but this bucket of water, how does the water "know" that it is rotating? Rotating with respect to what? Would the water's surface still curve? Newton would answer that of course it would, since the bucket knows that it is rotating with respect to absolute space. Mach disagreed. To Mach, motion was inconceivable except in relation to other matter. The relative motions of matter determined acceleration. If there is nothing but the bucket, it cannot rotate with respect to anything, and rotation therefore has no meaning or significance.

The Foucault pendulum provides another example. If we place a Foucault pendulum at the North Pole, it will swing in one plane with respect to the fixed stars, while the Earth turns underneath it. Newton associated the realm of the fixed stars with absolute space; hence, in the Newtonian view, the pendulum is moving inertially, with respect to absolute space. Now suppose that the "fixed stars" disappeared, leaving a universe consisting only of the Earth and the pendulum. How would the Foucault pendulum move then? According to Newton, the pendulum would be unaffected because the "fixed stars" merely serve as convenient markers in absolute space. The Earth would continue to turn underneath the pendulum as it swung. Mach, however, would claim that the pendulum would now swing in a plane fixed on the Earth, since the Earth would be the only other matter present. In the absence of the remainder of the universe, there is nothing to define rotation for the Earth, and inertial motion for the pendulum would be motion in a constant plane with respect to the only other matter in the universe, namely, the Earth.

It may seem hopelessly unrealistic to contemplate the consequences of thought experiments involving an empty universe, but if Mach is correct, there are observable effects even in our matter-filled universe. For example, the Foucault pendulum's motion, while dominated by the matter in the rest of the universe, must still show some influence from the nearby presence of the Earth. The rotating Earth must have *some* say in what constitutes a local inertial frame, so the local inertial frame must share in the Earth's rotation to some slight degree. This constitutes a kind of "dragging" of inertial frames by the rotating Earth, an influence which could, in principle, be measured. Thus, while the debate may at first have seemed to be about some esoteric, unresolvable issue, we see that there are real physical consequences and differences between the two points of view.

Mach's ideas, particularly the suggestion that the overall distribution of matter in the universe determines local motion, heavily influenced Einstein's thinking. Einstein took the viewpoint that matter determines what trajectories will be free falling, and hence inertial. The problem facing Einstein was to determine *how* matter could establish inertial frames. First, we must consider how we can define such frames in the presence of gravity. We have already shown that no frame which feels an influence of gravity can be inertial, and yet we know that gravity emanates somehow from matter. A more careful examination of the equivalence principle might provide a clue.

In adopting the strong equivalence principle, we have expanded the domain of special relativity. Previously, inertial motion was always *straight-line* motion, whereas it now involves curves, such as the curve of an orbit. We have also narrowed its scope, however, because we must now restrict ourselves to *local* measurements in "small" laboratories. Free fall is determined by the presence of gravitating masses, and the universe contains multiple overlapping and spatially varying gravitational fields. Inertial reference frames must be finite, therefore, because observers can be free-falling together only if the gravitational field they experience is uniform. For example, if one person jumps from an airplane over the North Pole and another over the South Pole, both are in inertial, free-falling frames, but those two frames are accelerating with respect to one other. Similarly, two skydivers falling side by side toward the center of the Earth are moving inertially, yet they converge toward one another, even though they started off with what appeared to be completely parallel motion in one common inertial frame. These ideas hint at the basis for general relativity; matter exerts its influence through its effect on the *geometry* of spacetime.

An Introduction to Geometry

Before we continue with our study of general relativity, we must take a detour through geometry. But what is geometry? It is the mathematics that describes the relationships of space, volumes, and areas. Since we seek to understand inertial frames of reference, perhaps it makes sense that geometry might have something to say about this. It turns out, however, not to be the geometry that we study in high school, but something more general.

Most people have some exposure to geometry in their secondary school education. The typical course on this subject seems to consist mainly of carrying out proofs of geometric propositions: the congruence of angles, similarity of triangles, and so forth. The major purpose is not so much to teach the applications of geometry, but to teach the process of drawing logical deductions from a set of *postulates*. Like all mathematical and logical systems, geometry is built upon a set of "obvious" assertions, which we call postulates or axioms. These postulates/axioms cannot be proved themselves, but they have consequences that we can deduce. All of the system is contained within the postulates. We have already seen an example of this when we studied how the special theory of relativity was derived from two simple assertions.

The geometry of our high-school days is based upon a set of five postulates systematized by the Greek geometer Euclid. The resulting geometry is called, appropriately enough, Euclidean, and its postulates are as follows:

1. It is possible to draw a straight line from any given point to any other point.

Note that we have defined neither "point" nor "straight line." Do these concepts seem obvious? Like an axiom, the concept of a "point" is not definable within the

Euclidean system, although we may define a "straight line" as the shortest distance between two given points. But now we have not defined what we mean by "shortest." Does that also seem obvious? We shall find that it *does* require a definition, and we shall soon provide one, although Euclid himself, and his contemporaries, probably took it as another self-evident concept.

2. A straight line of finite length can be extended indefinitely, still in a straight line.

3. A circle can be described with any point as its center and any distance as its radius.

4. All right angles are equal.

5. Given a line and a point not on the line, only one line can be drawn through that point that will be parallel to the first line.

For centuries, mathematicians were suspicious of the fifth postulate. It seemed as though it should be a provable statement, not an axiom, and some very distinguished mathematicians attempted to find proofs. All were flawed, but the struggle continued until the nineteenth century. The final acceptance that the fifth postulate is, indeed, a postulate, came from the independent demonstration by Carl Friedrich Gauss, Janos Bólyai, and Nicholai Lobachevsky that perfectly consistent geometries could be constructed if the fifth were replaced by some other postulate. These geometries are said to be *non-Euclidean.*

If we accept the fifth postulate, then we can prove numerous geometrical theorems, of which two will serve as examples. These are that the interior angles of a triangle sum to $180°$, and that the circumference of a circle is equal to $2\pi R$, where R is the radius of the circle. Both these theorems are familiar to nearly everyone, and most of us take it for granted that they are *facts.* Yet they are valid only for Euclidean geometry. Euclidean geometry is **flat**; it is the geometry of a set of planes. What happens if we reject the fifth postulate and substitute something different? We obtain new geometries which describe *curved* spaces. These spaces may have properties quite different from those of the flat Euclidean space. At first, these properties may seem so strange as to be unimaginable; and to many people, what they cannot imagine must be impossible.[4] But the non-Euclidean geometries are just as "real" as Euclidean geometry; it merely requires more reflection to think about them because they are unfamiliar. To study general relativity, we must abandon our prejudices for flat space and grant equal status to curved space.

It is easier to think about geometry if we start by considering only two-dimensional geometrical surfaces. An example of a curved geometry, one that should be easily imaginable, is the surface of a sphere. Where might we apply such a **spherical geometry** in our everyday life? To the surface of the Earth, of course, which is a sphere to a good enough approximation for most purposes. On a sphere, the equivalent of the straight line is the *great circle*, an arc of a circle whose center coincides with the center of the sphere. If you were to slice through a sphere along a great circle, you would exactly bisect the sphere. The circumference of a great circle that encompasses the entire sphere is the familiar $2\pi R$, where R is the radius of the sphere. The equator is a great circle, but other lines of latitude are *not* great circles. The circumference of a line of latitude depends on the location of the line. For example, you can walk along the entire length of a line of latitude by walking along a little circle centered on the North Pole. (Visitors to either of the poles sometimes do this and then claim that they have walked around the world. You could make a similar claim with nearly the same validity by performing a little pirouette in your backyard, since lines of latitude are arbitrary coordinates upon the sphere.)

The great circle is truly the equivalent of the straight line, in the sense that the shortest distance between any two given points on the surface of the Earth follows a great circle. This is why airplanes fly along great-circle routes whenever possible. Suppose you were to fly from Los Angeles to London. You would not fly directly east from Los Angeles; your route would take you to the north, over Montana and Canada, across the North Atlantic and into London via Scotland. If you plot this route on a flat map, which is the *projection* of the surface of the sphere onto the plane, and is therefore always distorted, this does not seem like the shortest route at all. If the Earth really were a plane, it would, indeed, not be the shortest distance. But if you plot the same route on a globe, you will easily see that it is the best one.

Consider any two of these globe-girdling great circles, as shown in Figure 8.5. With a little thought, you will quickly realize that they intersect *twice*. There are no parallel lines, in the sense of Euclid's fifth postulate, on the surface of a sphere; all "straight" lines intersect twice.[5] That is to say, one way in which we can define the geometry of the sphere is by retaining the first four Euclidean postulates and replacing the fifth with the statement:

5. Given a line and a point not on the line, NO lines can be drawn through that point that will be parallel to the first line.

Another property of spherical geometry is that the sum of the interior angles of a triangle drawn on the surface of the sphere is greater than 180°. For example, imagine a triangle made up of the portion of some line of longitude between the equator and the North Pole, another such line exactly 90° from it at the pole, and the equator. The interior angles of the resulting triangle are all 90°, for a sum of 270°. Now pick a point on the surface of the sphere. Locate all the points that are an equal distance r

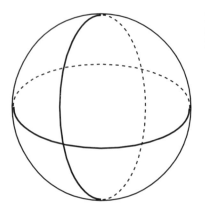

Figure 8.5 Great circles upon a sphere. In spherical geometry, any two great circles intersect at exactly two points.

from that point. By connecting those points you have drawn a circle, á la Euclid's third postulate. But what is the circumference of that circle, and how is it related to the distance r from the center point? On a flat plane the circumference equals $2\pi r$, but on a sphere the circumference is *less than* $2\pi r$. This can be demonstrated on a familiar kind of sphere; on a rubber playground ball, imagine drawing a circle along points equidistant from the inflation valve, as measured along the surface of the ball. Cut along the circle. You now have a little cap of rubber. If you try to press it flat, thereby forcing the spherical geometry circle onto a flat geometry, the rubber will stretch or tear; there is not enough material to make a flat circle. Finally, this spherical geometry is finite, yet has no edges. If you travel along any great circle (that is, along a "straight line") long enough, you will end up precisely where you began. In Euclidean geometry, straight lines have infinite length. If you are about to protest that the sphere does have a thickness in the up-down direction, remember that we are talking about the geometry of the two-dimensional surface of the sphere, as we would talk about the two-dimensional surface of a plane, not the three-dimensional space in which the sphere sits, or, as geometers say, in which it is embedded.

So far we have obtained two different geometries by assuming that either no, or one, line parallel to a given line, can be drawn through a given point. What if we assume that more than one parallel line can be drawn through such a point? This turns out to be the same as allowing an *infinite* number of parallel lines to be drawn through a point. Such a geometry may seem very strange to you, too strange to imagine, and indeed this **hyperbolic geometry** cannot be constructed, even in its two-dimensional form, in three-dimensional Euclidean space. It is extremely difficult, or even impossible, to visualize this geometry. Yet it is as self-consistent as spherical or flat geometry. The properties of this geometry are in some respects exactly opposite to those of the spherical geometry just discussed: interior angles of a triangle sum to less than $180°$, there is an infinite number of parallel lines through a point, and the circumference of a circle is greater than $2\pi R$. This geometry is also, like Euclidean geometry, infinite, but in some sense it is still "bigger" than a Euclidean space. In a three-dimensional hyperbolic space, there is more volume contained within a given radius than is contained in the corresponding radius within a Euclidean space.

Although the hyperbolic geometry cannot be visualized, a saddle exhibits some of its properties near the *saddle point* at the center; it can be employed as an aid to the imagination. At the saddle point, the surface curves "up" in one direction and "down" in the other. Mentally draw, and cut out, a little circle of some small radius around the center of the saddle, such that the circle contains material going both "uphill" and "downhill."[6] If you try to crush this circle flat, you will find you have too much material; there is overlap. This shows that circles in hyperbolic geometry are larger than the corresponding circles in Euclidean geometry. The saddle, a two-dimensional surface embedded in a three-dimensional flat space, has this property only at the saddle point. The hyperbolic geometry exhibits this property at *every* point.

In discussing the spherical and hyperbolic geometries, we have used two-dimensional examples. Both these geometries have three-dimensional forms as well, just as Euclidean geometry has two-dimensional (planar) and three-dimensional versions. We used the example of the surface of the Earth, a two-dimensional spherical geometry, because it is familiar to everyone and because we can visualize it. If we extend the spherical geometry to three dimensions, it retains all the properties described above, with appropriate additions for the third spatial dimension, but we can no longer visualize it. (Mathematically, a three-dimensional sphere is *not* a three-dimensional ball. It is the surface of a four-dimensional ball.) Similarly, the hyperbolic space can be described in three dimensions, but since we cannot even adequately visualize a two-dimensional hyperbolic surface, we have no chance of imagining the appearance of the hyperbolic space in higher dimensions. Nevertheless, the mathematics is essentially the same, regardless of how many dimensions we use.

Most of us probably think of geometries in terms of two-dimensional surfaces that exist within the three-dimensional Euclidean space of our experience. This can be misleading. All geometries have some properties that are intrinsic; these properties do not depend upon any higher dimensional entity. A sphere has an intrinsic curvature that is a property of the geometry itself and does not depend upon the sphere existing within, or being embedded in, a three-dimensional Euclidean space. The surface of the Earth constitutes a two-dimensional sphere, to a good approximation. If we were two-dimensional creatures, we would be incapable of visualizing the third dimension, so we would have no direct knowledge of the embedding of our sphere within any higher dimension; yet we could, by local measurements, determine that our geometry is curved. Furthermore, the existence of curvature in a geometry does *not* require a higher dimension in which the geometry curves. Mathematicians showed long ago that a geometry need not be embedded in anything at all, but can exist independently. In classical general relativity, the universe consists of a four-dimensional curved space-time geometry; it is not embedded in something of even higher dimension.

We have discussed three types of geometry: spherical, flat, and hyperbolic. These three examples are special cases from a whole host of possible geometries. What makes these geometries special is that they have the same properties everywhere, that is, they are all *homogeneous*; and they have no special directions, that is, they are all *isotropic*. One point in Euclidean flat space is the same as any other. One point on a sphere is like any other. (On Earth we regard the North and South Poles as special locations because the Earth is rotating, and the poles lie along the axis of rotation; this is a physical, not a geometrical, property.) Similarly, the hyperbolic geometry is the same at all locations. General relativity is not restricted to these specific geometries, but they have a special role to play in cosmology, for they are possible geometries for a homogeneous and isotropic universe. For now, however, we must discuss a few formal mathematical considerations of generalized geometries, as a way of understanding how general relativity works.

The Metric Equation

Suppose we wish to measure the distance between two points in one of these generalized geometries. We already know how to do that in Euclidean space; on the plane, the distance is given by the Pythagorean theorem as

$$\Delta r^2 = \Delta x^2 + \Delta y^2, \tag{8.1}$$

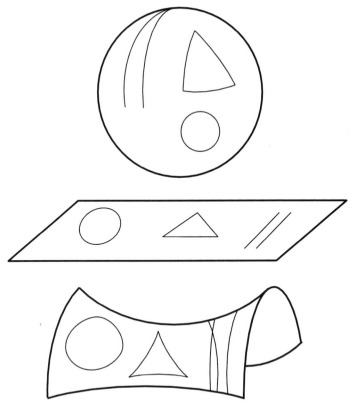

Figure 8.6   The three types of isotropic, homogeneous geometries and some of their properties. From top to bottom, these are the spherical geometry, flat geometry, and hyperbolic geometry.

where Δx and Δy are the distances given in perpendicular coordinates (x, y), laid out like a grid on our flat plane. For general geometries, we must write the analogous formula as

$$\Delta r^2 = f\,\Delta x^2 + 2g\,\Delta x\,\Delta y + h\,\Delta y^2, \tag{8.2}$$

where Δx and Δy are still our grid of coordinates, but now the grid follows the folds and curves of the geometry. This requires that we introduce the additional functions f, g, and h, which depend on the geometry. This formula is called the **metric equation**, and the quantities f, g, and h are the **metric coefficients**. We have expressed the distance in these formulae as Δr in order to emphasize that it represents the distance between two points, and thus can be taken as a small increment itself, as signified by the Greek letter Δ.

Since the metric coefficients depend upon the geometry, the general metric equation written here gives the distance between two points only in the case that those points are very close together, so that the values of the coefficients change little from one point to the next. Ideally, the points become so close together that their separation is infinitesimal. In order to compute the distance between two arbitrary points, we must know not only their coordinates and the metric equation, but also the path along which we wish to find the distance. This should not be surprising, since we all know from our everyday experience that the distance between two points depends upon the path we take from one to another! We compute the total distance by summing the small incremental distances along the path, or, in the language of calculus, by integrating along the path. The metric equation is useful for more than just the distance between two points, however; it can also be used to calculate more complicated quantities that specify the curvature intrinsic to the geometry. The metric equation and the coordinates together describe the properties of the geometry.

In Euclidean space, there exists a unique path between any two points whose distance is the shortest possible; it is the straight line which connects the two points according to Pythagoras' theorem. For two-dimensional Euclidean space, the length of this straight line is given immediately by equation (8.1), provided that we use the so-called Cartesian coordinates (x, y) that are everywhere perpendicular to each other. Cartesian coordinates happen to be the "right" coordinates for Euclidean space, in the sense that their corresponding metric equation is the simplest possible, but other coordinate systems may be used, with appropriate changes to the metric equation. If we choose to employ some coordinate system other than the usual (x, y) system in our ordinary Euclidean space, the metric coefficients will vary from one point to another, and computing distances is more awkward; regardless, the distance does not depend upon the choice of coordinates. When we extend our concept of distance to more general spaces and coordinates, the metric coefficients will usually depend upon the location for all choices of coordinates, due to intrinsic curvatures in the geometry.

Most geometries do not have a metric as simple as equation (8.1). For example, on the spherical surface of the Earth, we almost always use latitude and longitude as coordinates. To find the distance between two points on the Earth, we must use a more complicated metric equation, given by

$$\Delta r^2 = R^2 \Delta\theta^2 + R^2 \cos^2\theta \, \Delta\phi^2, \qquad (8.3)$$

where R is the radius of the sphere, θ is the latitude from the equator of the sphere, and ϕ is the longitude. In this example, the metric coefficients are $f = R^2$, $g = 0$, and $h = R^2 \cos^2\theta$. The h metric coefficient tells us, for example, that traveling 20° to the west in longitude is considerably farther at the equator, where $\theta = 0$ and $\cos\theta = 1$, than it is near the North Pole, where $\theta = 90°$ and $\cos\theta = 0$.

The distance between two points is a real physical property relating those points, and it does *not* depend upon the coordinates; the role of the metric equation is to describe how to compute this distance, given a particular set of coordinates. Does any of this sound familiar? It is very similar to our previous discussion of the spacetime interval, which also was an invariant physical property, independent of the coordinates Δt and Δx. We can write a general *spacetime* metric equation in the form

$$\Delta s^2 = \alpha c^2 \Delta t^2 - \beta c \Delta t \Delta x - \gamma \Delta x^2, \qquad (8.4)$$

where, for simplicity, we have have expressed only one of the spatial coordinates, x. In the present discussion, we have been restricting ourselves to spatial relations only, but we shall soon make use of the spacetime interval in curved spacetimes and general coordinates.

The Structure of General Relativity

We now have in place all the parts we need to complete the description of general relativity. We have learned that the equivalence principle implies that masses define inertial trajectories. We have seen how it is possible to construct geometries other than our usual flat Euclidean geometry and hence to use geometry to define the equivalents of straight and parallel lines. Now we must complete the task by showing how mass determines geometry, and geometry determines inertial trajectories.

To begin, recall that special relativity showed us how to relate observations made in one inertial frame to those made in any other inertial frame. We found that there is an invariant quantity, the spacetime interval. The spacetime interval between two events along any particular world line, which need not be inertial, corresponds to the proper time measured by a clock traveling on that world line. All observers will agree about this proper time, although they may not agree about the rate of ticking of the clock. In special relativity, our inertial observers move at constant velocity along a straight line through both space and spacetime. How can we define an inertial observer in the more general case of a curved spacetime? We can do so by recalling that in special relativity, the proper time interval between two events always has its greatest possible value along the world line of an inertial observer. In going to curved spacetimes, we must generalize this idea somewhat. Any path between two points that is an *extremum*, that is, is the longest or shortest possible, is called a **geodesic**. In special relativity, the geodesic is a straight line through spacetime, and always has the *maximal* value of the spacetime interval or proper time. (The straight line is the shortest distance in a Euclidean *space*, but in Minkowski *spacetime*, because of the negative sign in the definition of the spacetime interval, a straight line defines the *longest* proper time between two events. This is another way in which our intuition can be tripped up by special and general relativity.) We can immediately generalize this concept to curved spacetimes. Any observer traveling along a geodesic in spacetime is an unaccelerated, inertial observer. In general spacetimes, these geodesic world lines need not be the straight lines of the Minkowskian geometry; they are determined by the curvature of the spacetime geometry.

All well and good so far, but how do we determine the geometry of spacetime, so that we may compute the geodesics and find our inertial frames? This is where the equivalence principle shows the way. The equivalence principle states that inertial frames are identified with freely falling frames; they are completely equivalent. We know from experiment that free fall is determined by the distribution of masses producing what we have called gravity. Mass determines gravity, and gravity defines the inertial reference frames, or the geodesics, of the spacetime. Thus mass by its very presence causes spacetime itself to curve. Einstein's great contribution was to work out how this is accomplished through the geometry of spacetime.

How might this happen? We can construct a model that helps to visualize the idea. Imagine a sheet of rubber stretched flat, suspended between supports. The geometry on this sheet will be, of course, Euclidean. Draw straight, perpendicular coordinate lines upon the rubber surface. Now imagine that you scatter heavy steel ball bearings across the sheet. As you would expect, the bearings will distort the surface of the rubber. Lines that were straight on the flat sheet (the geodesics) now become curved.

On the surface of the sheet, some of those geodesics will twist around the spheres (the "orbits"), while others will be deflected by the spheres, and, far away from any ball bearings, where the rubber is still almost flat, the geodesics will once again become straight lines. The ball bearings determine the geometry of the rubber sheet, even quite far away from them.

In our model, the rubber sheet is filled with ball bearings of various sizes, causing the geodesics to curve in complicated and elaborate ways. Particles moving without friction across such a surface naturally follow geodesics along that surface. (You can experimentally confirm this by constructing a smooth curved surface, like a rubber sheet, and allowing marbles to roll along it.) The curvature determines the inertial (geodesic) motion at any point. Thus heavy ball bearings (matter) determine inertial motion of free particles through their effect on the geometry of the rubber sheet.

In general relativity, as in the rubber-sheet analogy, masses alter the geometry of four-dimensional spacetime, causing geodesics to be curved paths. The idea that masses determine inertial motion is similar to Mach's principle, although Mach never developed any formal way for matter to accomplish this task. Einstein himself intended to incorporate Mach's principle into the general theory of relativity, but it is present more in its spirit than in practice. The equivalence principle makes, in many cases, different predictions from Mach's principle. The mean distribution of matter in the universe does establish the geodesics, but within an inertial frame of reference, the laws of physics feel no effect of any matter whatsoever. General relativity walks a middle ground between Newton and Mach. We cannot say for sure which picture is absolutely correct for our universe, but experimental evidence supports the equivalence principle.

Once Einstein had decided that the geometry of spacetime fixes the inertial frames of reference, he had to establish the specific mathematical connection between geometry and gravity. This was the most difficult aspect, and it took him several years to find the right way. We will not trace the full development of mathematical general relativity, but we can outline some of the steps Einstein took.

Let us begin by returning to the spacetime of special relativity. It consists of a three-dimensional Euclidean space with one time dimension; it is called Minkowski

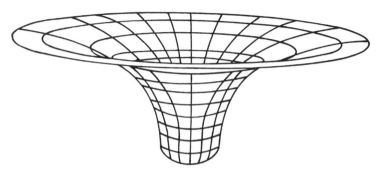

Figure 8.7 A small portion of a "rubber sheet geometry," showing its distortion by a massive "ball bearing." Away from the ball, the sheet becomes increasingly flat. Near the ball, the sheet is distorted by the presence of the mass. A small "test particle" rolling around on the sheet would be attracted to the ball due to the distortion of the rubber sheet.

spacetime. We know how to compute geodesics and find inertial frames in the Minkowski spacetime. Special relativistic inertial frames extend to spatial (and temporal) infinity. At each event, light cones can be constructed that divide spacetime into timelike, spacelike, and lightlike intervals.

We have found, however, that Minkowski spacetime is not adequate to describe spacetime in the presence of gravity. The equivalence principle tells us that inertial frames in curved spacetime must be freely falling. What else can we say about such inertial frames in the presence of gravity? We can easily deduce that they must be restricted in their extent by considering **tidal forces**. You are familiar with "tides," but what do we mean by a tidal force? Such a force occurs due to differences in the gravity acting over an extended body. For example, your feet are closer to the center of the Earth than is your head. By Newton's law of gravity, the gravitational force is not constant, but decreases over the distance from your feet to your head. Therefore, your feet feel a stronger attraction toward the center of the Earth than does your head. The difference in the attraction goes approximately as the ratio of your height to the cube of the radius of the Earth. Luckily, this ratio is quite small, and for all you care, you live in a constant gravitational field. If the ratio were not small, you could be torn apart by the tidal force.

Tidal forces do, in fact, ultimately cause the water tides on Earth; hence their name. The Moon can be on only one side of the Earth at a time, so its pull on the Earth differs significantly across the diameter of the Earth, as illustrated by Figure 8.8a. This causes the Earth to bulge on *both* sides. The Earth is mostly solid and has limited ability to change its shape in reaction to the changing and differential attraction of the Moon; but the waters of the oceans certainly can, and do, flow in response. Therefore, tides occur roughly on opposite sides of the Earth at the same time. The exact behavior of water tides depends in quite complicated ways upon the topography and shapes of the basins in which the oceans are contained, and so for the seas, this picture is greatly oversimplified, but it describes the basic driving force. The Sun also causes tidal forces, but although the Sun is many, many times more massive than the Moon, it is also much farther away, and tidal forces diminish even more rapidly (like $1/R^3$) than does the gravitational force with increasing distance. Consequently, the Sun's influence upon tides is only about half that of the Moon.

Tidal forces can be quite strong. The more rapidly the gravitational field changes over a given distance, the greater the tidal force. If an extended body in a gravitational field is to hold itself together in the face of strong tidal forces, some internal cohesive forces must be present. In most cases, at least in our solar system, intermolecular or self-gravity forces, or both, are adequate to keep the body intact. However, some comets have been observed to be broken apart by the tidal forces encountered in passing close to the Sun, or to one of the major planets such as Jupiter.

Tidal forces are an intrinsic property of gravitational fields. By the equivalence principle, a freely falling elevator above the Earth would be a local inertial frame, were it not for the presence of tidal forces. Since everything in the elevator is falling toward the very center of the Earth, a ball on one side of the cabin and another on the other side will accelerate toward one other in the elevator frame, as their free-falling trajectories converge. In free fall, both balls follow geodesics in the region of spacetime they occupy. Yet these geodesics are not parallel, because the balls approach one another as both fall toward the center of the Earth. From the point of view of geodesics, tidal forces result from the fact that geodesics in a curved geometry need not remain at some fixed separation. An extended body cannot travel on a single geodesic, and if the nearby geodesics, on which various parts of the extended body travel, should diverge, then stresses will result that would tend to pull the object apart. Conversely, converging geodesics could compress an extended body.

Do tidal forces invalidate the equivalence principle? Tidal forces seem to provide us with a mechanism for distinguishing between an elevator falling toward the Earth, and another floating in deep space. Actually, the equivalence principle remains valid, but the existence of tidal forces means that any inertial frame we might hope to construct must be "small," in the sense that the tidal forces within it must be zero or very small. Thus inertial frames in general relativity are *local*; that is, they are valid only in the immediate vicinity of the freely falling observer. A single inertial reference frame cannot be defined to cover all space and time when gravitating masses are present. Hence spacetime, in general, cannot be the special Minkowski spacetime. Within small, free-falling inertial frames, however, we know from the equivalence principle that the local geometry of spacetime must be Minkowskian, that is, flat. Thus no matter what the overall geometry of spacetime may be, locally it must reduce to a flat spacetime, for sufficiently small regions. Regardless of how spacetime may curve, it must be possible to consider a region small enough that the curvature can be ignored. The surface of the Earth behaves analogously; although the Earth is spherical and hence has a curved surface, it appears flat when observed locally, such as in a Kansas wheat field.

Fortunately, within the realm of all conceivable geometries, only a very few special geometries have this property of local flatness. Mathematicians tell us that the most general geometries that are locally flat are those studied by the German mathematician Georg F. B. Riemann (1826–1866). Such geometries are called **Riemannian geometries** and are characterized by invariant distances (e.g., the spacetime interval)

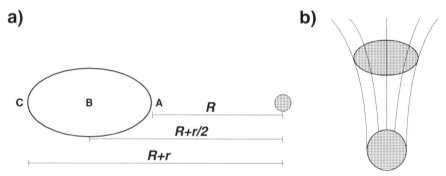

Figure 8.8   Tidal forces, from two points of view. (*a*) The Newtonian explanation. The gravitational force due to a distant mass at point A is GMm/R^2, whereas the gravitational force at B is $GMm/[R + (r/2)]^2$, and at point C it is $GMm/(R + r)^2$. The force at A is greater than the force at B, which in turn is greater than the force at C. Thus the planet is pulled into an ellipsoidal shape. (*b*) Tidal forces from the relativistic point of view. The divergence of adjacent spacetime geodesics pulls apart an extended body; the greater the divergence over a specified distance, the greater the tidal force.

that depend at most on the squares of the coordinate distances (e.g., Δx or Δt). For small enough regions, the metric equation must reduce to the familiar Pythagorean rule in space, or to the Minkowskian spacetime interval in a relativistic spacetime. Thus, *all possible* spacetime geometries can be represented by the form given schematically in equation (8.4).

This is a powerful notion because it eliminates most of the infinity of possible geometries and restricts our candidates to Riemannian geometries. Matter creates geometries that may be very complicated, but they cannot be arbitrary; they must be of Riemannian type. The mathematicians had already worked out a full set of equations describing these geometries. Any deviations from flatness in a Riemannian geometry are specified by mathematical expressions for the curvature. Similarly, tidal forces describe the gravitational deviations from flatness in a local frame. Thus, the geometrical curvature must provide a mathematical measure of the physical tidal forces in a gravitational field. All we must do now is write an equation that connects gravity, and the mass that produces it, to geometry.

Another important clue that guided Einstein in his search was the requirement that his new law of gravity reduce to Newton's law of gravity, for those cases in which velocities are much less than the speed of light and gravitational forces are weak. We are quite sure that Newton's law of gravity works very well for describing the trajectories of the planets and of spacecraft, so any new theory would have to be consistent with the old law under appropriate conditions. Although Einstein was developing a radically new theory, he was still stringently constrained by the success of Newton.

As we know, Einstein did succeed in deriving equations of general relativity that satisfied these requirements. The first complete publication of the general theory was a 1916 paper in *Annalen der Physik*, "Die Grundlage der allgemeinen Relativitätstheorie" (The foundation of the general theory of relativity), although portions of the equations had appeared earlier. In their most compact form, those equations are

$$G^{\mu\nu} = \frac{8\pi G}{c^4} T^{\mu\nu}. \tag{8.5}$$

No doubt this seems quite remarkable, to write down the entire universe in one line, but what do these symbols mean? The notation we are using here is very compact and comes from a branch of mathematics called *tensor analysis*; this single line represents ten complicated equations. But we can gain some insight without delving into the mathematics. The term on the left, $G^{\mu\nu}$, comes from mathematical geometry and describes the curvature properties of a four-dimensional Riemannian geometry. It actually represents ten different components; μ and ν are not exponents, but labels for the various space and time components of the geometry term. The term on the right-hand side, $T^{\mu\nu}$, has corresponding components in space and time; it is called the *stress-energy tensor*, and it contains the description of the matter and energy densities, pressures, stresses, and so forth with which spacetime is filled. The constant factors on the right are required for consistency of the units; our old friend, the gravitational constant G, appears in a prominent role even in general relativity.

The mathematics may be complicated, but in their essence, these equations state that *geometry = matter + energy*. Thus if matter or energy exist, they act as a *source* for the geometry. This is not all that different from Newtonian gravitation, with the notable exception that now *energy*, in any form, is also a source of gravity. (Special relativity has already taught us that mass is just a form of energy, so we should have expected a result such as this.) But let us go further and suppose that no matter or energy is present, so that we are left with $G^{\mu\nu} = 0$; this is still a valid equation. Geometry exists regardless of the presence of matter. Gravity itself turns out to be a form of energy, so not only does matter create gravity, that is, curvature, but gravity acts back on itself to create gravity. Spacetime curvature can exist and even act dynamically, without the presence of any matter or nongravitational energy. This is one of the ways in which Einstein and Mach part company.

Einstein's equations finally overcame one of the problems with Newton's law of gravitation by incorporating time and a finite propagation speed into gravity. This leads to a surprising consequence of the Einstein equations: moving masses can generate waves of curvature, or **gravitational radiation**. If the matter side of Einstein's equations changes, then the geometry will change as well. Thus a gravitational field which varies in time can produce a wave in the curvature of spacetime itself; this is a **gravitational wave**. Gravitational radiation propagates away from its source at the speed of light. To return to the analogy of the rubber sheet, imagine that we allow the ball bearings to roll, sending small ripples throughout the sheet. Gravitational waves are ripples in spacetime, like water waves that disturb the surface of a pond.

By now you should be unsurprised to learn that the Einstein equations are very difficult to solve. In other equations you might have solved, you were able to start with the set of rules associated with the geometry, and you permitted to choose your coordinates for your own convenience. Here those things are *part of the solution*. Einstein's equations are also highly *nonlinear*. This means that if you have found solution A, which could be, for example, the gravity around a spherical star, then the gravity around two spherical stars is *not* A + A. Solutions do not simply add together, or *superpose*; the solution of a full system is more than the sum of its individual parts. Consequently, Einstein's equations have been solved exactly only for a few simple cases.

Tests of General Relativity

It is a wonderful thing to have a beautiful theory of gravitation. But Einstein's equation (8.5) is not a *proof*; it is a hypothesis. In science, theories must be tested. The general theory of relativity satisfies our requirements for a scientific theory, since it makes many predictions that are useful for testing. However, all the gravitational fields which we have handy in the neighborhood of the Earth are extremely weak, in the sense that the curvature of spacetime in our vicinity is not very far from flat. Unfortunately, or perhaps fortunately, no sources of very strong, and thus especially interesting, gravitational fields are immediately available to us for direct measurements. The differences between Newtonian and Einsteinian gravity are most pronounced for strong fields, such as that of the black hole, a topic we shall treat in the next chapter. The weakness of our local fields complicates our efforts to test general relativity, since deviations from the predictions of Newtonian theory are small, and since agreement in the weak-field limit may not automatically extrapolate into the realms of strong fields. Even so, some ingenious experimental tests have been performed for the general theory.

One of the first predictions put to the test was that mentioned early in our discussion of the equivalence principle, namely, the bending of a beam of light in a gravitational field. We did not state so explicitly, but this would be predicted even within Newtonian theory, since we used nothing but the equivalence principle to obtain this result. However, general relativity goes further by recognizing that spacetime near a massive object is itself curved, not flat. Hence the bending of light, for example from a distant star as it passes close to the Sun, is greater than would be predicted from Newtonian gravitation, since the light both falls in the gravitational field, and travels through a curved spacetime. Specifically, the total bending of a light ray around the Sun would be *twice* as great in general relativity as it would be in a flat spacetime with Newtonian gravitation.

How might we measure the bending of starlight by the Sun's gravitational field? Normally, one cannot see stars close to the Sun, but during a total eclipse, such stars become visible. It is possible to make a careful determination of the location of the image of those stars during the eclipse. Those apparent positions can then be compared to the positions in the sky of the same stars during the part of the year when they are visible at night, when their light does not pass by the Sun. This experiment was performed by the British astronomer Arthur Eddington during the total solar eclipse of 1919, and the result was found to be consistent, within experimental error, with the prediction of general relativity. This experiment caught the public fancy and was responsible, more than anything else, for the elevation of Einstein to the exalted status of a popular hero.

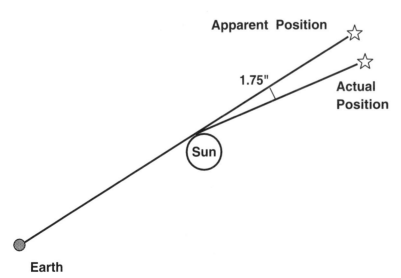

Figure 8.9 Bending of starlight as it passes through the gravitational field of the Sun. The angle is only 1.75 arcseconds; in the figure this angle is greatly exaggerated.

General relativity also predicts that the orbits of the planets will differ from the predictions of Newtonian gravity in flat spacetime. It had long been known that there was a discrepancy in the orbit of Mercury of 43 arcseconds per century, after its motion was computed and corrected for perturbations due to other planets. This is a very small residual, but it was well within the limits of the measurements of the late nineteenth century, and the cause of this discrepancy could not be easily explained within the context of Newtonian theory. One of the first problems that Einstein tackled with his new equations was the effect of curved spacetime on orbits. To his delight, Einstein found that the curvature of spacetime near the Sun accounted for the mysterious deviations in Mercury's orbit. General relativity predicted that the difference in the motion of Mercury due to the curvature of spacetime would be precisely 43 arcseconds per century. This does not prove that general relativity is correct, but it provides a simple explanation that accounts exactly for an observed datum; that fact by itself provides a powerful motivation for accepting the theory.

This type of measurement has been greatly refined with the help of space technology. Radar waves have been bounced off Venus and Mercury, determining their positions and orbits to great accuracy. Communications with spacecraft, particularly with the *Viking Lander* while it was on the surface of Mars, made it possible to measure the distance to that planet to within centimeters. These very exact measurements of planetary orbits make it possible to map the gravitational field of the Sun to extremely high precision; the results are all consistent with the predictions of general relativity.

Gravitational redshift and time dilation also provide a means to check the theory. The effect in the extremely weak field of the Earth is quite small, and clever experimentalists and very sensitive instruments are required, but the predictions of the equivalence principle have been verified. It is also possible, barely, to use the light from white dwarfs to test this aspect of the theory. The gravitational field of a typical white dwarf is sufficiently strong that the gravitational redshifting of photons departing its surface can be observed, with, of course, great difficulty; the shifts agree with the predictions of general relativity. The applicability of the theory for objects at astronomical distances is yet another confirmation that the physics we develop on Earth is valid for the universe as a whole. White dwarfs, many of which are near enough and bright enough to be observed easily, thus provide some quite tangible evidence for general relativity.

Recently, a very interesting test of general relativity has become possible by the discovery of binary pulsars. General relativity predicts that two compact and massive objects in orbit about one another will give off gravitational waves. The loss of energy to those waves will make the orbits decay and gradually spiral inward. Since a pulsar acts like a very precise transmitting clock, we can follow its motion in its orbit very closely and can determine the orbital period to almost fantastically high precision. Over years of watching one such binary pulsar, astronomers have found that the orbit is changing in exactly the way predicted by general relativity. The binary pulsar thus serves as an indirect detection of gravitational radiation. It was for the discovery of the first binary pulsar and its very significant consequences that Joseph Taylor and Russell Hulse won the 1993 Nobel Prize in Physics.

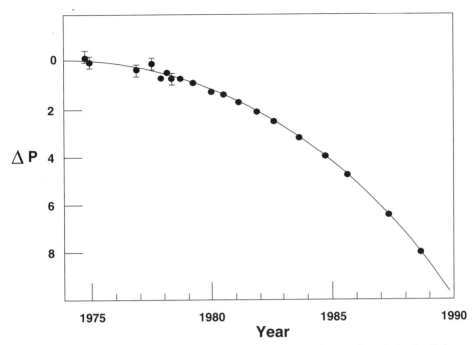

Figure 8.10   The binary pulsar $1913 + 16$ provides evidence for the existence of gravitational radiation. The loss of orbital energy to gravity waves causes the orbital period to change with time. The change in period, ΔP, is given in seconds. The solid line is the prediction from general relativity; the dots, the observations. (Adapted from J. Taylor and J. Weisberg, 1989.)

Of course, we would prefer to have a direct detection of gravitational waves, and it is possible, in principle, to detect such waves. Since spacetime curvature is changing locally in a gravitational wave, a tidal force is induced in any physical object through which the wave passes. One promising technology for detecting these tidal forces is the *laser interferometer*, a device in which a beam splitter creates two light beams that are sent for a round trip along perpendicular paths. If the distance in either of the two directions changes in time due to a stretching or compression by a gravitational wave, the interference pattern changes when the light beams are recombined. This is, of course, the Michelson-Morley experiment adapted to the search for gravitational radiation. The experiment is currently just at the limits of technology, as the effect at the scale of any laboratory-bound system is incredibly tiny, on the order of one part in 10^{20}.

Is such a small effect measurable? If we could build an apparatus with a very, very long baseline, then we would have a much better chance of detection. Such a device is currently under construction, the Laser Interferometer Gravitational Observatory (LIGO). LIGO consists of two laser interferometers. These are Michelson-Morley interferometers, each of which consists of perpendicular pipes through which laser beams travel. The current design of LIGO calls for each of these pipes to be approximately 5 km in length, but the light will be bounced back and forth down the pipe many times, creating an even longer effective baseline. Because such a system will be subject to many types of noise, for example, vibrations in the Earth, it is necessary to have at least two widely separated interferometers. One experiment would compare

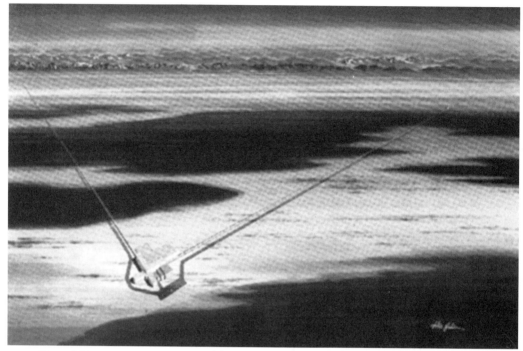

Figure 8.11   Artist's conception of the LIGO gravity wave detector. LIGO, basically a huge Michelson-Morley interferometer, consists of two 5 km long evacuated pipes at right angles. Gravity waves can cause small differences in light-travel time down the pipes, which can be detected as shifts in the interference fringes when the beams recombine. (Courtesy of the California Institute of Technology.)

the signals from both interferometers to look for common effects that might be due to gravity waves.

What might LIGO see? We have found indirect evidence for gravitational radiation from a binary pulsar. This radiation is not directly detectable by LIGO, but the universe must contain many binary pulsars. Eventually, a binary pulsar's orbit will shrink so much that the two neutron stars will spiral into one another. Such a cataclysmic collision would generate very strong gravitational radiation, producing effects which, although still small by ordinary standards, should be detectable on Earth, depending, of course, on the distance to the binary. Needless to say, the collision of two neutron stars is not an everyday occurrence in the neighborhood of the Sun, so we still need great sensitivity in order to be able to sample a very large volume of space, possibly to a radius of as much as 1 billion light years.

General relativity has shown us how matter and the geometry of spacetime are related. It has provided us with new insights into the workings of the universe and has predicted some remarkable new effects. General relativity has a profound impact on cosmology, fundamentally altering our view of the relationship of space and time. Yet how does that affect the humble stars and galaxies? Light may bend, gravitational waves ripple about, but it might appear that the mechanisms work pretty much as they always did in Newton's grand clockwork. But we cannot make a radical change in the underlying paradigm of the universe without finding unexpected consequences for things once thought to be quite ordinary. Before returning to cosmological models, we will examine the black hole, one of the most extreme consequences of general relativity, whose properties are almost beyond our imaginations.

KEY TERMS

free fall	weak equivalence	strong equivalence
gravitational redshift	principle	principle
spherical geometry	Mach's principle	flat geometry
metric coefficient	hyperbolic geometry	metric equation
Riemannian geometry	geodesic	tidal force
	gravitational radiation	gravitational wave

Review Questions:

1. What is "special" about special relativity, and what is "general" about general relativity?

2. Make a table of the properties of the three homogeneous and isotropic geometries we have studied: spherical, flat, and hyperbolic. Include (1) finite or infinite? (2) How does the circumference of circles relate to the radius of the circle? (3) What is the sum of the angles inside a triangle? (4) Given a line, how many parallel lines are there through another point not on that line (parallel-line postulate)?

3. State in your own words what Mach's principle is all about by considering the following thought experiment: The universe contains two observers who are initially at rest with respect to each other. Newton says that if the first observer accelerates away from the second, the first observer will feel a force, while the second will not. Why would Newton say this? What would happen according to Mach? What would Newton and Mach say if these two observers were the only objects in the universe?

4. What is the difference between the Newtonian version of the equivalence principle (the weak form) and the Einsteinian equivalence principle (the strong form)?

5. Consider the following experiment (which you can actually perform): Obtain a spring scale (e.g., a typical bathroom scale), place it in an elevator, and stand on it. Note the *exact* value when the elevator is at rest. Now ride up several floors. As the elevator starts up, there is an acceleration upward. Note how the reading on the spring scale changes. Next ride down. When the elevator starts down, note how the reading changes. Once the elevator reaches a constant velocity up or down, note the reading of the scale. What do you predict these various readings would be?

6. A space station in deep space is spun like a giant wheel to produce centrifugal force so the occupants have artificial gravity of one g. How does a clock out at the rim of the spacestation compare with one residing at the hub? What does this say about the behavior of a clock sitting on the surface of a planet with a surface gravity of one g?

7. Define *geodesic*. What does this mathematical quantity have to do with frames of reference?

8. (More challenging.) Imagine a point source with the mass of Jupiter, 1.9×10^{27} kg, at a distance of 4.2×10^8 m from a spherical object that has a diameter of 3640 m and a mass of 8.9×10^{22} kg. Consider a location on the surface of the sphere that is closest to the point mass. Now consider the location on the sphere that is exactly opposite the first location. (Refer to Figure 8.8a for a diagram of the situation.) What is the *difference* in the gravitational force between these two locations? How does this tidal force compare, as a percentage, with the gravitational force between the sphere (more precisely, the center of the sphere) and the point mass? You may use strictly Newtonian gravity for your answers. The figures in this problem correspond approximately to Jupiter's moon Io. Io is the only body in the solar system other than Earth which is known to have active volcanoes. (Many other objects have extinct or at best dormant volcanoes, but eruptions have been photographed on Io by spacecraft.) The ultimate source of the energy of the Earth's volcanoes is radioactive decay of uranium and thorium deep in the planet's core. Io probably lacks such a source. Does this problem suggest to you a possible energy source for the vulcanism of Io?

9. Describe two distinct experimental tests of general relativity. Explain how the results distinguish between Newtonian gravity and general relativity.

10. Explain in your own words what a gravitational wave is. At what speed do such waves propagate?

11. A spaceship is coasting in orbit. A second spaceship sits motionless on the launch pad. The two ships define frames that are accelerated with respect to each other, yet both might be regarded as inertial frames. Explain.

12. Under what circumstances is Newtonian mechanics valid? Does the development of general relativity mean that Newtonian theory is useless, or is an unacceptable scientific theory? Why or why not?

Notes

1. Science fiction movies of the 1950s often showed objects beginning to float around a spaceship's cabin after the craft had "left the Earth's gravitational sphere." The genre was aptly named, since this is completely fictional science.

2. You cannot save yourself in a falling elevator by "jumping up" at the last moment. Since you would have a large downward velocity, jumping upward would decrease this by only a small amount. If your legs were strong enough to provide the required deceleration, they would also be strong enough to absorb the impact at the bottom of the shaft, as that too is just a deceleration.

3. The movie *Apollo 13* filmed its scenes of weightlessness aboard NASA's airplane.

4. Conversely, many people are inclined to believe that whatever they can imagine, must occur. Neither attitude is defensible.

5. Lines of latitude are often called "parallels," because when they are drawn on a flat map, they are parallel to the equator. But this is merely a conventional expression that has nothing to do with the geometry of the sphere. Lines of latitude are not great circles, although lines of longitude are.

6. A *Pringles* brand potato chip provides a reasonable experimental model, if you would like to try this.

CHAPTER 9

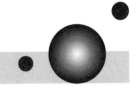

Black Holes

There's always a hole in theories somewhere, if you look close
enough.

—MARK TWAIN, *Tom Sawyer Abroad*

Schwarzschild's Solution

The death of supermassive stars must result in collapse; no known force can resist
gravity in such stars once their nuclear fires have died. The result of this inevitable
collapse is known as a **black hole**. The term "black hole" is nowadays bandied
about so much, in science-fiction novels and movies, as well as in the occasional
general-science articles of newspapers and magazines, that it would be difficult for
any reasonably literate person to be unaware of the expression. Yet few understand
why black holes exist, or what their properties really are. In popular thought, they
are often viewed as some sort of monstrous and voracious maw, devouring anything
that comes too near, even light and energy. While accurate in some respects, such
a notion falls far short of a complete description of black holes and their remarkable
properties.

Although the black hole is an extreme consequence of Einstein's theory of general relativity, the possibility of something like a black hole exists even within the
Newtonian theory of gravity. Any star or planet has a gravitational acceleration at its
surface. Escape from that star or planet requires a velocity large enough to overcome
the gravitational pull. This velocity is known, appropriately enough, as the **escape
velocity**. In Newtonian gravity, the escape velocity from a spherical object of radius
R and mass M is

$$v_e = \sqrt{2GM/R}. \tag{9.1}$$

On Earth, the escape velocity is about 11 km/s. Now imagine that we increase the
escape velocity until it is equal to the speed of light. If there existed a star with such

a huge escape velocity, no light could leave its surface; it would be a dark star. Light shining from the surface of this star might climb up, but like a ball thrown into the air, it must eventually turn around and fall back down. When this was first proposed, it was not known that the speed of light *in vacuo* is the ultimate speed limit, but with that additional fact, it is easy to conclude that *nothing* could escape from such a star. Setting $v_e = c$ in equation (9.1) and solving for the radius gives $R = 2GM/c^2$. For a star with the mass of our Sun, this radius is about 3 km; the Newtonian dark star is very compact indeed.

The story of the general relativistic black hole begins shortly after the appearance of the final version of Einstein's general theory of relativity late in 1916. Despite the great complexity of the Einstein equations, the German astronomer Karl Schwarzschild found one of the first solutions, almost immediately after Einstein published his results. Schwarzschild assumed a perfectly spherical, stationary ball of mass M, surrounded by a vacuum, that is, empty space. This is not a bad approximation to a star; the Sun, at least, rotates slowly and is very close to spherical, and, as far as we know, the Sun is a typical star. Schwarzschild then solved Einstein's equations to compute the spacetime curvature in the *exterior* of the star. Such a solution consists of a specification of the geometry of spacetime; this description can be encapsulated in the metric coefficients, as indicated by equation (8.4).

Schwarzschild's assumptions greatly simplified the mathematics required. First, he was solving for the gravity in a vacuum outside the mass. This meant that he could set the stress-energy term $T^{\mu\nu}$ in Einstein's equation equal to zero and work only with the geometry term. Since he was considering the space around a spherical mass, Schwarzschild employed spherical spatial coordinates, consisting of a distance R from the center of the mass, as well as the inclination from the origin expressed in terms of two angles, such as altitude θ and azimuth ϕ. (The precise definition of the radial distance is slightly more complicated than this, but the details need not concern us here.) The gravity arising from such a star must be spherically symmetric; that is, it should depend only on the distance from the star. Thus it was possible to ignore the angular terms, another considerable simplification for Schwarzschild. Finally, the star and its gravitational field are unchanging in time, which implies that the metric terms cannot depend on time, assuming that the time coordinate is chosen appropriately. The time coordinate Schwarzschild employed was a very sensible choice; it corresponds to the time measured by an observer very far away from the central mass, where gravity's effects diminish toward zero. With all these simplifications, Schwarzschild obtained his metric,

$$\Delta s^2 = (1 - 2GM/c^2 R)c^2 \Delta t^2 - \left(\frac{1}{1 - 2GM/c^2 R} \right) \Delta R^2 - R^2 (\Delta \theta^2 + \sin^2 \theta \Delta \phi^2).$$

$$(9.2)$$

This, in its glory, is a full general relativistic metric, or spacetime interval. The Schwarzschild metric is similar to the familiar flat spacetime interval of special relativity, as written in spherical coordinates, but it is modified by the appearance of the metric coefficients, which vary only with R. These new functions affect only the time and the radial measurements; the angular terms are unchanged from ordinary flat space, and need not concern us further here. We can interpret these metric coordinates in terms familiar from our previous study of general relativity; the Schwarzschild coefficients of Δt^2 and ΔR^2 respectively account for gravitational time dilation and length contraction. Keep in mind that this solution is valid in a vacuum outside *any* spherical body of mass M and radius R. It does not, however, hold in the interior of the body.

The combination

$$2GM/c^2 \equiv R_S \tag{9.3}$$

appears in both of the new metric coefficients. This expression is very important and will turn out to be intimately linked with many of the unusual properties of black holes. It has units of length and is called the **Schwarzschild radius**, R_S. Because c^2 is large, R_S will be extremely small unless M is also very, very large. For the mass of the Earth, R_S is equal to about 1 cm. Does this imply that the matter within 1 cm from the center of the Earth is inside Earth's Schwarzschild radius? No, because the Schwarzschild solution applies only *outside* a mass. Inside the Earth there is matter, and where there is matter it is necessary to solve the Einstein equations with the stress-energy term present.

Because the radius of the Earth is so much larger than the Schwarzschild radius, the metric expression $2GM/c^2 R = R_S/R$ will be very tiny for the gravitational field surrounding the Earth. This means that the modifications to ordinary flat space and spacetime will be equally small. Consequently, space and spacetime in the vicinity of the Earth are curved very little, although this small curvature still accounts for the gravitational field we experience. The major effect on spacetime around the Earth, the Sun, or any other spherical object that is large compared to its Schwarzschild radius, occurs through the metric coefficient of the time coordinate, because of the presence of the speed of light in the expression $(1 - R_S/R)c^2\Delta t^2$. The contribution of the radial coefficient is much smaller, with a correspondingly very small curvature to the *space* around the Earth; thus space in our vicinity remains very nearly the familiar Euclidean. In the relativistic view, we can say that Earth's gravity is mainly due to "time curvature."

What about an object whose radius R is comparable to $2GM/c^2$? The coefficient of Δt shrinks toward zero, while that of ΔR becomes huge. The general relativistic properties of such a compact star become increasingly evident. As the simplest example of relativistic behavior near the Schwarzschild radius, imagine that you establish a series of clocks lined up at fixed distances from the star, after which you retreat far from the star with the last clock. We have learned the general rule that a clock in a stronger gravitational field runs slower than an identical clock at a location where the field is weaker; how can we make this more exact? How do the time intervals measured on the clocks compare along the line?

It is the metric that enables us to relate the rates of ticking of all these clocks. Suppose that you consider two events, the instant when your clock reads 1:00, and the instant when that same clock reads 1:01. Your distance, call it R_∞, from the compact star is very much larger than R_S; hence the quantity $(1 - R_S/R_\infty)$ is essentially equal to one. Your clock then keeps the proper time of an observer located infinitely far from the compact star and experiencing zero gravity. As you observe clocks lying at distances R closer and closer to the Schwarzschild radius, $R \rightarrow R_S$, and the factor $(1 - R_S/R)$ drops toward zero. Now let us examine what this means for the Schwarzschild metric. Since the clocks are assumed to be at *fixed* spatial locations, we have $\Delta R = 0$ (as well as $\Delta\theta = \Delta\phi = 0$); thus only the time coordinate remains in the metric equation. Consequently, the product of Δt and its metric coefficient must always equal the invariant proper time interval Δs, or, mathematically, $(1 - R_S/R)c^2\Delta t^2 = \Delta s^2$. Although the rate of ticking of the clocks, given by Δt, depends upon any particular clock's location in the gravitational field, the spacetime interval Δs must be the same for *all* observers. If the product is to remain constant while one factor shrinks toward zero, then, obviously, the other factor must become ever larger. Thus we see that the closer the clock to R_S, the larger must be the time interval Δt, in comparison to the time coordinate of you, the distant observer, whose clock keeps proper time. In other words, the clocks nearer the Schwarzschild radius tick more slowly than your clock. We have found time dilation in the metric.

The Schwarzschild metric affects not only time, but also space. What happens to a standard length, that is, a "ruler," in the Schwarzschild metric? By an argument very similar to the one used for the clocks above, we can see that stationary rulers must contract, as measured by the observer at R_∞, such that $L_\infty = L_0/\sqrt{1 - R_S/R}$, where L_0 is the rest length of the ruler located at distance R from the center of attraction, and L_∞ denotes the length measured by the distant observer. This is a purely gravitational length contraction.

The metric affects not only spacetime, but also anything traveling through spacetime, including light. One of the most important consequences of the effect of the metric upon the propagation of light is the *gravitational redshift*, which is a consequence of the equivalence principle. Now that we have a specific metric, we can compute an explicit formula for the corresponding gravitational redshift. Redshift is *defined* to be

$$z = \frac{\lambda_{\text{rec}} - \lambda_{\text{em}}}{\lambda_{\text{em}}} = \frac{\lambda_{\text{rec}}}{\lambda_{\text{em}}} - 1, \tag{9.4}$$

where λ_{rec} is the wavelength of the light received at the detector, and λ_{em} is the "standard" wavelength, that is, the wavelength measured in the rest frame of the emitter. Wavelength is a length and will be contracted by the Schwarzschild gravitational field in the same way as any other length. Thus the gravitational redshift of a photon emitted at a distance R from the center of a compact object and received at "infinity" is simply

$$z = \frac{1}{\sqrt{1 - R_S/R}} - 1. \tag{9.5}$$

Since the Schwarzschild solution is valid only outside of a star, this formula holds when R is greater than the star's radius. Although we derived this result from consideration of length contraction, the identical result could be obtained from time dilation, because longer-wavelength radiation has a lower frequency, by the relation $\nu = c/\lambda$, and frequency is simply an inverse time interval.

Gravitational length contraction and time dilation occur in any gravitational field. But what if there existed a star whose radius was equal to the Schwarzschild radius? At $R = R_S$ the coefficient of Δt^2 becomes zero and that of ΔR^2 becomes infinite.

Does this mean that the metric equation has broken down? For a long time after Schwarzschild presented his solution, this was the prevailing opinion. It was believed that the solution simply was not applicable for so small a radius, and therefore no physical object could ever be smaller than its Schwarzschild radius. It took quite a while for scientists to realize that the solution does not fail. Instead, at the Schwarzschild radius, what fails is our choice of radial coordinate R. It is an artificial failure, similar to what would happen if we decided to measure temperature in terms of the inverse of degrees Celsius. On such a temperature scale, the freezing point of water still exists, but we have chosen a particularly inappropriate way to measure it, a marker that becomes infinite at this particular point. Once it was realized that the metric did *not* break down at the Schwarzschild radius, it became necessary to consider the consequences of a star which had collapsed to such an extent. These are the objects now known as black holes.

Properties of Black Holes

It is rather coincidental that the Schwarzschild radius is the same as the radius derived previously for the Newtonian dark star. Perhaps, since Newtonian gravitation is valid to a good approximation, we should have expected something not too far from its prediction. But the black hole is a much more interesting and exotic object than is the dark star, and insistence upon thinking about the black hole as if it were a Newtonian dark star will lead to misunderstanding of the essential properties of the black hole.

Why is every spherical object not a black hole? Because the Schwarzschild radius lies well within the outer surface of any "normal" object, even a neutron star. For example, the Schwarzschild radius of the Sun is approximately 3 km, compared to a solar radius of over 1 million kilometers. The Schwarzschild radius of the Earth is less than 1 cm. As emphasized above, the Schwarzschild solution applies *only* in the empty space to the exterior of the sphere; if the Schwarzschild radius is less than the radius of the body, then it is irrelevant within the body. The metric inside a star is not a Schwarzschild metric, but a different metric that includes the presence of the matter which generates the gravitational field. Only if the object has collapsed completely and disappeared beneath its Schwarzschild radius can a black hole be formed.

At the Schwarzschild radius, the coefficient of the time interval Δt in the Schwarzschild metric goes to zero. Therefore, the time interval itself, which is the proper time divided by this coefficient, becomes infinite; clocks stop. Similarly, radial intervals fall to zero, the ultimate length contraction. These effects are a consequence of our choice of coordinates, and coordinates themselves are not absolute even in Newtonian physics. Nevertheless, the time dilation, length contraction, and other relativistic effects that depend directly upon the metric coefficients, are real physical phenomena and can be measured with sensitive instruments. As is true for any massive object, the gravitational field near the black hole is stronger at small radius than it is far away, and so light climbing from close to the object suffers a gravitational redshift. In the case of a black hole, any light sent from the Schwarzschild radius is *infinitely* redshifted. The sphere defined by the Schwarzschild radius thus represents a surface from which light cannot travel to an outside observer. An observer outside this horizon can never see within it; the interior of the black hole is forever cut off from communicating with the rest of the universe. Events inside the black hole can have no causal contact with events to the exterior. This causal boundary between the inside and the outside of a black hole is an **event horizon**. It is the point of no return, from which no light or other signal can ever escape. The Schwarzschild radius marks the event horizon of the black hole.

From outside a black hole, the event horizon seems to be a special location. What would happen if an advanced civilization were to launch a probe toward a black hole? To the observers watching from a safe, far distance, the infalling probe's clock slows down; radio signals from the probe come at increasingly longer wavelengths due to the gravitational redshift. The probe approaches closer and closer to the horizon, but the distant observers never see it cross over into the hole. Time seems to come to a halt for the probe, and the redshift of its radio beacon goes to infinity, as measured by the faraway astronomers. At some point the last, highly redshifted signal from the probe is heard, and then nothing more. The probe disappears forever.

Does this mean that the probe is destroyed upon reaching the horizon? No; these strange effects, such as the freezing of time for the probe, are artifacts of the space and time coordinates of the *external* observers. The Schwarzschild radius is not a true **singularity** in the metric, a place where tangible, physical quantities such as pressure or density reach infinity, but rather it is a **coordinate singularity**, a point at which our choice of coordinate system fails. However, only the coordinate system defined by the observers at infinity fails; a coordinate system falling freely with the probe remains valid and indicates no changes in time or length values. Time and space seem normal to the probe, even at the horizon.

Extreme time dilations and length contractions are not unique to general relativity; an example from special relativity would be a spaceship accelerating toward the speed of light. To an observer at rest, the relativistic spaceship's clocks would seem to slow toward a halt, while meter sticks aboard the spaceship would shrink toward zero length.

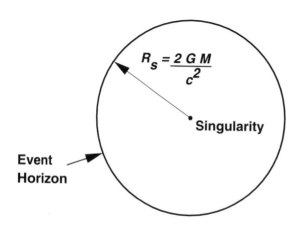

The Schwarzschild Black Hole

$$R_s = \frac{2\,G\,M}{c^2}$$

Singularity

Event
Horizon

Figure 9.1 Components of a black hole. The point of no return, the event horizon at the Schwarzschild radius, defines the size of the black hole. The singularity at the center is the point at which all ingoing world lines end, and matter is crushed to infinite density.

Yet the occupants of the spaceship would see nothing strange occurring. Similarly, to the ill-fated probe approaching the horizon of the black hole, nothing unusual occurs; physics continues to appear perfectly normal. This follows from the nature of spacetime and the equivalence principle. Even near a black hole, a sufficiently restricted, freely falling frame must be equivalent to any other inertial frame.

But there are other effects near the Schwarzschild radius that might be of consequence to an infalling probe. We have overlooked another real physical phenomenon: tidal forces. Since the probe is an extended body, it might well be in danger near the event horizon. Recall that tidal forces result when the gravitational force varies over a body. The gravitational field near a black hole increases so rapidly as the event horizon is approached, that the part of the probe closest to the black hole might experience a substantially larger gravitational force than more distant parts. The curvature of space close to the black hole is comparable to the scale of the hole's horizon, so tidal forces near a black hole will be very large for bodies whose size is not extremely small in comparison to the radius of the horizon. If your spaceship were to fall into a stellar-sized black hole, with a Schwarzschild radius of approximately 3 km, you would be torn apart by tidal forces. For such a small black hole, any spaceship that could accommodate human bodies would occupy a spatial volume that cannot be approximated by flat spacetime in the vicinity of the event horizon. But a black hole need not always chew its food before swallowing it; if you fell into a black hole a million times more massive than the Sun, with a horizon radius correspondingly a million times greater than that of the Sun, you would scarcely notice your passage across the event horizon. In such a case, the volume surrounding your spaceship would be fairly well approximated locally by a flat spacetime, and you would not feel strong tidal forces. An even larger black hole, such as might occur from the collapse of the core of a galaxy, would have reasonably small tidal forces even at the scale of an object as large as a star.

Thus, if you fell into a sufficiently large black hole, you would feel no ill effects as you crossed the horizon. What would you find inside the horizon of the black hole? Although an infalling probe could never return data from the inside, the Einstein equations still hold and still describe spacetime. If we continue to use the time and space coordinates appropriate for an observer at a great distance from the hole, as we have in equation (9.2), we find that within the Schwarzschild radius, the metric function behaves in a peculiar manner. When $R < R_S$, the metric coefficient of the time coordinate becomes negative, while that for the radius becomes positive. Outside the horizon, the signs of these functions are reversed. This suggests that within the black hole, the time and space coordinates (as defined by the external observer) exchange their roles. Recall that material particles must have world lines for which $\Delta s^2 > 0$. Outside the horizon, a world line can be fixed in space, with $\Delta R = 0$, as it advances forward in time. On the other hand, no world line could be fixed in time, while moving through space. Within the hole, in contrast, if the particle's world line remained at a fixed radius from the center, that is, $\Delta R = 0$, then the spacetime interval would become negative, or spacelike, which is not allowed for a particle world line. Therefore, inside the Schwarzschild radius, it is impossible for a particle to orbit at a fixed radius; its orbit *must* change. In fact, its radius must constantly decrease. The future, as it were, lies inward.

Let us consider this in terms of light cones. At any event (R, t) in spacetime, we can construct a light cone, as illustrated in Figure 9.2. For example, we could position a particle at some location and let it emit a pulse of light at some instant in time. Far from the black hole, the light cones are just as they would be in Minkowski spacetime. Nearer to the black hole, however, geodesic paths, including the null paths followed by light rays, point more and more toward the hole, that is, toward $R = 0$, because light falls in a gravitational field. If our light-emitting particle were near the event horizon, a large fraction of the light emanating from it would fall into the black hole. More and more of the particle's possible future world lines, which we know must be contained within its future light cone, point toward the hole. In other words, the light cones begin to tilt toward the hole. At the horizon, *all* of the particle's future will lie inward; one edge of the light cone will coincide with the horizon. This edge would describe a light beam directed straight outward, but frozen forever exactly at the horizon. Once inside the horizon, the light cone is even further tipped over. The future is directed toward smaller R, the past toward larger R. This is another way to look at the interchange of time and space coordinates; "out here" we may say that the future lies with greater values of time t. "In there" the future lies toward smaller values of the radius R. Interestingly enough, a world line inside the hole could move in the $+t$ or $-t$ direction, but that still does not permit time travel, because a world line can never emerge outside of the horizon, where t is once again the "usual" time coordinate.

A useful way to visualize this phenomenon is to imagine that spacetime is like water; a black hole is analogous to a drain. Objects falling radially toward the black hole are like boats floating unpowered in the water, moving with the current. Far from the black hole, our boat drifts very slowly toward the horizon, but the water, and hence the boat, gain speed the closer we approach the hole. If we wish to avoid falling down the drain, we must turn on our motor and aim away from the hole. There

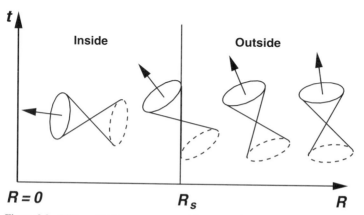

Figure 9.2 Tilting of light cones as a particle approaches a black hole. As the horizon nears, more and more of the future lies downward, toward the hole. The horizon itself coincides with a lightlike (null) surface, and the future lies entirely inward. Inside the horizon, the external coordinates t and R have swapped roles, and the future lies in the direction of decreasing radius.

comes a point, however, at which the water is flowing inward faster than the motor can drive the boat.

In this analogy, we can think of spacetime itself drawn inward at an increasing rate by the gravitational pull of the black hole. At the horizon, spacetime flows inward at the speed of light, so that a light ray emitted against the flow can, at best, stand still. Inside the horizon, spacetime flows at a speed faster than that of light, so even light cannot move outward anymore. (In case you are about to protest that nothing can move faster than light, remember that this is an analogy; moreover, the "motion" of spacetime, as we are describing it, carries no information and hence its superluminal motion cannot violate causality.)

Once inside the horizon, the radius of any particle's orbit must inexorably decrease, drawing the particle into the center. At the center, once again, our metric equation fails, here because $R = 0$. This time, however, the failure is real and unavoidable by a change of the coordinate system. At the center of the hole lies the true singularity, the point at which density becomes infinite. Any particle that crosses the event horizon is doomed, since it must fall toward the infinite crush at the center. The exact time required for infall to the singularity depends upon several factors, such as the path taken, but it is approximately equal to the time for light to travel a distance equal to the Schwarzschild radius. The larger the black hole, the longer this time is. If you fell straight into a solar-mass black hole, you would reach the singularity in roughly 10 microseconds. Infall from the horizon of a black hole of mass 10^8 $M_\odot$, such as might inhabit the cores of many galaxies, would take only 16 minutes.

But is the singularity truly infinite? Many scientists do not believe that infinite density can exist in the physical universe. We know, for example, that the general theory of relativity has never been made fully consistent with quantum mechanics, the other triumph of the physics of the first half of the twentieth century. The required theory would provide an explanation of **quantum gravity**, but no such theory has yet been developed. It is likely that there is a point at which Einstein's equations break down as a suitable description of the universe, and it may be that quantum effects prevent a literal singularity. In any case, it is probable that our current notions of "particles," and perhaps even our conceptions of space and time themselves, fail at a singularity. Even if such an effect occurs, however, the center of a black hole represents the highest density possible in the universe.

Whatever may be going on at the singularity of a black hole, it does not matter to the external universe. The singularity is surrounded by the event horizon, and hence no information or signals from the singularity can ever emerge. In fact, if you were to venture inside a black hole, giving up your life in order to see the singularity, your sacrifice would be in vain. Light rays cannot move toward *any* larger R within the horizon, including your position as you fall, so the singularity is invisible even from inside the horizon.

We can prove mathematically that it is impossible to observe light rays traveling from the singularity in the Schwarzschild solution. But what about singularities that might exist in other solutions, including those that we have not yet discovered? Might some solutions contain *naked singularities*, "bare" singularities which have no event horizons to shroud them? The conjecture that no singularities can ever be seen because they must be surrounded by event horizons is known as the **cosmic censorship** hypothesis: there are no naked singularities. This proposition holds that whenever a singularity forms, it will do so within the confines of a shielding horizon; thus whatever the properties of a singularity may be, they can have no effect on the rest of the universe. Although no exceptions are known, the cosmic censorship hypothesis has not yet been proven; it is based on experience to date with the Einstein equations, and on our beliefs about how the universe should work.

Perhaps it is because of the singularity that black holes are sometimes said to be the "densest things in the universe." Black holes may be very dense, but they need not be so. The black hole is not the singularity per se, but the volume of space surrounded by the event horizon. Just as for any other spherical object, a black hole's density is proportional to its mass, divided by its radius cubed. The radius of the black hole is itself proportional to the mass of the hole; hence the volume of the hole is proportional to the cube of its mass. Thus the *average* density of a black hole goes as the inverse of the mass squared; the more massive the black hole, the less dense it is. Specifically,

$$\rho_{bh} = \frac{3M}{4\pi R_S^3} = \frac{3c^6}{32\pi G^3 M^2} \propto \frac{1}{M^2}. \tag{9.6}$$

The black hole's density ρ_{bh} tells us by what factor a mass M would have to be squeezed to create a black hole. For example, to transform the Sun into a black hole, it would have to be compacted to a radius of 3 km. Since the Sun has a mass of 2×10^{30} kg, its density as a black hole would be about 10^{19} kg per cubic meter (kg/m^3). This is indeed fabulously dense, far beyond the imagination of any of us and considerably greater than the density even of an atomic nucleus, which is typically about 10^{17} kg/m^3. On the other hand, a black hole 100 million times as massive as the Sun, with a radius proportionally larger, would have an average density approximately the same as that of water, hardly an unusually dense substance. If a black hole were created from the Milky Way Galaxy by collapsing all its stars together, the entire Galaxy would be contained inside its horizon with the stars still well separated. In the most extreme limit, if the entire visible universe were within a black hole, its average density would be close to what is actually observed, about 10 atoms per cubic meter. Thus, you could be located inside such a high-mass, low-density black hole without your immediate surroundings appearing in any way exotic. However, if you were inside any black hole, you would have a limited (proper) time left to live, since nothing can stop the inevitable collapse into the central singularity.

Now let us return to the outside of the black hole and ponder a few more of its external properties. One of the most common misconceptions about black holes is that they possess some sort of "supergravity" power to draw distant objects into them. In reality, away from its immediate vicinity, a black hole has no more and no less gravitational pull than any other object of equal mass. Well away from a black hole, we would not notice its presence in any unusual manner; its gravitational field is not qualitatively different from the gravitational field of any other object in the universe. All massive objects produce curvature in spacetime. The unusual aspect of the field of the black hole is the strength of the curvature very near the event horizon. Far from the horizon, the gravitational field of the black hole is indistinguishable from the field of any other object of the same mass M.

You may recall that Newton showed that the gravitational field outside a spherically symmetric body behaves as if the entire mass were concentrated at the center. This holds true in general relativity as well; moreover, **Birkhoff's theorem** states that the gravitational field outside any spherical object, whether black hole or ordinary star, cannot be affected by purely radial changes in the object. If the Sun were to collapse suddenly to a black hole overnight, uniformly toward its center, we would certainly notice the absence of light, but its gravitational field at the distance of its planets would not change; the Earth would continue to orbit exactly as it does now. Indeed, the gravitational field would be unchanged right down to the former radius of the Sun. The bizarre effects of black hole gravity would manifest themselves only in the new vacuum region between this radius and R_S.

One such effect alters the properties of orbits around a black hole. In classical Newtonian gravity, it is always possible to orbit a gravitating body indefinitely, and arbitrarily closely to the body's surface, provided that no energy is lost to dissipation in an atmosphere. It is merely necessary to travel at a high enough speed, in a direction perpendicular to the radial direction, in order to balance the centrifugal and gravitational forces. In relativity, on the other hand, there is an ultimate speed limit, c; nothing can orbit at a speed greater than that of light. Close to a black hole, gravity becomes so intense that there is a minimum radius within which no material object can orbit fast enough to resist infall. At distances smaller than this radius, there are no stable circular particle orbits at all. For a Schwarzschild black hole, this point occurs at 3 times the horizon radius,

$$R_{min} = 3R_S. \tag{9.7}$$

Inside this point, a massive particle may fall in or fly out, but it cannot remain in orbit.

Spacetime curvature also affects the path of light beams; near a black hole, the bending of light becomes extreme. At 1.5 times the Schwarzschild radius, the path of a light beam passing the hole on a trajectory perpendicular to the radial direction is so strongly bent that the beam turns back and traces out a circular orbit. This is called the *photon orbit*, and it occurs at

$$R_{photon} = 1.5R_S. \tag{9.8}$$

Notice that the radius of the photon orbit is within the radius of the last stable particle orbit. The sphere defined by this radius R_{photon} is called the **photon sphere**. Within the photon sphere, not even light can remain in orbit but must move radially inward or outward. Although a spaceship cannot orbit at the photon sphere, it can hover at this distance from the black hole, albeit only by firing its rockets downward, toward the hole. If you were aboard such a spaceship, you could look along the photon orbit and see the back of your own head, by the light curving around the hole!

The intense gravitational field near the black hole can produce many other interesting effects due to the bending of light. Any gravitational field causes light to deviate from the straight trajectories it would follow in an empty spacetime. The effect is significant even for the field of the Sun, but is far more pronounced in the strong curvature of spacetime around a black hole. If a black hole were to cross the line of sight from the Earth to a distant galaxy, the light rays from that galaxy would be strongly bent and deflected as they passed through the field near the black hole. The

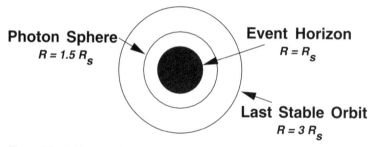

Photon Sphere
$R = 1.5\,R_s$

Event Horizon
$R = R_s$

Last Stable Orbit
$R = 3\,R_s$

Figure 9.3 Orbits around a black hole. The photon sphere is the distance from the event horizon at which light beams are bent into a circular orbit. The last stable orbit is the closest possible circular orbit for material particles.

bending and focusing of light by a gravitational field is called *gravitational lensing*, and the object that creates the image, such as a black hole (although any sufficiently massive object will do), is called a **gravitational lens**. The gravitational bending of light is different from the bending in ordinary glass or plastic lenses. In these, light rays are bent by refraction, the change in the speed of the waves when they pass from one medium (air) to another (optical glass or plastic). Nevertheless, the bending of light rays by a gravitational field can, under the right conditions, also cause an image to form. Many examples of gravitational lensing have been seen, although none is definitely associated with a black hole. Most observed gravitational lensing is due to the presence of a massive galaxy located between us and an even more distant galaxy or quasar. Examples include the striking lensing of distant galaxies by the galaxy cluster Abell 2218 (Fig. 9.4).

Figure 9.4 The gravitational field of galaxy cluster Abell 2218 produces gravitational lens images of an even more distant galaxy cluster. The lensed images appear as arcs circling around the massive core of Abell 2218. (W. Couch, University of New South Wales; R. Ellis, Cambridge University; STScI/NASA.)

Rotating Black Holes

Now, here, you see, it takes all the running you can do, to keep
in the same place.

—Lewis Carroll, *Through the Looking Glass*

So far we have been discussing black holes in terms of only one property, their
mass. What other properties can black holes possess? For example, nearly every
object in the universe rotates, so we would expect that any precursors to black holes
would likely rotate. What happens when a rotating object collapses to a black hole?
What if the star had a magnetic field, or an electric charge? What if the star is oddly
shaped, or, as an even more exotic possibility, what if it were made of antimatter rather
than matter? How would any of these things affect the black hole that is formed by
the collapse of such a star?

As remarkable as it may seem, no matter how complex the object that forms
the black hole, the resulting hole is very simple. The powerful singularity theorems
of Roger Penrose and Stephen Hawking showed that asymmetries or irregularities in
the collapse will not prevent the formation of a singularity. It can also be proved
that the only aspects of the precursor that are "remembered" by a black hole are its
mass, electric charge, and *angular momentum*. The spacetime around a black hole will
always settle down to a smooth, spherically symmetric configuration; any properties
other than these three will produce nonspherical components of the field, which will be
radiated away as gravitational waves. This theorem about the final state of the black
hole is known among relativists as the **no-hair theorem**, from the saying "black
holes have no hair." This does not mean that they are giant eight balls in space, but
rather that they have no detailed structure, or "hair," emerging from the horizon, since
that would break the perfect spherical symmetry of the spacetime surrounding them. If
more matter falls into a black hole, the mass of the hole changes, but the gravitational
field adjusts to maintain the spherical shape.

Electric charge is also preserved in the collapse to a black hole, if the precursor
had any. The electric field lines emerge from a uniformly charged object in purely
radial directions. Hence the electric field is spherically symmetric and is remembered
by the spacetime geometry outside the black hole. However, it is unlikely that any
black holes would actually maintain a net electric charge for very long. If they had
any net charge, they would rapidly attract opposite charges until they were neutralized.
Magnetic fields, on the other hand, are not spherically symmetric, and therefore any
stellar magnetic field will be radiated away as electromagnetic waves. In this respect,
a black hole is quite different from a neutron star. Neutron stars probably possess
enormous magnetic fields, which account for a significant portion of the emissions
from pulsars. Black holes have no magnetic fields of their own.

Angular momentum (spin) is another quantity that is preserved in the formation
of a black hole. If a black hole forms from a rotating object, it will remember the
precursor's original angular momentum. The Schwarzschild metric cannot describe a
rotating black hole; for that we need a more general solution, the **Kerr metric**, so
called for Roy Kerr, a New Zealand physicist who published it in 1963. The Kerr

metric is an *exact* solution to Einstein's equations for a rotating sphere, and it reduces to the Schwarzschild metric when the rotation is zero. When the rotation is not zero, however, the surrounding spacetime is endowed with several new properties.

One of the more remarkable properties of a rotating black hole is the "dragging of frames." This phenomenon was discovered very early in the history of general relativity by J. Lense and H. Thirring, and it is often called the *Lense-Thirring effect* for the general case of any rotating object. Frame dragging means that free-falling geodesics directed initially toward the center of the black hole will not fall straight along a purely radial path but will spiral in the direction of the spin of the hole. In other words, the inertial frames near such a body partake of its rotation. If you fell straight down toward the equator of a Kerr black hole from a great distance, it would feel to *you* that your path was straight and you were not rotating, but a far-off observer would see you spiraling inward as you neared the horizon. To you, on the other hand, it would seem that the distant stars would begin to rotate around you. Like a leaf sucked into a vortex at the bottom of a waterfall, you would be dragged into a spiral by the flow of spacetime. This is reminiscent of Mach's arguments; indeed, the effect is present for *any* rotating body, although it is very, very tiny for anything but a Kerr black hole. The definition of what constitutes an inertial frame is influenced by the rotation of a nearby, dominant mass. Inertial frames are at least somewhat determined by the distribution of matter. In its own vicinity, a rotating body vies with the overall matter distribution of the universe to establish what constitutes a local inertial frame.

The rotation of a black hole also alters the event horizon. What was a single event horizon for a Schwarzschild black hole now splits into two surfaces. The innermost surface, which is spherical and lies inside the usual Schwarzschild radius, is an event horizon, and is similar to its Schwarzschild counterpart in that it represents the "point of no return" for an infalling particle. The radius of the event horizon of a Kerr black hole is given by

$$R_K = \frac{G}{c^2} \left(M + \sqrt{M^2 - a^2} \right), \tag{9.9}$$

where a is a measure of the angular momentum in units of mass. If the hole is not rotating, then the angular momentum is zero, and the Kerr radius is equal to the Schwarzschild radius. As the spin of the black hole increases, the radius of the horizon shrinks. Even more interesting, this equation shows that there is a limit to the rotation speed even of a black hole. A black hole can exist only for $a \leq M$; a hole for which $a = M$ is said to be *maximally rotating*. At this limit, the Kerr horizon is rotating at the speed of light and has shrunk to half the Schwarzschild value.

The outer surface is called the **static surface**. It is oblate and touches the event horizon at the rotational poles of the black hole; it coincides with the Schwarzschild radius at the equator of the hole. This surface is called the static surface because at or inside this point nothing can remain static, that is, motionless, with respect to the spatial coordinates. If a spaceship slowly descends straight toward the hole while attempting to remain aligned with a distant star, the crew will find that in addition to firing their engine toward the hole to combat the pull of gravity, they must also aim against the direction of the hole's rotation, in order to maintain their position relative to the distant star. The effect becomes stronger as they near the static surface. Finally, at the static surface, it is necessary to move at the speed of light opposite the hole's direction of spin to avoid being dragged with the rotation of the hole; that is, it is necessary to move as fast as possible, just to stand still! The black hole compels free-falling trajectories to participate in its rotation; freely falling light cones tip increasingly toward the direction of rotation of the hole as the static surface is approached. Inside the static surface, the rotation of spacetime is so great that not even light can resist being dragged around the hole. Unlike the event horizon, however, the static surface is not a "one-way membrane"; it is possible to pass through it from the outside, and return to tell the tale.

The region between the horizon and the static surface is called the **ergosphere**. In principle, it is possible to extract energy from the ergosphere, a property from which its name is derived (from the Greek *ergos*, work or energy). An advanced civilization might accomplish this by sending spaceships into an appropriate orbit just inside the static surface. The spaceships would eject something—nuclear waste, perhaps—into the hole, in the opposite sense to its rotation. (That is, if the hole were rotating clockwise, as seen from its north pole, the spaceship would dump its load of waste into the horizon with a counterclockwise spin.) The waste would disappear forever into the hole, while the spaceship would acquire a "kick" of energy, leaving its orbit with more energy than with which it entered. If this is done inside the ergosphere, the energy thus acquired can actually exceed the rest energy of the waste sent down the hole. From where did the extra energy come? It came from the rotational energy of the black hole. Because material was sent into the black hole with opposite angular momentum, the hole is left with slightly less spin as a result of this encounter. In principle, a great deal of energy could be extracted from a Kerr black hole in this manner, but, lest you think that this would be an infinite source of energy, remember that as rotational energy is removed, the black hole must slow down. Eventually, all the rotational energy would be gone, and the Kerr black hole would become a Schwarzschild black hole. A classical Schwarzschild black hole is truly "dead" in the sense that no energy can be removed from it, not even by perturbing it.

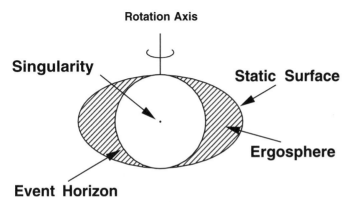

Figure 9.5 Components of a rotating Kerr black hole. The ergosphere (*hatched region*) of a Kerr black hole is located between the static surface (*outer curve*) and the event horizon (*inner curve*).

Hawking Radiation

Until now, our discussion of black holes has dealt with the consequences of the classical theory of general relativity. When we try to add the strange world of quantum mechanics to that of general relativity, we find that it is not quite true that Schwarzschild black holes never lose energy. Recall that Einstein's equations imply that gravity itself possesses energy. In principle, then, the spacetime curvature around a black hole could be tapped as a source of energy, even for the case of a nonrotating, stationary, Schwarzschild black hole. The Schwarzschild hole is "dead" only in the *classical* universe, that is, a universe without quantum mechanics. Although no one has succeeded in fitting gravity completely into a quantum-mechanical description, it is possible to carry out quantum calculations on a "background" of a smooth, curved spacetime (the *semiclassical* approach). Stephen Hawking of Cambridge University performed such a calculation and found something remarkable: black holes are not completely black. They actually *emit* radiation, although the amount is extremely tiny for most black holes.

What is the source of this **Hawking radiation**, if nothing can ever escape from a black hole? Its existence depends upon the quantum mechanical effect known as the Heisenberg **uncertainty principle**, which states, among other things, that energy need not be strictly conserved for short times, provided that it is conserved, overall, for longer time intervals. The greater the violation of energy conservation, that is, the more energy "borrowed," the more quickly it must be repaid. In the quantum universe, even the purest vacuum is filled with a "sea" of **virtual particles,** which appear as particle–antiparticle pairs and then disappear in the fleeting interval of time permitted by quantum mechanics. Can one of these virtual particles ever become real? Yes, if that particle can acquire the energy, it might be said, to pay off its loan of energy before it comes due.

This effect can be demonstrated in the laboratory. Set up two parallel conducting metal plates separated by an empty gap. Onto these parallel plates place opposite electrical charges, creating a voltage difference and a strong electric field running from one plate to the other. Now place the apparatus in a vacuum chamber, and increase the voltage across the plate. In the vacuum between the plates, negatively charged virtual electrons and positively charged virtual positrons are popping in and out of existence. But because they are doing so in the presence of an electric field, and because the field exerts a force upon charged particles that has opposite directions for opposite signs, a particle could be accelerated away from its antipartner before they have a chance to annihilate and disappear. The virtual electron is accelerated toward the positive plate, and the virtual positron is accelerated toward the negative plate. The virtual particles thus can become real, and we can measure this flow of electrons and positrons as a net electrical current from one plate to another. We see a current flowing in what we thought to be a vacuum! The energy for this process comes from the electric field, so energy is still conserved. Through a quantum-mechanical process, some of the energy stored in the electric field is converted into matter, in accordance with Einstein's law $E = mc^2$. This phenomenon is called *vacuum breakdown* and is an extreme example of a more general effect called *vacuum polarization*. The experiment demonstrates that the vacuum is not "empty," but is filled with virtual particles and fields. The virtual particles can also affect real particles in very small, but measurable, ways.

Near a black hole, virtual pairs are created and destroyed, just as they are everywhere. But near the horizon, the tidal forces are strong, and stress from the tidal forces can be utilized to bring a pair of virtual particles into real existence. One member of this pair of particle Pinocchios falls into the horizon, while the other escapes to infinity. The emergent particles are Hawking radiation. In analogy to the electric field discussed above, the energy to create the particles comes from the energy of the gravitational field outside the hole. As that gravitational energy is lost to the creation of particles, the strength of the gravitational field is diminished, and the hole shrinks. Eventually, it *evaporates* and disappears from the universe. The final moments in its evaporation produce an intense burst of very high energy particles and gamma rays. Since the radiation originates with virtual particle-antiparticle pairs, we should expect equal amounts of matter and antimatter to emerge from the hole. In fact, the easiest particle-antiparticle pairs to create are photons, particles of light. Photons, which are their own antiparticles, can appear at any energy level, whereas massive particles must be derived from at least as much energy as their rest energy requires. It is quite remarkable that a black hole could be a source of *any* kind of particle, since the classical theory of relativity predicts that matter can only disappear beyond the event horizon. But now we find that, whatever the black hole originally consisted of, it emits Hawking radiation composed of photons, matter, and antimatter.

Most black holes are not really very efficient at this process, however. Since the energy to create Hawking radiation comes from the tidal stresses, there must be substantial tidal stress present on the typical scale over which the virtual particles move. Because these particles exist only for a tiny span of time, this is a very small scale indeed. The larger the hole, the weaker the tidal stress on a small length scale; therefore, Hawking radiation is significant only for small holes. The time for evaporation of a black hole is proportional to the cube of its mass:

$$t_{\text{evap}} \approx 10^{10} \left(\frac{M}{10^{12} \text{ kg}} \right)^3 \text{ yr.} \tag{9.10}$$

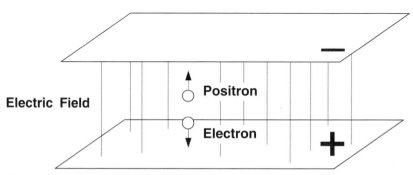

Figure 9.6 Virtual particles can become real particles if they can tap into a source of energy. Here an intense electric field between two charged plates accelerates a virtual positron and electron away from each other, endowing them with the energy necessary to become real. The energy is obtained from the electric field.

If you tried to watch a solar-mass black hole slowly evaporate as a result of Hawking radiation, you would have to wait about 10^{65} yr. Since we think the universe has an age of roughly 10^{10} yr, you can see that this is a very long time. Nevertheless, if the future of the universe extends to infinite time, as it does in some models, eventually all black holes will decay.

For Hawking radiation to be of much significance in the present universe, the hole must be very tiny, a "minihole." The only black holes that would be evaporating now would have masses of the order of 10^{12} kg, with a corresponding Schwarzschild radius of about 7×10^{-16} m; a "minihole" indeed. There is no observational evidence for the existence of such miniholes. Whereas large black holes can form from the collapse of ordinary astronomical objects such as stars, the only conditions under which miniholes could form would be inhomogeneities in the very early universe. There are severe constraints on the numbers of such tiny holes that could be produced in the big bang, making it doubtful that any such miniholes exist. For a black hole of stellar or greater mass, Hawking radiation would be essentially undetectable and would have no significant effect over most of the life expectancy of the universe. It is, however, a genuine phenomenon for even the largest black holes.

The fact that black holes radiate means that they have a *temperature*. An ideal emitter, or blackbody, radiates a continuum spectrum of photons, and that spectrum is uniquely determined by its temperature. The higher the temperature, the more energetic the spectrum. Remarkably, Hawking radiation turns out to be blackbody radiation. The temperature of this radiation, and thus of the black hole, is given by

$$ T_{bh} \approx 10^{-7} \left(\frac{M_\odot}{M} \right). \tag{9.11} $$

The radiation emitted by a solar-mass black hole is very small, so it has a very tiny temperature, only 10^{-7} K. Larger holes have even lower temperatures; a black hole with a mass of 10^6 $M_\odot$ has a temperature of only 10^{-13} K.

The realization that black holes emit radiation led Hawking to a wonderful unifying concept for black holes. Following an idea of Jacob Bekenstein, Hawking had already developed a theory of the merging of two black holes; such a merger forms a single black hole, with a surface area that is larger than the combined surface areas of the previous two separate holes. This was the law of black hole areas: regardless of anything black holes might do, whether they collide, gain more matter, or anything else, the result will always be a hole with a larger surface area than it had before. This is very reminiscent of the second law of thermodynamics, which states that the entropy, that is, the disorder, of an isolated system must always increase. If the size of a black hole can be equated with its entropy, then that implies that the black hole should be described by the laws of thermodynamics, which, in turn, means that it *should* have a temperature. Hawking radiation accounts for that black hole temperature, allowing Hawking to formulate all these ideas into the laws of **black hole thermodynamics**. We shall return to this topic later, for it has tantalizing implications for the evolution of the universe.

Black Hole Exotica

Hawking radiation seems quite odd, but there are even stranger things allowed by classical general relativity theory. One of these is the *white hole*, a kind of mirror image of the black hole. In a white hole, nothing can get *in*; it can only come *out*. Rather than spacetime flowing into the horizon of a black hole at the speed of light, spacetime flows out of the horizon of a white hole at the speed of light. Although this is intriguing, we know of no way in which a white hole could form. A black hole can be created in a straightforward manner, by the gravitational collapse of a star. A white hole would have to be inserted into the universe as an initial condition. And where would it get the matter and energy that would pour forth? Just because it is difficult to imagine does not mean that it cannot be, but in the absence of any evidence to the contrary, we can say with some confidence that white holes do not exist.

The mathematics of the Schwarzschild solution admits the possibility of another odd beast, the *wormhole*. The spacetime curvature produced by a Schwarzschild black hole can be envisioned as a kind of "funnel" in spacetime. A wormhole is a connection from one such spacetime funnel to another. Might the wormhole form a gateway from one point in spacetime to another? Could a spaceship travel through wormholes to reach very distant locations in both space and time? Unfortunately, this is not the case, at least not for the Schwarzschild wormhole. All paths through the wormhole that avoid the singularity are spacelike; that is, they can be traversed only at speeds greater than that of light. As we have seen, a massive particle cannot travel such a world line. Moreover, the full solution for Schwarzschild wormholes shows that they are dynamic and evolve. They turn out to be unstable; they pinch off, trapping anything within them at the singularity. Clearly, this is not a desirable feature for a transportation system.

Recently, Kip Thorne and colleagues at the California Institute of Technology have examined the wormhole solution in more detail. They have found circumstances under which it might be possible to construct a wormhole with a route that could be followed by a timelike path. This will most likely not help us find shortcuts from

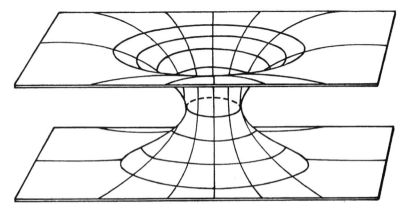

Figure 9.7   A wormhole is a solution of Einstein's equations that appears to connect two different universes or different regions of one universe. However, the classical wormhole is unstable and cannot be traversed by a timelike world line.

one galaxy to another, however; such a wormhole requires exotic conditions which are unlikely to exist in the physical universe. Moreover, it is still unclear whether it might yet violate physical laws, and even if it does not, it is probably too narrow for anything but an elementary particle to follow. It seems, then, that wormholes may be useful as a plot device in science fiction but have little, if any, relevance to the real universe. Why, then, do relativists study them? Aside from the intellectual pleasure of exploring such a unique topic, the study of wormholes and the possible quantum effects near them is a way of investigating the nature of quantum gravity. The odd properties of Hawking radiation, wormholes, white holes, and the like, provide insights to the properties of that as-yet-unknown theory.

The Role of Black Holes in the Universe

We have mentioned several exotic aspects of black holes, but always with the caveat that these effects are probably not important in the present universe. What, then, is the relevance of black holes now? Do they exist, and, if so, do they play any important roles, or are they merely mathematical oddities of the equations of general relativity?

Black holes must be the end stage of the evolution of very massive stars. If, upon consuming all its nuclear fuel, a star is left with a core mass greater than the upper limit for a neutron star, collapse is inevitable. Black holes might also form at the centers of dense clusters of stars, or in the cores of galaxies, perhaps as those galaxies are created. Because general relativity *requires* a black hole to form if the conditions are appropriate, the universe may well harbor countless black holes of varying sizes, ranging from modest black holes left behind after supernova explosions, to huge monsters residing at the center of galaxies. Yet by their very nature, black holes are *black*; they emit no light, and the tiny amount of Hawking radiation from any moderate-sized black holes would be completely undetectable. How, then, can we see them?

The answer is that while we cannot see the holes, we can infer their presence by their effects on the light and matter that we *can* see. We have already mentioned how a passing black hole can distort the light from a distant star or galaxy in a way we might be able to detect. But any object with mass can produce a gravitational-lens effect, and it would be difficult to distinguish a lens created by a black hole from one created by a dim, but otherwise normal, star. The easiest way to detect black holes is by their gravitational attraction on other matter. There are many possibilities, depending on the mass of the hole and the kind of matter that surrounds it. Stars and gas can fall into the gravitational well of a black hole. Stars can be torn apart by tidal forces; gas can be heated to enormously high temperatures, compressed, and shocked. Gas orbiting a black hole can be whipped around at extraordinarily high velocities in a very small region of space. We could observe such things with our increasingly sensitive telescopes, especially the *Hubble Space Telescope*. But evidence for the existence of compact sources of energy has been seen for over thirty years.

Where is the first place to go looking for a black hole? Since we believe that supermassive stars collapse to form black holes, there should be numerous stellar-mass black holes in our own Galaxy. Finding them would not be easy because, as we have noted, an isolated black hole will produce no luminosity. Hence our first candidate locations should be *binary* stars, stellar systems consisting of two stars orbiting one another. In a binary system, the gravitational effect of a black hole will influence its visible companion in a detectable way. In some rare cases, we can observe the wiggles in the motion of a star with an unseen partner and deduce from Kepler's laws that the mass of the invisible companion must be greater than that allowed for a neutron star. All such detections are still somewhat uncertain, and none has established the presence of a black hole beyond all doubt. However, if the two stars are sufficiently close, gas from one star can be pulled away from it and fall onto the black hole, becoming very hot and radiating high-energy photons before disappearing down the hole.

Astronomers had not really given this idea much thought until the early 1970s, when the X-ray satellite *Uhuru* detected powerful X-rays coming from the constellation Cygnus. This source, designated Cyg X-1, proved to be a binary system that emitted energetic X-rays, but irregularly; the X-rays "flickered" over a very short time interval, about 0.01 s. Observations made with optical telescopes determined that the system included a hot, massive star. Wobbles in its motion made it possible to surmise that it has a companion with a mass of about 5–10 times that of the Sun. Furthermore, this companion could not be detected by optical telescopes. The X-ray flickering is thought to occur in hot gas near the invisible companion. The rapidity of the variation is significant in establishing the size of the companion, because no object can vary in a systematic and regular fashion on timescales shorter than the time it takes light to cross it. As an analogy, imagine a huge marching band, spread out over such a large area that it takes 10 s for sound to travel from one end to the other. Now imagine that the musicians are blindfolded and must play from what they hear. Such a band could not play staccato notes in unison every half-second. The sound would arrive at a distant listener spread out over a 10 s interval. Since the speed of light is the fastest speed attainable, the largest region of an astronomical body that can be in causal contact over a time interval Δt, is of size $c\Delta t$. In the case of the unseen member of the Cyg X-1 binary, light can travel only approximately 3000 km in 0.01 s. If the radius of the "dark star" is indeed 3000 km or less, it is a little smaller than the size of the Earth. Existing stellar theory cannot accommodate such a large mass for such a small star, in any form other than a black hole. Neutron stars cannot have a mass more than about 3 times that of the Sun; thus the unseen companion is either a black hole or something unknown to current theory.

What is the origin of the X-rays from Cyg X-1? The X-rays are produced in gas which is lost from the normal star and drawn to the black hole. Because the gas possesses some angular momentum, it orbits around the black hole, flattening into a spinning disk of gas called an **accretion disk**.[1] Turbulence in the disk causes the gas to spiral slowly toward the black hole. As the gas falls into the gravitational potential well of the black hole, it loses gravitational potential energy in exchange for a gain in other forms of energy; half of the gravitational energy is converted into heat energy. Collisions between the infalling gas and the matter already occupying the region close to the horizon could also compress and heat the gas. A sufficiently hot gas will emit X-rays, just as a cooler gas might emit visible light.

How much energy might be thus liberated when gas falls into a black hole? There is no clear answer to this question, as it depends upon details of the behavior of the infalling gas, but theoretical estimates range from a few percent, to as much as 40%,

Figure 9.8   Model for a binary system consisting of a hot, massive star and a companion black hole. Gas is drawn from the normal star and forms an accretion disk around the black hole. (STScI/NASA.)

of the rest energy, m_0c^2, of the gas. Considering that nuclear reactions release at most about 1% of the rest energy of the matter, it is clear that "gravity power" is a far more efficient means of generating energy.

Accretion disks can make black holes detectable at great distances, if only indirectly. In the early 1960s, when radio astronomy was yet a young science, astronomers mapped the sky at radio wavelengths, finding an abundance of radio sources. Most could be identified with known objects. For example, the center of the Milky Way Galaxy, which is located in the constellation Sagittarius, was found to shine brightly in the radio. (We cannot see the center of our own Galaxy in optical wavelengths because of absorption by intervening clouds of cosmic dust.) Many other radio sources were associated with optical galaxies; one of the brightest radio objects in the sky is a galaxy which we see across a distance of 500 million light years, Cyg A. Cyg A throws out over 10 million times as much radio energy as an ordinary galaxy. But not all bright radio sources could be traced to optical galaxies; many such sources, when seen in optical wavelengths, were, in the 1960s, indistinguishable from stars. It was clear, however, that they could not be normal stars; normal stars are dim in the radio, and these were bright. More mysterious yet was the presence of unrecognizable emission lines in their optical spectra. Some scientists went so far as to propose that an unknown element existed in these objects.

The resolution of this puzzle came in 1963 when astronomer Maarten Schmidt recognized that these strange lines were, in fact, the usual lines of hydrogen, but redshifted so much that they appeared in a completely unexpected portion of the spectrum. Cosmic redshift, as we shall discover in chapter 10, indicates distance. The large redshifts that were measured for these objects implied fantastic distances, up to *billions* of light years. To be visible over such distances, the objects had to be almost unimaginably luminous. How luminous? The objects outshine even the brightest galaxies by factors of 100 or more. In some cases, a luminosity 10,000 times greater than that of an ordinary galaxy would be required for an object at such a great distance to appear so bright. Clearly these objects, whatever they were, were not stars. Since it was uncertain *what* they might be, and because on photographic plates they appeared as unresolved starlike points, they were called **quasi-stellar objects**, which is often shortened to **quasars**; they are also sometimes referenced by their acronym QSOs.

It was soon discovered that the light output of many quasars fluctuates considerably over short intervals of time. As for Cyg X-1, the distance light travels over the interval of the variations sets an upper limit to the size of the source. Hence, changes in the appearance of quasars over times of days or less means that the light must be coming from a region less than about a light day in size. All that energy is coming from a region comparable to the solar system in extent.

The rapid oscillations and the starlike appearance point to a very compact energy source for quasars. Stars alone could never provide so much energy; what could? The best theory available today requires supermassive black holes, holes with masses from several million up to a billion times that of the Sun. Such holes have Schwarzschild radii as large as the orbit of the Earth around the Sun, and would be surrounded by hot gas spiraling into the hole through a huge accretion disk. The whole system would be comparable in size to our solar system and could process each year several solar masses' worth of gas. If a black hole could release just 10% of the rest energy of this gas via the accretion process, then the consumption of 1 $M_\odot$ of gas per year would provide enough energy for a luminosity roughly 100 times that of a garden-variety

Figure 9.9 *Hubble Space Telescope* photograph of a quasar. The *HST* resolves the quasar 1229+04 into a point source surrounded by an ellipsoidal figure, perhaps like the bar of a barred spiral galaxy. The image clearly shows at least one spiral arm, with a bright spot suggesting hot gas or recent star formation at one location on the arm. (J. Hutchings, Dominion Astrophysical Observatory; STScI/NASA.)

galaxy. A typical spiral galaxy might shine with the brightness of 10^{11}–10^{12} $L_\odot$, while an average quasar emits 10^{13} to 10^{14} $L_\odot$.

Quasars are not the only place where we might find supermassive black holes. The center of a galaxy represents another place where gravitational collapse might occur; the larger the galaxy, the more prone to collapse its core might be. A small minority, about 1%, of the galaxies we observe are **active galaxies**; that is, they emit more than just ordinary starlight. Active galaxies occur in diverse shapes and sizes. One category is known as *Seyfert galaxies*; these appear to be typical spiral galaxies but have abnormally bright centers with bright emission lines from hot gas. *BL Lacertae* objects are distant elliptical galaxies with a rapidly varying, unresolved point of nonstellar emission in their cores. *Radio galaxies* are giant elliptical galaxies that produce large amounts of radio energy in their central regions; examples include Cyg A, mentioned above, and the famous galaxy M87 in the constellation Virgo.

The common feature of all of these galaxies is that they emit copious amounts of nonstellar energy from a relatively small region in their centers, or nuclei. These bright cores are thus called **active galactic nuclei**, or **AGN**s. The AGNs that are brightest at radio wavelengths show an even more remarkable feature: *radio jets*, beams of radio-emitting matter, probably in the form of an energetic gas, which appear to be shot from the very heart of the galaxy. Some active galaxies possess two symmetric jets, directed oppositely away from the center. Sometimes only one jet is seen to emerge from the galaxy, but maps of the radio energy reveal *radio lobes*, large regions of diffuse radio emissions, on both sides of the galaxy. The single jet almost always runs from the center of the galaxy to one of the radio lobes; the partner jet on the other side of the galaxy is believed to be present, but unseen because it is beaming away from our line of sight.

Some of the most powerful jets are gigantic, as much as 3 million light years in length. These huge jets have been powered over their lifetimes by enormous amounts of energy, as much as 10 million times the rest energy of all the matter in the Sun. Some jets are moving so fast that they can be observed to change over short timescales; anything that changes on a human timescale is astoundingly fast by astronomical standards. Some jets appear to move faster than light, but this can be explained as an illusion caused by the beaming toward us of a jet whose gas is moving at speeds near that of light. We can thus infer that the most energetic jets consist of gas moving at relativistic speeds.

Jets require a compact energy source in the center of the galaxy. It must be a source that is capable of beaming huge quantities of energy in a specific direction for a very long time; it must also be capable of processing millions of solar masses of matter into energy over the jet's lifetime, at high efficiency. Again, the best candidate for such a powerhouse is a supermassive black hole. Gas would be squirted out from an enormous accretion disk in the two directions perpendicular to the disk, along its axis of rotation. Perhaps if the supermassive hole is a Kerr hole, its rotational energy might be tapped somehow. A spinning black hole represents a huge reservoir of available energy, if that energy can be extracted. One suggestion involves magnetic fields, generated in the surrounding accretion disk and connecting it with the black hole. As the field lines are wound up, they accelerate and focus the outflowing jets. This picture is somewhat speculative, but plausible. The study of AGNs, jets, accretion disks, and supermassive black holes is one of the most active areas of research in astronomy,

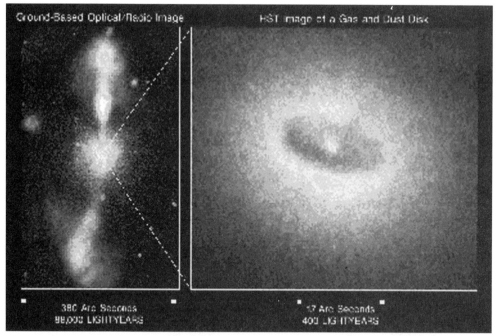

Figure 9.10 *Hubble Space Telescope* observation of the core of an active galaxy, NGC 4261. On the left is a photograph combining data from ground-based optical and radio telescopes, showing powerful jets emanating from the core of the galaxy. The right-hand photo is an *HST* image of the very central region of the galaxy, possibly showing an accretion disk. (H. Ford, Johns Hopkins; W. Jaffe, Leiden Observatory; STScI/NASA.)

both in theory and in observations. New observations continue to provide better data, by which theories can be tested, but improved data also present us with new mysteries.

The giant elliptical M87 is a famous example of an active galaxy. It has been known for many years that this galaxy is special. It is unusually large, even for a giant elliptical; its volume is nearly as great as that of the entire Local Group. It sits at the apparent center of a very large cluster of galaxies, the Virgo Cluster. Its core is prodigiously energetic, and a well-defined jet shoots from the heart of the galaxy. This jet is bright not only at radio wavelengths but at optical and higher wavelengths as well. The jet is observed to emit *synchrotron radiation*, a well-defined pattern of wavelengths characteristic of electrons spiraling around a magnetic field. It was long suspected that M87 might harbor a black hole, but it is some 50 million light years distant, and even the best ground-based telescopes could not clearly resolve the motions in the innermost regions of the galaxy. Once the *Hubble Space Telescope* was repaired, M87 was one of its first assignments for spectroscopy. The results were spectacular; gas in the central 60 lt yr of M87 rotates much more quickly than can be easily explained by any hypothesis other than orbits around a supermassive black hole. At last the black hole has emerged from the pages of texts on general relativity and shown itself to be as much a resident of the real universe as the stars.

In the last few decades, many observations have suggested that quasars are themselves extreme examples of active galaxies. They are so far away that we see them

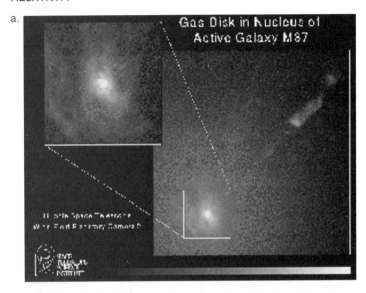

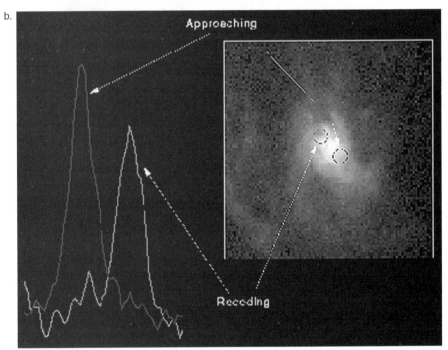

Figure 9.11 (*a*) The giant elliptical galaxy M87 (cf. Fig. 4.15) as seen from the *Hubble Space Telescope*. The right-hand picture shows the jet emerging from the core of the galaxy, which is seen in closeup in the upper right. (*b*) Emission lines from the central core of M87 show Doppler shifts. Gas that is approaching us is blueshifted, while gas that is receding from us is redshifted. The sharp peaks of the two spectral lines, and their wide separation, indicating a large velocity difference, can be easily explained only by the presence of a massive black hole at the center of the galaxy about which this gas is orbiting. (H. Ford, Z. Tsvetanov, A. Davidsen, and G. Kriss, Johns Hopkins; R. Bohlin and G. Hartig, STScI; R. Harms, L. Dressel and A. K. Kochhar, ARC; Bruce Margon, University of Washington; STScI/NASA.)

when the universe was much younger than it is now; studying them provides clues to the history of the universe. Quasars are often found in association with other objects which definitely have the appearance of galaxies, and which have similar redshifts. It has been possible to detect the faintest wisp of spirals around some QSOs themselves. If quasars/QSOs are indeed active galaxies, then they join the lineage as its most energetic members.

All these objects have some things in common: an apparently very compact energy source and an astonishing output of energy. If all are powered by black holes, they can be explained by differing scales: the larger the central black hole, and the more gas available for its appetite, the greater the energy it would generate. If this hypothesis is correct, active galaxies in general, and especially quasars, could not sustain such an outpouring of energy for long periods. We might see active galaxies during an explosive stage of their existences. They blaze for only a short time, on cosmological timescales, then, when the black hole has devoured all the matter readily available in its vicinity, the galaxy settles into quiescence. Perhaps it will sit placidly for the remaining lifetime of the universe, or perhaps another source of gas will replenish the accretion disk and cause a new outburst. We can see only a snapshot of the universe as it is during our short lifespans; we cannot watch the evolution of an active galaxy. It is as if we could visit a family reunion only once, seeing various members of a family each at a single age; from that information, we could try to construct a hypothesis of how a given individual would age and change throughout life. We do much the same with active galaxies, making our best effort to understand them with the data available to us.

KEY TERMS

black hole	escape velocity	Schwarzschild radius
event horizon	singularity	coordinate singularity
quantum gravity	cosmic censorship	Birkhoff's theorem
photon sphere	gravitational lens	no-hair theorem
Kerr metric	static surface	ergosphere
Hawking radiation	uncertainty principle	virtual particles
black hole	accretion disk	quasi-stellar object
thermodynamics	active galaxy	(QSO)
quasar		active galactic nucleus
		(AGN)

Review Questions:

1. A neutron star is very compact and dense, but it is not a black hole. If a typical neutron star has a mass of 2.5 $M_\odot$, what is its Schwarzschild radius? If the actual radius of the neutron star is 30 km, how does this compare to the Schwarzschild radius?

2. Excited atoms of hydrogen emit light with a wavelength of 1216 $\times$ 10^{-10} m (that is, 1216 Å). Suppose that you detect this line in emissions coming from a very compact

source within the Milky Way Galaxy, but you measure its wavelength to be 1824×10^{-10} m. What might account for the change in wavelength? If the light originated from near a black hole, from how close to the Schwarzschild radius, expressed as a fraction of R_S, was the radiation emitted?

3. Define and distinguish *singularity*, *coordinate singularity*, and *event horizon*.

4. You are the commander of an exploratory mission to a black hole. You launch a robotic probe on a trajectory that will take it into the black hole. The probe has an internal clock and sends a radio pulse back to your ship at a fixed interval, in the reference frame of the probe. What effects do you observe in the signals from the probe as it approaches the black hole? What might you observe if the hole is rotating?

5. You observe an X-ray source to vary on a timescale of 0.001 s. What is the upper limit for the size of the X-ray emitting region? What is the mass of a black hole with a Schwarzschild radius of this size?

6. The Earth is a rotating body whose gravity is not as strong as that of a black hole. Does the Earth exhibit any of the effects we discussed for black holes, such as gravitational redshifts, frame dragging, or gravitational time dilation? How are the Earth and a black hole alike, and how are they different?

7. For a rotating (Kerr) black hole, define and distinguish the *static surface*, the *event horizon*, and the *ergosphere*. How might energy be extracted from a Kerr black hole? Is this an infinite source of energy for some advanced civilization?

8. How massive would a black hole have to be in order for it to evaporate due to Hawking radiation in only 1 year? How big is that mass compared to some object with which you are familiar? (On the surface of the Earth, 1 kg $\approx$ 2.2 lb.)

9. Discuss the leading model for X-ray emissions from a binary system that might include a black hole. If nothing can escape from inside a black hole, from where is the energy coming?

10. Discuss the "unified theory" for AGNs. Include topics such as the possible identity of the "central engine," the origin of jets and radio lobes, and the range of activity.

11. If the Sun collapsed and formed a black hole, how would the orbit of the Earth be affected? Would any gravitational radiation (gravitational waves) be produced?

12. The Milky Way Galaxy may have a total mass of around 10^{12} $M_\odot$, or one trillion times the mass of the Sun. What is the Schwarzschild radius for the Milky Way? Divide the mass of the Galaxy by the volume of such a black hole ($\frac{4}{3}\pi R_S^3$) to obtain the density of such a black hole. (The mass of the sun $M_\odot = 2 \times 10^{30}$ kg, and the Schwarzschild radius of the Sun is 3×10^3 m.) How does that density compare with water, which has a density of 10^3 kg /m^3?

13. You plan to take a spaceship down to the photon sphere and hover above the black hole to observe the back of your head. What sort of acceleration will you experience as you hover at this point? (Answer qualitatively, e.g., small, comparable to one g, several times g, much bigger than g, incredibly huge.) Does your answer depend upon the mass of the hole?

Note

1. See also chapter 5 and Figure 5.7.

PART IV
The Big Bang

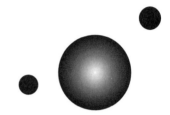

CHAPTER 10

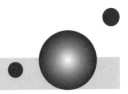

The Expanding Universe

For at least as long as written history has existed, humanity has set its sights upon understanding the shape, scope, and history of the universe. To this task we bring our senses, our experience, and our reason. This was as true for ancient cosmologists as it is for modern scientists. Today, however, our senses are augmented by powerful tools, we benefit from the accumulated and recorded experience of many generations, and we have developed mathematical languages that provide an efficient means to systematize our reasoning. In this chapter, we shall see how these advances led to one of the greatest discoveries in history: the expanding universe.

The biologist or the geologist is accustomed to gathering data in the field; the chemist, to the direct manipulation of molecules in the test tube; the physicist, to the construction of apparatus to measure a particular phenomenon. The astronomer, in contrast, must be content to look. The only exceptions are the occasional meteorite, and the several hundred kilograms of moon rocks returned to Earth by the Apollo astronauts. And although we have sent robotic investigators to other planets to do our experiments and sampling remotely, these efforts are confined to our own solar system and will remain so for the indefinite future. Almost everything we know about the universe at scales larger than that of our solar system comes from the electromagnetic radiation we collect. Even what knowledge we might obtain about the nonluminous matter in the universe must be inferred from its effect upon the matter we can see. Despite this fundamental limitation, the astronomer can learn a great deal through careful observations of the light which reaches the telescope.

Although we tend to think of light as just something that illuminates our surroundings, there is much more to it than our eyes can see. Light, or electromagnetic radiation, can be found in a full range of energies. This distribution of energies makes up the electromagnetic spectrum; what we call visible light is just one small range of light's energy spectrum. Complete knowledge of an astronomical object requires observing across the full spectrum. As a possibly more familiar example, consider the spectrum of sound frequencies heard while listening to music; different notes of music correspond approximately to different frequencies of sound wave. You would probably agree that a Beethoven symphony is best appreciated by registering the arrival of individual notes and distinguishing among the different sound frequencies. There would be much less benefit in summing the total sound energy of all the notes arriving at a microphone during the performance! Similarly, while it can be quite useful to measure the total luminous flux from an astronomical object, a far greater wealth of information is obtained by performing *spectroscopy*, the measurement of the quantity of light energy at each wavelength. The technique of spectroscopy was developed and applied to astronomy late in the nineteenth century. The significant advances that made this possible were the recognition that different elements have unique spectral signatures; the development of photographic techniques that not only could make a permanent record of a star's spectrum but also were much more sensitive than the eye; and finally, the construction of substantially larger telescopes that made it possible to collect enough light to perform spectroscopy on faint objects.

So much useful information is derived from the spectrum of electromagnetic radiation that telescopes devote most of their time to spectroscopy. In analyzing a spectrum, an astronomer considers many issues. Which lines are present? Are they emission or absorption lines? Are they shifted from their laboratory-based standard positions? What is the overall distribution of energy in the spectrum? The particular lines present in the spectrum of an astronomical object can identify the composition of the emitter, while other characteristics of the spectrum give clues to the temperature of the object, its internal motions, and the processes occurring within it. From this we learn such things as how the light was emitted, which elements are present and in what abundance, the velocities of the gas that emitted the light, and what population of stars a galaxy contains.

When an astronomer compares the spectrum of a star or a galaxy to laboratory standards, the emission or absorption lines associated with individual molecules and elements are typically not located at exactly the same wavelengths as the standards. Since we now have a good understanding of the elements, we would not be inclined to hypothesize the existence of new elements. The discrepancy, instead, is explained by an overall shift of the spectrum. A shift to longer wavelengths and lower energies is called a **redshift**, while a shift to shorter wavelengths and higher energies is a **blueshift**. These are the generic terms for these shifts, even if the radiation detected is not near the "red" or "blue" part of the visible spectrum. Because the relative *spacing* between the lines of a given element never changes, and a red- or blueshift occurs for the spectrum as a whole, it is still possible to identify elements in a shifted spectrum by comparison to measurements in our Earthly laboratories.

How might the spectrum of light be blue- or redshifted? The most mundane, and prevalent, process is the ordinary Doppler effect, which is a consequence of the relative motion of the source of the light and our detector. Doppler shifts are easily detected and are an important source of information about motions in the universe. The formula for the nonrelativistic Doppler shift is

$$z = \frac{\lambda_{\text{rec}} - \lambda_{\text{em}}}{\lambda_{\text{em}}} = \frac{v}{c}. \tag{10.1}$$

The shift is symbolized by z; this quantity is also often called the "redshift" whether it actually represents a redshift or a blueshift. The shift z is a positive quantity when light is redshifted, that is, when λ_{rec} is greater than λ_{em}, and is negative when it is

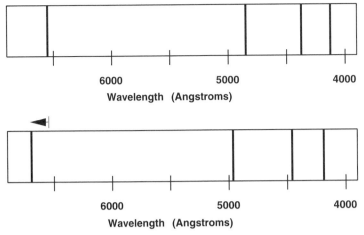

Figure 10.1 *Top*: a spectrum of the element hydrogen, showing four emission lines from the so-called Balmer sequence. *Bottom*: these four hydrogen emission lines shifted to the red by a fixed amount corresponding to a redshift of about 2%.

blueshifted. This equation must be suitably modified when dealing with relativistic velocities, as we showed in our discussion of special relativity.

Even though the Doppler shift almost always tells us only about the relative *radial* motion of the source,[1] it still provides abundant information. For nearby objects, an overall Doppler shift indicates whether they are approaching or receding from Earth. Many objects show both a red- and a blueshift, sometimes superposed upon an overall shift; this indicates that part of the object is approaching and part receding, perhaps relative to a bulk motion of the object as a whole. Such spectra reveal that the object is rotating and can even enable astronomers to measure its rotation rate. In a few cases, the spectrum of a star is found to shift back and forth at a regular interval, indicating that the star is in orbit around another object whose light cannot be resolved over the glare of its brighter companion. Sometimes careful searches for the partners of these *spectroscopic binaries* fail to find the companion; such a system may be a candidate for a neutron star or even a black hole. The Doppler shift is without doubt one of the most important tools of astronomy.

Another source of spectral shifting is the gravitational redshift. Gravitational redshifts occur when photons climb out of a strong gravitational field to a point where the field is weaker. (There is also a gravitational blueshift that occurs when light falls from a weaker to a stronger point in a gravitational field, such as when light travels from a satellite in orbit to the surface of the Earth.) The gravitational field of the Earth is extremely weak in comparison with the fields of stars or galaxies. Since the photons from those objects were emitted from a much stronger gravitational field than that of Earth, we observe only redshifts from astronomical sources. These gravitational redshifts are almost always extremely tiny and difficult to measure, unless they originate from compact objects such as neutron stars or white dwarfs. Detection of the gravitational redshift from gas near a black hole would be important, and recently the first such observation has been reported. A group of astronomers led by Y. Tanaka used an X-ray satellite to observe the core of a Seyfert galaxy; they found a severely redshifted X-ray emission line of iron. The redshift was consistent with light originating within a distance of approximately 3–10 times the Schwarzschild radius of a black hole. Further observations of this type may soon provide unambiguous direct evidence for supermassive black holes in active galaxies.

The Schwarzschild gravitational redshift provides an example of how spacetime curvature, as described by a metric function, can affect light as it moves through spacetime. Other solutions to Einstein's equations must also have the potential to produce redshifts and blueshifts. In this chapter we will introduce the **cosmological redshift**, which is produced by the overall metric, that is, the geometry, of the universe. As we shall discover, the cosmological redshift is a consequence of the fact that the universe is not static and stationary. It is a dynamic, changing spacetime: an *expanding* universe.

The Discovery of the External Universe

Today we are accustomed to thinking of the Milky Way Galaxy as merely one among billions of galaxies in the universe, and not a particularly significant galaxy at that, except, perhaps, to us. But before the twentieth century, few imagined that external galaxies might exist. One of the first original thinkers to grasp this idea was Thomas Wright of England who, in 1750, published the suggestion that the Earth and Sun lay within an enormous shell of stars. He pointed out that our view through this shell would appear as the Milky Way, the band of light that circumscribes the heavens. Wright's book stimulated the philosopher Immanuel Kant to modify and extend the hypothesis. Kant realized that the appearance of the Milky Way in the sky could be explained if it was shaped like a disk, with our Sun somewhere within the disk. In 1755 he published a book describing a universe inhabited by a finite Milky Way surrounded by many similar Milky Ways, all clustering in groups of ever-increasing size.

Such a prescient view of the universe was a minority position. If we were surrounded by other Milky Ways, where was the evidence? Astronomers had cataloged many fuzzy patches of light, called **nebulae** from the Latin word for "cloud," but the true nature of these objects was unknown. It was suspected that the nebulae were blobs of glowing gas spread among distant stars. Such gas clouds do exist in space. They contain mostly hydrogen and glow with light from atoms energized by hot stars embedded within the nebula. Most such clouds that are easily observed are very near the solar system; a well-known example, visible to the naked eye in moderately dark skies, is the Orion Nebula, the faint patch in the "sword" below Orion's belt. Not all nebulae, however, are so obviously gas clouds.

The first of the great catalogs of nebulae was compiled in 1780 by Charles Messier, primarily as an aid to astronomers searching for comets. Although many more nebulae have been discovered since Messier's time, the objects described in his catalog are still known today as the Messier objects and are designated by the letter "M" followed by a number; an example is the great elliptical galaxy M87 in the constellation Virgo. The list of nebulae was soon expanded by William Herschel (1738–1832), a professional musician turned astronomer who almost single-handedly created the field of galactic astronomy. Herschel set about a detailed study of the distribution of stars in the Milky Way, using telescopes of his own design and construction. In 1785 Herschel published the first diagram of the Milky Way, which he called a "detached nebula." He suggested that many of the mysterious nebulae could be similar agglomerations of stars.

As telescopes improved, some nebulae revealed an overall structure. The first direct evidence in support of Kant's view of the universe came from the observations of the Irish nobleman Lord Rosse in 1845. He observed that some nebulae had a distinct spiral structure, suggesting to him that they could be "island universes" similar to our own Milky Way. The nebulae that resembled whirlpools of light were designated *spiral nebulae*. Astronomers remained divided over the nature of these spiral objects. Some agreed with Rosse and Kant that they were external galaxies, while others believed them to be spiral conglomerations of stars within the Milky Way, as globular clusters are spherical associations of stars within and around our Galaxy. Some argued that the whirling appearance of the spiral nebulae suggested that these were new stars and solar systems, caught in the early stages of formation.

The difficulty in elucidating the nature of the spiral nebulae was compounded by the lack of a good determination of the size of the Milky Way, and our location within it. Herschel had described the Milky Way as a somewhat small, amorphous disk of stars, with the Sun near the center. What was not appreciated at the time was that dust

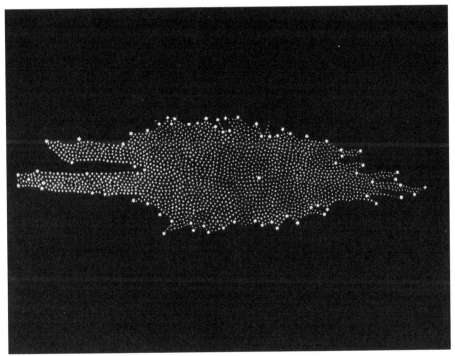

Figure 10.2 Herschel's depiction of the distribution of stars in the Milky Way. The Sun was thought to be near the center of a somewhat irregular distribution of stars. Courtesy of Yerkes Observatory.

within the Milky Way blocks our view through the Galaxy itself. In particular, this effect led Herschel to underestimate considerably the size of the Galaxy. Obscuration by dust also made the spiral nebulae appear to be preferentially located out of the plane of the Milky Way, suggesting to nineteenth-century astronomers that the distribution of these nebulae was somehow related to the Galaxy, and hence that they must be associated with it. On the other hand, there was some evidence that individual stars were present in the spirals; if so, their faintness would argue for a great distance, well beyond the boundaries of the Milky Way. With no knowledge of the actual size of the Milky Way, it was difficult to determine whether the nebulae were located inside or outside our Galaxy. If the Milky Way was as big as some believed, the small apparent diameters of the spirals meant that they would be fantastically remote, if they were comparable to the Milky Way in size. And if that were the case, then no individual stars could possibly be seen if they were only as bright as known stars in the Milky Way.

One piece of evidence came from the observation of novae in some of the nebulae. A **nova**, from the Latin *nova stella* or "new star," is an abrupt increase in the brightness of a star, due to an enormous flare-up. Novae tend to reach a maximum at a fairly uniform brightness. Since they do not represent the end of a star, but rather a large and temporary increase in its energy output, they are fairly common, and, in fact, they can repeat themselves. This is in contrast to the much brighter, but rarer, **supernova**, which does result from the explosion of a star. Although it can be seen to much greater distances, the observation of a supernova is chancy and was especially so before the days of systematic searches for them. In 1885, astronomers observed what they thought to be a nova in the Andromeda Nebula; although it did not appear to be unusually bright, it was nevertheless comparable in luminosity to the rest of the nebula. Clearly, it seemed to them, the nebula could be neither too far away nor composed of billions of unresolved stars. Ironically, what had been observed was actually a supernova, which *does* rival the brightness of an entire galaxy. The relative faintness of the supernova was evidence for a substantial distance. But with no distance reference, and little understanding of the distinction between a nova and a supernova, astronomers assumed that what they had seen was an ordinary nova.[2] Such preconceptions can confuse a scientific question for years, as the history of astronomy vividly illustrates. The nineteenth century passed with no resolution of these issues in sight.

New and important data were introduced in 1912 by Vesto Slipher, who measured the spectral shifts, and hence the radial velocities, of some of the spiral nebulae. He found that many of them had velocities much greater than is typical for stars within the Milky Way. In fact, some of them had velocities that might be so great as to exceed the escape velocity from the Milky Way, a finding which certainly argued in favor of the "island universe" model. The matter began to become clearer in 1917, when Heber Curtis found three faint novae in spiral nebulae. Based on this, he correctly rejected the Andromeda nova as anomalous, and employed the dimmer novae to conclude that the spiral nebulae must be millions of light years away.

During the first twenty years of the twentieth century, the nature of the "spiral nebulae" remained one of the major scientific controversies. Among the more influential observations of the time were those of Adriaan van Maanen of the Mount Wilson observatory, who claimed in 1916 to have directly observed rotational motion in the spiral nebula M100. If visible transverse motion could be observed in only a few years' time, then the nebula could not possibly be very far away, else the implied rotational speed would be in excess of the speed of light. Although it was not realized at the time, van Maanen's observations were simply erroneous. Acceptance of his results led many astronomers to consider them the final blow against the island-universe hypothesis. Even today, it is unclear how van Maanen, a highly competent and experienced astronomer, could have committed such a gross error. Perhaps his interpretation of his data was affected by his beliefs.

At the same time that Curtis and van Maanen were carrying out their research on the nebulae, a young astronomer named Harlow Shapley, also working at the Mount Wilson Observatory near Los Angeles, set about to make a careful study of the size and extent of the Milky Way. He focused his attention on globular clusters, the gigantic, spheroidal agglomerations of stars that orbit the Milky Way. Shapley's work on the spatial distribution of the globular clusters showed that they occupied a roughly spherical region. He postulated, correctly, that the center of the sphere was the center of the Milky Way—and the Sun was nowhere near it. Before Shapley attacked the problem, the Milky Way was believed to have a diameter of 15,000–20,000 lt-yr, with the Sun at its center. Shapley concluded that its true diameter was nearly 300,000 lt-yr, with the Sun located near the edge. Unfortunately, various errors led him slightly astray; although the Milky Way is indeed much larger than anyone dreamed at the turn of the century, Shapley's estimate of its diameter was too large by roughly a factor of 3. However, his somewhat erroneous conclusion led him to the belief that the Milky Way was so enormous and grand that the "spiral nebulae" must be mere satellites about it, spiral counterparts of the globulars.

These issues crystalized in 1920, when a formal debate was held on the subject at the National Academy of Sciences in Washington. Representing the "local hypothesis" was Harlow Shapley; the "island universe" hypothesis was championed by Heber Curtis. In all fairness, Shapley was mainly concerned with establishing the size of the Milky Way. Yet it was felt that his success would defeat the island universe hypothesis *en passant*, since the distances required for the spiral nebulae would simply be too unimaginably great, if the Milky Way were as large as Shapley believed. Although Curtis, an experienced public speaker, mounted a more focused argument during the 1920 debate and was, in the formal sense, generally judged the "winner," Shapley proved more persuasive in the larger scientific discussion, on the basis of his masterful calculations of the size of the Milky Way. Interestingly, Shapley's determination of the diameter of the Milky Way, although an overestimate, was nevertheless fairly accurate in comparison with the small diameter fashionable at the time; yet Curtis' view on the nebulae ultimately proved correct. In truth, they were both right; the fact was that nobody was then quite ready to conceive of the true vastness of the cosmic distance scale.

The Cosmic Distance Ladder

How would an astronomer have gone about measuring distances in 1920? It is a difficult problem, as it remains today, because the methods available to measure very remote distances are not necessarily very accurate. The procedure that is used has come to be known as the **cosmic distance ladder**, because each successive distance scale

Figure 10.3 Harlow Shapley (1885–1972). Shapley's measurements of globular clusters enabled him to determine the size of the Milky Way and the location of the Sun within it. Courtesy of Yerkes Observatory.

depends on accurate measurements at the earlier stages (or "rungs") of the process. The first of these rungs is the most direct and accurate method, but it is useful only for nearby stars. This is the method of **parallax**, the measurement of the apparent shift of a star's location on the celestial sphere due to the motion of the Earth in its orbit. Parallax was sought in vain by early astronomers, but the stars are too distant for these shifts to be detected with the naked eye. Today we can measure these shifts, and once the very small corresponding angle has been observed, triangulation is used to determine the distance; it is very similar to the method employed by a surveyor to determine the distance to a mountaintop by measuring its angle from two different positions. Since the resulting triangles are, in the astronomical case, extremely long and thin, we may use the formula

$$D = \frac{2 \text{ A.U.}}{2p} = \frac{\text{A.U.}}{p},$$ (10.2)

where an A.U. (astronomical unit) is the mean distance of the Earth from the Sun, p is the measured parallax angle, and D is the desired distance to the star. (Notice that the parallax angle is defined in terms of half the total baseline, as illustrated by Fig. 10.4.) If D is to be determined in absolute units, such as meters, p must be expressed in radians. However, p is usually measured in arcseconds, 1/3600 of a degree. From this, we derive the unit of length called the *parsec*, which is that distance producing 1 arcsecond of parallax over the baseline of the Earth's orbit. A parsec (pc) corresponds to 3.26 lt-yr.

The parallax angles of even fairly nearby objects are incredibly tiny, and the distances of the stars remained beyond the reach of astronomers until technological improvements in telescopes and their mountings made it possible to determine star positions with great accuracy. After centuries of futile attempts by many observers, the famous German mathematician and astronomer Friedrich Wilhelm Bessel announced in 1838 that he had measured the parallax of the inconspicuous star 61 Cygni. Without the aid of photography, Bessel found a parallax angle of 0.3 arcseconds, from which he computed a distance for this star of 10.9 lt-yr, a figure very close to the modern result of 11 lt-yr.

Parallax is the only *direct* method of determining interstellar distances; it requires no knowledge of the structure or brightness of the star, nor does any intervening matter affect the result. Unfortunately, parallaxes can be detected even with current technology only for those stars within a radius of some 100 pc from the Sun. Parallax measurements require extremely accurate determinations of an object's position at different times of the year. For very distant stars, the blurring of the star's image by the optical distortion that is inevitable in any telescope swamps the minute shifts in its apparent position. Therefore, beyond our immediate stellar neighborhood, we must turn to more indirect means.

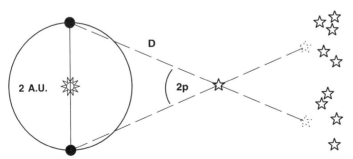

Figure 10.4 The geometry of parallax, showing the angular shift in a nearby star's position due to Earth's orbit. The figure is enormously exaggerated. The largest parallax angle observed for any star is less than 1 arcsecond.

The most common indirect approach is the method of **luminosity distance**. This approach depends upon the fact that if the intrinsic luminosity of an object is known, then a measurement of its *apparent brightness* makes it possible to deduce the distance traveled by the light. The amount of light received at a particular location from the source is reduced with increasing distance, because the energy emitted must spread out over a larger and larger sphere as it travels outward into space. (Fig. 4.13 illustrates the geometry involved.) The surface area of a sphere is given by $A_s = 4\pi r^2$; hence the brightness, which is the energy per unit time per unit area, must diminish by the inverse square of the distance from the source. The apparent brightness received at the surface of the Earth from a star at a distance d_L is thus simply $L/4\pi d_L^2$; in this expression, L is the star's *luminosity*, its total output of energy per unit time. We may rearrange this formula to obtain

$$d_L = \sqrt{\frac{L}{4\pi b}}, \tag{10.3}$$

where b is the apparent brightness obtained from the light falling upon the telescope.

As a specific example, consider the energy radiated into space by the Sun. The Sun has an intrinsic luminosity of $L_\odot \approx 4 \times 10^{26}$ watts. Suppose an astronomer on a distant planet observes the Sun with a telescope whose effective mirror radius a is 2 meters, in Earthly units. The surface area of the mirror is thus $\pi a^2 = 4\pi$ m^2. Suppose further that the alien astronomer's telescope collects from the Sun an energy per unit time of 4×10^{-8} W. The apparent brightness measured at this distance is thus $10^{-8}/\pi \approx 3 \times 10^{-9}$ W/m^2. Assuming the astronomer knew, or could estimate, the luminosity of stars of the Sun's type, the astronomer could then apply equation (10.3) to find that the distance of the Sun from the alien planet is 1×10^{17} m, or approximately 10.6 lt-yr. As suggested by this example, the quantity of energy received from even a nearby star is very small. As enormous as the total energy output of a star may seem, this energy spreads out through the vastness of space; only a tiny fraction arrives at Earth and impinges upon your retina, or upon the focus of a telescope. In comparison, a typical flashlight emits about 5 W of energy from a lens of approximately 15 cm^2, for a brightness of roughly 333 W/m^2. It should be obvious why astronomy demands huge telescope apertures, sensitive detectors, and dark skies.

A major weakness of the method of luminosity distance is its reliance upon a knowledge of the absolute luminosity of the target object. The first step in obtaining absolute luminosities depends on determining distances to nearby star clusters, particularly the Hyades cluster in Taurus, by other, more direct, means, such as parallax, or by determinations of the inherent proper motions of the stars of the cluster. By measuring the apparent luminosities of the stars in the same cluster, we can then work "backward," using our knowledge of distance and apparent luminosities to compute the intrinsic luminosities of the different kinds of stars found in the cluster. Once the luminosity of a particular type of star is known, it becomes a **standard candle**, a term referring to any object of known luminosity. As it turns out, stars of a given mass and age vary little in luminosity, so if we find another star of the same type that is too far away for its parallax to be measured, we can obtain its luminosity distance.

The method of luminosity distance has a further set of difficulties. One particularly important confounding effect is the presence of intervening dust, which reduces the apparent brightness of a standard candle beyond that explained by distance alone. This phenomenon is known as **extinction**, and it was a significant source of systematic error in the 1920s, when the existence of interstellar dust was not yet recognized. This phenomenon specifically led Shapley to overestimate the size of the Milky Way and confused the study of spiral nebulae for several decades.

In the early debates over the nature of the spiral nebulae, the greatest problem was that ordinary stars were the only well-established standard candles; but in the nebulae, individual stars were simply too faint to detect and too small to resolve. Even today, it is difficult to observe individual stars in galaxies, and it was nearly impossible with the technology available in the 1920s. If it is difficult to see a star, it is even more difficult to determine its type, since accurate spectra are required, and spectroscopy demands the collection of quite a lot of light. The farther the galaxy, the more severe this problem becomes. Heber Curtis attempted to use novae as a standard candle in the Andromeda Nebula, but the presence of a supernova complicated the issue. Beyond that, novae themselves are not precisely consistent in their luminosities. A new and better standard candle was required.

The Hubble Law

Edwin Hubble, one of the astronomers most instrumental in changing our view of the universe, now enters the story. In the 1920s, Hubble undertook a systematic survey of spiral nebulae. The critical breakthrough occurred in 1924, when Hubble detected a star of the type called a **Cepheid variable** in the Andromeda Nebula, which, to reveal the outcome of this story, is now known as the Andromeda Galaxy. Cepheid variable stars vary in brightness with a fixed periodicity. The periodicity varies from star to star, over a range of about 3–50 days, but, as was discovered by Henrietta Leavitt in 1912, the periodicity is a function of the star's maximum luminosity. Specifically, the brighter the star, the longer the period. Herein lies the importance of Cepheid variables. By measuring the star's apparent brightness along with its period of variation, a relatively straightforward operation, Hubble was able to compute the Cepheid's luminosity distance, and hence the distance to the nebula in which it resided. Hubble's discovery that the Andromeda Nebula was very remote, far beyond the reaches of the Milky Way, settled the debate once and for all; the spiral nebulae were external galaxies. We now know that the distance to the Andromeda Galaxy is approximately 2 million lt-yr (700 kpc). Andromeda is the nearest large galaxy to the Milky Way, to which it is mutually gravitationally bound.[3] In modern astronomical usage the term "nebula" is reserved exclusively for those objects which truly are clouds of gas and dust; external star systems are always called "galaxies." Some texts continued to refer to both objects as "nebulae" until well into the 1950s, however.

It is useful to pause and reflect upon the discovery of other galaxies. This revelation was another blow to humanity's anthropocentric cosmological point of view. Copernicus removed Earth from the center of the universe. Centuries elapsed during which the heliocentric theory was grudgingly accepted, but humans stubbornly retained their sense of specialness by shifting the center of the universe to the *Sun*. Again, appearances conspired to make it seem so: the band of the Milky Way has a nearly

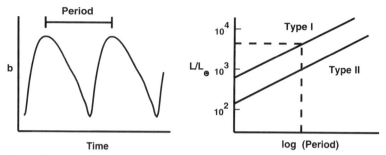

Figure 10.5 The apparent brightness of a Cepheid variable changes with time. By observing the variation one can easily measure its period. The luminosity of the Cepheid can then be inferred from the period-luminosity relationship: period increases with luminosity. There are actually two populations of Cepheids, type I and type II; the type II Cepheids are not as luminous.

uniform brightness all around the sky, implying that we are at its center. Astronomers did not realize that dust obscured their view of the true center of the Galaxy. Then, with his study of globular clusters, Harlow Shapley proved that the Sun is not at the center of the Galaxy. But the question remained whether or not the Milky Way constituted the bulk of the matter in the universe. Finally, Hubble showed that the Milky Way is not the only galaxy in the universe. We now know that the Milky Way is a good-sized, although typical, spiral galaxy and is a member of a rather insignificant group of galaxies falling toward a much larger cluster of galaxies. And even that

Figure 10.6 Edwin Hubble (1889–1953). Best known for his discovery of the expanding universe, Hubble also was the first to measure the distance to the Andromeda Galaxy, proving that the so-called "spiral nebulae" were external galaxies. Courtesy of Hale Observatories, AIP Emilio Segrè Visual Archives.

attracting cluster is not one of the biggest of all clusters. The universe is filled with numberless galaxies, organized into huge structures stretching over millions of parsecs. In humbling humanity, twentieth-century astronomy has outdone even Copernicus.

By establishing their true nature, Hubble created a new branch of astronomy, the study of galaxies. Hubble was a pioneer on this new frontier. Using the recently completed 100 inch telescope on Mount Wilson, Hubble developed a classification scheme for galaxies, based upon their appearance, which is still widely used today. Galaxies are grouped into two morphological classes, spiral and elliptical. Spiral galaxies have a disklike shape with a central bulge; the disk contains spiral arms of greater or lesser prominence. Elliptical galaxies are, as their name suggests, ellipsoidal conglomerations of stars. They exhibit little or no substructure such as spiral arms or flattened disks. Hubble subdivided these groups further by developing classification criteria for galaxies of each type, based on details of their overall structure. The ellipticals were grouped on the basis of their overall ellipticity, from the nearly round E0 type, to the highly elliptical E7. Spirals were divided into two groups, those with prominent stellar bars extending across the nuclear region (the "barred spirals") and those without. These groups were further classified by the tightness of the spiral arms and the compactness of the nuclear bulge. Hubble also created another category, the irregular galaxies, for galaxies whose appearance is, as the name implies, irregular. The Large and Small Magellanic Clouds are two nearby dwarf irregular galaxies.

As valuable as these discoveries were, Hubble's preeminent contribution to cosmology lay elsewhere. At the time of the Shapley-Curtis debate, it was known that the majority of the spiral nebulae showed redshifted spectra; by 1922 Vesto Slipher had found that the spectra of 36 out of a sample of 41 of these objects were redshifted. (The Andromeda Galaxy is an example of a "spiral nebula" that is blueshifted; it and the Milky Way orbit one another, and currently Andromeda is approaching us.) Following his triumph with the Andromeda Galaxy, Hubble, aided by Milton Humason, obtained dozens of galactic spectra with the Mount Wilson 100 inch telescope.

The galactic spectra provided redshifts, but redshifts alone gave little information, other than that the galaxies were nearly all receding. Hubble and Humason went further and combined their redshift data with the distances to the galaxies. Redshifts were easily measured; obtaining distances was the challenging part. Hubble began with Cepheids, but they quickly become too dim to function as standard candles for more distant galaxies. Beyond the "Cepheid limit" other calibrations must be used. Hubble and Humason used the best standards known to them, beginning with the brightest supergiant stars of a galaxy. They assumed that the brightest stars in any galaxy all have about the same luminosity. Stars cannot be infinitely bright, so there must be a cutoff; for most large galaxies, the brightest stars seem to be of similar luminosity. This hypothesis was checked using galaxies whose distances were found via Cepheids, and it worked reasonably well. The greatest difficulty with this approach is that the brightest *object* in a galaxy need not be its brightest normal *star*; for many galaxies, small clouds of extremely hot hydrogen gas may be the brightest pointlike object. At cosmological distances, it can be very difficult to distinguish such emission regions from a star. Finally, at great enough distances, even the brightest stars fade into the general glow of the galaxy, and other techniques must be brought to bear. At very large distances, Hubble and Humason were reduced to using the apparent luminosity of the galaxy as a whole. They knew that the intrinsic luminosity of galaxies probably varied a good deal, but they hoped to limit the variation by comparing galaxies of the same Hubble classification. Despite the many potential sources of error, Hubble and Humason found that when redshift was plotted versus distance, the points were not randomly scattered about, but lay very close to a single straight line. In Hubble's own words, he had found a "roughly linear relationship between velocities and distances."

In fact, what Hubble had found directly was a linear relationship between *redshift* and distance ℓ, symbolized $z \propto \ell$. The distinction is subtle, but significant. What Hubble measured was the shifts in the spectra. A redshift can be caused by several factors, the most obvious of which is radial motion. In the cosmological case, however, the redshift is caused by the relativistic expansion of the universe itself, but this was not understood at the time that Hubble and Humason were compiling their data. Hubble interpreted the extragalactic redshifts in terms of the familiar Doppler shift. As it happens, the cosmological redshift and the Doppler redshift behave the same way for nearby galaxies and small redshifts, up to approximately $z \leq 0.1$. For small redshifts, then, one may employ equation (10.1), in which the redshift is directly proportional to a velocity. Under these conditions, redshift can be equated to a velocity, and the

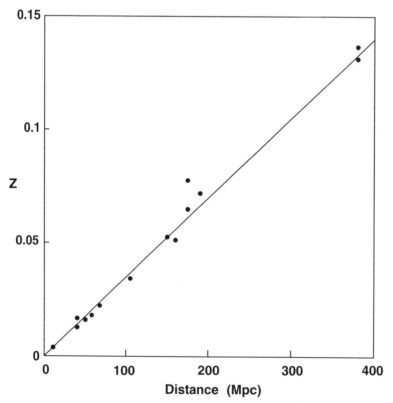

Figure 10.7 A Hubble diagram obtained using redshifts and distances to galaxy clusters out to a redshift of roughly 0.1. The derived Hubble constant for these data is 105 km/s/Mpc.

graph of velocity versus distance will also be a straight line. A straight line through the origin, that is, zero relative velocity at zero distance, implies a relationship of the form $v = H\ell$, where H, the slope of the line, is now called the **Hubble constant**. The general relationship

$$v = H\ell \tag{10.4}$$

is called the **Hubble law**. The Hubble law is a *theoretical* description of the behavior of the universe. Hubble's observed redshift-distance relationship provides the experimental evidence for this law.

The Hubble constant in equation (10.4) must be determined by observation. It is usually expressed in units of kilometers per second per megaparsec. For example, if $H = 100$ km/s/Mpc, then a galaxy with a recessional velocity of 3000 km/s, corresponding to a measured redshift of 0.01 ($z = v/c$), would be located at a distance of 30 Mpc from Earth. If $H = 50$ km/s/Mpc, then the same galaxy, with the same redshift and inferred recessional velocity, would be 60 Mpc distant.

The current measured values for H fall mostly between 40 and 100 km/s/Mpc, the large range representing the continuing uncertainties in the measurements of distances. Hubble's original value for this constant was around 500 km/s/Mpc, quite different from today's value. The reason for this is instructive, for it illustrates why it is still difficult to determine H. Hubble did not realize that there are actually two classes of Cepheid variable stars; this was not discovered until 1952. Hubble was actually observing a brighter class of Cepheid in external galaxies, a type of star now designated type I Cepheids; these are the "classical" Cepheids, like the prototype star δ Cephei, and they are intrinsically very bright stars. However, the period-luminosity relationship used by Hubble had inadvertently been calibrated for distance using another kind of variable star, now called type II Cepheids, which are dimmer than type I Cepheids by approximately a factor of 4. As a result of this confusion, as well as other factors such as failure to take extinction sufficiently into account, Hubble systematically underestimated distances to the type I Cepheids by more than a factor of 2. This is an example of a type of error which can throw off a set of measurements by a much larger factor than that indicated by formal experimental uncertainties. Stated uncertainties generally do not include such *systematic errors*, but only the random errors that are inevitable in experiments. Because of systematic errors, Hubble believed that galaxies were, on the average, much smaller than the Milky Way. This should have been a clue that something was wrong, if we adhere to the Copernican principle; when more correct distances are employed in our size estimates, we find that the Milky Way is a typical, perhaps even smallish, spiral.

Today, technology has improved the longest distance measurements. One of the most important recent developments was the launch of the *Hubble Space Telescope (HST)*. Above the distortions caused by the Earth's atmosphere, the *HST* can resolve individual stars to a much greater distance than most ground-based telescopes. Cepheids have now been detected as far as the Virgo cluster of galaxies, at a distance of 17 Mpc. However, beyond the point at which stars can be resolved even by present-day optical instruments, other standards must be employed. One of the most reliable is the **Tully-Fisher relationship**, named for its discoverers R. Brent Tully and J. Richard Fisher. The Tully-Fisher relationship is a correlation between the width of one particular emission line of hydrogen, that at a wavelength of 21 cm in the rest frame of the hydrogen, and the luminosity of the spiral galaxy from which the emissions are observed. The main reason that there is a width at all to the 21 cm line is the rotation of the galaxy. Some of the gas is approaching, while some is receding, but at the resolution of ordinary radio observations, only the collective photons from all the gas in the galaxy are detected, causing the line to smear into a band. The broader the band, the faster the rotation of the galaxy and the greater its luminosity.

The Tully-Fisher relationship is based upon observations of many galaxies, not upon theory; however, there is a simple, qualitative explanation that accounts well for it. A brighter galaxy obviously has more mass in stars than does a dimmer galaxy; we would thus expect the overall mass to be greater as well, including any nonluminous matter that might be present. If the ratio of total mass to luminous mass is roughly constant for most spiral galaxies, then the brightness of any spiral galaxy's emissions should be related to its total mass. From Kepler's third law, we know that a more massive galaxy would rotate faster. The width of the 21 cm line indicates the rotation rate of the galaxy, which in turn depends upon the total mass; since more massive galaxies are also brighter, the 21 cm line can thus stand as an approximate proxy for the total luminosity. The most remarkable aspect of this relationship is how good it is. It has proved to be one of the most useful distance indicators for very remote galaxies, although even it has its limitations; beyond approximately 200 Mpc, the width of the line becomes difficult to measure accurately.

The measurement of cosmological redshifts, while considerably easier than that of distances, is not without its sources of uncertainty as well. Most of the scatter in a typical Hubble diagram is not due to simple instrumental error. Rather, the observed redshift is a composite of red- or blueshifts due to all velocities and gravitational effects. Most galaxies are members of clusters and interact gravitationally with other galaxies. Gravity is a long-range force, so even the clusters can be influenced by other clusters. For instance, the Milky Way is primarily in a mutual orbit with the Andromeda Galaxy, but also, along with other members of the Local Group, orbits the Virgo Cluster. The intrinsic motion of an object due to its particular responses to forces such as local gravitational attractions is called the *peculiar motion* of the object; its velocity due to such movement is called its **peculiar velocity**. This name does not mean that the velocity is due to anything strange or unusual. It is simply the velocity that results in the ordinary, classical Doppler shift due to the unique motions of a given galaxy, as distinct from the overall Hubble effect. The net redshift is a superposition of the peculiar Doppler shift upon any cosmological redshift.

For nearby objects, it can be very difficult to extricate the Hubble law from the Doppler redshifts. However, because peculiar velocities will tend to be in all directions, both toward and away from us, their effect on the Hubble diagram will average to zero if we use a large number of galaxies at a given distance. Furthermore, peculiar velocities will all be less than some certain maximum amount, perhaps a few hundred kilometers per second, whereas the Hubble effect increases with distance. It follows that the relative importance of peculiar velocity is itself a function of distance. The closest objects, such as the other members of the Local Group, are dominated by their peculiar motions and are essentially unaffected by the Hubble law.

A little further out, matters become quite complicated. For example, peculiar motions play a significant role in attempts to determine the Hubble constant from observations of the Virgo Cluster. The Virgo Cluster is, in cosmological terms, quite near. It is easiest to measure the distances of the closest objects accurately, such as by finding Cepheid variables in a galaxy. If the galaxy M100, a member of the Virgo Cluster whose distance has recently been determined to very high accuracy by the *HST*, showed only a cosmological redshift, then we could immediately determine the Hubble constant, since redshifts can be measured to very, very high precision; this is especially true for nearby objects whose spectra can be easily obtained. Unfortunately, the galaxies of the Virgo Cluster are executing complicated internal motions. More than that, the cluster is also a center of attraction for the Milky Way and its companions, increasing further the Doppler shift due to relative peculiar motions. All of this creates Doppler redshifts that are not insignificant compared to the cosmological redshift. Redshifts measured for the galaxies of the Virgo Cluster thus do not reveal an unambiguous cosmological redshift. In order to obtain a value for the Hubble constant from the measurement of the distance to M100, a model for the motions of the galaxies of the Virgo Cluster, and of the motion of the Milky Way toward them, must be employed to interpret the data. The distance to M100 serves mainly to establish a rung on the cosmic distance ladder so that more distant galaxies may be used in the determination of the Hubble constant.

For distant objects, those well beyond the Virgo Cluster, the cosmological redshift is so large as to completely swamp any peculiar velocities. The systematic increase of redshift with distance is the strongest argument that the cosmological redshift *is*, in fact, cosmological. If it were due to peculiar motions of the galaxies, the redshift would show no tendency whatsoever; indeed, it might be expected that just as many distant galaxies would show blueshifts as redshifts, which is emphatically not observed. Some have argued that the large redshifts of quasars are due to peculiar velocities, mostly by appeal to a few anomalous cases that could easily be misleading. However, nearly all cosmologists agree that the data present overwhelming support for the interpretation that the major contribution to the redshifts of distant objects is the cosmological redshift, due not to any peculiar motions but to the *expansion of space itself.*

This is a new and challenging concept. Usually, when we speak of a Doppler shift's implying a certain recession velocity, we mean that the shift is due to the inherent motion of the source relative to the receiver. But regarding the cosmological redshifts in such a manner could lead to a picture of galaxies streaming away from us. Such a picture implicitly places the Milky Way in the center of some great explosion, a point of view that is quite clearly inconsistent with the cosmological principle. From the Copernican principle, we should expect that we are not at the center of *anything*, much less some universal cataclysm. This means that the recession velocity in the Hubble law is very different from the kind of velocity to which we are accustomed. The cosmological redshift is due to the properties of *space itself.* Since we observe that all galaxies that are not gravitationally bound to the Milky Way show a cosmological *redshift*, and never a blueshift, they must be receding from us. What this observation implies is that space is *expanding everywhere.* Every galaxy sees every other galaxy expanding away from it. The overall motion of galaxies away from one another, due to the general expansion of the universe, is called the **Hubble flow**. In the next section we shall consider how such a strange notion arises from Einstein's theory of general relativity.

The Theoretical Discovery of a Dynamic Universe

In the first decades of the twentieth century the astronomers, mostly in the United States, were enlarging the Milky Way, discovering external galaxies, and collecting the first hints of an overall cosmological redshift. Meanwhile the theorists, mostly in Europe, considered cosmology to be too speculative, almost metaphysical, and thus hardly worthy of serious scientific contemplation. But now and then a physicist or an astrophysicist dabbled in cosmology. Albert Einstein was among the first to investigate cosmology from a firm theoretical basis. Shortly after Einstein had completed the correct formulation of the equations of general relativity, he turned his attention, with his usual scientific audacity, to their implications for the entire universe. The new theory of gravity seemed to have properties which could solve some of the age-old questions of the universe. By admitting the possibility of spacetime curvature, it was at last possible to construct a universe that was comfortably finite, thus avoiding the disturbing prospect of infinite space, yet without invoking an equally unfathomable edge. All that was necessary was to insert enough matter-energy into the universe to force space to curve back upon itself, forming a spherical geometry that was both homogeneous and isotropic. Such a universe has a pleasing Machian property about it: the overall distribution of matter exactly determines the shape and size of space.

Einstein found that constructing such a model was easier imagined than done. The difficulty was that even in its relativistic form, gravity remains an attractive force. The tendency for a distribution of mass to undergo gravitational collapse, a problem that plagued Newton's clockwork universe, is not alleviated by confining the universe to a finite spherical domain. Quite the opposite, relativity enhances the propensity of matter-energy to collapse. Einstein's equations predicted that his spherical universe, left to its own devices, would come crashing down on itself in about the amount of time it would take light to complete one circuit through the universe. Of course, a model in which the universe almost instantly collapsed upon itself was not very satisfying. It was, however, possible to adopt an alternative picture in which the universe was expanding, rather than contracting; the expansion would tend to counteract the pull of gravity. The important point was that, due to the omnipresent gravity, the universe could not remain still. Much like a ball thrown into the air, it had to move. But this was still 1917, a time when, to astronomers, the universe consisted entirely of the Milky Way. Astronomers may not have known too much, but this they knew: the Milky Way was not contracting or expanding. Stars were moving about, but there seemed to be no systematic expansion or contraction.[4]

Einstein was faced with a quandary. His equations predicted a dynamic universe, not the static universe that everyone believed to exist. Hence he decided, probably reluctantly, that the equations must be wrong. But how could they be wrong when they had worked so well for explaining the corrections to the orbit of Mercury? The only possible answer was that there must be a term that is important only on the cosmic scale. How could such a term be accommodated without destroying the mathematical properties that had taken so long to establish? In formulating the Einstein equations, there is a point at which, essentially, an integral is performed. It is always possible, when integrating, to add a constant term; Einstein had initially set that constant of integration to zero, consistent with the notion that the force of gravity drops toward zero at large distances. But what if that constant were not zero? There could exist a term that is immeasurably small on the scale of the solar system, but which nevertheless creates a repulsive force at large scales, a force just sufficient to counteract the attraction of the matter in the universe. Such a force would have interesting properties. It would be zero on small scales but would *increase* as a function of distance. Consequently, any nonzero repulsive cosmic force, no matter how small, would ultimately dominate on the largest scales.

Einstein called his new term, designated by the Greek letter Λ, the **cosmological constant**. It may seem at first glance to be completely ad hoc, but it enters the equations of general relativity as a perfectly legitimate possible contribution to large-scale gravity. Looking back at Einstein's equation (8.5), we see that there are two ways in which to regard Λ. Einstein's equation states that the geometry of spacetime equals mass-energy. The Λ term could be a constant of integration added to the geometry on the left-hand side of the equation, that is, it might be regarded as a mathematical correction. However, it could also arise from physical phenomena. Whereas geometry, lurking on the left-hand side of the equation, is well defined in terms of established mathematical quantities such as the metric, the right-hand side is much less well established and could conceivably include new effects, or additional properties of matter and energy that are still unknown. The cosmological constant can thus also be interpreted as a contributor to the mass-energy of the universe that produces a repulsive force, a kind of "negative energy" term. In this case, the cosmological constant would correspond to a negative energy associated with the vacuum of space itself. Λ will reemerge in this guise much later, when we examine the so-called inflationary cosmological model.

Einstein added Λ to his equations to provide a balance between the attractive force of gravity and the repulsive force of Λ. Unfortunately, as it seemed at the time, the effort failed. It was not any artificiality of the cosmological constant that ultimately proved fatal to Einstein's static universe. Instead, the model proved to be *unstable*; while it could be set up in a static equilibrium, it simply did not *remain* static. The cosmological constant was introduced in order to provide a delicate balance between gravity and a repulsive force. The Λ force increases linearly with increasing distance, while ordinary gravity diminishes as the inverse square of the distance. Thus, if such a balanced universe were to expand just a bit, the repulsive force would grow, while the attractive gravitational force would decrease. But this would mean that the universe would expand even more, leading to further decreases in gravity and more expansion. Conversely, if the universe contracted a little, the force of gravity would increase, over-coming the cosmological repulsion. Either way, the universe was destined to move. It is analogous to a marble sitting on the top of an inverted bowl. The marble may be at rest (static), but any slight perturbation would cause it to roll away from the top.

Again the static model failed, and this time there was no saving it. Happily, the changing view of the universe eliminated the need for such a model. Edwin Hubble's data indicated that the universe, finally realized to be much larger than the Milky

Way alone, was *expanding*. With Hubble's observational evidence of the reality of the expansion of the universe, Einstein attempted to retract his cosmological constant, going so far as to call it his "greatest blunder." This is probably too harsh, although one can certainly sympathize with his chagrin. Had he believed what his equations, in their original form, were telling him, he could have predicted the expansion of the universe before it was observed. Theory and observation are often at odds, and theory should remain mutable in the face of the experimental results. However, the problem here lay with the perennial assumption that humanity's observations, up to a particular point in history, provide a sufficiently representative sample of the universe from which to draw cosmological conclusions; this dates back at least to the first person who gazed across a wide open field, and concluded that the Earth was flat.[5]

The cosmological constant can be used to create many interesting alternative models of the universe, the simplest of which was published by Willem de Sitter in 1917, the same year as Einstein's static universe appeared. In contrast to the Einstein model, the **de Sitter model** contains no matter, just spacetime and the repulsive Λ. In this model the geometry of space is flat, but the repulsive force causes space to expand exponentially, although this was not recognized when the model was first published. If a few stars were sprinkled into this otherwise-empty universe, they would recede from one another with velocities proportional to the distance. Given the obvious limitations of the de Sitter model, it is not surprising that it was not taken very seriously. But Slipher continued to publish redshift observations; for a while, the overall recession of the galaxies was even called the "de Sitter effect." The clues were emerging, but the synthesis had not yet occurred. A decade passed after the publication of the de Sitter model before the realization dawned that the universe is expanding.

A Metric for an Expanding Universe

Once it is accepted that the universe need not be a static entity, and, indeed, is *expanding*, it becomes necessary to construct models that are consistent with this new reality. What difficulties might we encounter in undertaking so ambitious a task? First of all, our knowledge of physics has been obtained from experiments carried out on one tiny planet in one infinitesimal region of the universe, in relatively small ("human-sized") laboratories. Do the laws we derive hold for the universe at large? We cannot know for certain, but if we accept the cosmological principle, they do, although it is possible that they also change with time. A more subtle question is whether there might be laws governing the large-scale behavior of the universe that simply have too small an effect to be seen at the scale of our Earth-bound laboratories. An example would be Einstein's cosmological constant Λ, which clearly does not affect gravity even on the scale of our Galaxy, but might have a profound influence on the universe as a whole. Ultimately, we can only attempt to observe large-scale phenomena, then fit them into our physics. If they cannot be made to fit, then additional physical laws may be required. But unless we are *forced* to think otherwise, we should assume that our existing physics can explain the universe as a whole. We should seek complicating factors that obscure the action of known laws, before we postulate new physical principles.

Another significant problem is that the universe is, by definition, unique; we cannot observe other universes. Ideally, we seek a theory that explains everything that happens in this one universe solely in terms of the universe; the theory should describe everything that happens and must not allow anything that does not occur. We do not have, and may never find, such a theory. In the meantime, we are free to construct simplified models of the universe, to derive their predictions, and to see how well a given model is supported by the observations we can make.

How do we derive a model of the universe from Einstein's theory of general relativity? It is simple enough to describe what must be done: calculate the total matter-energy content of the universe and find the spacetime geometry consistent with that distribution. In practice, this is far from easy. Before we delve into the details of cosmological solutions to the Einstein equations, let us begin by investigating and clarifying what we mean by "expanding spacetime," and how such a concept can be incorporated into a spacetime metric. By adopting the cosmological principle, which immediately tells us that the universe is homogeneous and isotropic, we place a considerable constraint on the permitted appearance of the metric. Specifically, the metric coefficients must be the same everywhere; they cannot depend either on spatial location or on direction. The usual flat spacetime Minkowski metric from special relativity,

$$\Delta s^2 = (c\Delta t)^2 - (\Delta x^2 + \Delta y^2 + \Delta z^2),$$

provides a trivial example of a homogeneous and isotropic metric, since the Minkowski metric coefficients are constants. This is, however, a static metric; it does not change with time. We can generalize it by including an arbitrary scale function and allowing that function to vary with the time coordinate:

$$\Delta s^2 = (c\Delta t)^2 - R^2(t)(\Delta x^2 + \Delta y^2 + \Delta z^2). \tag{10.5}$$

The coefficient $R(t)$ is known as the **scale factor**. The notation means that R is some function that depends only on time t. For example, $R =$ constant $\times t$ would provide a scale factor that increased linearly with time.

We can make space expand (or contract) by adjusting the scale factor function. It is important to understand what we mean here by "expansion." If the scale factor increases with time, then two particles, separated spatially at time t_1 by some distance and both at rest with respect to the cosmic frame, are separated at a later time t_2 by a greater distance. The expansion of the universe occurs everywhere and is manifested by an overall increase in the separation of particles; it is not an enlargement into some predetermined, larger entity. Thus it is that a space which may be already infinite can still be expanding.

Let us examine the properties of the cosmological metric (10.5) in some detail. In this form of the metric, the coordinates (x, y, z) are said to be **comoving coordinates**; that is, they remain fixed, and the distance between coordinate locations can be scaled up (expansion) or scaled down (contraction) through the scale function. The surface of a balloon functions well as an analogy. Suppose that we paint lines of "latitude" and "longitude" on the surface of the balloon and then paste some paper disks at various positions on the balloon. Now we inflate the balloon; as it expands, the painted lines expand along with it, as illustrated in Figure 10.8. The coordinates of the disks, relative to this grid, do not change. However, the distances between the disks, measured along the balloon's surface, do increase. The paper disks themselves do not expand; only the surface of the balloon enlarges. Similarly, the spatial coordinates in the cosmological metric do not change with time; the time variation has been taken entirely into the scale factor $R(t)$. In comoving coordinates, a cluster of galaxies, like the paper disks affixed to the balloon, would always keep the same location as time evolves, although the physical distance between one cluster and another would change according to the scale factor. Comoving coordinates are a conceptual aid to help us think about the expanding universe; for real measurements, astronomers would use some coordinate system fixed to a convenient location, such as the center of the Earth or of the Milky Way, and these physical coordinates would change in time. The *metric* measures physical distances; but the *coordinates* which mark locations are arbitrary and may be chosen for our convenience.

The metric scale factor $R(t)$ then provides all the information we need in order to describe how the universe changes with time. A given model of the universe can be illustrated by a plot of the scale factor versus cosmic time; some examples are shown in Figure 10.9. The vertical axis represents the scale factor R in arbitrary units. In practice, specific values of R are never given. All that matters is how R changes, that is, the ratio of R "now" to R "then." Figure 10.9a shows a scale factor that increases with time, but at an ever-slowing rate. Figure 10.9b illustrates a scale factor that increases as a simple linear function of time. Figure 10.9c shows a scale factor increasing with time, but at an accelerating rate. The final frame, 10.9d, displays a scale factor *decreasing* with time. Such a universe is contracting rather than expanding. Graphs such as these are invaluable for clarifying the characteristics of a particular model of the universe.

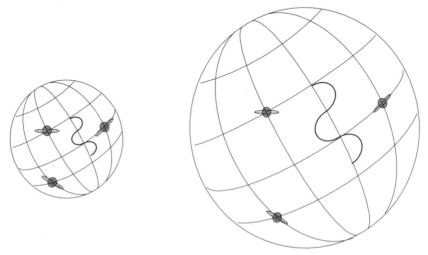

Figure 10.8 Schematic illustration of a two-dimensional expanding spherical universe, analogous to the surface of a balloon. There is no center of expansion; all points move away from one another. The undulating line indicates the redshifting of light by the expansion of spacetime. Comoving coordinates are given by the "latitude" and "longitude" lines; these scale up with the expanding sphere.

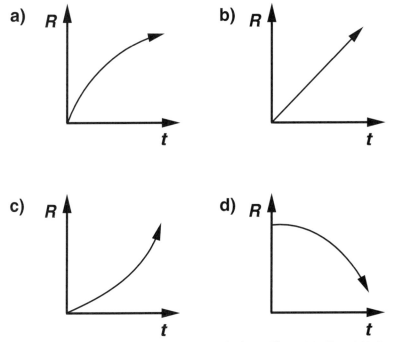

Figure 10.9   Graphs of scale factor R versus cosmic time t. Figures (a), (b), and (c) show expanding universe models, with the rate of expansion slowing, held constant, and increasing with time. Figure (d) shows a model that is contracting at an ever-increasing rate.

Perhaps you wonder why the scale factor has any effect on anything, if everything in the universe simply scales up uniformly. If the universe is expanding, then does each galaxy expand as well, and the solar system, and atoms, so that the expansion would be unobservable? There are several ways to answer this question. First, we must emphasize that in cosmology we consider the largest scales. The standard cosmological solutions to the Einstein equations are obtained by assuming some *averaged* matter or energy distribution. The solutions then correspond to the behavior of the overall gravitational field of the universe. A galaxy, a star, the Earth, you, all reside in local gravitational fields created by the presence of concentrations of mass in a given vicinity; these details are not taken into account in a cosmological solution. Second, recall that in general relativity the metric determines geodesics, and geodesics are the free-falling world lines. You can regard the geometry of spacetime near the Earth, or another massive object, as resulting in a "gravitational force"; similarly, the geometry of the universe provides an "expansion tidal force." Objects following the cosmological geodesics move apart. But this "expansion" tidal force is incredibly weak on the scale of the solar system, or even the galaxy. It only builds up to significance over millions of parsecs. It is very easy for galaxies, or the Earth, or the solar system, to hold themselves together in the face of the expanding universe.

The metric (10.5), which we have introduced to describe the dynamic universe, is really only appropriate for flat spacetime geometry. We need to introduce a means for the other types of homogeneous and isotropic geometries to be included. H. P. Robertson and A. G. Walker independently constructed such a metric in 1936; they showed that the most general spacetime metric for a dynamic, homogeneous, and isotropic universe can be written in the form

$$\Delta s^2 = (c \Delta t)^2 - R^2(t) \left(\frac{\Delta r^2}{1 - kr^2} + r^2 \Delta \theta^2 + r^2 \sin^2 \theta \Delta \phi^2 \right), \qquad (10.6)$$

where we use *comoving spherical coordinates*, with r a radial distance. (Recall the use in chapter 9 of spherical coordinates in describing the Schwarzschild metric.) As was the case for the metric (10.5), the scale factor $R(t)$ is a function of cosmic time only, but it can change with time in a yet-unspecified manner. This metric is known as the **Robertson-Walker metric**, and its most prominent feature that we have not already encountered is the constant k. This **curvature constant** specifies the curvature of the three-dimensional *spatial* part of the spacetime. In our earlier discussion of geometry, we stated that there are three homogeneous, isotropic geometries: flat, spherical, and hyperbolic. These correspond to three values of k: $k = 0$ describes a flat geometry; $k > 0$ gives a spherical geometry; and $k < 0$ represents a hyperbolic geometry. We have not yet asserted that our space corresponds to any particular one of these three, only that it is one of them. The Robertson-Walker metric can accommodate all three types of homogeneous and isotropic geometries.

A remarkable feature of this metric is that its geometry is *uniquely* determined by the sign of the constant k. No matter how we adjust our coordinates, the sign of k can never change, although we are able to change its numerical value by a coordinate transformation, if we wish. The sign of the curvature constant specifies the type of the geometry, as we have indicated above. It is convenient, therefore, to label the three types of geometry with specific values of k: $k = +1$ corresponds to spherical geometry, $k = 0$ to flat, and $k = -1$ to hyperbolic. It can be shown that the scale factor R is fundamentally related to the magnitude of the curvature of the three-dimensional, spatial part of the metric. If $k = 0$, the spatial geometry is flat, and this length is irrelevant to the geometry, although its change with time controls the behavior of many physical quantities. For the spherical and hyperbolic universes, R indicates the characteristic curvature of space. In the case of spherical spatial geometry, $k = 1$, this scale is easily interpreted as the radius of the spatial part of the spacetime at any fixed cosmic time t. For the hyperbolic geometry, which has negative constant curvature, visualization is not possible, but still R sets the length scale.

Now let us verify that the Robertson-Walker metric is indeed consistent with the cosmological principle. Is a universe described by such a metric homogeneous? The curvature constant k is the same everywhere; the universe has the same geometry at all points. Every point has the same expansion factor; that is, spacetime is evolving in exactly the same way at all spatial locations. We have implicitly included this characteristic by requiring that the scale factor depend only on time, and not on the spatial coordinates. Therefore, this metric describes a homogeneous spacetime. Homogeneity makes it possible to define a "standard clock" for the universe, which can be said to keep **cosmic time**. In the form in which we have presented this metric, the time coordinate t is a cosmic time coordinate, determined solely by the rate of expansion of the universe as a whole. Any clock, anywhere in the universe, which is always instantaneously at rest with respect to the average matter distribution of the universe, will keep cosmic time. If such a universe starts with a big bang, then cosmic time indicates the time elapsed since the big bang.

Is this universe also isotropic? The expansion is the same not only everywhere in space, but in all directions, so the metric is isotropic. An isotropic expansion is sometimes said to be *shape preserving*, because shapes do not change, but merely scale up: a square becomes a larger square, a sphere becomes a larger sphere, as the expansion progresses. If the expansion were anisotropic, that is, if the expansion factor were different in one or more spatial directions, then a sphere would be converted into an ellipsoid, and in principle such a transformation is detectable. An anisotropic expansion would also be observable in the redshift pattern; the redshift-distance plots of galaxies in different directions would obey different Hubble laws. As best we can tell from current data, the Hubble constant is the same in all directions, so the expansion of the real universe appears to be isotropic.

If the universe *is* isotropic and homogeneous, then it follows that the Hubble law will describe any expansion or contraction. The Hubble law depends only on distance, not on direction; hence it is isotropic. The Hubble law is a simple linear relationship between distance and velocity; a straight line looks the same at all of its points. In fact, the Hubble law is the *only* law for which the expansion is the same in all directions and in all places.

We can illustrate this with a simple example. Start with a uniform distribution of galaxies, each of which is initially separated from its nearest neighbors by a distance d. Label the galaxies along any one line as A, B, C, and so forth, as indicated by Figure 10.10. Now scale everything up by doubling the distances during some time interval Δt. After the expansion, A is separated from B by $2d$, from C by $4d$, from D by $6d$, and so forth. Construct a velocity for each galaxy, relative to A, by computing the distance moved and dividing by the time interval. Thus B moves away from A at a recession velocity of $(2d - d)/\Delta t = d/\Delta t$; C recedes at $(4d - 2d)/\Delta t = 2d/\Delta t$; similarly D moves at $3d/\Delta t$. The recession velocity of any galaxy relative to A increases linearly with the distance from A, which is the Hubble law. However, our

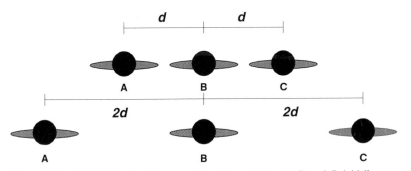

Figure 10.10 The Hubble law demonstrated by three galaxies, A, B, and C, initially separated by distances d. After a time Δt this distance has doubled to $2d$, and the distance between A and C has increased to $4d$. The recession velocity of B from A is $d/\Delta t$, and of C from A is $2d/\Delta t$. The Hubble Law predicts just such an increase in recession velocity with distance.

choices of reference galaxy and line of sight were completely arbitrary; we could have centered our reference on any of the galaxies and looked in any direction. From this we see that a homogeneous and isotropic expansion is described by the Hubble law.

The Hubble law is a statement of how the universe expands, and the Hubble constant is a measure of how fast it expands. In Chapter 11, we will learn how the Hubble constant can be related to the change with time of the scale factor. For now, we shall accept that the Hubble constant indicates the speed of expansion. An extremely useful byproduct of this fact is that the Hubble constant can tell us something about the length of time over which the galaxies have been separating. From the defining equation (10.4), you can see that the unit of the Hubble constant is inverse time (s^{-1}). Therefore, the inverse of the Hubble constant is an interval of time. This interval is called the **Hubble time** (or Hubble period, or Hubble age). Mathematically, the Hubble time is simply

$$t_H = \frac{1}{H},\qquad(10.7)$$

where H must be expressed in consistent units, such as inverse seconds. The Hubble time provides an *estimate* of the age of an expanding universe. More exactly, it is the age of an idealized universe that expands from zero size at a constant rate given by the value of H in question. We shall soon discover that the Hubble "constant" is not, in general, truly a constant, but varies over the history of the universe; thus the Hubble time does not give the age of the actual universe. Nevertheless, it does usually provide a good first approximation to the true age.

The Hubble time is akin to an estimate of the time required for a journey. If you were to take a car trip, at any point in your trip you could use the instantaneous speed on the speedometer, plus the distance traveled as measured by the odometer, to estimate how long you had been driving. If your speed had been changing, then obviously this could be a rather poor estimate, whereas if the speed were relatively constant, it would be a reasonably good estimate. If you wanted to compute from

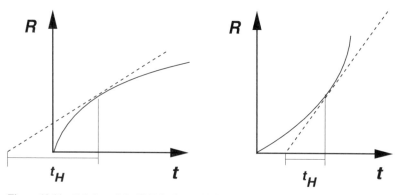

Figure 10.11   Relation of the Hubble time, which is a linear extrapolation from "now" back to $R = 0$, as indicated by the straight broken lines, to the actual age of the universe, obtained from the scale factor. On the left, a universe for which the rate of expansion is decelerating; in this case, the Hubble time is longer than the true age of the universe. On the right, a universe whose expansion is accelerating; its Hubble time is shorter than its actual age.

your speedometer readings the *exact* time you had been driving, you would have to know in detail how your speed varied during your entire trip. Similarly, to find the exact age of the universe, we must know how the Hubble constant has changed with time. Given a model, we can compute the precise age of this universe. However, we do not yet have sufficient data to decide which model best describes our physical universe, so the Hubble age provides a convenient timescale by which we may judge various theories.

Equation (10.7) shows that the higher the expansion speed, that is, the larger the Hubble constant, the shorter is the Hubble time. This should make intuitive sense, as well. If the universe is expanding rapidly (large Hubble constant), then it would reach its present scale more quickly than it would at a small expansion rate (small Hubble constant). For reference, remember that if the Hubble constant is approximately 100 km/s/Mpc, then the Hubble time is roughly 10^{10} yr; if $H = 50$ km/s/Mpc, then $t_H = 2 \times 10^{10}$ yr. Because the relationship between the Hubble constant and the Hubble age is a simple inverse proportionality, Hubble times for other values of the Hubble constant can be easily computed by using these reference values.

A related concept is the **Hubble length**. This is the distance that light travels in one Hubble time,

$$D_H = ct_H = \frac{c}{H}. \tag{10.8}$$

For a Hubble constant of 50 km/s/Mpc, this length is 1.5 trillion parsecs; it is huge. The Hubble length has several interpretations. Perhaps the easiest way to understand its significance is to consider a sphere centered on the Milky Way, of radius equal to the Hubble length; this delimiter is called the **Hubble sphere**. The volume enclosed by the Hubble sphere is an estimate of the volume of the universe that could possibly be within our past light cone. The Hubble sphere thus defines the limit of our *observable* universe, containing all that could have affected us, or been affected by us, up to the present time. The Hubble sphere is not the whole universe; *every* point in the universe has its own Hubble sphere. The Hubble sphere corresponds, roughly, to the past light cone at a specific place and time. Thus the existence of the Hubble sphere does not imply a "center" at any special location. The Hubble length can change with cosmic time if the Hubble "constant" changes; the exact time behavior depends upon the variations in the Hubble age at different epochs. If the expansion rate is slowing, then the Hubble age increases with cosmic time; hence the Hubble sphere will become larger with time.

Another interpretation of the Hubble length is that it is the distance at which the speed of recession from our vantage point is equal to the speed of light. Again the Hubble length represents the greatest distance to which we can possibly see, regardless of any other limitations under which we might operate. Just as the Hubble time is only an approximation to the age of the universe, with the actual age dependent upon the exact model, so the Hubble sphere is only an approximate measure of the size of the observable universe. The true size must be computed using a specific model $R(t)$. The Hubble sphere is still a good approximate length scale for conceptualizing the universe and serves as a perfectly adequate estimate for most purposes.

The Cosmological Redshift

Now that we have developed the Robertson-Walker metric, we can understand the origin of the cosmological redshift. Because of the change in the cosmic scale factor, the wavelength at the time the light was emitted is different from the wavelength at the time it is detected. The cosmological redshift is a consequence of the effects of gravity upon spacetime, as specified by a general relativistic metric. As lengths increase with time in the expanding universe, the wavelengths of light moving through space also increase. Specifically, since wavelength λ is just a length, and we know from the scale factor $R(t)$ how lengths change in a Robertson-Walker metric, we can write for the wavelength "now," λ_{rec}, the formula

$$\lambda_{rec} = \left(\frac{R_{now}}{R_{then}}\right)\lambda_{em}. \tag{10.9}$$

But the redshift, from whatever cause, is *defined* as the change in wavelength divided by the standard value of that wavelength, that is, $z = (\lambda_{rec} - \lambda_{em})/\lambda_{em}$, where λ_{em} is the wavelength as measured at the emitter. Thus we find

$$1 + z = \frac{\lambda_{rec}}{\lambda_{em}} = \frac{R_{now}}{R_{then}}. \tag{10.10}$$

The redshift provides a direct measurement of the ratio of the scale factor "now" to the scale factor "then," when the light was emitted.

Relativity has shown that anything causing a change in length (length contraction) must also affect time as well (time dilation). Wavelength times frequency always equals the speed of the wave; light traveling through a vacuum obeys the equation $\lambda\nu = c$. Thus both the wavelength and the frequency ν change with the expansion, since c is a constant for all observers and all times. Frequency is measured in cycles per second (cycles/s), so the inverse of a frequency is a time. Hence the expanding universe produces a cosmological time dilation, as well as a redshift. For example, suppose that a clock attached to a distant, high-redshift quasar measures the frequency of a particular ultraviolet light beam, emitted somewhere within the quasar, as 10^{15} cycles/s. By the time that light reaches us on Earth, it has shifted into the infrared, with a frequency of 10^{14} cycles/s as we measure it. Thus it takes 10 of our seconds for us to see the quasar clock tick off one quasar second. In other words, we see the quasar clock running very slow. We can derive this using the wavelength-frequency relationship, $1/\nu = \lambda/c$, in combination with the redshift formula, equation (10.10), to obtain

$$\frac{\nu_{em}}{\nu_{rec}} = \frac{\lambda_{rec}}{\lambda_{em}} = \frac{R_{now}}{R_{then}} = 1 + z. \tag{10.11}$$

Notice that whereas an expanding universe *increases* wavelengths, it *decreases* frequencies. The frequency formula is, as you would have expected, very similar to the formula for wavelength; it can be used to indicate the time dilation of very distant sources relative to our clocks. In the above example, the quasar has a redshift $z = 9$, a huge redshift, much bigger than has ever been observed for any quasar. A more typical quasar might have a redshift of $z = 1$. An observer on Earth would find that this quasar's clocks run slow, relative to Earth clocks, by a factor of 2.

We have seen how two important observable properties of the universe are related to the scale factor: the cosmological redshift and the Hubble law. These effects contain subtleties that lend themselves to misunderstandings. One point of confusion is the belief that the cosmological redshifts are ordinary Doppler shifts resulting from the motions of the galaxies through space. After all, cosmologists talk about "expansion velocities," and the Hubble law relates just such a velocity to a distance. An analogy with Doppler shifts can be useful, particularly for nearby galaxies where the redshifts are small, but the cosmological redshift is really more akin to the gravitational redshift discussed in connection with black holes. Both cosmological and gravitational redshifts arise directly from a metric. The cosmological redshift is produced by photons traversing expanding space. It does not occur because of relative motion at the moment of the emission or reception of the light, as is the case for the conventional Doppler shift, but as the light *travels* through space. The formula $z + 1 = R_{now}/R_{then}$ tells us only the *ratio* of the scale factor today, when the light is received, compared to the scale factor at the time the light was emitted. It tells us nothing about how the expansion (or contraction) proceeded with time.

As an illustration, suppose that the universe is not expanding at some arbitrary time t_{then}. A distant galaxy emits light toward the Milky Way at t_{then}, while it is at rest with respect to the Milky Way. Write the scale factor at that time as R_{then}. Suppose further that after the light is emitted, the universe begins to expand rapidly. As the light crosses the space between the emitting galaxy and the Milky Way, it will be redshifted because of the expanding space. Finally, suppose that the expansion stops abruptly at a scale factor R_{now}, just as the light reaches us. When the light arrives, it still has a large redshift, in accordance with equation (10.10), despite its being both emitted and received while the source is at rest with respect to the Earth. It would have had exactly the same redshift in the more realistic case that the universe had expanded at a continuous rate from R_{then} to R_{now}. Again, the cosmological shift observed from some specific galaxy tells us only the relative scale factors, not the way in which the universe evolved from R_{then} to R_{now}.

This is not to say that we cannot derive such information about our universe, but it is a difficult business. Redshift alone is not enough; we need another piece of information. Consider equation (10.4), the Hubble law. We called it the "theoretical" Hubble law because it describes the state of the universe at one instant of cosmic time, and this is something that we *cannot* directly observe. When we make observations, we do not see the universe at a single moment in cosmic time. Because the speed of light is finite, when we look at distant stars and galaxies, we are seeing them as they were at the time in the past when that light left them. The farther we look into space, the farther back in time we look; the travel time of the light is called the **lookback time**. We see a redshift because the universe had a different scale factor when the light left the emitter. The redshift would give us the ratio R_{now}/R_{then}, and the lookback time would tell how far in the past the universe had the scale factor R_{then}. If we knew both the lookback time and the redshift for a large number of objects, we could construct a complete graph of $R(t)$.

In practice, the redshift is easy to measure, but the lookback time is not. To obtain the lookback time, we must measure the distance to the object emitting the light. This brings us right back to the same difficult question that confronted Hubble, namely, the impediments to seeing distinct standard candles at such great distances. When we compute the distances to remote objects, we must take into account the change of the Hubble constant with time, and the increasing distances between objects as time progresses. Only when we look at nearby galaxies can we ignore the complexities of a specific model. For such galaxies, the lookback time is relatively small, and the Hubble constant has not changed significantly since their light was emitted. In these cases, we can approximate distances directly by relating the Hubble law to the redshift. But this is only an approximation, and a good approximation only for nearby galaxies. For anything else, we need a complete model $R(t)$.

We have learned how the expanding universe was discovered observationally. We have seen how the expanding universe can be incorporated within a spacetime metric through the use of a homogeneous and isotropic scale factor. Finally, we have examined how some of the observed properties of the universe, such as the Hubble constant and the redshifts, can be related to the all-important scale factor. There remains the task of constructing specific models of $R(t)$; we take up this challenge in the next chapter.

KEY TERMS

redshift	blueshift	cosmological redshift
nebula	nova	supernova
cosmic distance ladder	parallax	luminosity distance
standard candle	extinction	Cepheid variable
Hubble constant	Hubble law	Tully-Fisher relationship
peculiar velocity	Hubble flow	cosmological constant
de Sitter model	scale factor	comoving coordinates
Robertson-Walker	curvature constant	cosmic time
metric	Hubble length	Hubble sphere
Hubble time		
lookback time		

Review Questions:

1. In retrospect, it seems obvious that the spiral nebulae are external galaxies. Discuss the reasons that this hypothesis was so slow in gaining acceptance. What finally proved it?

2. You are measuring distances to galaxies using a particular standard candle. At a professional meeting, another astronomer announces that your standard candle is actually twice as luminous as previously believed. If she is correct, how would you have to modify your derived distances?

3. Discuss the difficulties in measuring extragalactic distances. What phenomenon confused scientists for several decades, causing them to overestimate distances? What kinds of distance indicators can be used? Describe some potential sources of error with the methods.

4. Assume that the Hubble constant is 50 km/s/Mpc. Using the Hubble law and the nonrelativistic redshift formula ($z = v/c$, where z is the redshift), calculate the distance of a galaxy whose spectrum has a redshift of 1%. (The speed of light is 3×10^5 km/s).

5. Explain why the overall expansion of the universe does not make the solar system expand as well.

6. A Hubble constant of 50 km/s/Mpc corresponds to a Hubble time of about 20 billion years. What would be the Hubble time for a Hubble constant of 100 km/s/Mpc? 75 km/s/Mpc? If the rate of expansion of the universe is slowing down with time, will the Hubble time over- or underestimate the age of the universe?

7. The "Hubble sphere" forms a sort of edge to our observable universe. Why isn't this a "real" edge to the universe? Why doesn't this edge violate the cosmological principle or the Copernican principle?

8. In order to prevent his model from collapsing, Einstein added a term Λ to his model. A positive Λ resulted in a repulsive force to counteract gravity. What do you think would happen if a *negative* Λ were used instead?

9. Some quasars have redshifts of 4 or even greater. Redshifts can be caused by relative motions, by gravitational fields, or by the expansion of the universe. Consider the case of gravitational redshift. Using equation (9.5), compute how close to the black hole the emitting region would have to be in order to produce a redshift $z = 4$. Give your result in terms of the Schwarzschild radius. Why is it unlikely that the redshifts of quasars could be explained solely by gravitational redshifts?

10. Consider a quasar at a redshift $z = 2$. If the quasar's light output varies with a period of 1 day as we observe it, what is the period of variation in the quasar's frame? What does the quasar's "lookback time" mean?

11. Consider the universe at a redshift of $z = 2$. If two galaxies were separated by a distance ℓ at that time, what is their separation today? What is their separation in *comoving coordinates* today compared with then?

12. Draw a graph showing scale factor versus time using the following data for redshifts and lookback times: At $z = 0$, $t_{lb} = 0$; $z = 1/3$, $t_{lb} = 12$; $z = 1$, $t_{lb} = 21$; $z = 1.5$, $t_{lb} = 24$; $z = 3$, $t_{lb} = 28$; $z = 9$, $t_{lb} = 31$. The units are arbitrary. Locate these data points on the graph and then draw a freehand curve through them.

Notes

1. The exception is the transverse Doppler shift due to relativistic time dilation. See chapter 7 for more details.

2. See chapter 5 for a discussion of the properties of novae and supernovae.

3. The Andromeda Galaxy, also known as Messier 31 (M31), can be seen by the naked eye from a dark location. It is literally "as far as the eye can see."

4. In fact, there is a systematic motion: the galaxy rotates. But at the time, the Milky way was believed to be a somewhat amorphous disk of stars, with the Sun at the center. Coincidentally, 1917 was also the year that Shapley published his observations of globular clusters, indicating that the Sun was actually at the outskirts of the Milky Way, not the center.

5. It must be admitted that Einstein certainly had demonstrated the hubris to put his theory ahead of observation. When asked what he would have thought if the 1919 eclipse expedition had not observed the predicted shift of the stellar positions, he reportedly said "I would have had to pity our dear Lord. The theory is correct." In this case Einstein believed more in his theory's beauty, simplicity, and unity than in the ability of astronomers to measure the positions of stars accurately.

CHAPTER 11

Modeling the Universe

Two major developments, Hubble's observations and Einstein's theory of general relativity, moved the subject of cosmology out of the realm of the mainly philosophical and firmly into the arena of science. Now the task of developing an actual model of a dynamic universe lies before us. While at first glance it may seem an impossible undertaking, we are aided by the adoption of the cosmological principle. This is an enormous simplification, for it implies that the metric which describes space and time, and specifies their evolution, must be the Robertson-Walker metric. The real universe is complex, with many intricate structures and objects, so it may seem excessively crude to assume that the universe is perfectly smooth and homogeneous; but we must start somewhere, and it seems prudent to begin with the simplest case. As yet, however, we have gone no further than this, and until we determine the parameters $R(t)$ and k, we can say nothing more. How might we go about evaluating the scale factor and the curvature constant?

The construction of cosmological models corresponds to the task of solving Einstein's equations of general relativity, for an isotropic and homogeneous universe. Recall that Einstein's equations (8.5) state that the geometry of the universe is determined by its mass and energy content. The detailed mass and energy distribution of the universe is obviously very complicated. Galaxies and stars are scattered unevenly throughout space; interstellar and intergalactic gas float about in irregular patterns. In addition to this matter distribution, the universe is filled with photons, and possibly other more exotic particles, all contributing some amount of energy. In keeping with the cosmological principle, however, we assume that the characteristics of individual clumps of matter, such as the galaxies, are not important. Instead, we shall take all the galaxies, stars, and planets in the universe, grind them up into a fine dust, and distribute that dust evenly throughout space, so that at every location the average mass density takes a constant value, ρ. The universe may also be filled with energy from sources other than rest mass-energy; similar to the mass-energy, these other forms can be characterized by a uniform energy per unit volume, or *energy density*. By replacing the complex matter and energy contents of the universe with these constant average values, we greatly simplify the right-hand side of Einstein's equations. The variables describing the contents of the universe, such as ρ, are now independent of spatial location; at most they can depend only upon time, for consistency with the cosmological principle.

This is as far as we are going with Einstein's equations. Solving for the metric terms from the geometry term requires more differential calculus than we wish to demand here. Fortunately, we can gain considerable insight into the nature of the solution by considering Newtonian physics in a flat space. Newtonian physics is an adequate description of the universe as long as the distances we consider are much less than the radius of the Hubble sphere, and the expansion velocities are much less than c. In a sufficiently small region, we can safely ignore any curvature of space, so the assumption of a flat space is not a severe limitation. Moreover, remember this: in a homogeneous, isotropic universe, anything we learn locally tells us something about the way the universe works everywhere. Of course, although these may seem to be reasonable justifications for using Newtonian physics, cosmologists do find their solutions for the universe using Einstein's general relativity. Remarkably enough, the equations that result from the Newtonian analysis are almost exactly the same as those which emerge from the Einstein equations.

A Newtonian Universe

The universe envisioned by Newton when he developed his law of universal gravitation was infinite and unchanging, filled in all directions with stars acting under the mutual gravitational forces of all the other stars. The image is much like that of the air molecules in a huge room; the molecules fly around with random velocities and interact with each other, but overall the air in the room is still. There was one aspect of this model that was rather troubling. If the cosmos is filled with stars that attract one other by gravitational forces, should not all the stars pull themselves together into a lump? Newton reasoned that in an infinite universe, the forces would be balanced at each point by equal gravitational pull in all directions. But in a universe with an infinite number of stars, the forces in all directions would be infinite, so this would be a delicate balance indeed!

Rather than dealing with an infinite extent of stars, let us confine our attention to a large, but finite, spherical portion of the universe, with radius R. Assume that it is filled uniformly with matter. We shall focus our attention on a small bit of matter sitting on the edge of the sphere. Let this little "test particle" have a mass m_t, while the total mass of all the other matter within the sphere shall be M_s, as in Figure 11.1. There is nothing special about this test particle. Indeed, since R is arbitrary, as long as it meets the requirements for a Newtonian analysis to be valid, the test particle could actually represent *any* random gravitating particle within the distribution of matter. As long as any exterior matter is distributed uniformly, the gravitational effects of the matter *outside* the sphere cancel out, in a manner similar to Newton's explanation of the gravity of a sphere.[1] Therefore, the only contribution to the gravitational acceleration of our test particle comes from the matter within the sphere. We can now apply Newton's formula for the gravitational force,

$$F_g = -\frac{Gm_t M_s}{R^2}, \tag{11.1}$$

to obtain the force acting on the test particle. The situation we have described is exactly analogous to the force of gravity on you (a test particle) due to the mass of the Earth (a large mass-filled sphere). Just as for a test particle on the Earth, the force of gravity due to the sphere is the same as if all the mass of the sphere were concentrated at its center; thus a simple equation like (11.1) can still be used. The sign is negative because gravity is always attractive; the mass in the sphere attempts to pull the test particle into the sphere, toward smaller R.

Imagine now that the test particle must remain on the surface of the sphere, but the sphere itself can expand or contract, although its total mass M_s cannot change. Hence the quantity R is not only the radius of the sphere, but also specifies the location of the test particle. As R changes, the particle moves along with it at some velocity. As always, a velocity is the change of position with time, so we can write

$$v = \frac{\Delta R}{\Delta t} \equiv \dot{R}, \tag{11.2}$$

where we have used a new notation, $\dot{R}$, to represent the change in the location R with respect to time. Assume now that the particle and sphere are expanding with some outward velocity. Since there is a net gravitational force acting on the particle, there must be an acceleration, specifically, the usual Newtonian gravitational acceleration

$$g = -\frac{GM_s}{R^2}. \tag{11.3}$$

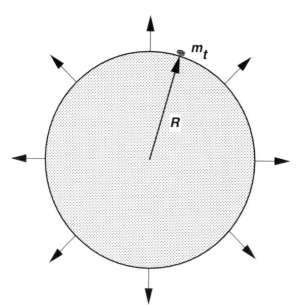

Figure 11.1 A Newtonian universe. A sphere of radius R contains uniformly distributed dust. A test particle with mass m_t, located at the edge of the sphere, feels the gravitational attraction of all this mass directed toward the center of the sphere. The whole system is expanding due to some initial outward velocity, but gravity causes that expansion to decelerate.

The gravitational acceleration is always directed toward the center. If the sphere expands outward, gravity will reduce its velocity, possibly leading to an infall and recollapse. This acceleration, or change of velocity with respect to time, is represented by the notation $\ddot{R}$.

If we consider the "test particle" to be simply any representative particle, our system consists of a sphere of particles moving in the radial direction under the influence of gravity. We have encountered such a situation before, in another guise. In chapter 9 we discussed the escape velocity, defined to be that velocity which just permits a particle to escape from the gravitational field of an object of a certain mass and radius,

$$v_{esc} = \dot{R} = \sqrt{\frac{2GM_s}{R}}. \qquad (11.4)$$

The concept of escape velocity has its most obvious application when considering the launch of a rocket from the Earth. If a rocket has less than escape velocity, it cannot escape from Earth, but falls back. A rocket launched radially precisely at escape velocity will be slowed by gravity, such that at any distance R from Earth, its speed equals the escape velocity appropriate to that R; the rocket travels indefinitely outward with a velocity that approachs zero as R goes to infinity. Finally, if the rocket's velocity exceeds the escape velocity, then it has more than enough speed to leave Earth's gravitational field. The rocket escapes to infinity, where it still has a positive velocity.

For the case of the expanding sphere, the question is whether or not the particles that make up this self-gravitating sphere are moving outward with sufficient velocity to avoid recollapsing due to their own gravity. If the sphere has precisely the correct velocity just to avoid recollapsing as R becomes infinite, the expansion speed will be equal to the escape velocity at every R. However, among all possible expansions, movement at escape velocity is a special case. We can account for this by adding to the expression for the expansion velocity an additional constant term, positive for velocity exceeding the escape speed, and negative for total velocity less than the escape speed.

A more precise treatment of this situation phrases it in terms of the energy of motion, or the *kinetic energy*, of the test particle. In the Newtonian physics we are considering for the moment, this energy is given by one half the mass of the particle times its velocity squared, $E_k = \frac{1}{2}m_t v^2$, and the kinetic energy per unit mass is simply this quantity divided by the mass m_t, that is, $\mathcal{E} = \frac{1}{2}v^2$. The square of the escape speed gives an escape energy per unit mass, and the generalization of equation (11.4) can be written in the form

$$2\mathcal{E} = \dot{R}^2 = \frac{2GM_s}{R} + \text{constant}. \qquad (11.5)$$

The constant allows for more or less energy than the escape value. By letting the radius go to infinity, causing the "matter" term to vanish, we find that this constant is equal to twice the kinetic energy per unit mass remaining when the sphere has expanded to infinite size. We shall denote this quantity by $\mathcal{E}_\infty$. We can thus express equation (11.5) as

$$\dot{R}^2 = \frac{2GM_s}{R} + 2\mathcal{E}_\infty. \tag{11.6}$$

Energy as defined here can be zero, positive, or negative.

Given an expanding Newtonian sphere, we consider each of the possibilities in turn:

1. If the energy per unit mass at infinity is negative, $\mathcal{E}_\infty < 0$, then the sphere has net negative energy. It will stop expanding at some point *before* it reaches infinite radius. It will then recollapse.

2. If the energy per unit mass at infinity is zero, $\mathcal{E}_\infty = 0$, then the sphere has zero net energy. It has exactly the right velocity to keep expanding forever, although the velocity will drop to zero as time and radius go to infinity.

3. If the energy per unit mass at infinity is positive, $\mathcal{E}_\infty > 0$, then the sphere has net positive energy. It will keep expanding forever at a rate that is faster than necessary to prevent it from stopping and recollapsing and will reach infinite radius with some velocity remaining.

Now we must make the great leap from the analogy of the sphere to the universe. Instead of regarding the sphere as an isolated ball of mass, we can regard it as a typical volume of space, filled with a smooth distribution of dust at a constant density. Next, we will rewrite our equations in terms of density, not total mass, because the average density of the universe is locally measurable, whereas the total mass is not. Density is mass per unit volume; thus we shall replace the total mass within the sphere by its volume multiplied by its density, $M_s = 4/3\ \pi R^3 \rho$. Performing this substitution in the velocity equation (11.6) we obtain

$$\dot{R}^2 = \frac{8}{3}\pi G\rho R^2 + 2\mathcal{E}_\infty. \tag{11.7}$$

What is the limit of this equation as R becomes large without bound? Although we have now written the equation in terms of density, mass is still conserved, meaning that the quantity ρR^3 remains constant. Equations (11.6) and (11.7) behave in exactly the same way as the scale R becomes infinitely large. Thus, like the Newtonian sphere, the universe too can have positive, zero, or negative net energy. If the universe has negative energy at infinity, it cannot "escape" its own gravity, and it will eventually cease to expand, subsequently collapsing back on itself. If the universe has positive energy at infinity, it has sufficient energy to expand forever. If the universe has zero energy at infinity it will expand forever, but the expansion velocity will drop to zero as R becomes infinite. The constant $2\mathcal{E}_\infty$, then, is related to the fate of the universe.

So far our discussion has been in terms of Newton's equation of gravity. Of course, from our study of general relativity, we know that Newtonian gravity cannot apply to the universe as a whole. We were justified in applying it to investigate a spherical region carved out of the universe, as long as the sphere was relatively small compared to the Hubble length and any velocities within it were nonrelativistic.

Under such conditions, Newtonian gravity is a good approximation for gravity in a flat space. Of course, to extend our analysis to the entire universe and to curved spaces, we must return to general relativity. If we had worked out the fully relativistic equations for a universe with uniform matter density, we would have obtained the equation

$$\dot{R}^2 = \frac{8\pi G}{3}\rho R^2 - kc^2, \tag{11.8}$$

as the relativistic equivalent of the "escape energy" equation (11.7); here k is the same curvature constant that appears in the Robertson-Walker metric. Since we can always choose our coordinate values to take whichever scale we wish, we shall adjust them so that k will take one of the three values 0, +1, or −1. These correspond to the three isotropic and homogeneous geometries: flat, spherical, and hyperbolic, respectively.

Equation (11.8) is nearly the same as that which emerged from our Newtonian treatment of the universe, except that now R is the scale factor rather than the radius of some arbitrary sphere. In other words, the gravity of the mass in the universe acts on the spacetime scale factor in much the same way that the gravity of the mass inside a uniform sphere acts on its own radius R. There are important additions due to relativity, however. First, we have replaced the $\mathcal{E}_\infty$ term in the Newtonian equation with the curvature term. This term retains its significance as an energy at infinity, but is now tied into the overall geometry of space. Second, relativity requires that we must include *all* forms of energy, not just rest mass, in our definition of ρ; mass and energy are equivalent, and both contribute to the force of gravity. This relativistic equation (11.8) is called the **Friedmann equation**, after the Russian physicist Alexander Friedmann.

Because of the complicating addition of energy to the source of the gravitational field, we cannot solve the Friedmann equation until we have specified how the total mass-energy density changes with time. This requires two more equations; the relativistic equation for the *conservation* of mass-energy density, and an *equation of state*, a relationship between matter density and energy density. We shall not derive these equations; instead, we shall content ourselves with describing some important qualitative features. In our present universe, the only significant contribution to nonmatter energy density is from photons left over from the big bang. Early in the history of the universe, the photon energy density was dominant, but their energy has been reduced by redshifting. The energy density these photons now contribute is almost negligible compared to the ordinary mass density, and we shall ignore them when we describe the evolution of the universe today. In the case of zero energy density, the quantity ρR^3 remains constant; as R increases, ρ decreases precisely as in the previous Newtonian analysis.

Universe models of this general type, that is, those that are determined by a Robertson-Walker metric, obey the conservation of mass-energy, admit no cosmological constant, and contain some specified total energy density, are the **standard models**. The standard models are not the only possible cosmological models, but they are those that follow from our minimal set of assumptions, specifically, a universe in which only gravity operates and the cosmological principle holds. These models all decelerate, since gravity acts to pull matter together and to slow down the expansion. We know from observation that the expansion factor is increasing at the present time, that is, $\dot{R}$ must be positive; we have also just argued that $\ddot{R} < 0$ for the standard models. From these facts, we can conclude that the function R(t), whatever its form, must have reached $R = 0$ at some time in the past. We may always adjust our cosmic time coordinate such that this occurred at $t = 0$; thus we obtain, as the initial condition of the universe, that

$$R(0) = 0.$$

Since the scale factor gives the separation of comoving points at a particular cosmic time, it obviously must be related to the density. But if comoving points have zero separation, then the density must be infinite. This initial state of infinite density is what is meant by the **big bang**. All standard models begin with a bang. The big bang is *not* a point in space; the scale factor is zero at all spatial locations, in accordance with the cosmological principle and the Robertson-Walker metric. In other words, the big bang happened everywhere.

If we briefly consider the big bang limit of equation (11.8), we see something interesting. The closer the approach to the initial time, the less the geometry of the model matters. As R shrinks, the first term on the right-hand side of the equation dominates over the constant contribution due to the curvature. This is a valuable simplification, since it means that we can describe the earliest stages in the universe without worrying too much about what its true curvature might be.

The ultimate fate of the universe, on the other hand, depends very strongly on the spatial curvature. Our discussion of the Friedmann equation and its Newtonian equivalent has suggested that the energy density of the universe is related to its spatial curvature. Our energy arguments enable us to predict the future behavior of the models corresponding to each of these geometries: the flat and hyperbolic models will expand forever, while the spherical model will recollapse into something conventionally called the **big crunch**. The spherical model both begins and ends with a bang; it recollapses back to $R = 0$. The flat and hyperbolic models, by virtue of their endless expansion, end not with a bang, but a whimper.

Hubble's Law and the Scale Factor

The Friedmann equation governs the evolution of the scale factor $R(t)$ for the case of a universe described by the Robertson-Walker metric; that is, a universe which is isotropic and homogeneous. A solution to this equation, for a given choice of curvature k and density ρ, is a model of the universe. The scale factor itself is not *directly* observable; the observables are various quantities derived from it. A model specifies the time behavior of the scale factor, from which observable quantities can be computed. The predictions of the model can then be compared with actual measurements, to evaluate how well that model fits the data. Hence we must obtain theoretical expressions for the observable parameters in terms of those quantities that are specified by the model itself, such as the scale factor, the curvature constant, and any cosmological constant. One of the most important observables is, of course, the

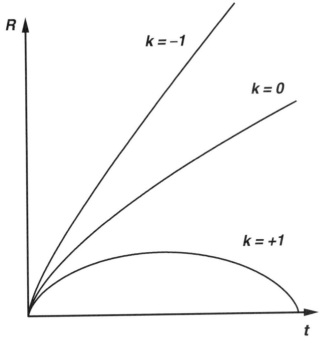

Figure 11.2 The behavior of the scale factor for the three different geometries of the standard models. All begin with a big bang. The $k = +1$ spherical model expands to a maximum size, and then contracts to a big crunch. While both the flat ($k = 0$) and the hyperbolic ($k = -1$) universes expand forever, the hyperbolic universe expands at a faster rate than the flat universe.

Hubble constant H, so let us first investigate how the Hubble constant is related to the scale factor.

The Hubble law itself can be derived directly from the Robertson-Walker metric. Although a rigorous proof requires some basic calculus, a simplified demonstration can be visualized from Figure 11.3. At a time t, two galaxies are separated by the *comoving* distance D, corresponding to a *physical* distance $\ell(t) = R(t)D$. Since the

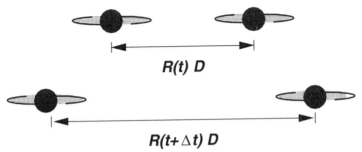

Figure 11.3 Two galaxies are separated by comoving distance D. At time t their physical separation is $R(t)D$. A time Δt later their separation is $R(t + \Delta t)D$. The scale factor increases, but the comoving distance remains constant, by definition.

change in separation of these "test" galaxies is a consequence only of the change in the scale factor, we know that at some later time $t + \Delta t$, their physical separation is $R(t + \Delta t)D$, or, approximately, $(R + \Delta R)D$. (Recall that the comoving distance does not change, by definition.) Thus after an interval of time Δt, the separation of the galaxies has changed by a distance $(\Delta R)D$. Dividing the change in distance by the time interval gives a velocity, $v = \dot{R}D$ (compare eq. [11.2]). Since the initial separation was $\ell = RD$, comparison with the theoretical Hubble law $v = H\ell$ leads to the conclusion that

$$H = \frac{v}{\ell} = \frac{\dot{R}}{R}.$$ (11.9)

Because R is a function only of time, its rate of change also depends only on time, and thus the Hubble constant is itself a function of cosmic time alone. Of course, this means it is not truly a *constant*. Homogeneity requires only that it be the same at every point in space at a given instant of cosmic time. Referring to it as a constant derives from its mathematical role as the "constant of proportionality" in the Hubble law, and so the term "Hubble constant" persists. The general symbol for the Hubble constant is H, without a subscript. Its value at the present time, that is, "now," is denoted H_0, pronounced "H-naught." This is the value we could determine through measurements of galaxies that are near to us.

Observing the Standard Models

The average mass density of the universe is another potentially measurable quantity. All that is necessary, in principle, is to add up all the mass seen within a large, representative volume of space around us. Needless to say, this is easier said than done. The most obvious stumbling block is that we can detect only luminous matter; the presence of any nonluminous, or "dark," matter can be inferred only indirectly. Furthermore, it is not so easy to find a representative volume. We require a volume large enough to be regarded as homogeneous; but how large would a truly homogeneous volume be? Galaxies cluster, and clusters themselves cluster, complicating our determination of the average matter density, as well as of any local deviations from it. Finally, just seeing a galaxy does not tell us its mass, whether in luminous or in dark matter; we must somehow deduce its mass from its brightness, or its motions, or from its interactions with its neighbors.

We will discuss the measurement of mass in the universe in more detail in later chapters. It turns out, however, that the average density of the universe has great cosmological significance, making an accurate measurement of this quantity especially important. To show this, we begin with the Friedmann equation. Substituting our new expression (11.9) for the Hubble constant, and rearranging the equation to solve for the curvature term, we obtain

$$\frac{kc^2}{R_0^2} = \frac{8\pi G\rho_0}{3} - H_0^2.$$ (11.10)

Although this equation holds for all times, we have written it to be evaluated at time "now," as indicated by the subscripts. Thus we can determine whether the curvature is positive, negative, or zero by measuring the present values of both the Hubble constant and the average density. Through comparisons with observations, we can determine which of the three geometries the actual universe most resembles.

We may draw some interesting conclusions from a careful examination of the version of the Friedmann equation specified by (11.10). First, suppose the universe were empty, that is, $\rho = 0$. The square of the Hubble constant is always positive, so the curvature (the left-hand side) must be negative in that case. Hence if there were no matter at all, space would have hyperbolic geometry. A flat or spherical geometry can occur only in the presence of matter. On the other hand, if we set $k = 0$, corresponding to the spatially flat universe, then the matter density must equal a very specific **critical density**, which we denote by ρ_c. The critical density is a very important parameter in cosmology. Returning to equation (11.10) but dropping the subscripts and setting $k = 0$, we obtain a definition of the critical density

$$\rho_c = \frac{3H^2}{8\pi G}. \tag{11.11}$$

This is the density required for a flat spacetime, the critical case that demarcates the open, infinite universe from the closed, finite universe. For a Hubble constant of 100 km/s/Mpc, the value of this critical density is $\rho_c \approx 2 \times 10^{-26}$ kg/m^3. This corresponds to approximately 10 hydrogen atoms per cubic meter of space, a quantity which might not seem particularly dense by Earthly standards. Since the critical density scales with H^2, a Hubble constant of 50 km/s/Mpc produces a value one-fourth as large.

That the density and the Hubble constant are so intertwined should not be surprising. The Hubble constant is, after all, a measure of the velocity of the expansion of the universe; it is this expansion which must be overcome by the gravitational force of the matter in the universe. The critical universe is precisely balanced, so for a given mass density, the Hubble constant must have a very specific value, which we might call the "critical Hubble constant." Despite the difficulties in measuring distances, however, the Hubble constant should be the easier of the two quantities to measure, so we generally speak of a critical density implied by the measurement of H_0, rather than a critical Hubble value implied by a measured density.

From the critical density, we can define another important parameter for observational cosmology. This quantity is called the **density parameter**, or **omega** (Ω), and it specifies the ratio of the actual mass-energy density of the universe to the critical density:

$$\Omega = \frac{\rho}{\rho_c}. \tag{11.12}$$

We can use this definition to rewrite the Friedmann equation once again as

$$\Omega = 1 + \frac{kc^2}{H^2 R^2}. \tag{11.13}$$

Equation (11.13) shows that the value of Ω is intimately linked to the geometry of the universe; in particular, if $k = 1$, then $\Omega > 1$, whereas if $k = -1$, then $\Omega < 1$.

In the spherical universe, the density is greater than the critical density, whereas in the hyperbolic universe, the density is less than critical. A spherical universe contains sufficient mass for gravity to overwhelm its expansion and cause recollapse, whereas a hyperbolic universe has too little mass to allow gravity to overcome its expansion. The critical universe is exactly balanced; there is sufficient matter to halt the expansion, but only after infinite cosmic time has elapsed. Thus the density of the universe directly determines the geometry of spacetime in the standard cosmologies.

We can also see from equation (11.13) that for the spherical and hyperbolic universes, the density parameter Ω changes with time, because the denominator $H^2 R^2$ in the curvature term on the right-hand side is a function of cosmic time. Only in the flat universe is the density parameter a constant; in this critical case, $\Omega = 1$ for all times.

We have established that the universe is presently expanding, and have stated that gravity slows the rate of expansion. Let us define a new parameter, the **deceleration parameter**, which relates any change in the expansion rate to observations. The rate of change of the scale factor, $\dot{R}$, specifies the expansion of the universe. The rate of change of the rate of change of the scale factor, conventionally symbolized by $\ddot{R}$, denotes how the expansion itself changes in time, and this gives us the acceleration of the universe. The negative of the acceleration is, as usual, the deceleration, so we might define this new parameter as simply $-\ddot{R}$; but it is more convenient to define it as a dimensionless quantity, to make it independent of whatever way we may have chosen to set the specific value of R. The standard definition is given by the formula

$$q = -\frac{\ddot{R}}{RH^2},$$ (11.14)

where the inverse factors of R and H perform the role of making q dimensionless. Physically, the deceleration parameter tells us whether the expansion rate ("velocity," if you prefer) of the universe is increasing or decreasing. By the name and definition of this parameter, one might think that there is a prejudice that the expansion of the universe is slowing down. This is the case for many models, because the attractive force of gravity always acts to pull together the matter of the universe, slowing the expansion. Even so, the deceleration parameter can indicate either a deceleration or an acceleration of the expansion by its sign; it is positive for a deceleration, and negative for an acceleration. Its value at the present time is denoted by a zero subscript, as q_0 ("q-naught"). All standard models are characterized by $q > 0$; *all* these models are decelerating.

If we combine various formulae developed above for the standard models, we obtain the remarkably simple result that the deceleration parameter is directly related to the density parameter,

$$q = \frac{1}{2}\Omega.$$ (11.15)

Since Ω and q are so simply related, the specification of a value of q also determines the geometry of space and hence the specific model.

Although at this point the deceleration parameter may still seem like an abstract concept, it can immediately tell us something about the difference between the actual age of the universe and the Hubble time. If the cosmic expansion is slowing down, then the Hubble constant we measure today will be smaller than it was previously. The Hubble time will thus overestimate the age of the universe. On a long car trip, suppose you drove from point A to point B, a distance of 100 miles, while slowly reducing your speed. If your speed were 20 miles per hour when you arrived at point B, then you would overestimate the length of your trip, if you merely divided 100 by 20 to obtain an elapsed time of 5 hours. Conversely, if you accelerated during your trip, then you would underestimate the time of travel, if you divided the total distance traveled by your instantaneous speed upon your arrival at point B. In the same way, in a decelerating universe with $q_0 > 0$, the age of the universe will be *less than* the Hubble time, because at earlier times it was expanding at a faster rate, whereas an accelerating universe with $q_0 < 0$ will have an age that is *greater than* the Hubble time. Only in a universe that expands at a constant rate $q_0 = 0$, will the age equal the Hubble time.

From these results we have complete limits on the standard models in terms of observables. If the density of the universe is greater than the critical density, the model is a **closed universe** with $q > 1/2$. In such a universe, the curvature constant $k = 1$, and the energy at infinity is negative. Sufficient matter is present to halt the expansion eventually, at which point the universe begins to contract. The universe then contracts to ever smaller size, ending in a big crunch at some finite time in the future. This universe has a spherical spatial geometry; it is finite in both space and time.

If the density is precisely equal to the critical value, we obtain a **flat universe** with $q = 1/2$. The curvature constant is $k = 0$, and the universe expands to infinity; the expansion would come to a halt only after the passage of an infinite interval of time. This is a special case because its spatial geometry is Euclidean, or flat, and because the parameters describing it can take only very specific values. The density parameter Ω is equal to one at all times. The universe is infinite in both space and time. (In fact, it is infinite in space *for all times*, not just at infinite time.) This very special case is also known as the **Einstein–de Sitter model**.[2]

If the density is less than the critical value, then $k = -1$, and we have the third standard cosmological model. This model also expands forever, but the expansion never ceases, even after infinite time. The geometry of this **open universe** is hyperbolic, and it is infinite in space and time. For this model, the deceleration parameter takes values $q < 1/2$. The density decreases faster than does the Hubble "constant," so the density parameter Ω approaches zero at large times.

The Friedmann equation provides the means to produce a complete cosmological model: we need merely solve for the function $R(t)$, then use it to evaluate the observable parameters. We have already discussed these solutions qualitatively, but of course there are precise mathematical solutions for $R(t)$. Since the Friedmann equation is a differential equation, its solution, while straightforward, requires calculus. The easiest solution to obtain is that for the flat model, also known as the Einstein–de Sitter model. It is relatively simple to show that

$$R(t) = R_i(t/t_i)^{2/3}, \tag{11.17}$$

where t_i is some nonzero, but otherwise arbitrary cosmic time at which the scale factor R has some value R_i. The behavior of this scale factor is illustrated in Figure 11.4.

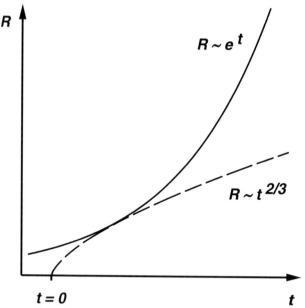

R

$R \sim e^{t}$

$R \sim t^{2/3}$

t = 0   t

Figure 11.4   The scale factor R as a function of time for a de Sitter universe (*solid line*), and the Einstein–de Sitter flat standard model (*dashed line*). The flat model begins with a bang at time $t = 0$. The exponential curve of the de Sitter model never goes to $R = 0$ so there is no big bang; this model is infinitely old.

The solutions for other models can be understood in comparison to the flat model. For example, the scale factor R in the closed spherical model increases less rapidly with time, reaches a maximum, and then falls back to zero. In the open hyperbolic model, R increases more rapidly than $t^{2/3}$. For the hyperbolic model with $k = -1$, when t becomes very large, the solution for R is nearly proportional to time. At this point the mutual gravitational attraction of all the mass in the universe is so weak as to have almost no effect upon the expansion of space.

From the explicit formula for the scale factor, the Hubble constant can be computed as a function of time. For the Einstein–de Sitter universe, this is given by the simple result $H = 2/(3t)$; as in all standard models, the Hubble constant decreases with increasing cosmic time. From this expression, it follows that the true age of the critical universe is 2/3 the Hubble time. Specifically, if the Hubble constant were presently 50 km/s/Mpc, the corresponding Hubble time would be 20 billion years, and the actual age of the flat universe would be $2/3 \times 20 = 13.3$ billion years. The Einstein–de Sitter solution provides an important dividing line for the standard models. Denser, closed models will have *smaller* ages than that of the Einstein-de Sitter model; open models will have *greater* ages than this flat model. If, for example, we could demonstrate conclusively that some star cluster was older than 13.3 billion years, this datum would rule out flat and closed standard models unless the Hubble constant proved to be less than 50 km/s/Mpc.

TABLE 11.1
Standard Friedmann Models

Model	Geometry	k	Ω	q_0	Age	Fate
Closed	Spherical	+1	> 1	> 1/2	$t_0 < \frac{2}{3}t_H$	Recollapse
Einstein–de Sitter	Flat	0	= 1	= 1/2	$t_0 = \frac{2}{3}t_H$	Expand forever
Open	Hyperbolic	−1	< 1	< 1/2	$\frac{2}{3}t_H < t_0 < t_H$	Expand forever

Models with a Cosmological Constant

Now that we have developed our standard models, we shall consider some variations. Einstein originally attempted to find a cosmological solution that was both *static* and spatially finite, with no boundary. He quickly discovered that such a universe cannot *remain* static, as we can appreciate from our analysis of the standard models; gravity's inexorable pull forces such a model to evolve. Rather than accept an evolving universe, however, Einstein added a term, conventionally symbolized by Λ and called the **cosmological constant**, to his equations of general relativity. This quantity, unlike the Hubble "constant," is a true constant; its value never changes for a given cosmological model. It acts over long distances; in Einstein's original formulation, it provides a repulsive force that counters gravity. In the Einstein static model, the effect of this parameter on the scale of the Earth is immeasurably small, yet on cosmic scales it just balances the tendency for matter to pull the universe toward collapse. Of course, Einstein's attempt to make a static universe ultimately failed, and this led him to recant on the cosmological constant. But the genie could not be put back into the bottle quite that easily. The cosmological constant can be used to create interesting alternative models of the universe, and cosmologists have done so.

We begin our study of the "nonstandard" cosmological models by modifying the Einstein equations to include the cosmological term. The relativistic equivalent of Newton's gravitational acceleration equation (11.3) is

$$\ddot{R} = -\frac{4G\pi}{3}\rho R + \frac{\Lambda R}{3}. \tag{11.18}$$

The first term on the right can also be written in the form $4G\pi(\rho R^3)/3R^2$, showing that it is just the familiar gravitational term. (We have implicitly assumed that any energy density other than mass is negligible. This is valid for most of the history of the universe.) The new term on the right can be interpreted as the force associated with the cosmological constant. With a positive sign, Λ provides a force opposite to that of gravity, that is, a repulsive force, which is just what Einstein needed to make his model static. However, the cosmological constant need not be repulsive; it could also be negative, that is, attractive. If the cosmological constant provides another attractive force, it supplements gravity.

An important property of the cosmological constant arises from the dependence of the Λ-force upon the scale factor. Whereas the gravitational force of the mass in the universe drops off as the inverse square of R, the Λ-force *increases* with R. This means that when the universe becomes "large enough," the Λ-force will inevitably dominate over the usual gravitational force. As one consequence of this, if Λ is negative, providing an additional attractive force, the universe will slow down and recollapse, no matter how rapid its present expansion or how small Ω might be. Hence it is possible to have a recollapsing universe that does not have a spherical geometry.

The presence of Λ means that we must return to the Friedmann equation and recompute the values of the Hubble constant, the critical density, and the deceleration parameter in light of this new term. With the cosmological constant, the Friedmann equation becomes

$$\dot{R}^2 = \frac{8}{3}\pi G \rho R^2 + \frac{\Lambda R^2}{3} - kc^2. \tag{11.19}$$

If we combine this with equation (11.18) and use the definitions of our various cosmological parameters, we find the following relationships:

$$\Lambda = 4\pi G \rho - 3H^2 q, \tag{11.20}$$

and

$$\frac{kc^2}{R^2} = 4\pi G \rho - H^2(q + 1). \tag{11.21}$$

The most important consequence of the inclusion of Λ that is apparent from these equations is that the simple relationships characterizing the standard models are now a little more complicated. The critical density is no longer directly related to the Hubble constant, and a measurement of the deceleration parameter does not immediately determine the model.

We can use these relationships to compute the special value of Λ that Einstein used to derive his static, spherical universe. Setting H to zero (static), and q to zero (no acceleration, i.e. no net force), and using equation (11.20), we find that the Einstein critical value of Λ is

$$\Lambda_c = 4\pi G \rho. \tag{11.22}$$

From equation (11.21), we further obtain $kc^2/R^2 = \Lambda_c$. Thus, in the Einstein static universe, the cosmological constant has a positive value that is determined by the average density of the universe. It is also just equal to the positive spatial curvature; this model corresponds to a spherical geometry.

A very different cosmology, which contains a cosmological constant but is devoid of matter, is the eponymous **de Sitter** model. The de Sitter universe has flat spatial geometry ($k = 0$), zero density ($\rho = 0$), and a positive cosmological constant. From equation (11.21), we immediately see that $q = -1$; that is, this universe is accelerating. From equation (11.20), we find that the Hubble constant is truly a constant, and

$$H = \sqrt{\Lambda/3}. \tag{11.23}$$

Since H is a genuine constant in this case, the time dependence of the scale factor can be determined by elementary calculus from the equation $H = \dot{R}/R$, with the result that

$$R(t) = R_i e^{Ht/t_i}, \tag{11.24}$$

where R_i is the value of the scale factor at some arbitrary time t_i.[3] The behavior of this *exponential* solution is dramatic; the increase in the scale factor is very rapid.

Another feature of the exponential curve is that its appearance is the same everywhere. That is, any section of the curve can be overlaid on another section, with just a change in scale. As one consequence of this, the de Sitter universe is infinitely old and never goes through a big bang. Running backward in time, $R(t)$ shrinks to smaller and smaller values but never reaches zero. All this may seem so strange that you might think it could not have anything to do with the physical universe, but the de Sitter universe describes the behavior of several important models. One is a possible inflationary phase in the early universe, a cosmology which we shall describe in more detail in chapter 15. Another example is the late evolution of any open, expanding universe with matter and a positive cosmological constant. As the universe expands, ρ drops toward zero; as it does so, the model behaves more and more like the empty de Sitter universe. Yet another example, though a rather more extreme one, is the steady state model.

The **steady state model** obeys an idealization known as the *perfect* cosmological principle. The perfect cosmological principle holds that not only is every point in space representative of the universe as a whole, but each point in time is representative of the entire history of the universe as well. In other words, the steady state universe is isotropic and homogeneous in time as well as in space. As in the de Sitter universe, the Hubble constant is truly a constant, the same for all times, and again this yields an exponential expansion. Moreover, the steady state model also has a deceleration parameter $q = -1$. This model expands without a big bang, and continues to expand in the same manner forever. The universe has always existed and always will exist.

An important difference between the de Sitter model and the steady state model is that the latter may contain matter. But if this universe is expanding, should not the matter density decrease as the third power of the expansion factor? If the model is to adhere to the perfect cosmological principle, this cannot occur. The steady state cosmology thus requires the existence of a "creation field" that creates mass at exactly the correct rate to balance the expansion. This creation field is also responsible for the acceleration; ordinary mass could only produce a deceleration. The creation field, in other words, produces the cosmological constant Λ, with a value given by equation (11.23).

The steady state model demands the introduction of new physics, the creation field, beyond the equations of general relativity and other standard laws of physics. That alone need not rule this model out; but the steady state cosmology has been fairly emphatically rejected by observations. For one thing, the perfect cosmological principle requires that the average appearance of the universe remain unchanged for all time, yet observational evidence shows that the universe has evolved considerably over the past several billion years. The universe during the time of the quasars was a very different environment from what we see around us at the present epoch. Another test of the steady state cosmology is to measure the deceleration parameter q. Although such attempts have not been completely successful, the data, while still inconclusive in choosing among standard models, do not seem to be at all consistent with the value $q = -1$. The most serious blow to the steady state model was the detection of the cosmic microwave background radiation. This provides direct evidence for the hot initial state we call the big bang, as the next chapter will explain.

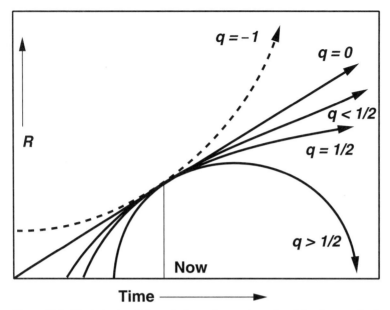

Figure 11.5 The behavior of the scale factor for a variety of models, all constrained to pass through the time *now* with the same slope. The value of the deceleration constant determines both the model's future and the age of the universe. The larger the value of q, the shorter the time back to the big bang. The exponentially expanding de Sitter model, with $q = -1$, never intersects $R = 0$.

The steady state model was never particularly popular with cosmologists or astronomers, although it has been advocated by a vocal minority as an alternative to the big bang. Most astronomers regard the steady state cosmology as an interesting and perhaps enlightening model which is ruled out by observations.

The de Sitter universe, the Einstein static universe, and the steady state model are all quite different from our standard models, and all three are inconsistent with observations. However, they are by no means the only possible models that contain a cosmological constant. Combining Λ with more realistic assumptions can provide interesting alternatives. A dramatic example is obtained by the addition of a *negative* Λ term. Because a negative Λ augments gravity and becomes stronger with increasing scale factor, it will cause *any* geometry ultimately to recollapse, even the flat and hyperbolic models. With the right balancing of terms, the flat and open models can exist for an indefinite period, but the return to the big crunch is inevitable. Adding a negative Λ to the spherical model, which was already fated to recollapse, causes the end to come just that much sooner.

Other, more interesting, possibilities are created by the addition of a positive (repulsive) Λ term. A positive Λ will not change the ultimate fate of either the flat or hyperbolic models, but it will change their behavior as they expand. As the Λ-force begins to dominate over gravity, the universe will start to accelerate, and q will become less than zero. The acceleration will increase until the models are expanding exponentially. Eventually, they will behave like empty de Sitter universes, as the gravitational attraction of the increasingly low matter density becomes utterly irrelevant.

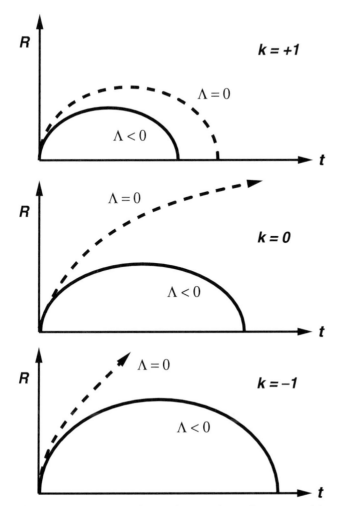

Figure 11.6 The addition of a negative (attractive) Λ force to any of the standard models, regardless of geometry, results inevitably in recollapse. The evolution without the Λ term is shown as a dashed line for reference.

Another interesting example is the modified standard model with a spherical geometry that is prevented from recollapsing by the repulsive force due to Λ. Such a model was devised by the Belgian physicist and cleric Georges Lemaître; it is therefore called the **Lemaître model**. The fate of this spherical model depends on the exact value of the cosmological constant. Recall that Einstein introduced his Λ term to provide a precise balance with gravity. Thus the Einstein critical value Λ_c determines whether a spherical model with a cosmological constant will recollapse or will expand forever. If $\Lambda < \Lambda_c$, models with spherical geometry will recollapse, although the closer Λ is to the critical value, the longer the model lasts before the big crunch.

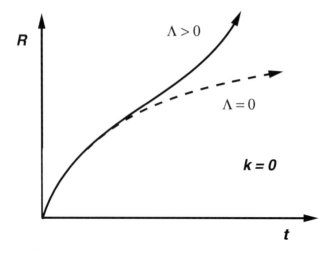

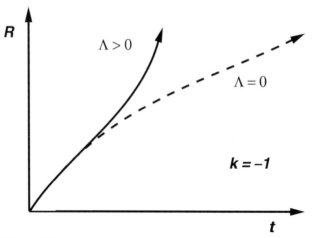

Figure 11.7   When a positive Λ term is added to models with flat or hyperbolic geometry, they expand more rapidly, eventually at an accelerating rate with $q < 0$.

If the cosmological constant is barely larger than Λ_c, the universe may spend a very long time with gravity and the Λ force in near balance, in a lengthy, nearly static, hovering period. After this hovering interval, the universe begins to expand again, now at an accelerating rate. There would be interesting observable consequences if we lived in an accelerating Lemaître universe. For instance, the age of the universe would be much greater than the Hubble time, which is one reason why this particular model once enjoyed some popularity among cosmologists. Hubble's initial measurements gave a very large value of H_0, which was only slowly revised downward. These early observations indicated a Hubble time close to 2 billion years, much less than the age of the Earth! In the Lemaître model, the Hubble age would provide only a rough estimate of the time back to the hovering period; the universe could be considerably older than that. Nowadays the Hubble time and the ages of the constituents of the universe are in better agreement, and the Lemaître model has lost its brief popularity.

Another observable effect in the Lemaître model is that R would remain nearly constant for a long period during the past. Because of this, there could be many objects at the special redshift associated with the "hovering" interval. This was of some interest in the 1970s, when there appeared to be an excess of quasars at a redshift of approximately 2. However, the excess turned out to be largely a selection effect; the technique used to find quasars favored those that had a redshift close to 2. Since then, better observations have reduced the apparent "bulge" in the quasar distribution considerably. Today we recognize that there was a great age of quasars when the universe was younger, but they are distributed over a range of redshifts. Nothing about the number of quasars at any particular redshift now indicates that there was ever any sort of hovering period.

One final consequence of the Lemaître model is particularly interesting. In any closed spherical geometry, it is possible for light to travel all the way around the cosmos. For example, we could look far into space and see our own Milky Way forming. In the standard spherical model, the transit time for light around the universe is the same as the entire lifetime of the universe, so that by the time we can "see ourselves," we are caught in the big crunch. This limitation is overcome in the Lemaître model, with its static period. In this case, the universe might be sufficiently old for some photons to have had time to travel completely around the universe. The light from a distant quasar exactly halfway around the universe could arrive from two opposite directions at the same time; we would thus observe the same quasar in opposite directions in the sky. Searches have been made for such "mirrored" quasars, but none have been detected.

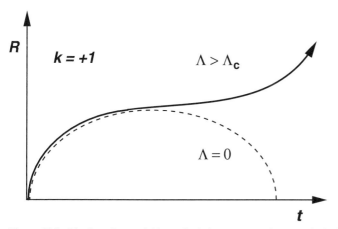

Figure 11.8   The Lemaître model has spherical geometry and a cosmological Λ value slightly greater than the Einstein critical value Λ_c. This model features a hovering period during which the scale factor remains nearly constant over a long time interval. Following the hovering period, expansion continues at an accelerating rate with $q < 0$.

A historically important aspect of the Lemaître model is that it begins from a state of large or infinite density. Lemaître seems to have been the first to take seriously such an initial state, which he called the "primeval atom." It can be said with justification that Lemaître paved the way for acceptance of the later big bang models.

In the final analysis, how should we regard the cosmological constant term Λ? Is it just a free parameter, a "fudge factor," thrown in to adjust the models as needed, or does it have a serious role to play in cosmology? Before dismissing Λ too quickly, bear in mind that it is not completely arbitrary. It is allowed by the equations of general relativity; it even has a physical interpretation as an overall cosmic contribution to gravity by space itself. It is possible, in principle, to measure its value, although as is usual in observational cosmology, such measurements are difficult. For the present epoch we know that Λ, if present at all, must be *close* to zero; but even a tiny value, whether positive or negative, can have significant effects.

While there is considerable evidence that the standard models do provide an excellent description of the universe, over nearly all of that portion of its history to which we have direct observational access, the actual universe is not restricted to obey our simplest models. As we shall see in chapter 15, the cosmological constant can be associated with a physical quantity called the *vacuum energy density*. It may happen that, due to processes from particle physics, the vacuum energy density is different at different times in the history of the universe. In such circumstances, Λ could change as the universe undergoes transitions from one state to another. Such a universe could not be approximated by a single model over its life but would move from one model to another. An understanding of the Λ models is essential to comprehending the frontiers of cosmology.

The Life of a Universe

The destiny of a standard universe is predetermined at its birth. A spherical universe must remain spherical for all times, if the Einstein equations are not to be violated. Similarly, an open, hyperbolic universe will be open forever; it cannot suddenly jump into a closed geometry. A flat universe will be eternally flat. The standard models

TABLE 11.2
Cosmological Models

Model	Geometry	Λ	q	Fate
Einstein	Spherical	Λ_c	0	Unstable: collapse or expand
de Sitter	Flat	> 0	-1	Exponential expansion
Steady state	Flat	> 0	-1	Exponential expansion
Lemaître	Spherical	$> \Lambda_c$	< 0 after hover	Expand, hover, expand
Closed	Spherical	0	$> 1/2$	Big crunch
Einstein–de Sitter	Flat	0	$1/2$	Expand forever
Open	Hyperbolic	0	$0 < q < 1/2$	Expand forever
Negative Λ	Any	< 0	> 0	Big crunch

are derived from the most basic set of assumptions: general relativity describes the behavior of gravity on the large scale, and the universe conforms to the cosmological principle. Add to this the observational fact that the universe is expanding, and we conclude that the universe began a finite amount of time in the past and was once much more compact, dense, and hot than it is now. This hot, dense, initial phase of the universe is known as the big bang.

If we admit additional cosmological models by adding a cosmological constant, then the universe need not have begun with a bang. The steady state universe is an example of a model with an infinite past. Even stranger things are possible. With the appropriate positive value of Λ, we can create a model that is infinitely old and that spent the first half of its infinite existence in a state of collapse. In such a model, the Λ force causes a turnaround at a minimum size, after which the universe begins expanding again, heading back out to infinity. This is an "inverted" closed universe, in a sense. Such models are rather artificial and without any compelling basis in observation. The two guiding observational facts are the cosmological redshifts indicating that the universe is expanding, and the accumulating evidence that there was a big bang.

Given that the universe almost certainly had a beginning, let us turn our attention to the question of its end. There are two general classes of end possible: with a bang or a whimper. What would these two different fates be like?

The spherical standard universe ends in a big crunch, the return to a very dense state such as was present at the beginning. The gravitational attraction of the matter in the universe is sufficiently great that it eventually halts, and then reverses, the expansion. If we lived in such a universe, the first sign of the contraction would be that relatively nearby galaxies would no longer show redshifts. Instead, we would observe only blueshifts in these galaxies as the collapse began. As galaxies fell together, the transition from redshift to blueshift would move to greater and greater distances, that is, to larger and larger lookback times. At first glance, it might seem that the evolution of the universe would run in reverse; however, because things have changed over the lifetime of such a universe, conditions during the collapse will not be identical to those during the expansion. Inchoate matter at the beginning organized itself into stars, and the stars into galaxies. Plenty of hydrogen existed to make new stars; the galaxies were the lively places that we know today. But as time passes, the gas is used up; the stars cease to shine, with massive stars dying first in violent explosions, while low-mass stars slowly burn out as cinder-like white dwarfs. At the end, the contents of a closed universe will be dominated by stellar ashes such as dead white dwarfs, neutron stars, and black holes. In fusing hydrogen to heavier elements, the stars converted mass into energy. In collapsing to black holes, the smooth, initially uniform, spacetime continuum developed numerous singularities at the centers of these holes. These conditions, quite different from those prevailing during the expansion, are carried into the collapse and the new overall singularity.

The future history of the universe is quite a bit different if it ends in a whimper, as for the flat and open universes. These universes exist forever; time begins, but never ends. But as time increases into the infinite future, the average matter and energy densities drop continuously, and all temperatures decrease toward zero. Stars burn out, white dwarfs cool as much as they can. Black holes, drawn together by gravitational radiation, merge with each other and with other, less exotic, remnants.

But even the largest black holes, over a fantastically huge but not infinite time, decay via Hawking radiation into particles and photons. In some theories of particle physics, the proton, the basic constituent of matter, turns out to be unstable, albeit with an enormous lifetime. If this is true, all matter will break down into more fundamental forms. Eventually, it will no longer be possible to extract useful energy from anything; the universe will become ever colder and increasingly disorganized. These universes evolve toward a **heat death**, a state of minimum temperature and maximum disorder, or **entropy**.

It is characteristic of the universe that its entropy increases as it evolves. What, then, is entropy? There are several ways in which to think about it, but fundamentally, it is a measure of the *disorder* of a system. Suppose, for example, that you contrived, by exerting work with a piston upon the air in a room, to confine all the air into a small volume. That state would be reasonably well ordered, as gases go. Now release the air; it fills up the room and the molecules mix more or less randomly. As it expands, the initially compressed air could do some work, such as turning a turbine. But once the air is thoroughly mixed, it can no longer do any work. The motions of the molecules are random, not systematic, and cannot become sufficiently organized to turn a turbine. We say that this new state, with air spread throughout the room, has more disorder, that is, higher entropy, than the old.

As another example, the gasoline that fuels your car is in a relatively ordered state, containing substantial available chemical energy. When the engine burns the gasoline, the vapor is converted into various gases, some of which push the pistons that ultimately cause the wheels to turn; after each stroke, these combustion products are vented through the exhaust system. The waste gases are in a much more disordered state than was the original gasoline, so the final entropy is much higher than the initial. You cannot run this process in reverse; the exhaust cannot be collected and pushed backward through the engine to create gasoline. You have released energy at the expense of creating quite a lot of entropy. Thermodynamics is the science of energy and entropy, and the **second law of thermodynamics** states that entropy at best remains the same, and usually increases, in any process. It is possible to reduce entropy and create order locally by means of the expenditure of energy; but such a process always results in an overall increase in the entropy of the universe. You are an ordered system, for example, but to exist you must utilize considerable energy from the food you eat. The energy is available in low-entropy chemical form (e.g., sugar). Most of that energy turns into higher-entropy random heat and is radiated from your body, unavailable for further use by you. You maintain your relatively ordered state at the expense of producing greater disorder in your environment.

Nearly all the energy on Earth is, or was, ultimately provided by the Sun. Energy from the Sun drives the atmospheric and oceanic motions that control the Earth's climate. Photosynthetic plants and bacteria capture energy from the Sun's light; some of this energy they use to manufacture organic compounds for their own use, while the rest is stored in chemical form. When other organisms consume plants, the stored energy becomes available to them. Fossil fuels such as petroleum and coal are the remains of ancient plants; these fuels are, in a sense, frozen sunshine. Overall, however, the total energy on Earth is in balance. Except for relatively brief intervals of climate shift, the Earth neither gains nor loses energy over periods of roughly a year or longer. Energy not used to maintain the *status quo* is reradiated in the form of heat and returned to space.

Heat engines are devices that use a temperature (and entropy) differential to perform work. They take energy from some high-temperature source, and exhaust higher-entropy heat at a lower temperature. The Earth system, including the biosphere, is a huge heat engine, receiving usable (lower entropy) energy from the Sun, extracting work, which goes into such purposes as sustaining low-entropy entities such as living creatures, and finally reemitting the unused energy in the form of high-entropy, lower-temperature heat. The high temperature of the radiation beaming from the Sun onto the much colder Earth makes life possible.

The Sun is a ball of gravitationally compressed gas that contracted from a diffuse, cold cloud of interstellar gas and dust. How can we reconcile the presence of stars, which form spontaneously, with the second law of thermodynamics? After all, stars might seem to be more ordered than a swirling gas cloud. However, gravity complicates the picture. Stars do not consume energy as they form. Whether they ever reach a state of nuclear fusion or not, protostars release energy as they contract; if they do ignite nuclear reactions, considerably more energy is produced. Our previous examples have suggested that release of energy tends to be accompanied by an *increase* in entropy. In a gravitating system, the higher-entropy states are those that are contracted. The more clumped the matter, the higher the entropy. The ultimate is the black hole, which is maximally contracted and is in a state of very high entropy. This may seem counterintuitive, because it is directly opposite to the case of a gas in which gravity is negligible; in such a gas, the more diffuse states have higher entropy. Yet with gravity, the diffuse gas has the potential to release its gravitational energy, and possibly to perform work, by contracting, a capacity that is reduced the more the gas clumps. Hence a star actually has higher entropy than the gas from which it formed, and we can easily account for the spontaneous creation of stars. The clumping of gas into stars, of stars into galaxies, and of galaxies into clusters obeys the second law; entropy increases.

Entropy and the second law of thermodynamics play a key role in defining the direction of the **arrow of time**. The laws of mechanics, and of special and general relativity, and even of quantum mechanics, show no obvious asymmetry in time; they do not disclose why time has a preferred direction. Yet a broken bottle never spontaneously reassembles itself. The air in a room does not abruptly coalesce into one corner, leaving a vacuum in the rest of the room. Left to their own devices, things evolve from order into disorder. So it is for the universe. Why is a contracting universe not simply an expanding universe running backward in time? The direction of time appears to be defined by the direction of increasing entropy, and this is independent of the behavior of the scale factor; the universe always moves from an ordered state to a state of greater disorder. The universe winds down.

What is the entropy of the entire universe? One component of the total entropy is measured by counting photons. As we shall see in chapter 13, at the beginning of the universe, both matter and antimatter existed in nearly equal amounts. Most of the photons in the present universe were produced by the mutual annihilation of this matter and antimatter. These photons are in thermal equilibrium; that is, they represent a blackbody. Since they are all in equilibrium, there are no variations in the temperature of the photons that could be used to perform work. If it had been possible to separate the matter and antimatter into different boxes in the very early universe, the matter-antimatter reaction could be used now for some application such as powering spaceships. Had this been the case, there would be fewer photons in the universe and the entropy of the universe would be lower. Another component of the entropy of the universe is contained in black holes. Black holes are the ultimate in gravitational compaction, and have very high entropy. The formation of black holes over the lifetime of the universe is another manifestation of the inexorable increase in entropy.

Speculating on the ultimate fate of the universe is a particularly stimulating pastime. In the flat and open universes, entropy increases indefinitely, and the opportunities for further extraction of useful energy diminish. The universe becomes dominated by high-entropy photons, all at some incredibly low temperature. Eventually, there is too little energy available to be converted into appreciable amounts of work; flat and open universes simply fade away. Such a heat death does not lie in the future for the closed universe, but there too entropy may have important implications. It is a popular notion that if the universe returned to a big crunch, it might bounce to a new big bang and start again. This would be a cyclic universe, a sort of modified steady state model. The universe continues into the indefinite past and the indefinite future, but it has individual manifestations, each separated by the infinite crunch of a singularity. If it turned out that the universe could somehow pass through the singularity at the big crunch and rise from its own ashes, it seems probable that, unlike the phoenix, it does not return as simply a younger version of itself. The big crunch would represent a state of higher entropy than the big bang. Even if the universe "tunneled" through the singularity, reemerging in another big bang, the new universe would likely "remember" its entropy. If so, the next universe would have more photons, fewer particles, and higher entropy than the old, and, with less matter density, would expand to a larger size than in its previous incarnation. After some number of cycles, the matter density contained in the new universe would be too low for the formation of galaxies and stars.

There is a greater difficulty inherent in the cyclic model, however. The big crunch would consist of many merging black holes, separate singularities coming together into one final universal singularity. A singularity created from the mergers of many black holes is quite different from the initial singularity out of which our universe emerged. If the universe is to reappear in a new manifestation, can it find its way back to the smooth state such as apparently existed during "our" big bang? This is a question for which we have no answer, although it seems somewhat improbable. In the absence of a quantum theory of gravity, we cannot know much of how singularities behave. Even so, there is really nothing to suggest that a big crunch could produce another big bang.

So long as there is no definitive observation that determines the fate of the universe, we are free to state an emotional preference for its ultimate destiny. Would you rather see it all end in fire, or in ice? Some people, including some cosmologists, become emotionally distraught by such a choice, and opt for a steady state model. Some seek a compromise in a closed universe that goes through endless cycles. But, others counter, is eternal repetition really preferable? Nothing ever really changes in the steady state model. In an infinite steady state universe, presumably there are an infinite number of Earths, with all possible histories, but still very much the same. In the standard models, there is an end. If we pass this way but once, we must make the most of it.

KEY TERMS

Friedmann equation	standard models	big bang
big crunch	critical density	density parameter/omega
deceleration parameter	closed universe	flat universe
Einstein–de Sitter	open universe	cosmological constant
model	steady state model	Lemaître model
de Sitter model	entropy	second law of
heat death		thermodynamics
arrow of time		

Review Questions:

1. Using the Friedmann equation along with the definition of q, show that the deceleration parameter q is equal to 1/2 for a $k = 0$ universe. (This requires some algebra.)

2. Describe the expanding steady state model. Describe an observation that would test the predictions of the perfect cosmological principle. Is the steady state model in conflict with present observations of the universe?

3. Suppose it were discovered that the Hubble time was 17 billion years, and the oldest stars were 15 billion years old. Among the standard ($\Lambda = 0$) models, which would be acceptable? What possibilities open up if a cosmological term Λ is added to the model?

4. We have observed quasars with redshifts as large as $z = 4$. How big was the universe then compared to now? A useful quantity is the lookback time, the travel time required for light with a certain cosmological redshift to reach us. The actual value of the lookback time depends on the specific model. For the flat Einstein–de Sitter model it is

$$t_{lb} = \frac{2}{3H_0} \left[1 - \frac{1}{(1+z)^{3/2}} \right]$$

for redshift z. Using this formula, what is the lookback time to the $z = 4$ quasar, if the Hubble time is 20 billion years? If the universe were the closed spherical model, would the lookback time be larger or smaller than that for the flat model?

5. How does the density parameter Ω depend on the Hubble constant? If the universe were found to have a density ρ just equal to the critical density for a universe with a Hubble constant of 50 km/s/Mpc, what would be the Ω corresponding to this same density for a Hubble constant of 100 km/s/Mpc? For $H_0 = 25$ km/s/Mpc?

6. How would nonzero cosmological constants alter the evolution of each of the three standard Friedmann models? Illustrate with diagrams for $\Lambda < 0$, $0 < \Lambda < \Lambda_c$, and $\Lambda > \Lambda_c$, where Λ_c is the Einstein critical static universe cosmological constant.

7. (More advanced.) Demonstrate by direct substitution that the relation $R \propto t^{2/3}$ satisfies the Friedmann equation for the case of the matter-filled, zero pressure, flat universe. (This requires a little calculus.)

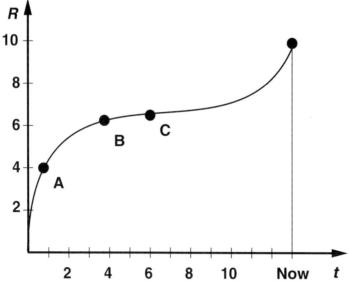

Figure 11.9 (Exercise 11.10)

8. What is the ultimate fate of the closed universe? Of the open and flat? Explain what is meant by the *heat death* of the universe. What fundamental law of physics predicts such an ultimate state? Why is a heat death an inevitable consequence of this law for certain universes? Why does this same law appear to make a cyclic universe unlikely?

9. Briefly describe the history $R(t)$ of a cosmological model you like, giving the value of its defining parameters (e.g., Ω, k, Λ). Explain why you prefer it; your reasons may be philosophical, theoretical, or observational.

10. Consider the figure of a "hovering universe." Labeled on it are several points. Answer these questions: what sign does the deceleration parameter have NOW? What sign did it have for quasar A? What redshift do quasars B and C have? Is the NOW-measured Hubble period greater or less than the actual age of the universe?

Notes

1. Chapter 3 describes Newton's proof that a sphere attracts gravitationally as if all its mass were at its center.

2. Be careful not to confuse the Einstein–de Sitter model with either the Einstein static model, or with the empty and expanding de Sitter model. The Einstein–de Sitter model is another name for the flat, matter-filled, standard universe. Obviously, Einstein and de Sitter between them made many early contributions to cosmology.

3. For those unfamiliar with it, e represents a special number which occurs frequently in mathematics. It is the base of the so-called natural logarithms, and it has many other interesting properties. Like π, it is a transcendental number; it happens to be given by $e \approx 2.718\ldots$.

CHAPTER 12

A Message from the Big Bang

Darkness and the Expanding Universe

When the Sun goes down, the night sky darkens, lit up only here and there by points of lights we now know to be distant suns. But have you ever wondered *why* the sky is dark at night? You might be inclined to argue that the sky is bright during the day because the Sun is so close, and therefore its light fills the sky when it is visible; at night, conversely, the distant stars cannot brighten the sky. But this argument is inadequate, a fact which Kepler was one of the first to recognize. If the universe is infinite, and contains an infinite number of stars that live forever, then every line of sight must end on a star. It is true that a star's light diminishes as the distance squared, but as distances become greater, the volume of space sampled increases by exactly the same factor. Thus the night sky should be everywhere as bright as the average surface of a star; both night and day would be ablaze. Yet we do not observe this, and this has important cosmological implications.

Kepler was certain that this paradox required that the universe be finite. This was a satisfactory solution to him, but it later ran aground in the Newtonian universe. Newton was concerned with balancing gravitational attractions equally in all directions, in order that his universe not collapse. To Newton, this required an infinite space, filled uniformly with an infinite number of stars. Newton believed, or perhaps hoped, that this arrangement solved the problem of gravitational collapse; about this he was quite wrong. Not only is the infinite Newtonian universe gravitationally unstable, but it also exacerbated the problem of the night sky. Newton's friend Halley tried to banish the paradox by attributing darkness to the remoteness of most of the stars, but this argument is incorrect. Even if an individual star were invisible to the eye because of its great distance, its light, combined with the light from an infinite number of other imperceptible stars, would still accumulate to light the sky. Indeed, were it not for the fact that nearer stars completely block the light from those behind them, the sky would be infinitely bright in Newton's universe.

As early as 1744, a Swiss astronomer, Jean-Phillipe Loys de Chesaux, attributed the darkness of the night sky to absorption by a fluid he imagined to permeate all space.

Nearly 80 years later, the German astronomer Heinrich Olbers repeated this argument. For whatever reason, the name of Olbers stuck to this awkward difficulty, and it has become generally known as **Olbers' paradox**. But the explanation was incorrect, as John Herschel showed in the middle of the nineteenth century. Any fluid that filled the universe and absorbed starlight would, according to the laws of thermodynamics, heat up until its temperature was equal to the average temperature of the stars; it would then radiate just as much light as if it were itself a source of starlight. Herschel himself favored a hierarchical view of the universe, in which stars clump into clusters, the clusters bunch into larger clusters, and so forth ad infinitum. In a hierarchical universe, or for that matter in any nonuniform distribution of stars, there do exist lines of sight that are empty; this is the salient feature which distinguishes uniform from nonuniform. But if the universe is to be isotropic and homogeneous on the large scales, as the modern view requires, then a strictly hierarchical model is ruled out.

Olbers' paradox hung over cosmology well into the twentieth century. With the discovery of the expanding universe, many cosmologists immediately accepted it as the answer. The light from the most distant stars is so redshifted that it contributes no appreciable energy to lighting our skies. Some authors have gone so far as to assert that the darkness of night is sufficient proof that the universe is expanding. However, cosmologist E. R. Harrison has emphasized that the resolution of the paradox does not require expanding space. The crucial flaw in the traditional argument was the assumption that the stars could shine forever. Of course, with our modern understanding of energy conservation, this could not possibly be the case. Light carries energy, and thus stars must liberate energy in order to shine. Stellar lifetimes are very finite. When we look sufficiently far out into space, and therefore back in time, eventually we look to a time before any stars existed. Moreover, in any universe that is not infinitely old, or that expands, the size of the *observable* universe is finite, because of the finite speed of light. The finite volume of the observable universe contains a finite number of stars, so most lines of sight never intersect the surface of a star at all. Even if multiple generations of stars live and die, the sky will still be dark. The number of stars is too small, and stellar lifetimes are simply too short, to fill the vastness of space with light. The darkness of the night sky quite elegantly rules out the simplest naive model of an infinite universe filled with infinitely old stars.

However, there is more light in the universe than that which originates with the stars. The **cosmic background radiation (CBR)** fills the sky in all directions, yet it is invisible; its wavelengths lie below the range that our eyes can see. The expanding universe *does* have something to tell us about this version of Olbers' paradox. As is true for the light from distant quasars, the expansion of the universe has caused the cosmic background radiation to redshift and lose energy on its long journey across the universe. But if the light now has lower energy, was there a time when its energy was high? Was the volume of space once small enough that the universe could have been filled with hot photons of visible light? If this were the case, the cosmos was once ablaze throughout but has become dark due to the cosmological redshift. The fact that we cannot see the CBR demonstrates that the universe is expanding. The darkness of the night sky, a simple fact of life which most of us have taken for granted since childhood, is now seen to yield an important clue to the structure of our universe. In cosmology, the most innocuous phenomena can sometimes prove to be very profound.

Noise from the Sky

In 1964, Arno Penzias and Robert Wilson of Bell Laboratories were searching for the source of some weak noise observed in the signal detected by a sensitive radio antenna in Holmdel, New Jersey. The antenna had originally been built for communications via the satellite *Echo*, but Penzias and Wilson, who were radio astronomers, planned to study radio emission from the Galaxy. In order to map such an extended source, they had to characterize all potential sources of noise in their receiver so as to be able to subtract that noise from the desired signal. They began their calibration with a wavelength much shorter than the radio wavelengths anticipated to originate from Galactic sources, expecting that any noise in the microwave band would be due to their receiver, or to the Earth's atmosphere. Accordingly, they chose a wavelength of 7.35 cm for their initial tests. Penzias and Wilson felt confident that this would enable them to evaluate any noise due to the antenna's electrical circuits or to radiation from the atmosphere, but to their surprise, a persistent excess noise remained after they had accounted for every source they could identify. In their determination to find the source, they went so far as to dismantle part of the receiver in the spring of 1965, cleaning it thoroughly and removing the residue from a pair of pigeons that had been nesting in it. Yet despite their best efforts over many months, the excess noise remained.

Radio astronomers describe their signals by, roughly speaking, fitting the radiation to a blackbody spectrum, regardless of whether the original source emits blackbody radiation or not. The use of such an *antenna temperature* enables them to standardize

Figure 12.1 Penzias and Wilson and the horn antenna with which they discovered the cosmic background radiation. (Courtesy of Lucent Technologies.)

the signals and provides a reference for comparison purposes. The enigmatic noise discovered by Penzias and Wilson corresponded to an antenna temperature of approximately 3.5 K, a relatively small signal, but still larger than they had expected for electrical noise. If the atmosphere had been responsible, the amplitude of the noise would have varied from the zenith to the horizon in a predictable way, since the emissions would be proportional to the thickness of the atmosphere being observed. Atmospheric emissions coming from near the horizon would, because of the low angle, come from a much thicker layer of atmosphere than would those originating directly overhead.[1] But the background noise was found to be independent of the direction in which the antenna was pointed, ruling out an atmospheric cause.

Neither did it vary with the time of day or year, which was evidence against an origin in the Galaxy, or any other anisotropic celestial source. Given that the noise could not be attributed to the antenna circuits themselves, such constancy in space and time indicated a *cosmic* origin, but in 1965, few scientists anticipated that the cosmos itself might hum with microwave energy. It happened, however, that P. James E. Peebles, a young theorist at Princeton university, only a few miles from Holmdel, had just carried out a calculation that predicted low-temperature radiation from the early universe. He and his colleagues Robert Dicke, P. G. Roll, and D. T. Wilkinson were in the process of constructing a receiver specifically to look for this background when they learned of Penzias and Wilson's discovery. The mysterious emissions were soon identified with this overall cosmic background radiation. The discovery of the CBR was the most significant cosmological observation since Hubble's results and earned for Penzias and Wilson the 1978 Nobel Prize in physics.

From Hubble to Holmdel

The two greatest cosmological observations of the twentieth century were the discovery of the expansion of space, and the discovery of the cosmic background. In both cases, the discoveries were astonishing and revolutionary; but we can also see, in retrospect, that cosmological theory was prepared for them. Einstein's theory of general relativity, and the difficulties in obtaining a static model, provided an immediate theoretical interpretation for Hubble's finding. Similarly, the existence of the CBR was anticipated by cosmologists investigating the early history of expanding models. The explanation of the background radiation as a relic from a hot, dense phase in the history of the universe was sufficiently persuasive to create a coalescence in cosmological theory; indeed, it is a primary piece of the evidence that makes the "standard" models standard.

Prior to 1965, however, little was known with certainty, and the case that could be made for any of the standard models was no more compelling than arguments for other models. During the era between the two World Wars, many astronomers developed models, based more upon philosophy than upon data, since hardly any data were available at the time beyond the bare knowledge that distant galaxies were receding. Some of these models were developed in Great Britain, whose scientists had long taken a more philosophical approach to their cosmology than had those in the United States. The British tendency was to avoid an explicit beginning for the universe. Arthur Eddington, a distinguished British astronomer who was one of the first to realize that nuclear processes must power the stars, devised his own model in the 1930s, in which the universe, although it had a beginning, emerged calmly and gradually from a nearly static initial state. The Eddington model was essentially an Einstein static model with positive cosmological constant that, after an unknown length of time, began expanding. Through this contrivance, Eddington avoided the question of an initial state in the finite past.

Another British astronomer and stellar theorist, E. A. Milne, rejected entirely any cosmological explanation in terms of general relativity; with a few exceptions such as Eddington, British scientists and mathematicians of the 1920s and 1930s were never very receptive to the general theory. Milne's model, which he derived in the 1930s and continued to defend until his death in 1950, was based on special relativity. There was no gravity at all on the cosmological scale. He adopted the point of view that the apparent expansion of the universe was simply the result of an infinity of galaxies expanding outward at ever-increasing velocities approaching the speed of light. The outer edge of this ensemble was identified with a sphere, expanding at the speed of light into flat Minkowski space. Within the sphere, increasingly distant galaxies were Lorentz-contracted by just the right amount to fit an infinite number of galaxies into the finite volume of the sphere. Although it may not be immediately obvious, such a universe obeys the cosmological principle. Because the speed of light can never be reached, the view from every galaxy is the same; that is, each observes surrounding galaxies moving according to the Hubble law. As it turns out, Milne's model is mathematically equivalent to the empty hyperbolic standard model, to which Milne's name is sometimes now attached. Milne himself recognized the equivalence but disliked the idea of curved space, preferring his own interpretation.

Eddington's and Milne's models represent interesting, but futile, attempts to explain the existing data of their time in a manner consistent with their philosophical prejudices. We should not immediately dismiss such efforts as foolish or old-fashioned; aesthetical considerations have continued to guide many cosmologists throughout the twentieth century. When few data are at hand, not much else is available to help with the construction of models. Better data from space-based, and improved ground-based, telescopes should continue to make it possible for cosmology to rely in the future more upon observations and less upon speculation, but all cosmological observations are very difficult and not always good enough to be of much help. Significant progress has often occurred because some scientists held stubbornly to a particular viewpoint in the face of apparently contradictory data which later proved to be wrong. Yet cosmologists must always be prepared to give up their preferred models if the weight of data refutes them. It is a fine line to walk, but there is no other option.

The British school of cosmology reached a pinnacle in the steady state model, a theory first advanced in 1948. Hermann Bondi and Thomas Gold, two Austrians living in England, developed one version of a steady state cosmology, while Fred Hoyle simultaneously worked out another. Bondi and Gold were uncomfortable with general relativity and based their model more directly upon Mach's principle, and particularly on the perfect cosmological principle. Hoyle, in contrast, developed a relativistic model that maintained a constant density by the introduction of a new physical phenomenon, a creation field. Both models required the spontaneous generation of new matter, usually assumed to take the form of hydrogen. Hydrogen atoms were postulated to appear as necessary to balance the expansion; however, the rate of production of new matter was so small, only about 10^{-24} protons/s/cm^3, that it could not possibly be directly observed. Although such a minute quantity of matter creation may seem like only a tiny violation of known physical principles, the creation of *any* amount of matter out of nothing within the physical universe would require a significant modification in our understanding of physical law. The proponents of the steady state model have never provided a physical theory to account for such a creation.

In contrast to the steady state and other "eternal" models, the big bang models must deal with the evolution of the universe from an initial state. The Belgian scientist Georges Lemaître was the first, around 1931, to advocate an explicit beginning, although initially his ideas were not well received. His model, which remained essentially unchanged after the 1930s, was a spherical geometry with both matter and a positive cosmological constant. It began with a dense initial state from which the universe rapidly expanded, followed by a lengthy hovering period, during which the cosmological constant nearly balanced gravity. During this time the universe looked much like an Einstein static universe, and the scale factor changed only very slowly. Finally, the cosmological term won out, and the universe resumed a rapid and accelerating expansion.

During the 1930s, Lemaître was one of the few scientists to take seriously the concept of a dense initial state. Hubble was an early convert, and some of Hubble's colleagues at the California Institute of Technology worked on the theory of an expanding universe: specifically, H. P. Robertson, whose name is attached to the metric he devised to describe such a model, and Richard Tolman, who developed the thermodynamics of an early universe dominated by the energy of radiation. Apparently, however, neither Robertson nor Tolman was particularly inclined to make the leap of asserting that the model was a valid description of physical reality. It was Lemaître who took the models as representative of reality and unhesitatingly explored their ultimate consequences. He believed that the universe began with a density comparable to that of an atomic nucleus, and then, in a process he likened to radioactive decay, the particles split apart to ever lower densities. Lemaître had little training in quantum physics, nor, apparently, much interest in it; he envisioned the beginning of the universe as a "fireworks" of radiation, which, he speculated, might provide the explanation for cosmic rays.

The big bang did have other early advocates, especially among some nuclear physicists in the United States. Novel and bold theoretical calculations, aided by improvements in the understanding of nuclear physics, provided new avenues of investigation. The first theoretical prediction of the cosmic background radiation came as early as the 1940s from the work of George Gamow and his collaborators, especially Ralph Alpher and Robert Herman. Gamow, Alpher, and Herman were primarily interested in **nucleosynthesis**, the creation of elements, in the early universe; their prediction of the cosmic background came as a byproduct. They postulated that the universe began as pure neutrons, some of which decayed to create protons, plus electrons and antineutrinos. All the elements were then built up via neutron capture. To prevent all matter from ending up as helium, they concluded that the early universe had to be hot; there must have been a large number of high-energy photons present for

Figure 12.2   George Gamow (1904–1968). Gamow's calculations of primordial nucleosynthesis led to the first prediction of the existence of the cosmic background radiation. Alpher and Herman created this image of Gamow rising like a genie from a bottle of "ylem," his name for the primordial stuff of creation. Courtesy of Dr. Ralph Alpher.

every nucleon. Gamow and Alpher described this model in the so-called "$\alpha\beta\gamma$" paper, published in 1948 in the *Physical Review* under the names of Alpher, Hans Bethe, and Gamow.[2] Gamow, Alpher, and Herman realized that this radiation would have eventually escaped from the primordial "ylem," as Gamow called the hot initial state; hence the relic radiation could still be present in the universe, although greatly redshifted in energy due to the overall expansion. Gamow initially predicted a present temperature for the radiation of about 10 K. Unfortunately, although Gamow and his coworkers' contributions to big bang nucleosynthesis were widely recognized, their prediction of a background of low-temperature radiation throughout the universe was not.

Alpher and Herman later repeated the calculations and corrected some arithmetic errors, reducing their estimate of the present temperature to approximately 5 K, not far from that observed 14 years later by Penzias and Wilson. Alpher and Herman, later working with James Follin, continued to develop the theory of nucleosynthesis; the trio published an important paper in 1953. Their later model was more modern, hypothesizing that the primordial mixture consisted of photons, neutrinos, and both neutrons and protons. They succeeded in predicting an abundance of helium of approximately 25% by mass, which agreed quite well with observations of the solar helium abundance. Unfortunately, they were stymied by the "mass gaps," which might be better termed "stability gaps," at atomic mass 5 and 8; no stable nuclides exist with those masses. Hence their nuclear progression was halted, and they were unable to explain the existence of elements heavier than helium.

If the big bang was incapable of producing all the elements, how did most atoms originate? Many scientists were convinced that all elements originated in the stars. Some of the theory of stellar nucleosynthesis was originally motivated by the steady state cosmology, which was at the time still a viable competitor to the Friedmann-Lemaître standard models. The steady state model never experiences a hot, dense phase in which nuclear fusion could occur, and therefore it must explain all elements beyond hydrogen as originating in stellar cores; theoretical work seemed to show that this was possible. Edwin Salpeter showed in 1952 how the "stability gaps" could be bridged in stars, but the mechanism proposed did not operate under the conditions of a hot big bang. In 1957, Geoffrey Burbidge, Margaret Burbidge, William Fowler, and Fred Hoyle wrote the definitive paper on the theory of nucleosynthesis in stars. The "stability gap" problem arises from the fact that the fusion of two helium nuclei produces an unstable isotrope of beryllium, ^{8}Be, which promptly decays back into two helium nuclei. How do stars "jump" over this gap in the elements to reach the stable isotopes further along in the periodic table? Within stars, the gap is overcome by the so-called "triple-alpha" process, through which helium can be converted to carbon. Although ^{8}Be is unstable, at the high temperatures and helium densities present in the cores of massive stars enough of this isotope will be present that occasionally a nucleus of ^{8}Be captures a helium nucleus, forming a ^{12}C nucleus. The early universe never achieves the densities and temperatures appropriate to this reaction; stars are the furnaces in which the heavy elements are forged.

The stars are the source of all the heavier elements of the periodic table, from carbon on up. The common isotopes of the elements between helium and carbon (beryllium, lithium, and boron) cannot be generated by ordinary stellar nucleosynthesis but are produced mainly by reactions involving *cosmic rays*. Cosmic rays are high-energy, relativistic particles, mostly protons, which are ejected from pulsars, supernovae, and other energetic sources. When these particles traverse interstellar gas, some collisions with the gas particles are inevitable. If the cloud has been enriched with carbon and oxygen by earlier generations of stars, a proton will occasionally strike a nucleus of one of these atoms; with so much energy, the proton literally knocks the nucleus apart, creating the light elements. The rarity of these formation processes accounts for the scarcity of these isotopes; they are by far the least abundant of the elements lighter than iron. Still, with one exception (^{7}Li), the formation of these isotopes depends heavily upon prior nucleosynthesis in stars, since carbon and oxygen must be present for the reactions to occur.

Despite the success of the stellar theory of nucleosynthesis, there remained the problem of explaining the large abundance of helium in the universe. The stars can create helium, of course; the Sun and other main-sequence stars obtain their energy by the fusion of hydrogen into helium. Even so, it was extremely difficult to demonstrate how the stars could create *so much* helium. It was already established by the 1950s that the mass of even the oldest of stars consisted of approximately 25% helium, well in line with the prediction of Gamow and his successors, but far more than could be easily accommodated by stellar theory. Not until the acceptance of the hot big bang model did it become clear that this was another instance in which two theories were *both* right. The big bang created helium, as well as trace quantities of a few other light isotopes; all others are the products of stars or stellar explosions. In retrospect, this seems like an obvious reconciliation of the mutual difficulties of the two models, but at the time it was widely believed that it must be one *or* the other. Science is not immune to philosophical prejudices and idiosyncratic blind spots; but over time, the pieces generally fall into place.

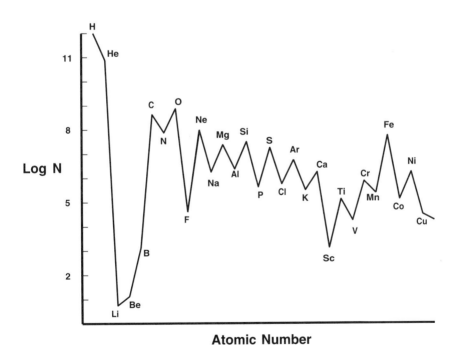

Atomic Number

Figure 12.3 The relative abundance of the first 30 elements. Hydrogen and helium are, by far, the most common elements. All of the hydrogen and most of the helium was created in the big bang. Stars produce carbon from helium directly through the triple-alpha process. Elements of higher atomic number are synthesized by further nuclear reactions in very massive stars. Note the deep minimum for the light elements between helium and carbon; the major source of these elements is infrequent collisions between cosmic rays and interstellar atoms of carbon and oxygen.

The expanding big bang model, in one form or another, gradually became sufficiently widely known that in 1951, Pope Pius XII officially approved big bang theory. Most scientists were unimpressed. After all, Christianity had, with much greater fervor, asserted for centuries that the only acceptable cosmology was the Aristotelian/Ptolemaic model. Even Georges Lemaître, who was a Roman Catholic priest, took pains to separate his science from his religion, at least publicly. (On the other hand, Gamow, who enjoyed tweaking other scientists, once cited the papal approval in a technical paper.) During the first 60 years of the twentieth century, there were essentially no observations that could choose among the big bang, the steady state, or other models, since all could explain the observed redshift-distance relationship, although in very different ways. The data at high redshift, which could, and later did, decide between steady state and big-bang models, were either lacking or unreliable until the 1960s. For example, Hubble and Humason's observations prior to the 1950s contained sufficient systematic error that a straightforward standard model would not fit; with $H_0 \sim 500$ km/s/Mpc, the Hubble time was simply too short to account even for the age of the Earth.

The observational situation improved slowly after the opening of the 200 inch reflector atop Mount Palomar, near Los Angeles. Walter Baade showed in 1952 that a misidentification of the Cepheid variable stars in the Magellanic Clouds, upon which much of the cosmic distance ladder was ultimately based, had led to an overestimate by a factor of 2 in the extragalactic distance scale. In 1958, Allan Sandage established that the scale was too large by at least another factor of 2 because many of the very bright "stars" used by Hubble and Humason for the most distant galaxies were in fact extended regions of very hot, ionized hydrogen gas. These and other new results dropped the value of Hubble's constant from the original estimate to the range of 50–100 km/s/Mpc, where it remains today. The "age problem" for the standard models became less critical.

The steady state model came under scrutiny in several other areas as well. One of the positive aspects of the steady state model is that it is so eminently falsifiable. It is so tightly constrained by the perfect cosmological principle that it makes very specific predictions on many fronts. One of these is that, on average, the universe should look today much as it ever has; there should be no overall change or evolution to the universe. Yet evidence for just such change and evolution accumulated slowly after World War II. The new science of radio astronomy provided important data in the late 1950s and early 1960s. Ironically, England, the home of the steady state model, led the world in radio astronomy, a field that ultimately traced its ancestry to wartime radar. Martin Ryle of Cambridge University claimed in 1955 that counts of radio sources as a function of redshift were incompatible with the prediction of the steady state model. The earliest data turned out to be inconclusive, but as the technique was refined, it became more and more apparent that the data were in conflict with the steady state model. Radio astronomers also discovered a new phenomenon, the quasars, mysterious objects that emitted huge quantities of radio energy, but on optical photographic plates appeared starlike. When Maarten Schmidt realized, in 1963, that the bizarre spectra of some quasars could be explained as the familiar spectrum of hydrogen, but redshifted far more than anyone had ever imagined to be possible, it was nearly fatal to the perfect cosmological principle. The environment of the quasars, so manifestly different from that of nearby galaxies, was a clear example of change in the universe. Quasars were abundant in the more distant past, but hardly exist at all in the recent epoch of the universe.

While this was all evidence *against* the steady state theory, it was not evidence *for* the big bang. Although such observations validated the idea of a universe evolving from an initial state, the evolution of specific structures such as galaxies and quasars could be explained in many ways without directly testing the big bang itself. Only a remnant of the very earliest era in the existence of the universe would be convincing. That evidence was the CBR; this background radiation was virtually impossible to explain within the steady state model but was a natural outcome of a hot big bang. The presence of the CBR, the lack of a physical theory to explain the creation of matter required in the steady state theory, as well as the lack of agreement with observations of galaxies and quasars, has led almost all astronomers to abandon the steady state model as a possible alternative to the big bang.

By the early 1960s, cosmologists had many revolutionary ideas before them: the expansion of the universe, the origin of the elements, the implications of the early dense phase of the Lemaître and standard models, and the evidence for evolution in the universe. These various lines of thought, both theoretical and observational, were merging to create a climate receptive to the discovery of the cosmic background radiation. Indeed, astronomers were just beginning to search specifically for it. Robert Dicke, the original designer of the Holmdel receiver, had arrived at the idea that relic radiation might be present from an early phase of the universe; in 1964, Dicke, Roll, and Wilkinson began constructing a microwave receiver on a rooftop at Princeton University. Meanwhile, Peebles was set to the task of computing the temperature of the anticipated background. Peebles has written that neither he nor his collaborators at Princeton were, at the time he began his work, aware of the earlier calculations of Gamow and his coworkers.

Dicke had based his expectations not upon a "hot big bang," but upon a cyclic model, in which the universe is expanding from an earlier state of collapse. He was looking for evidence of element *destruction*, not creation, although Peebles' eventual theoretical work amounted to a rediscovery of Gamow's, with more realistic conditions. In any case, they were on the right track and would surely have discovered the CBR, had it not been for the serendipitous, but timely, results of Penzias and Wilson. The Russian physicist Ya. B. Zel'dovich also improved upon Gamow's results in the early 1960s, about the same time that Peebles was performing his calculations independently; Zel'dovich, however, was aware of Gamow, Alpher, and Herman's prediction. A. G. Doroshkevich and I. D. Novikov even suggested in 1964 that microwave radiation might be used to check Gamow's theory, but the Russians never pursued the matter further. Better communication among theorists and observers, as well as among different groups of scientists, might have speeded up the discovery of the CBR somewhat. Regardless, it is clear that the scientific atmosphere was ripe by the time that Penzias and Wilson announced their results.

Given the importance of the CBR in establishing the standard big bang model, it is somewhat ironic that the CBR had been detected as early as 1941, even before the steady state model had been formulated or any work on nucleosynthesis had taken place. Interstellar gas clouds often contain molecules as well as atoms; and molecules possess discrete energy levels, just as do atoms. In general, the spectra of molecules are much more complex than those of atoms, since not only can electrons jump around, but the molecule as a whole can rotate or vibrate, or both; nevertheless, molecules can also be identified uniquely by their spectra. In 1941, W. S. Adams observed transitions of cyanogen (CN) in a molecular cloud between the Earth and the star Zeta Ophiuchus. From these data, A. McKellar found that one line in the spectrum of the cyanogen could be explained only if the molecules were being excited by photons with an equivalent temperature of approximately 2.3 K. At the time, no explanation could be found for this phenomenon, so it simply disappeared into the sea of scientific information. Only in 1965 did George Field, I. S. Shklovsky, and N. J. Woolf realize the significance of this observation. In 1993, K. C. Roth, D. M. Meyer, and I. Hawkins again took spectra of cyanogen in several clouds between the Earth and nearby bright stars in an intentional search for the CBR excitation, finding a temperature of 2.729 K, in excellent agreement with other measurements.

Studying the Cosmic Background

Why is the CBR so important to cosmology? The CBR provides evidence for the big bang itself; this alone qualifies it for the title of a "great discovery." The CBR is among the few things that can tell us about the conditions in the very early universe. When we observe the background photons, we are quite literally looking back to a time approximately a million years after $t = 0$.

The CBR is also the best evidence we have that the universe adheres to the cosmological principle. A truly homogeneous and isotropic big bang should produce a relic cosmic background that is a perfect blackbody in all directions, excluding any possible interactions with matter lying between its distant source and our radio antennas. But is the spectrum of this radiation really consistent with blackbody radiation? A blackbody spectrum is produced by a dense gas in perfect thermal equilibrium; that is, the energetics of all the gas is fully described by a single temperature. The specification of that temperature determines precisely what should be observed at every wavelength in the spectrum. Penzias and Wilson observed the background radiation only at a single wavelength. Although the energy measured at that wavelength was appropriate to a blackbody of around 3 K, they did not actually know whether the radiation they discovered was truly part of a blackbody spectrum or not. The shape of the spectrum could be determined only by taking data at many points.

The CBR has even more to say about the early universe. We know that the universe is not *perfectly* homogeneous on all scales today; galaxies and clusters of galaxies obviously contain more matter than exists in intergalactic space. Therefore, while the universe must have been highly homogeneous at the time the CBR was emitted, it cannot have been completely homogeneous even then. The CBR carries important information about the structure of the universe. Galaxies had their origins in slightly overdense regions of gas in the early universe. Any such irregularities in that gas will produce slight variations in the temperature of the CBR; thus it is important to discern how smooth the background radiation truly is. After Penzias and Wilson's discovery, astronomers faced two important questions: (1) Is the spectrum of the CBR truly a blackbody, and (2) Does the spectrum exhibit any anisotropies, that is, variations in temperature across the sky?

Scientists took up the challenge and began measurements of the intensity of the CBR over a full range of wavelengths and in different directions on the sky. Unhappily for astronomers, however, this is not an easy task from beneath the blanket of Earth's atmosphere. The atmosphere is very nearly opaque to several regions in the electromagnetic spectrum. To confirm a true blackbody spectrum, the CBR had to be measured into the infrared, at wavelengths shorter than those of the radio band. The infrared was an important part of the spectrum to observe, since the peak emission of a 3 K blackbody lies in this region. Unfortunately, in the particularly relevant range of the spectrum, water molecules in the atmosphere block almost all the radiation; the only way to observe at these wavelengths is to go above the Earth's atmosphere. For nearly 25 years, observers flew receivers to the top of the atmosphere on balloons and rockets, searching for the elusive and important infrared emissions. Such experiments were exceedingly difficult and complex, and subject to numerous systematic and instrumental errors. Figure 12.4 shows the state of the data from this research, as of the late 1980s. The infrared measurements seemed to suggest possible deviations from a blackbody, although the radio data indicate that the background is well fitted by a blackbody.

Clearly, the best hope of measuring the CBR with high accuracy was to place a receiver aboard a satellite, far above the atmosphere and its obfuscating effects. At last, the *Cosmic Background Explorer (COBE)* satellite was launched in 1989. *COBE* was able to measure the intensity of the CBR across a broad range of infrared wavelengths without interference from the atmosphere. The results were spectacular. The CBR was found to obey a perfect blackbody law to better than 0.03%, an impossible precision to achieve before the satellite observations became available. The temperature of the CBR was at last confirmed to be 2.735 K, with an uncertainty of 0.01 K.

What of the possible anisotropy in the CBR? In addition to absorbing much of the electromagnetic radiation impinging upon the Earth, the atmosphere also varies in density and other properties, complicating comparisons from one direction to another. Balloon and rocket data again indicated that the CBR had the same temperature, to within the experimental errors, at all points in the sky. Such measurements were valuable, but were, obviously, very prone to errors. Careful comparisons of the temperature of the sky in different directions in the radio bands had found no variations down to about one part in 10^5, but only better data from a satellite could settle the question. *COBE* included a special instrument, called the Differential Microwave Radiometer, that simultaneously compared the radiation coming from two directions at three different wavelengths. This device mapped the full sky and found that the temperature of the CBR was nearly the same in all directions, to a level of precision unattainable in earlier experiments. This provides the best confirmation available that on the largest scale, the universe is very isotropic. (We may then conclude that it is also homogeneous, from the Copernican principle.) The observed isotropy of the CBR puts stringent limitations upon any cosmological model we might construct; no matter what, it must always accommodate this fact. This evidence justifies our assumption that the Robertson-Walker metric provides a good first approximation to the universe, and we can continue to work out the details within the context of models based upon this metric.

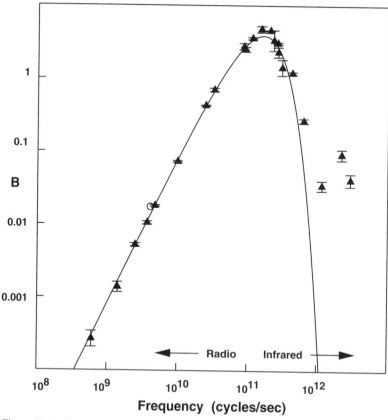

Figure 12.4 Observations of the brightness B (in arbitrary units) of the CBR at various frequencies. These observations were obtained before the launch of the *COBE* satellite. The radio observations are well fitted to a blackbody with a temperature of 2.7 K, but Earth's atmosphere is opaque in the interesting infrared region where the spectrum should turn over. Observations in this part of the spectrum require balloon- or rocket-borne telescopes. In the 1980s these observations suggested excess infrared radiation was present. Penzias and Wilson's original measurement is shown as an open circle.

COBE also found the limit to this isotropy. A more detailed statistical analysis of the *COBE* data indicated the presence of very small anisotropies, near the level of sensitivity of the instrument. The temperature fluctuations are quite small, showing that the variation in temperature from one part of the sky to another is less than about one part in 10^5. The spatial scale corresponding to these anisotropies is much smaller than the overall scale of the universe, although it is larger than the scales of the superclusters of galaxies. The anisotropy data are also very important in constraining theories and may have something to tell us about the origin of the large-scale structures we observe.

Where Are We Going?

The raw data from *COBE* do show some significant nonisotropic effects that are well understood. For example, the plane of the Milky Way itself stood out in the *COBE*

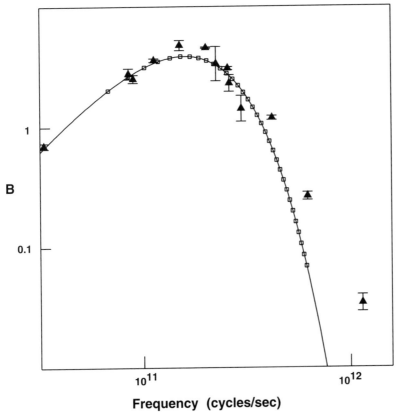

Frequency (cycles/sec)

Figure 12.5   Observations of the brightness B (in arbitrary units) of the CBR, measured over the critical range of infrared frequencies by the *COBE* satellite. The *COBE* data are the squares and they are fitted by a blackbody curve with a temperature of 2.735 K. The observations show no evidence for any deviation from a perfect blackbody; in fact, the squares used here are substantially too large to represent the true error bars. The pre-*COBE* data is included, for comparison.

data; dust within the Galaxy emits considerable infrared radiation. Nevertheless, it was simple to subtract away this effect, leaving only measurements of the CBR.

Another large and systematic anisotropy is seen in the CBR itself. This is due to the Earth's motion through space, which creates a Doppler shift in the CBR. This anisotropy is said to be *dipole* because it has two well defined, and opposite, points: the point of largest blueshift indicates the direction in which we are heading relative to the CBR, while the point of greatest redshift is immediately opposite. Between the two extremes, there is a smooth and systematic transition over the sphere of the sky. This is a well-known, well-understood phenomenon that is easily accounted for when searching for the more subtle anisotropies. These smaller anisotropies are distributed over the entire sky and are indicative of small variations in matter density present in the early universe; one of *COBE*'s chief tasks was to measure the scale of any such variations. The dipole anisotropy is of less intrinsic interest, since its origin is not mysterious; however, the data from *COBE* enabled scientists to determine the Earth's motion through the universe with unprecedented accuracy.

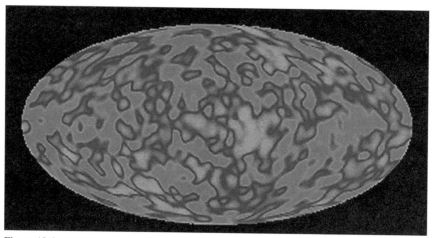

Figure 12.6 *COBE* map of the sky showing temperature variations in the CBR. The temperature variations are very small, less than 20 millionths of a degree. (Goddard Space Flight Center/NASA.)

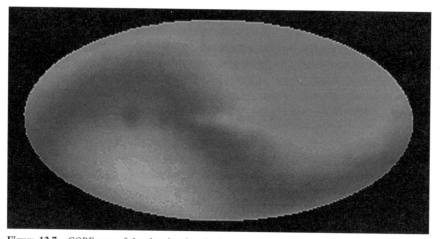

Figure 12.7 *COBE* map of the sky showing the large scale temperature variation in the CBR due to the motion of the solar system with respect to the cosmic background. The temperature difference corresponds to a velocity of about 390 km/s. (Goddard Space Flight Center/NASA.)

If we wish to study the motions of external galaxies, we must first determine our own peculiar velocity. The *COBE* dipole data showed that we are moving at about 390 km/s in a direction toward the region of the sky that happens to be assigned to the constellation Leo. There are many components that contribute to this overall motion. The Earth orbits the Sun with an average speed of about 30 km/s. The Sun itself orbits the center of the Galaxy, with a speed of about 220 km/s. The Milky Way, in turn, moves through space due to gravitational interactions with the other Local Group galaxies, and due to the infall into the Virgo Cluster of galaxies. Finally, the Virgo Cluster itself may have a systematic motion. The net resultant of all these yields the Doppler shift relative to the CBR.

Does this motion with respect to the background radiation imply that the CBR constitutes a "special" frame, like the ether of the nineteenth century? In a sense, it is true that the CBR defines a special frame. If we had no peculiar motion, but were simply being carried away from other galaxies solely by the Hubble flow, we would see no Doppler shift in the spectrum of the CBR. This standard of rest in turn provides a convenient definition of the cosmic time, as in the Robertson-Walker metric. Yet the existence of cosmic time might seem itself to violate the equivalency of frames. Does this somehow repudiate special relativity?

The first response to this question is simply that we can define any convenient frame we wish, and we can always note that we are moving *relative* to such a frame. More important, the cosmic rest frame defined by the CBR results from *general*, not special, relativity, and special relativity can be valid only in localized regions of spacetime if gravitation is present, as it always is. The cosmic rest frame represents the inertial frame of observers who are freely falling in the large-scale gravitational field of the universe; that is, observers who are moving only with the Hubble flow. The "specialness" of the frame of the CBR should be a consequence of the overall distribution of mass in the universe and does not conflict with the special theory of relativity.

The assertion that the frame of the CBR defines the cosmic rest frame is a testable question: Does the frame in which distant galaxies recede isotropically coincide with the frame of the CBR, as obtained by subtracting away the dipole anisotropy? If both phenomena are the result of an isotropic expansion, and if there are no intrinsic large-scale anisotropies in the CBR, then this should certainly be the case. There is no particularly good explanation for any possible skewing of the background radiation with respect to the Hubble flow, but a few suggestions have been advanced, so we should check to see whether our fundamental assumptions will hold up. Unfortunately, the answer to this question is not so easily obtained. The *COBE* measurements determine the *net* motion of the solar system relative to the CBR, but that net motion is composed of many subcomponents. We must somehow extricate the overall motion of the Milky Way, the other galaxies in the Local Group, and the motions of nearby clusters, such as the Virgo Cluster, all relative to the most distant galaxies. Only after we can subtract the peculiar motion of the Milky Way can we determine the frame of the Hubble flow.

Like so many cosmological observations, this measurement is quite difficult. Imagine you were riding an amusement-park ride which simultaneously rotated about an axis and traveled in some arbitrary direction: you would have a difficult time sorting out the motions if everything in the background was moving as well. All the galaxies in our immediate vicinity, and indeed for a distance of at least several megaparsecs, have their own, ill-determined peculiar velocities. The project is further complicated by the fact that, while we can measure redshifts for the galaxies, they provide only the radial component of the velocity relative to us. That is, we can determine whether a galaxy is approaching or receding along our line of sight, but we cannot observe its complete, three-dimensional motion. We must simply do our best to sort out all these motions.

How do astronomers proceed with this task? First, it is easy to account for the motion of the Earth around the Sun. Next, consider that the Sun travels along a nearly

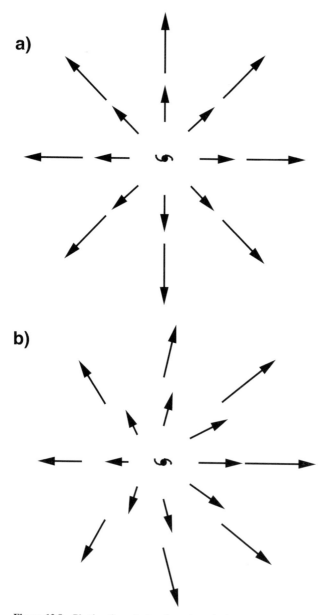

Figure 12.8 Plotting the velocity of nearby galaxies surrounding the Milky Way as vectors illustrates the appearance of (*a*) simple Hubble flow, and (*b*) Hubble flow distorted by the peculiar velocity of the Milky Way. The actual situation is more complicated since every galaxy has its own peculiar motion.

circular orbit in the Galactic disk. Since the total mass of the Galaxy within the orbit of the Sun is not independently known, we cannot apply Kepler's third law directly to compute this orbit. (In fact, observations are used to go the other way: from the measured velocity, we compute a Galactic mass from Kepler's law.) Astronomers must carefully analyze the velocities of globular clusters to find the center of the

Galactic motion. The Sun's motion is determined by measuring the motions of nearby stars, which travel almost on the same Galactic orbit as the Sun. The result is that the Sun orbits the Galactic center with a speed of about 220 km/s, very nearly on a circular orbit in the plane of the disk. This enables us to correct for the motion of the Sun within the Galaxy, in order to pick out the inherent motion of the Milky Way.

The next step is to accumulate redshifts of, cosmically speaking, nearby galaxies. It is then possible to plot a map, with vectors drawn in for the radial velocities, corrected for the local solar motion. Scrutiny of such maps, in principle, enables us to deduce the motion of the Milky Way, and even of the Local Group as a whole. It should not be surprising that it is difficult to carry out such a program; nearly every group of astronomers who have attempted it have obtained somewhat different results. A recent development gives an example of the importance of this task. Several analyses carried out since 1988 have indicated a bulk motion toward something called the "Great Attractor," a hypothetical mass concentration in the direction of the constellation Centaurus, beyond a large galaxy cluster called Hydra-Centaurus. A 1994 study from *HST* data may indicate peculiar motions on even larger scales, a result that would be difficult to explain within the context of current models, should it hold up.

At some point all the galaxy motions should blend into a background due only to the Hubble flow, but until we can determine the peculiar motions of our and other galaxies, it is not so easy to extract the pure Hubble flow. We can plot velocity-redshift diagrams for distant galaxies located in all directions and attempt to shift the data until the scatter of the points is minimized; but so far this approach is not really definitive. For the present we shall simply state that the data are not inconsistent with the coincidence of the CBR frame and the Hubble-flow frame, but the problem is still unsolved.

Traveling Photons

The CBR provides the most direct data we have from the early era of the universe's existence. It stringently constrains the permissible models for the formation of the universe and its constituent structures. It tells us immediately that the universe was once very much hotter than it is today, which is by itself convincing evidence for the hot big bang. The fidelity of the CBR to a blackbody spectrum is a powerful vindication that the radiation comes from the universe itself and not, as some rival theories have proposed, from a general background of stellar and gaseous emissions. It would be impossible to obtain such a perfect blackbody spectrum from a combination of dim sources, none of which would be expected to be at the same temperature as any of the rest, nor even, by itself, exactly a blackbody. The sum of many discrete sources could be a blackbody only if the emitted photons were brought somehow into *thermal equilibrium* by various interactions. Models that invoke such effects are, however, contrived and artificial.

In contrast, the cosmic explanation of the blackbody spectrum is simple, elegant, and fits the data extremely well. The hot big bang model proposes that the universe was in thermal equilibrium in its early stages of existence; thus it would naturally have been filled with blackbody radiation appropriate to its temperature. As the universe expanded it cooled, eventually to the point at which the matter in the universe became transparent to light. Then the state of thermal equilibrium between the photons and the matter ceased, and the blackbody radiation streamed freely through space. Since then the universe has been transparent to photons, and the cosmic radiation has traveled unimpeded through space, until finally a few of its photons struck receivers located on a small planet in the Milky Way galaxy. The background radiation has been redshifted by the universal expansion, just like photons from distant galaxies, so the CBR photons now have energies much lower than they had upon their emission. This would not affect the *shape* of the spectrum, however. All blackbody spectra have exactly the same shape, differing only by their amplitudes and their peak wavelengths, both of which depend only upon the temperature of the radiation. Therefore, any process which affects all photons in the spectrum equally, as does the universal expansion, cannot change the shape of a blackbody spectrum but can only shift it. Hence we see the spectrum today still as a blackbody, but at a very low temperature, since the CBR photons have all been equally redshifted to low energies.

To understand how the redshift works, imagine carving out a typical region of the early universe and watching as it expands. By the cosmological principle, this volume must be completely representative of the universe as a whole, as long as it is sufficiently large for true isotropy to prevail. Early on, the matter density and temperature were very high, and the universe was opaque: matter and radiation constantly exchanged energy. As the universe expanded, cooled, and rarefied, there came a point at which the matter ceased to interact with the radiation, and the universe became transparent. Since this decoupling, the cosmic radiation has evolved independent of the matter. These cosmic photons are neither created nor destroyed; they simply stream through space in all directions. On average, there will be some number of photons per volume; this is simply the photon number density, which we shall denote by n. We can characterize each photon's energy by specifying its wavelength λ and using the equation $E = hc/\lambda$, where h is Planck's constant. Our representative region will thus contain some amount of energy due to the photons, which we would strictly obtain by summing the energy of each of the n photons in this volume. But this is just an arithmetical average, and h and c are constants for all space and time, so we can always write the total energy per unit volume, that is, the energy density, as $\mathcal{E} = nhc/\lambda_a$, where λ_a is the representative wavelength of all the photons in the volume. (More precisely, it is the wavelength corresponding to the average frequency of the spectrum.)

How does this energy density change with expansion? First, the wavelength of any individual photon is redshifted by the overall expansion. Since the scale factor R is the same for all photons at any cosmic time, it follows that $\lambda \sim R$ for any wavelength, including the representative wavelength λ_a. This accounts for one power of inverse wavelength in the energy density. There is one more effect to consider: as the volume expands, the fixed number of photons occupies a larger and larger region; the photons become more and more diluted. Hence the *number* density of photons decreases due to the increase in volume from the expansion. Since the volume increases as R^3, we have $n \sim 1/R^3$, just as for matter density. Unlike the matter density, however, the photon *energy* density is also affected by the aforementioned redshift. Combining these two effects, we find that the energy density in the photons of the cosmic radiation decreases as the *fourth power* of the scale factor, that is, $\mathcal{E} \sim 1/R^4$.

We can carry this simple calculation further. It can be shown that the energy density of blackbody radiation is proportional to the fourth power of the temperature, $\mathcal{E} \propto T^4$. Substituting our previous formula for the behavior of the energy density as a function of the scale factor, we find the remarkably simple result that the temperature of the cosmic radiation diminishes as the inverse of the scale factor,

$$\frac{T_{\text{then}}}{T_{\text{now}}} = \frac{R_{\text{now}}}{R_{\text{then}}} = (1 + z). \tag{12.1}$$

This formula asserts that the temperature of the CBR increases as we look backward in time.

This prediction has been put to the test; in 1994, a group of astronomers used the Keck telescope to search for excitations in the atoms of nebulae at high redshift. The approach is exactly like the studies of cyanogen in nearby clouds that have looked for molecular perturbations due to the background radiation, but the astronomers found a transition in atomic carbon that was more appropriate for their measurement. They obtained spectra from two clouds lying close to a distant quasar, at a redshift of $z = 1.776$. The measurement was difficult, since the clouds were remote and spectra demand the collection of considerable light; fortunately, the Keck, with its 10 m primary mirror, is capable of collecting such huge amounts of light. The astronomers used a very long exposure and an extremely high resolution spectrograph. Their results were gratifying; the excitation of one cloud corresponded to a temperature of 10.4 K ± 0.5 K, while that of the other indicated a temperature of 7.4 K ± 0.8 K. The predicted temperature of the background radiation at the redshift employed is 7.6 K (cf. eq. [12.1]). Much of the difference between the measured and the theoretical temperatures is most likely due to the interference of molecular collisions and other phenomena within the clouds themselves, which muddy the interpretation of the data. By itself, this measurement, or any other single measurement, cannot "prove" the big bang, but it is another datum lending its support. In principle, the method could be extended to clouds even farther away.

Where Has the Energy Gone?

If the background radiation has redshifted to lower temperatures throughout most of the history of the universe, with a corresponding decrease in the energy of each photon, what happened to all that photon energy? Isn't energy, taken in all its forms, conserved? If the CBR photons once had a high temperature, and thus high energies, but now have a low temperature and low energies, where has that energy gone? In fact, what happened to the energy lost by *any* redshifted photon traveling through the universe?

We can understand *how* the photons lose energy by an analogy to an expanding gas. You may be aware that when a gas expands, it cools. The thermal photons from the big bang are behaving in exactly the same manner; as the universe expands, the temperature of the photons is reduced. Although this line of reasoning seems perfectly sound, it fails when we try to extend it to account for the lost energy. When an ordinary container holding a gas, which might be a "gas" of photons, expands, then something must cause it to do so. If the gas particles themselves cause the expansion, such as in the cylinders of an internal-combustion engine, then some of the internal (heat) energy of the gas must be converted into the work required to expand the container. If an external agent, such as a motor pulling on a piston, causes the expansion, then similarly some work is exerted, so that overall energy is still conserved. However, there is no such external agent in the cosmos, nor is there a boundary against which the photons push. The photons themselves are certainly not driving the expansion of the cosmos. We have learned from general relativity that any energy density, including that due to photons, would contribute to gravity and thus would tend to make the universe *collapse*, not expand. In any case, the background radiation is nearly perfectly isotropic, and thus there are no pressure changes from one point to another to create any "photon push."

What about the gravitational field itself? In the case of a photon climbing from a Schwarzschild field around a massive object, we could understand the redshift qualitatively by imagining that it was "using up" its intrinsic energy to gain gravitational potential energy, much as a tossed ball rising in a gravitational field loses kinetic energy. Unfortunately, this conceptualization does not work for cosmological models. The gravitational potential energy is a consequence of the change in a gravitational field in *space*. The universe as a whole is spatially homogeneous and isotropic at all times in its existence, so there is no spatial change in the field; indeed, in the case of a flat universe, there is not even a spatial curvature, but there is still an expansion redshift.

So what are we to make of this? Does the universe violate the conservation of energy? The principle of conservation of energy that is familiar to physicists is a *local* statement, known to hold only for finite regions. Cosmological energy is, by definition, quite nonlocal. A major impediment to our understanding is that we currently do not even have a consistent *definition* of total, cosmological energy. We cannot formulate a conservation law for a quantity we cannot define. It may not even be possible to define such a cosmological energy, in which case there is no reason to expect that any corresponding law of energy conservation will exist. If this is true, we need not be concerned with the lost energy of the redshifted photons. On the other hand, this rather glaring exception to an extraordinarily fruitful principle may tell us something. We know that general relativity is an incomplete theory, because as it stands, it cannot be melded with the other three fundamental forces of nature. Those three forces all do conserve energy. In the theories that explain the other three forces, the conservation of energy arises because their laws and equations are indifferent to whether time runs forward or backward; these theories are *symmetric* under time reversal. In contrast, the universe does not appear to be time symmetric; there is an arrow to time in the evolution of the cosmos, running from the low-entropy big bang to the high-entropy heat death of the future. If time symmetry is required for energy conservation to hold, then the universe as a whole simply may *not* conserve energy. Perhaps, if we achieve the "final theory" which unites all four fundamental forces, we will find that the grandest laws of physics are not symmetric under time reversal. Perhaps then we shall be able to formulate some more complete conservation law, or else we shall understand why this is not possible. Unless and until we reach this summit, the missing energy of the redshifted photons must remain unexplained.

Testing the Models with Observations

Even with the restriction to isotropic, hot–big bang models demanded by the most natural interpretation of the CBR, we still find a plethora of models from which to choose. We have the three standard models—closed, flat, and open—plus all the variations created by the introduction of a nonzero cosmological constant. The task, then, is to determine which model best agrees with the actual universe in which we live. Each specific model predicts relationships among the various parameters, such as H, q, Ω, and Λ. In principle, many of these cosmological parameters are observable. Even the geometry itself, as specified by the curvature kc^2/R^2, is measurable. Our program thus seems simple: measure the parameters, then compare the data to the predictions of various candidate models in order to deduce which model best describes our universe. If it were only so simple, of course, we would know right now whether the universe is open, flat, or closed; we would know its age to perhaps a few million years; and we would know a great deal about its matter content.

As things now stand, we have no definitive answers to any of those questions. Fortunately, the immediate prospects for improvements in our knowledge are quite good. Observations have improved dramatically recently, especially with the repair of the *HST* and the construction of giant ground-based telescopes, such as the Keck telescope in Hawaii. Beginning with Galileo, history has shown that the introduction of new observational capabilities inevitably leads to significant progress, as new portions of the universe are opened to scrutiny for the first time. Nevertheless, we still have far to go before any firm statements can be made about the values of the important cosmological parameters, and what the data tell us about the structure of our universe. There are many avenues we might explore in order to pin down the most appropriate model, and many different kinds of measurements we might carry out.

The Hubble Constant and the Deceleration Parameter

A good place to begin is with a continuation of Hubble's original program of measuring redshifts and distances for a large ensemble of galaxies. Such a program would measure the Hubble constant and the deceleration parameter, both their current values, H_0 and q_0, and how they change with time, $H(t)$ and $q(t)$. The values "now," or, more precisely, good approximations to them, are given by the redshifts of fairly nearby galaxies. The time dependence over longer intervals is observable because we are looking back in time as we look out into space. For example, if we determine the Hubble constant for a selection of galaxies with a very high redshift, we are measuring the Hubble constant from a time long ago, when the light from those galaxies first started on its way to us. A solution for the scale factor $R(t)$ can be used to derive a theoretical formula for $H(t)$, which in turn can be rewritten as $H(z)$, giving us the Hubble constant as a function of redshift. (Recall that $z+1 = R_{now}/R_{then}$.) If we repeat this calculation for all candidate models, we can assemble a set of curves of predicted redshift versus distance for each type of universe, as shown in Figure 12.10. Then, assuming we know a good **standard candle**, we can measure the apparent brightness as a function of redshift for a large sample of these candles. Superimposing the

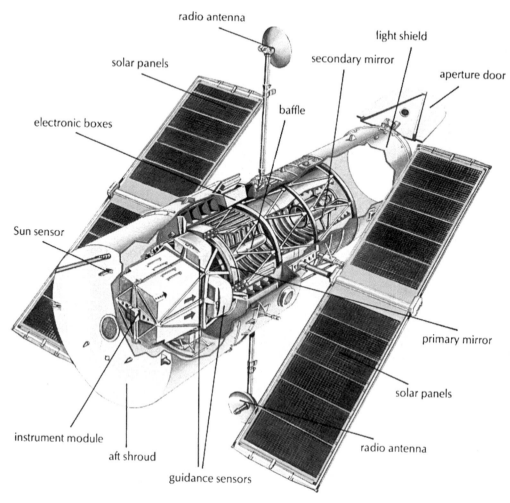

Figure 12.9 Diagram showing the components of the *Hubble Space Telescope*. The telescope is designed to carry out observations in the visible and ultraviolet region of the spectrum while in near-Earth orbit. The unprecedented resolution of the telescope is possible because it is above Earth's atmosphere, which blurs and distorts ground-based observations. The primary mirror of the *HST* is 2.4 m in diameter. (STScI/NASA.)

observed data upon a graph of the theoretical curves would tell us in which type of universe we live.

The Hubble constant is measurable *in principle*; in practice it is very difficult, the greatest difficulty being the accurate determination of distances. The uncertainty in various stages of the distance ladder leads to the current disparity in measured values of the Hubble constant, from around 50 to 100 km/s/Mpc. While we may be fairly confident that we know the present Hubble constant H_0 to within a factor of approximately 2, the deceleration parameter is much more sensitive to measurement errors, especially the distance determination. Different q_0 values result in different

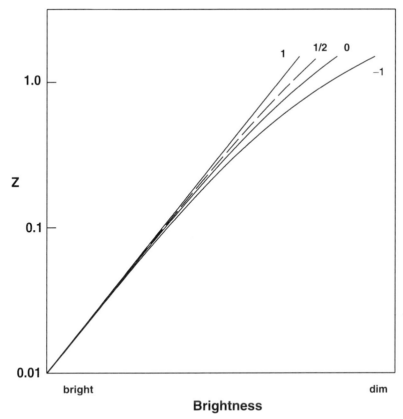

Figure 12.10   Redshift versus observed brightness for a standard candle for various cosmological models, labeled by deceleration parameter q_0. The Einstein–de Sitter model, with $q_0 = 1/2$ is a dashed line. For most of the graph the open, flat, and closed models are indistinguishable. At high enough redshifts where measurements could discriminate among the models, accurate data become increasingly difficult to obtain.

curves on a theoretical plot of redshift versus the brightness of standard candles, but the deviation of the curves becomes significant only for very large redshifts and correspondingly large distances. At large redshift it is difficult to observe anything, much less find good standard candles. The absorption of light by matter in intergalactic space and within our own Milky Way is a further source of uncertainty. There are other, more subtle, complications as well. One of the most important is the possible evolution of the source with time. As we look deeper into space, galaxies and other potential standard candles will be younger and younger. Their brightness may well change with time, rendering them quite unsuitable as a "standard candle." The redshift itself may also change a galaxy's appearance. Astronomers can observe in only one band (visible, radio, etc.) at a time, but observations in different regions of the electromagnetic spectrum emphasize different attributes. For example, at higher and higher values of z, detectors sensitive to visible wavelengths will actually be observing light emitted, in the galaxy's own frame, from increasingly bluer portions of that galaxy's spectrum; the image will not show the normal *optical* appearance of a galaxy, but will be skewed toward objects bright in blue or ultraviolet light.   All these sources of uncertainty are larger than the spread produced by variations from one model to another in the theoretical deceleration parameter q_0. Just where precision measurements become particularly important to the determination of q_0, our ability to make them becomes very poor.

Recent progress provides new optimism that we may finally be narrowing down the range of possible values, at least for the Hubble constant. One of the most important developments was the repair of the *HST*. By installing corrective optics, NASA brought the resolving power of the telescope back to its original design specifications, which exceed anything possible from the ground. Using these improved optics, a research team lead by astronomer Wendy Freedman of the Carnegie Observatories detected classical Cepheid variable stars in galaxies located in the Virgo Cluster, the nearest large galaxy cluster. By calibrating the distance to the Virgo Cluster, the cosmic distance ladder previously established can then be used to obtain the value of the Hubble constant. This team measured a distance to one member of the cluster, the galaxy designated M100, of 17.1 ± 1.8 Mpc. From this the researchers computed a Hubble constant of $H_0 = 80 \pm 17$ km/s/Mpc. Not all the progress has been attributable to the *HST*, however. Sophisticated instruments and imaging techniques have now made possible ground-based observations of ever more distant Cepheids. Such an observation has been carried out by M. J. Pierce and colleagues using the Canada-France-Hawaii telescope. They observed Cepheids in the galaxy NGC 4571, another member of the Virgo cluster. These astronomers obtained a distance of 14.9 ± 1.2 Mpc, from which they derive a Hubble constant of $H_0 = 87 \pm 7$ km/s/Mpc.

These new values of H_0 still contain uncertainties, of course. Even a very precise distance to the Virgo Cluster does not unequivocally pin down the Hubble constant. The Virgo Cluster cannot be used to determine the Hubble constant directly, since it is sufficiently close to be interacting gravitationally with the Milky Way and its neighbors. The Local Group, to which the Milky Way belongs, is falling toward the Virgo Cluster, and the Virgo Cluster itself may be moving with respect to the "Great Attractor," thus distorting the local Hubble flow considerably. Furthermore, we do not know exactly where an individual galaxy is located with respect to the center of the Virgo cluster; is it on the near or far side of the cluster, as seen from the Milky Way? The cosmic distance ladder can be tied to the distance to the Virgo Cluster, but it also depends upon a good estimate of the motion of the Milky Way relative to the Virgo Cluster. Another group of researchers, led by Allan Sandage and G. A. Tamman, used the same data with a different estimate for the Virgo infall and obtained a Hubble constant of approximately 65 km/s/Mpc.

Cepheids are not the only possible standard candle; Sandage and others have attempted to use a particular type of supernova, the Type Ia resulting from the explosion of a white dwarf, to measure luminosity distances. Here the controversy is whether this type of supernova is sufficiently uniform in its peak luminosity; astronomers are still divided over the issue. Supposing that these Type Ia supernovae are well calibrated, various groups have computed the Hubble constant to lie between 50 and 67 km/s/Mpc. An advantage to the use of supernovae is that if we could discover a firm relationship between their luminosity and their characteristics, then we would be able to leap over several rungs of the cosmic distance ladder, since supernovae can be seen for enormous distances.

All this shows that the distance ladder is improving, but it is still shakier than we would prefer. While the stated errors in these measurements do not, of course, take into account *unknown* sources of error, the astronomers have attempted to allow for systematic errors in their error estimates. In any case, most of the newest measurements are consistent with other measurements over the past decade, many of which have bounced around in the range of approximately 65 to 80 km/s/Mpc. Perhaps it is not overly optimistic to suggest that we are finally closing in on the elusive value of H_0.

There are other, more subtle, approaches to measuring H_0 that bypass the distance ladder entirely and obtain distances directly. One of these makes use of small-scale distortions in the CBR caused by its interactions with clouds of X-ray emitting hot gas in remote galaxy clusters. Microwave photons are scattered by the hot gas, and in the process they gain some energy, a phenomenon known as the *Sunyaev-Zel'dovich effect*. Thus the CBR spectrum is slightly distorted; the radio region of the spectrum loses energy, while the shorter wavelength microwave region gains. The magnitude of the effect depends on the size of the X-ray emitting region in the cluster. A comparison of the physical extent of the cluster, as inferred from the distortion in the CBR, with its size on a photographic plate, gives the distance. The redshift of the cluster, always relatively easy to measure, then provides H_0. This method has become feasible only recently, and results from it are still tentative. Some such measurements have tended to favor lower values of the Hubble constant, perhaps $H_0 \sim 50$ km/s/Mpc. As usual, more data are needed before much can be said definitively.

Another clever, recent method for measuring H_0 exploits gravitational lensing. Certain configurations of object and lens produce multiple images that are fairly widely separated on the sky. If the object is a quasar whose light output varies, as is true for many quasars, then it may happen that one image brightens before the other, since the length of the light path, and the gravitational time dilation, can vary between the two images. In principle, measurements of the time delay can determine the ratio of the distance from Earth to the lensing galaxy, compared to the distance from the lens to the quasar. This ratio depends upon the Hubble constant, but not upon the absolute values of the distances; thus it is entirely independent of the cosmic distance ladder. This technique is not without its own problems, of course, not the least of which is the paucity of suitable candidates for the multiple images required. A few measurements using this approach have also indicated lower values of the Hubble constant, $H_0 \sim 50$ km/s/Mpc. However, some later efforts have found larger Hubble constants, in line with the more direct methods. Obviously, not all sources of error have been determined in this, or any, attempt to snare the slippery quarry, H_0.

The Age of the Universe

The higher values of the Hubble constant, indicated by distance-ladder methods such as Cepheid variables, create some knotty problems for cosmological theory. The biggest potential embarrassment is the age of the universe. The Hubble constant implies a Hubble time; for any decelerating universe, the Hubble time is the upper limit to the actual age of the universe. Each specific standard model predicts an age for the universe that is some fraction of the Hubble time. For the Einstein–de Sitter model, the age of the universe is two-thirds the Hubble time:

$$t_{\text{flat}} = \frac{2}{3H_0} = \frac{2}{3}t_H. \tag{12.2}$$

For the open model, the age lies between $\frac{2}{3}t_H$ and t_H, depending upon the density; the lower the density, the closer the actual age is to the Hubble age. For the spherical model, the age is *less than* $\frac{2}{3}t_H$. From this it is obvious that the spherical universe might be subject to a very stringent test if the Hubble time is not comfortably larger than the age of objects in the universe; a spherical model might well be simply too young to accommodate such evidence.

If we could obtain the age of the universe by independent means, we could compare it to the Hubble time and determine the correct model. Not surprisingly, it is difficult, if not impossible, to find the exact age of the universe independent of a model. The best we can do is to obtain the ages of various constituents of the universe. Obviously, the universe must be *at least* as old as its oldest components. Well-established radioactive dating techniques indicate that the solar system is about 4.5 billion years old; this sets a certain lower bound on the age of the universe. Models of the Sun are consistent with this, indicating that it is approximately 4.6 billion years old. Of all astronomical objects, stars are probably the best understood; their ages can be estimated with good confidence from stellar models.[3] We can seek a lower bound for the age of the universe by determining the age of its most ancient stars. The oldest stars of which we are aware are located in globular clusters; their inferred ages range from perhaps 10 to around 18 billion years old, with the stellar modelers favoring something more toward 15 billion years. If that is so, then stellar ages are a fairly severe constraint on cosmological models. For example, if the Hubble constant is 50 km/s/Mpc, then the age of the universe in the Einstein–de Sitter model is about 13 billion years. If stars prove to be older than this, the open model is the only acceptable standard model.

The constraints become even tighter if you favor a larger value of H_0, as many of the recent measurements discussed above indicate. If $H_0 = 75$ km/s/Mpc, the Hubble time is only about 13 billion years, meaning that even the empty open model cannot accommodate some estimates of stellar ages. In fact, the cosmological imperative is the only reason that some stellar modelers are seeking mechanisms to reduce the derived ages of the globular clusters. Otherwise, there would probably be near agreement that the globular clusters are at least 14 billion years old.

The situation is sufficiently acute that many theorists, especially those who favor the Einstein–de Sitter model, have suggested that something is wrong with the cosmic distance scale. Such an objection is not entirely without merit; when Hubble first determined a value for the constant now named for him, the number he reported would have indicated a universe with a Hubble age of only 2 billion years, much less than the known age of the Earth! This is one of the reasons that the Lemaître model had a period of interest, and why the steady state cosmology enjoyed popularity for as long as it did. The systematic errors that caused this overestimate of H_0 were not corrected until the 1950s. It is possible that some unknown systematic error may yet be distorting our modern results. It is true that many different measurements have recently yielded $H_0 \sim 75$ km/s/Mpc; however, some of the techniques that avoid the "distance ladder" problem do tend to point toward a somewhat lower value for the Hubble constant. The controversy will probably continue for quite some time yet, even with improved data.

The Geometry of the Universe

If distances are difficult to obtain, especially for high-redshift galaxies where the differences among the models are particularly striking, what alternatives are there? One is to measure the density of the universe directly, since in the standard models, the density is closely related to the geometry. These measurements can be carried out in various ways, all of them uncertain. One technique involves measuring the masses of galaxies and galaxy clusters and counting the number of galaxies and clusters in a known volume of space. Multiplication of the average mass by the number density of galaxies gives the mass density for that volume.

But how can a galaxy or a cluster of galaxies be "massed"? The only way is to use Newton's laws to interpret detailed observations of their gravitational interactions. We can determine the rotation rate of some individual galaxies by measuring the Doppler shift at varying distances from the center; the resulting curve indicates the rotational speed as a function of radius. Kepler's third law then provides an estimate of the total mass within that radius. At increasingly large radii from the center, we can hope to reach something like an edge for the galaxy, so that the overall rotation rate can indicate the total mass of the galaxy. It is somewhat more complicated to weigh a cluster of galaxies. Within a cluster, we can observe the motions of the galaxies themselves under the influence of the gravitational field of the cluster as a whole. This approach does not yield a mass directly, but statistical studies of the motions of the cluster members can determine the overall mass of the cluster. We will examine these dynamical estimates of mass further in chapter 14.

An entirely different approach to obtaining the mass density of the universe uses the abundance of the heavy hydrogen isotope, deuterium, to infer the density during the early phases of the big bang. We shall defer a complete discussion of this method to chapter 13, but, to put it briefly, it makes use of the fact that the rate of production of deuterium in the early universe is very sensitive to the density of potential reactants. The data from these two quite disparate techniques tentatively indicate that $\Omega < 1$, that is, the universe is open.

Although in the standard models the density parameter is directly related to the geometry of space, this is not true if there is a nonzero cosmological constant. For example, one possible way out of the "age of the universe" dilemma, should it remain with us in the future, is to include a cosmological constant in our model. If Λ is not zero, the models have additional freedom, and the independent measurement of cosmological parameters becomes crucial. For example, it can be shown that the cosmological constant can be expressed in terms of observable parameters as follows:

$$\frac{1}{3}\Lambda = H_0^2 \left(\frac{1}{2}\Omega - q_0 \right). \tag{12.3}$$

In the standard models the left-hand side is zero; measurement of Ω determines q_0. But if we could *independently* measure the density of the universe and the deceleration parameter, we could determine at least the sign of the cosmological constant. Measurement of these quantities by similar means would not do, since the errors would then tend to go in the same direction, distorting the result of the subtraction. With the possibility of a cosmological constant, it becomes especially important to infer directly the geometry of space from independent measurements. How might we measure k itself?

An interesting attempt to measure the geometry of the universe exploits the dependence of distance upon geometry. It must be kept in mind that when we speak of distances in an expanding universe, we must make a distinction among different ways of describing "distance." The complication arises from the fact that looking out into space means looking back in time. The quasar light that we see today was emitted long ago, when the quasar was much closer to our Galaxy. Hence there is a "distance then," or **emission distance**, when the light was emitted, and a "distance now," or **reception distance**, when we receive the light. Since they are distances, they must scale according to the rule for distances in an expanding universe, as given by the Robertson-Walker metric (10.6). It must be emphasized that *both* the emission and the reception distances depend individually upon the cosmological model, but their ratio is a simple function of the redshift:

$$\frac{d_{now}}{d_{then}} = \frac{R_{now}}{R_{then}} = 1 + z. \qquad (12.4)$$

Notice that the emission distance is smaller than the reception distance by precisely the ratio of the scale factor "then" to "now." A galaxy at a redshift of $z = 2$ was one-third as far from us when the light we receive today was emitted.

In ordinary Euclidean geometry the **angular size**, that is, the angle occupied upon the sky, of a *known* length is a direct way to measure distance. If we symbolize the proper length of the standard by ℓ, its angular size by θ, and its distance by d, then the Euclidean formula relating these quantities is simply $\ell = \theta d$. But expansion and non-Euclidean geometry introduce new effects. At first, as we look to greater and greater distances, the galaxies' angular sizes become smaller and smaller as usual; they are farther and farther away, after all. But eventually there comes a point at which the lookback time becomes important. At significant lookback times, the universe was appreciably smaller. Very distant galaxies were much closer to us at the time of their light's emission; in fact, for very large redshifts, they would have practically loomed over us. Since we see such a galaxy as it was then, it appears larger on the sky than it would had it been located at that distance all along. Therefore, beyond some turnaround point at which emission distances become significantly smaller than reception distances, the apparent size of a galaxy actually *increases* as a function of redshift. This effect provides a way to relate distance to geometry; it is most significant for the spherical universe, since of the three, that geometry has the least volume. Several such tests have been carried out, using various "standard yardsticks" such as the largest spirals in a cluster, or the lengths of radio jets emerging from remote radio galaxies. Unfortunately, the data collected so far do not unambiguously select *any* of the standard models.

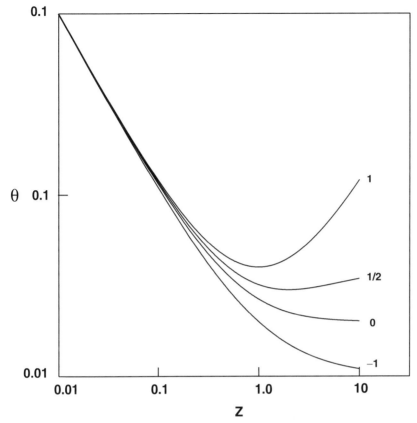

Figure 12.11 Angular size θ as a function of redshift z for an object of a given proper length. The lines are theoretical curves for different cosmological models labeled by the deceleration parameter q_0.

One of the difficulties with the angular-size tests is that data from relatively high redshifts must be obtained before substantial differences in the standard models become apparent. Obtaining spectra requires a lot of light, and that is hard to collect at large redshifts, where the galaxies are very faint. Moreover, redshifts must be obtained galaxy by galaxy, a long and tedious project. In principle, it is possible to determine the geometry of the universe without measuring redshifts at all, simply by counting the number of objects seen as a function of apparent brightness. Recall that the different geometries predict different relationships between geometrical quantities such as the radius and circumference of a circle, or the radius and volume of a sphere. If galaxies were equally spaced throughout the universe, then we could use them to "measure" the size of circles, or the volume of spheres, with a given radius. In practice it becomes rather complicated, but it is possible. One approach is to count the number of sources at a given optical or radio brightness for some area in the sky. If galaxies were all identical, then the apparent brightness of any galaxy's image would be immediately related to its distance. Astronomers could then count the number of galaxies at each distance, thus obtaining a rough estimate of the volume of space at that distance. The different geometries predict different number counts, with the hyperbolic geometry giving more sources, while the spherical geometry gives fewer. In principle, this method could determine the curvature without the need for accurate distance measurements.

Unfortunately for astronomers, galaxies are not all the same, but vary in their intrinsic brightness; they may also change in brightness over their lifespans. In order to compute the theoretical curves required for comparison of the data with the model, we must make assumptions about the nature of the sources we are studying; specifically, we must either assume that the galaxies do not change over the large times required for their light to reach us, or we must develop a model of the evolution of these sources. There is little doubt that the first approach rests on a very dubious assumption. Looking back in time, we can see changes in the appearance and apparent properties of galaxies and quasars. There are no, or at best very few, nearby quasars, but many at high redshifts. Normal galaxies too change with time, growing brighter as their stars form, or as they collide and merge, and becoming dim as their stars age and burn out. How can we develop a model for this complicated evolution? At the present, we cannot account for all observed phenomena, but must make our best model based upon what we know. If we can develop a good model for the distribution of the inherent brightness of a large population of galaxies, and if we can somehow account for possible changes in their brightness over time, then we can apply corrections for these effects to our counts of number at a given brightness. The brightness of any

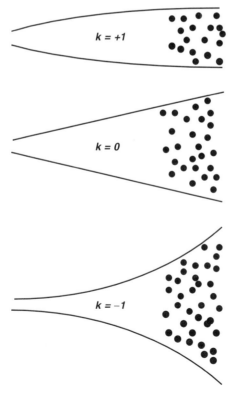

Figure 12.12 In principle, number counts of galaxies can distinguish between the different geometries. If galaxies were distributed equally throughout space, different geometries would show different numbers of galaxies as a function of distance. Spherical geometry has the least volume, hyperbolic the most; hence the closed universe would show the lowest number of galaxies at a given distance, while the open universe would contain the highest number.

individual galaxy may not be a function of its distance alone, but with the corrections for galactic variability, it is possible to arrive at an overall figure which stands as a proxy for the distance *distribution* of the galaxies under study.

Another weakness of this approach is that sources, such as galaxies or quasars, are not equally spaced throughout the universe, but cluster. Further corrections must be applied in order to separate the effects of clustering from the effects of geometry. Another confounding effect is the possibility that the spatial distribution of galaxies, not just their separation, might change over the history of the universe due to factors unrelated to the geometry of space. For example, the number of galaxies per unit of volume might increase as galaxies are formed, or the number density might decrease when galaxies merge.

Like the density estimates, number counts tend to support either a flat or an open universe, but again, like the measurements of the density parameter, they are as yet inconclusive. All of the complications mentioned above, such as changes in the brightness or the clustering properties of galaxies, could have an effect upon the number count that is far greater than that due to the geometry of space itself. Problems such as these leave the determination of the curvature of the universe still unsolved. The most that can be said with near certainty is that the results are inconsistent with the perfect cosmological principle and the steady state model; the universe *is* evolving.

We have described several methods for measuring by independent means the cosmological parameters. The potential for errors in all these measurement techniques is large. However, in the standard models the parameters H, q, Ω, and k, as well as the measured age of the universe, are not independent. Even with large error uncertainties in all values, the accumulated evidence may point toward a particular type of model. Independent and reliable measurements of the Hubble constant, the density, the deceleration parameter, and the geometry, along with limits on the age of the universe, would pick out the specific model that best describes our universe. The capabilities of both ground and space-based observatories are improving so rapidly that there is real hope that we will have these answers in the near future.

KEY TERMS

Olbers' Paradox	cosmic background	nucleosynthesis
standard candle	radiation (CBR)	reception distance
angular size	emission distance	

Review Questions

1. Briefly describe how the cosmic background radiation (CBR) was first discovered (with an antenna, not theoretically).
2. Why do you think Gamow and his collaborators' predictions were ignored for so many years?

3. Explain the significance of the measurements of the *COBE* satellite. What is the shape of the CBR spectrum? Does it vary from one direction to another, in either shape or temperature? What anisotropies were found by *COBE*, on both large and small scales, and what is the significance of these anisotropies?

4. Why is the CBR considered to be the most definitive evidence available for the big bang?

5. If the CBR came from the early moments of the universe, when conditions were much hotter, why is its temperature so cold today? What happened to the energy lost by the photons?

6. What is the actual age of an Einstein–de Sitter model whose Hubble time is 12 billion years? If the oldest stars are found to be 14 billion years old, what is the maximum possible value for the Hubble constant in an Einstein–de Sitter universe?

7. What is the difference between "reception distance" and "emission distance"? What kind of information would you need, in addition to redshift, in order to determine these quantities exactly for objects such as quasars?

8. Consider the universe at a redshift of $z = 2$. If two galaxies are separated by a distance ℓ at the time corresponding to this redshift, what is their separation today?

9. Explain how the measured angular size of galaxies as a function of redshift could be used, in principle, to determine the geometry of the universe.

10. What phenomenon accounts for most of the darkness of the sky at night? How does the expanding universe contribute to darkness at night?

11. Supporters of the steady state model have argued that the gradual creation of matter out of nothing in empty space is philosophically preferable to the big bang, which features creation of everything out of nothing at $t = 0$. Do you agree? How are these two types of creation physically and logically distinct?

Notes

1. A similar effect accounts for the reddening of the Sun at sunrise and sunset; light emitted near the horizon traverses more atmosphere and thus has more of the blue photons scattered out of it.

2. Bethe's name, which is pronounced "beta," was added for humorous effect. Gamow even tried to induce Herman to change his name to "Delter."

3. Chapter 5 discusses some of the ways in which stellar ages can be determined.

CHAPTER 13

The Early Universe

The recognition that the universe is expanding leads naturally to the question of its origins. From what might the universe have expanded? What might have happened when the universe was much smaller? What does "smaller" mean? Within the standard models, a straightforward projection to earlier time leads to the conclusion that the universe was once much denser and more compact than it is today. Indeed, taking this to its ultimate limit, the universe was *infinitely* dense at that cosmic time when the scale factor R was zero.

If the universe began with a "big bang," what was this event like? Can we learn anything about it today, or is it too far from our experience to try to understand? How can we even begin to think about the universe near its beginning? You might believe that the earliest stages of the universe are so unimaginably complicated that we would not even have the capability to describe them. Yet there are good reasons to believe that near its beginning, the universe was in many ways much simpler than it is today. Complex objects such as stars and galaxies had not yet formed. The universe consisted of a soup of elementary particles, interacting with one another in relatively simple ways.

The further back in time we go, the smaller the scale of the universe, the greater its density, and the higher its temperature. Because of the high density, elementary particles of all kinds constantly exchanged energy and momentum; the particles coexisted in an equilibrium defined by a single set of statistical properties. A state completely characterized by the statistical quantity we know as temperature is called **thermal equilibrium**. The simplest model we can study assumes that the early universe may be described by a state of thermal equilibrium; from the big bang, until the universe cools sufficiently that the photons no longer interact with the massive particles, the temperature alone will be our guide to the many particles and their interactions. Like any model, this is subject to testing; it finds its justification mainly in the uniform blackbody spectrum of the cosmic background radiation. It does introduce some problems of its own; we shall examine these failings in later chapters. But first let us see how far we can go with this simple assumption and how well the data support it. We shall find that it will go very far indeed.

The Radiation Era

Although photons have no rest mass, they possess energy proportional to their frequency.[1] As we have learned from general relativity, both energy and mass create gravity. Just as the universe contains a rest-mass density, it also contains an energy density, the latter being defined simply as the energy per unit volume. In the present universe, the energy density is mostly due to the cosmic background photons, which number more than a billion photons for every particle of ordinary matter. However, because each photon has lost much of its energy in the overall expansion, the energy density in the CBR is minute in comparison to the mass density of ordinary matter. Thus, we say that the universe today is **matter dominated**.

Early on, however, conditions were quite the opposite. For the first several thousand years of the universe's existence, the radiant energy density provided the most important contribution to the gravity; therefore, this stage in the history of the universe is called the **radiation era**. Since the rest-mass density of the matter was entirely negligible in comparison to the energy density of the radiation, the universe is said to have been *radiation dominated*.

How must the Friedmann equations be changed to account for photons? Obviously, we must include the photons' energy density in ρ; but in this case, the pressure due to the photons is also a significant contributor to the universe. The pressure in the early universe was not a gas pressure, since contributions from the kinetic and other energies of the massive particles were largely insignificant, but was a *radiation pressure* due to the energy density of the photons. It may not be immediately obvious to you that a massless particle can exert a pressure, but it can. Photons carry energy and momentum; thus when they impinge upon a surface, some momentum can be transferred to it. Radiation pressure is not something that often appears in everyday life, but its effects can be seen in the sky; the tails of comets point away from the Sun in part because the specks of dust that make them up are buffeted by the pressure from the Sun's photons.

Previously, we have investigated the expansion of the universe when pressure is assumed to be negligible. We found its behavior to be very much like the prediction from ordinary Newtonian gravity in an expanding background, with some corrections for relativistic effects. A pressureless universe was easy to understand in fairly familiar terms. What happens when we introduce a pressure? In general, a pressure can be associated with any energy density other than pure rest-mass energy. The distinction is that energy density is, as its name implies, an energy per unit volume. Pressure, on the other hand, arises when the energy contained in a volume is able to exert a force on some surface, which could be a surface within the volume and need not be a boundary. An *equation of state* is a relationship which connects pressure (force per unit area) with energy density (energy per unit volume); the ideal-gas law discussed in chapter 5 is an example of such an equation.

One counterintuitive result of this new effect is that any ordinary positive pressure actually *increases* the rate of contraction by increasing the gravitational force. Pressure does not "blow up" the universe, it accelerates its collapse! However, some reflection should make it clear that this is not so surprising after all. In general relativity, not only energy density, but also *stress*, of which pressure is one form, participates in gravity. From this the conclusion follows that pressure too must actually contribute to gravitational attraction and thus possibly to a collapse. But you might still be unconvinced; does not pressure also push objects apart, and so should it not push the universe outward? No, because pressure only pushes when there is a *change* in pressure. Pressure pushes from regions of higher pressure to regions of lower pressure. In a homogeneous and isotropic universe, the pressure must be the same everywhere, so there is no net push. Pressure is left with nothing to do but increase the gravitational attraction.

Given that the evolution of the early universe was controlled by the energy density and pressure of the photons, we must determine the behavior of the photon energy density as a function of the scale factor. This problem can be approached from more than one direction; the equation for local energy conservation could be written and formally solved. We can also make use of the more intuitive argument developed in chapter 12. In either case, we find that the energy density varies like the inverse fourth power of the scale factor,

$$\mathcal{E}(t) = \mathcal{E}(t_1) \left[\frac{R(t_1)}{R(t)} \right]^4 , \tag{13.1}$$

where t_1 is some arbitrary point in cosmic time. The radiation energy density drops more rapidly than it would if it were decreasing due only to the volume expansion of the universe, since the redshift introduces an additional power of the scale factor.

The energy density of radiation in thermal equilibrium is proportional to the fourth power of the temperature, $\mathcal{E} \propto T^4$. Hence the temperature of the radiation in the universe is a simple function of the scale factor

$$T(t) = T(t_1) \left[\frac{R(t_1)}{R(t)} \right] . \tag{13.2}$$

Using the formula for redshift as a function of scale factor, we obtain the temperature at any past time t in terms of the present temperature and the redshift:

$$T(t) = T(t_0)(1 + z), \tag{13.3}$$

where t_0 is the time "now" and t is the time corresponding to the redshift z. We need not know the actual value of t in order to make use of this equation; neither do we need to know t_0, since we can measure $T(t_0)$.

As we go backward in time, corresponding to larger and larger z and smaller and smaller scale factor, we find a universe filled with photons of increasing temperature. The cosmic background radiation today has a temperature of about 2.7 K. At a redshift $z = 1$, this same background radiation had a temperature of 5.4 K. As we continue to look into the past, the temperature rises into the thousands of degrees. The background radiation, which today is mostly in the microwave band, becomes visible; the universe at this time was bright with light. Further back in time, the temperature rises to ever greater values. Indeed, if the universe began with a scale factor of zero, the initial temperature must have been infinite!

The behavior for the scale factor as a function of cosmic time in the early universe is computed by solving the Friedmann equation, under the assumption that the energy density arises only from radiation pressure. It is easiest to solve this equation for the case of flat space. In our studies of the present universe, we have often employed the Einstein–de Sitter model, the $k = 0$ solution for the case of pressureless "dust," for comparison purposes. As it turns out, $k = 0$ is a good approximation regardless of the actual curvature, at sufficiently early times. In the standard models, the density, from whatever source, becomes large as the scale factor shrinks; further, the energy density will change with R *at least* as R^{-3}. Therefore, as R becomes small, the first term on the right-hand side of the Friedmann equation (11.8), the density term, will become large, whereas the curvature term cannot change. Eventually the density term dominates so completely that the curvature term is irrelevant. Even at the present time, the curvature is not a dominant factor; this is one reason why it is so difficult to determine the curvature from observations. Early in the history of the universe, the curvature had almost no effect upon the evolution of the cosmos.

Thus we confidently proceed to ignore the curvature term in the early universe, at least for the purposes of gaining an overall picture. We can now solve the Friedmann equation to find that during the radiation era, the scale factor is given by

$$R(t) \propto t^{1/2}. \tag{13.4}$$

In comparison, the matter-dominated Einstein–de Sitter universe has $R(t) \propto t^{2/3}$. The age of a flat, radiation-dominated universe is only 1/2 the Hubble time. The deceleration is larger in such a universe, with $q = 1$. As we have stated, the presence of radiation pressure actually increases the gravitational force, thus braking the expansion more rapidly.

Matter and Energy

High temperatures and energies imply a drastically different universe from what we know today. The further we probe toward $t = 0$, the more exotic the universe becomes. At sufficiently early times, conditions were so extreme that even atoms could not have existed. The universe was like a tremendous particle accelerator, with high-energy particles zipping about at relativistic speeds, crashing into one another and interacting with photons. Our description of the very earliest moments of the universe must necessarily be somewhat tentative, as the theories of matter and energy under such extreme conditions are still rudimentary. But let us see how far we can go with the knowledge we have.

If the very early universe was filled with particles, from where did they originate? As has been amply demonstrated experimentally, Einstein's famous equation $E = mc^2$ means that matter can be converted into energy. What may not be so well appreciated is that it also goes the other way: pure energy can be converted into matter. We have seen an example of this phenomenon in our study of Hawking radiation from a black hole, but other such processes exist. In the early universe, creation of matter from energy was one of the most important effects. By Einstein's equation, the rest mass of any elementary particle is equivalent to some amount of energy, defined to be the rest energy of that particle. In the state of thermal equilibrium, the total energy is divided equally among all species of particle, including photons. At any cosmic time in the early universe, the temperature implies a mass scale, via this mean energy per particle. If the temperature of the early universe is at or above this *threshold* value for a given particle, two colliding photons can produce the particle and its antiparticle, a phenomenon known as **pair production**. The threshold temperature thus represents the minimum energy required for matter-antimatter partners of a specific mass to be created from the collision of two photons. Particles of a given species can still be produced at temperatures well above their threshold, of course; in this case, they are simply created with kinetic as well as rest energy. Pair production need not always result from photon collisions; if the temperature is at least twice the threshold, particle pairs appear directly from the energy of the fields.

It is simple to quantify how temperature determines the types of massive particle that could be present at any point in the early universe. In thermal equilibrium, the mean energy per particle is proportional to the temperature

$$\langle E \rangle = \frac{3}{2} k_{\text{B}} T. \tag{13.5}$$

As in the ideal-gas law, Boltzmann's constant k_{B} appears as a conversion factor between temperature and energy units. To find the threshold temperature corresponding to any particle mass, equate the rest-mass energy of the particle with the mean energy of the photons, and solve for the temperature:

$$T = \frac{2m_0 c^2}{3k_{\text{B}}}. \tag{13.6}$$

The higher the temperature, the more massive the particles that can be produced. As an example, if we wish to create a proton-antiproton pair by photon collision, the temperature must be approximately 10^{13} K.

Figure 13.1 The process of pair production. Two high-energy gamma rays collide and produce an electron-positron pair. Each gamma ray must have at least as much energy as the rest mass energy of the electron, according to Einstein's formula $E_0 = m_0 c^2$.

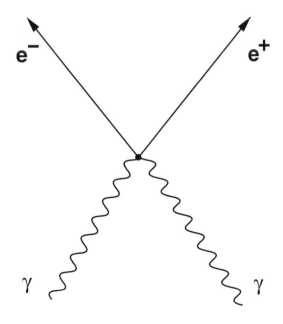

At temperatures above the threshold for a particular particle type, each reaction proceeds at a rate precisely equal to its backreaction; creation and destruction of any kind of particle must be exactly in balance. In pair production, for example, photons produce pairs, and the pairs almost immediately annihilate back into photons. Moreover, approximately as many particles were present as photons. Any excess of photons would create more particles, causing the number of particles to rise, whereas a shortage of photons would mean that particle creation would not be able to keep up with particle destruction. At any cosmic time in the early universe, all particles permitted to exist at the corresponding temperature were constantly colliding, materializing, and annihilating, such that the number of members of any particle species remained the same, and each particle was matched by an antiparticle. During epochs of particle creation, then, the contribution of matter to the energy density of the universe was not negligible. When the temperature dropped below the threshold temperature for a particular pair type, the annihilation rate exceeded the production rate, and the pairs of particle were rapidly destroyed.

In the "ordinary" matter creation from photons which we have described, both a particle and its antiparticle *must* be created, because in any such reaction, certain properties of the particles, as well as the total mass-energy, must remain the same. Nevertheless, matter and antimatter cannot have been in exact balance for all times, because the universe now is filled with matter even though the radiation temperature has dropped to 2.7 K, far below the threshold temperature for any of the constituent particles of atoms. If nucleons and antinucleons had occurred in equal numbers in the early universe, they would have annihilated, leaving only photons. Considerations of causality indicate that matter and antimatter could not have segregated themselves fast enough in the early universe to prevent this. Furthermore, there is no evidence for any significant accumulations of antimatter anywhere in the present universe. Galaxies and clusters do not exist in isolation; if an antimatter galaxy existed, it would undoubtably find matter in short order. Any such antimatter would annihilate whenever it encountered matter, creating a copious flux of characteristic gamma rays. Matter-antimatter reactions are particularly violent and can be seen over enormous distances if they occur. No tremendous cosmic flux of gamma rays from annihilation reactions is seen, and we can conclude that only matter exists in appreciable amounts.

Since our present universe is made of matter, at some point in the early high-temperature epoch, the symmetry between matter and antimatter must have been violated. The amount of leftover matter is rather small; there is one particle of matter per 1 billion photons, meaning that the excess of matter over antimatter was about one part in a billion. Yet that small quantity of matter makes up all that we can see, and all that we are. There is as yet no firm explanation of how this effect occurred. While it could have been simply an initial condition, built into the very beginning itself, physicists believe that it may be a consequence of the fundamental laws of physics; one leading possibility will be discussed when we delve into the chronology of the big bang.

The extreme conditions of the early universe require that our understanding of its history be inextricably linked with particle physics. Throughout this century, there have been occasional interactions between cosmology and other branches of physics; some of the most distinguished physicists of the first half of the twentieth century, such as Enrico Fermi, George Gamow, Robert Oppenheimer, and many others, made important contributions to cosmology. But for the most part, nuclear and particle physics advanced independently of cosmology. Particle physicists have sought to build ever larger accelerators, in order to study physics at higher energies. But such accelerators take an increasing toll in effort and resources. Our study of special relativity showed how difficult it is to accelerate even elementary particles to relativistic speeds; if we wish to create even more exotic states, such as significant quantities of antimatter, the engineering problems become considerable, even overwhelming. The Superconducting Supercollider was to have consisted of an evacuated ring, 54 miles in circumference, about which nearly infinitesimal particles would have been driven to ever higher energies by the magnetic field from superconducting magnets. The cost of this great machine proved prohibitive, however. And even the Supercollider could not have reached the energies for which the particle physicists ultimately yearn. In order to test the leading edge of particle physics to the utmost, much greater energies are necessary. With any realistically foreseeable technology, only the early universe itself could be an appropriate laboratory.

Fields of Dreams

The world of elementary particles is a realm controlled by **quantum mechanics**, the physics of the very small. In quantum mechanics, the sureties of our familiar, macroscopic world vanish, to be replaced by a physics in which only probabilities can be known. We cannot predict, for example, when a given atom of uranium will decay; not because of any lack of understanding about the decay process or ignorance of the initial state of the atom, but because it is *fundamentally* unknowable. The best we can do is to compute the probability that the atom will decay in any specified interval of time. If we have a large number, that is, an *ensemble*, of uranium atoms, then we can predict how many will have decayed after a specified time interval has elapsed, but we can say nothing definite about the fate of any individual atom.

Quantum mechanics also demands a blurring of the concepts of "particle" and "wave." According to quantum mechanics, each elementary particle can show both corpuscular and wave behaviors, although never both at once. We are already familiar with something that can show either wave or corpuscular behavior: light. The typical wavelength of a photon of visible light is about 5×10^{-7} m, which is greater than the size of molecules. This means that visible light often manifests itself in the manner of a wave as it interacts with surrounding matter. Yet we have often explicitly treated light as a particle, the photon, such as when we deal with quantum transitions in atoms. We are less familiar with the wave nature of things we call "particles" because their wavelengths are so small. In general, the wavelength of a massive particle depends upon its velocity, but high-energy particles moving at relativistic speeds manifest a wavelength known as the **Compton wavelength**,

$$\lambda_C = \frac{h}{m_0 c}.$$

(13.7)

For example, the Compton wavelength of a proton is approximately 2×10^{-16} m. Although our mental picture of entities such as the proton and the electron is firmly rooted in the concept of "particle," their wave nature is quite easily observed in high-energy physics experiments. Wave-particle duality is as real for the proton as for the photon.[2]

In modern theories of quantum mechanics and particle physics, a wave, and hence its allied particle, can be associated with a **field**. In physics, a field is a convenient mathematical representation of a quantity which is extended in space or time, or both. The gravitational and electromagnetic fields are familiar descriptions of the corresponding forces. What may not be so obvious is that these fields are associated with particles. The photon provides a reasonably concrete illustration; the photon, the particle of light, is linked with the electromagnetic field. In *quantum field theory*, this concept is further extended. Not only the photon has a field, but so does every particle. There is an "electron field" and a "neutrino field," and so on for each particle. A few of these particle/wave fields *mediate* forces.[3] The "photon field" *is* precisely the electromagnetic field. The field of an as-yet undiscovered particle, the **graviton**, is the gravitational field. The energy and momentum of a field is *quantized* into bundles, or *quanta*, which can, under appropriate conditions, manifest themselves as particles, just as the photon carries a discrete quantum of energy.

The strength of the field is related to the number of quanta that are present. The surface of the Sun spews forth so many photons each second that the electromagnetic field (the light) emanating from it seems continuous and thus obeys the laws of classical optics; but many photoelectric detectors, such as the charge-coupled devices (CCDs) used with modern telescopes, and even in some of the "digital cameras" now on the consumer electronics market, can, if sufficiently sensitive, detect electromagnetic fields so weak as to represent only a few photons. In fact, a CCD works very much like the retina of the human eye; nature discovered the principle long before humans incorporated it into modern technology. The retina is lined with cells containing special molecules that can, upon being struck by photons, change their configurations. The alteration of such a molecule rearranges its electrical charges and thus creates a weak electric current. Ultimately, after considerable amplification and processing by nerves, the brain interprets such currents as an image. The retina is a device for converting a quantum field into a pattern of electrical activity which a processor, the brain, can recognize! Yet when the light enters the eye, passing through the cornea and the lens, it behaves like a wave, and we can use classical wave optics to compute how much it will be refracted by the cornea and lens and where the focus will be; if necessary, we can then interpose an artificial lens between the source and the eye in order to shift the point of focus to the retina. In one part of the eye, light behaves like a wave; in another, the very same light is a particle. According to quantum mechanics, it is either, depending upon which behavior the "experiment" elicits. Quantum mechanical effects are not weird theoretical constructs, with no connection to reality; it *is* the way the universe works. Quantum mechanics, perhaps even more so than relativity, is very nonintuitive. But whether we are aware of it or not, it impinges in many ways upon our "classical," macroscopic world.

Field theories of one form or another are the foundation of most of modern particle physics. Some of the most important characteristics of the fields are their

symmetries, those quantities that remain invariant under specific transformations. We have already discussed invariance in relativity; there we can find the most intuitive forms of symmetry. The spacetime interval is unchanged when the coordinates change. The cosmological principle is a statement of the symmetry of the universe in both spatial location (homogeneity) and spatial direction (isotropy). Many conservation laws can be attributed to various symmetries. For example, energy conservation arises from symmetry with respect to translations or reversals in time. Energy can be defined to be a particular quantity that does not change when time changes, and the fact that such a quantity can be specified at all is due to the existence of the symmetry in time of fundamental laws of physics. Similarly, the conservation of linear momentum can be understood as resulting from symmetry under straight-line translations in space, while conservation of angular momentum is a consequence of symmetry with respect to rotations in space.

Particle physics itself can be characterized as a search for symmetry. The menagerie of particles can be divided into a few families, each with various internal symmetries, that is, symmetries that are properties of the field itself and not of the outside world. Electric charge, for example, represents a symmetry in the electromagnetic field under certain transformations of abstract coordinates. Three of the four fundamental forces of nature can be understood in terms of the symmetries of an appropriate field theory. (Gravity is the lone holdout, so far resisting all efforts to fit it into this picture.) If these forces share all symmetries, then they are indistinguishable from one other. This was the state of things in the earliest moments of the big bang. Today we see the forces as distinct because of the loss of symmetries at early times, once the temperature dropped below certain levels. Such a loss of symmetry is known as **spontaneous symmetry breaking**.

It may be difficult to visualize a spontaneous symmetry breaking of some abstract field theory, but we are all familiar with a very similar symmetry breaking; we call it "melting." A liquid, such as water, has higher entropy and much less symmetry than a crystalline solid. If the conditions are not right, the symmetry is lost, or "broken"; the ice melts. Similarly, under appropriate conditions, forces are united in a symmetry; but failure of those conditions, such as a lowering of temperature in the case of fundamental forces, breaks the symmetry. The melting, or freezing, of water is called in physics a *phase transition*; boiling of water from a liquid to a gas is also a phase transition. Remarkably, spontaneous symmetry breaking is not only similar to

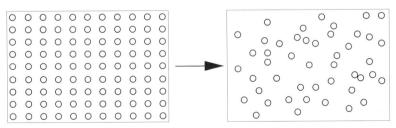

Figure 13.2   In melting, molecules go from an orderly, symmetric arrangement to a more random state. Thus, the phase transition of melting results in a loss of symmetry.

such mundane occurrences as an aid to the imagination; the two concepts are actually *mathematically* similar. It is thus quite justifiable to think of a spontaneous symmetry breaking as a kind of phase transition.

The ultimate goal, the Holy Grail of particle physics, is the "theory of everything," a theory that would encompass all particles and forces, showing them to be manifestations of an underlying simplicity. Like the Holy Grail of mythology, the final theory has proved elusive. Many have thought to grasp it, only for their vision to evaporate in the glare of data. It may be that a few have glimpsed its outline, but as yet no one has seen it clearly. But the quest continues, and it may someday be successful, for much has already been learned. As new and more powerful particle accelerators were constructed, particle physics contributed new ideas and discoveries to cosmology, until the universe seemed comprehensible down to the first hundredth of a second. Now it may be that cosmology can return the favor, by providing clues toward the understanding of conditions that may never be reproducible by humans.

The Beginning of Time

We would like our cosmological theory to describe the history of the universe all the way back to the big bang, with its soaring temperatures. But can classical general relativity apply as far as $t = 0$, the very initial state, with its extraordinary conditions? Certainly not. Gravity has not yet been fully incorporated into the other great theory of modern physics, quantum mechanics. For the conditions prevailing throughout most of the history of the universe, we can separate the two theories because the scales over which they dominate are so vastly different. Quantum mechanics rules the smallest scales, while gravity governs the largest scales. In the present universe, there is little overlap in the domains of these two theories. But as we approach $t = 0$, their regimes must merge together.

Gravity is by far the weakest of the four fundamental forces. Nevertheless, it controls the evolution of the universe because it is long range, and especially because it is *only* attractive; all the mass in the universe contributes, and the gravitational force is never partially cancelled. Even at quite early times in the history of the universe, gravity was much weaker than any of the other fundamental forces, so we may go to very early epochs before we need worry about quantum gravity. At the very beginning of the universe, however, the scales characteristic of quantum mechanics and of gravity were similar, and gravity was comparable in its immediate effects to the other forces. We must have a full theory of quantum gravity in order to describe the universe under such conditions. Since we have no such theory, the earliest moments of the big bang remain a mystery.

Rather than starting from time zero, then, we must pick up the story where classical general relativity gains control of the universe as a whole. This occurs at a cosmic time of 10^{-43} s, the **Planck time**. At this time, the characteristic length scale of the universe was $ct = 1.6 \times 10^{-35}$ m, the **Planck length**.[4] This length is much smaller than the Compton wavelength of any elementary particle. Indeed, the very idea of a "particle," at least as we currently conceive of it, must break down during this initial period. We can say essentially nothing about the behavior of the contents of the universe from the beginning until the Planck time, an interval which is often called the **Planck epoch**. During this epoch, all four fundamental forces of nature (gravity, electromagnetism, and the weak and strong interactions) composed a single force. At the end of the Planck epoch, the gravitons, the particles which carry gravitational force, and which so far have eluded detection, fell out of equilibrium with the other particles, and at the same time, gravity decoupled from the other forces. The gravitons then streamed out through the universe, forming a cosmic background of gravitational waves. This event occurred so early that the energy of these waves has been redshifted nearly away, and they are utterly undetectable today; the invaluable information that they could provide about the earliest moments is beyond our grasp for the foreseeable future. The decoupling of gravity was the first spontaneous symmetry breaking in the universe, the loss of the perfect symmetry and equivalence among all four forces with which the universe is thought to have begun. As strange as it may seem, at its very beginning the universe, with its exotic conditions we cannot yet comprehend, was in some ways as simple as it could ever be.

From the Planck time till about 10^{-35} s, the temperatures are so high that we still have little understanding of the nature of matter. This can be called the **unified epoch**, since during this stage, three of the four fundamental forces, electromagnetism, the weak force, and the strong force, were unified, that is, they made up a single, indistinguishable force. Although theories exist that apply to conditions during the unified epoch, they are still incomplete and not always consistent with experimental data. Nevertheless, they provide the beginnings of a framework to understand the behavior of particles and forces during this epoch. These theories are called **grand unified theories** (GUTs), because they attempt to explain the unification of the three forces. Unfortunately, the temperatures, and hence the energies, during this epoch are far beyond what we could ever hope to reach in the largest particle accelerator we could imagine. Physicists and cosmologists hope that the universe itself will provide experimental evidence for conditions during the unified epoch and thereby guide the development of GUTs.

GUTs deal with the interactions among various particles, specifically, the elementary particles called **hadrons**. The "elementary" fermions can be subdivided into two families, the hadrons and the **leptons**, according to whether or not they interact via the strong interaction. Hadrons are massive and participate in the strong interaction. They are not point particles, but consist of smaller particles called **quarks**, which seem to be pointlike. Six species of quark were predicted from theory; five were found fairly easily in high-energy particle physics laboratories, with reliable evidence for the sixth beginning to appear early in 1995. Based upon their construction, and the species of quarks present, the hadrons may be further broken down into the **baryons**, which consist of three quarks, and *mesons*, which are composed of a quark and an antiquark. Mesons are heavy particles with extremely short half-lives; we shall have little further to say about them. Baryons, on the other hand, are extremely important; by far the most common baryons are the nucleons, the proton and the neutron. "Baryon" is sometimes even loosely used as a synonym for "nucleon," although this is not quite correct. Many other baryons exist, but in the present universe they are rare and decay very quickly, eventually always becoming protons, since the proton is the least massive baryon.

Particle Families

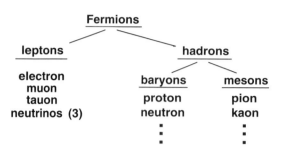

Fermions

leptons   hadrons

electron   baryons   mesons
muon
tauon   proton   pion
neutrinos (3)   neutron   kaon
                 .   .
                 .   .
                 .   .

Figure 13.3 Fermion families. The hadrons participate in the strong interaction, the leptons do not. Hadrons are composed of quarks and are subdivided into the baryons and the mesons. The proton and neutron, the constituent particles of atomic nuclei, are the most important examples of baryons.

Gluons are the particles that hold quarks together in hadrons. In the present universe, solitary quarks do not exist in nature but are only briefly created in high-energy particle collisions. During the unified epoch, however, conditions were so extreme that quarks had not condensed into hadrons; the universe consisted of a brew of highly relativistic particles, including quarks, gluons, and more exotic particles.

The most significant remnant of the unified epoch is the excess of matter remaining from it. Particles created from pure energy in "ordinary" processes must always be created in matter-antimatter pairs. Furthermore, when a particle and its antiparticle collide, they destroy one another, converting their rest masses entirely into photon energy. This corresponds to a rule called *conservation of baryon number*, where antiparticles of baryons are "negative baryons." For example, if a single neutron, a baryon, decays, only one proton can be created. If there were no baryons to begin with, then only particles which are not baryons, and thus have zero baryon number (e.g., photons), or else a pair consisting of both a baryon (any baryon has a baryon number of +1) and its antiparticle (baryon number −1) must be created. Since baryons are made of quarks, this also implies conservation of the number of quarks in a particle reaction, and under ordinary conditions, this is true. However, a common feature of most GUTs is that, under GUT conditions, this particular conservation law no longer holds; reactions take place which transform quarks into leptons, and vice versa, thus violating baryon conservation. These reactions occur in such a way that the result is always a very, very tiny excess of matter. The process by which matter was preferred over antimatter, creating the stuff of our universe, is called **baryogenesis**.

If baryogenesis had *not* occurred, whether by the GUT mechanisms or by some other means, no matter would now exist, for every particle would have eventually destroyed itself with its antimatter partner. Modern estimates find that there are approximately a billion (10^9) photons left over from the early seconds of the universe for every hadron; therefore, this asymmetry between matter and antimatter during the unified epoch must have been only about one part in a billion. Yet it is just this asymmetry that led to the creation of all the many forms of matter that we know today, including ourselves.

Baryogenesis in GUTs leads to another very important prediction: the proton is unstable. There is no particular reason that the proton *should* be stable. After all, it is not, in some sense, truly "elementary," but is a composite of three quarks. However, protons are one of the most important components of ordinary matter, so the question of their stability is of considerable significance for the stability of matter as a whole. We can immediately see that the proton must have an *extremely* long life expectancy, for if it did not, matter would disintegrate over the current age of the universe. If the life expectancy of the proton were short, comparable to the age of the universe, then given the vast number of protons within our Hubble sphere, we would expect to be able to see proton decay on a regular basis. In particular, the human body contains approximately 2×10^{28} protons. If the proton's life expectancy were close to the age of the universe, 10^{10} yr, then on average, roughly 10^{18} protons would decay in your body per year! Since the decay products would have considerable energy and would rapidly be converted into gamma rays, ordinary matter would be noticeably radioactive. Life as we know it probably could not exist in a universe with such a short proton lifetime. On the other hand, a very large proton lifetime, considerably greater than the age of the universe, would mean that very few protons would have decayed by now, explaining the absence of an observable effect.

As it happens, testing this prediction is an experiment that does not require a particle accelerator at all; it is one of the few aspects of GUT theories that can be *directly* tested in Earthly laboratories. It is easy to design an experiment to measure the life expectancy of the proton. Simply gather a huge number of protons (a large quantity of water will work nicely) and watch for decay products. For example, if you have approximately 10^{30} protons, and the proton's half-life is 10^{30} yr, then you can expect to see roughly one proton decay per year. Experiments of this nature have been performed, and the average lifetime has proved to be too large to measure, yet. A firm lower bound can be placed: the proton half-life is *at least* 10^{31} yr, probably closer to 10^{32} yr; it might even be stable. Unfortunately, this disagrees with the simplest GUT theory, which predicts a proton lifetime of about 10^{30} yr. This does not rule out all GUTs, but means that the simplest version cannot be correct. More sophisticated, and thus more difficult, theories are required. Even so, simple GUTs have some successes, such as explaining why matter exists. We may not yet fully understand the unified epoch, but there is good reason for optimism that it will become comprehensible in the near future.

If inflation, an enormous, exponential expansion of the universe, occurred, it must have taken place sometime during the unified epoch, probably around 10^{-37} s after the big bang. We shall discuss inflation in chapter 15.

Quarks, Hadrons, and Leptons

The end of the unified epoch came at 10^{-35} s, when the temperature dropped below the level required to maintain the grand unified symmetry, and the strong interaction decoupled from the others. What followed may be called the **quark epoch**. During the quark epoch, the universe consisted of quarks and gluons, along with the carrier particles of the combined electromagnetic and weak force, as well as more exotic heavy particles; plus, of course, the antiparticles of all. We still cannot say very much about this period. Although a theory of strong interactions, *quantum chromodynamics*, or QCD, exists, its equations are so complicated that very little is known about their solutions; this difficulty occurs precisely because the coupling between two hadrons is so strong. We know even without any equations that the strong force is exceedingly strong; after all, it holds together nuclei against their electrostatic repulsion. But the little information so far extracted from the theory has revealed a most curious property of this force: it actually becomes stronger with increasing distance. This is why free quarks are never found under natural conditions in the present universe; if any two quarks were somehow pulled apart, the force between them would increase until the energy in the strong field would create a new pair of quarks. It would be akin to trying to divide a magnet; when a bar magnet is split, the result is two smaller magnets, not two distinct poles. At extremely high energies and densities, however, the strong force becomes negligible, and the quarks are able to behave as if they were perfectly free particles. Thus there was no compulsion for the quarks to form particles until the temperature dropped far enough.

During most of the quark epoch, the weak and electromagnetic forces were unified. The theory of the unified weak and electromagnetic forces, the **electroweak interaction**, is well established; the 1979 Nobel Prize in physics went to Steven Weinberg, Abdus Salam, and Sheldon Glashow for their work in developing this theory. At sufficiently high temperatures, the weak and electromagnetic forces were of comparable strength. Instead of photons, there were two force-carrying bosons, both massless. Around a time of 10^{-11} s and a temperature of 10^{15} K, the weak interaction decoupled from the electromagnetic force, leaving all forces separated as they are today. During this transition, the carrier bosons of the unified electroweak force were transformed into four new particles: the W^+, W^-, and Z^0, which acquired mass, and the photon, which did not. The three massive particles are carriers of the weak interaction, whereas the familiar photon is the carrier particle of the electromagnetic force. Because of this, the weak interaction has a short range, while the range of the electromagnetic force is, in principle, infinite.

The masses of the W and Z particles are comparable to the masses of fairly heavy atomic nuclei. We have discovered that a large mass corresponds to a very high energy; hence it should be no surprise that it was difficult to create W and Z particles even in advanced accelerators. Indirect tests confirmed the electroweak theory, leading to the Nobel Prize for its developers, but the final proof had to await detection of the

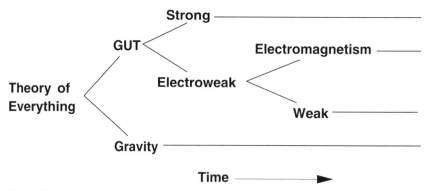

Figure 13.4 Force tree showing how the four fundamental forces we know today separated as time advanced in the early universe.

bosons themselves. Finally, the W and Z particles were found in experiments at the CERN accelerator near Geneva, Switzerland, in 1983, more than 15 years after the theory was first proposed.

The symmetry breaking of the electroweak force was, like other spontaneous symmetry breakings in field theories, analogous to a phase transition such as the melting of ice. One characteristic of phase transitions is that local conditions may affect quantities such as the rate or timing of the transition. Nearly everyone has seen a frozen creek or lake whose ice contains dark planes separating regions that froze at various rates or times, or even including pockets of liquid coexisting with the solid state. Similar phenomena could occur in spontaneous symmetry-breaking transitions; the universe may be divided into many domains in which the electroweak transition occurred differently. Such divisions between regions would be defects in the structure of spacetime, similar to the "defects" in a crystal that divide one ordered area from another. This could have significant ramifications for structure in the universe. If these defects exist, they could have attracted matter, providing seeds for later gravitational collapse.

After the separation of the fundamental forces, the next major stage occurred approximately 10^{-6} s after the big bang, when the quarks condensed into hadrons, ushering in the **hadron epoch**. The hadron epoch was brief, for the temperature soon fell below the threshold for protons. At this point, the asymmetry of matter remaining from the unified epoch was permanently frozen; all baryon-antibaryon pairs disappeared, leaving behind photons, while unpaired baryons survived. Those photons now make up most of the CBR. But as cosmic time has passed, the photons have lost their energy to the redshift. The baryons, on the other hand, retained their rest mass unchanged. This leftover bit of matter, the one-part-in-a-billion survivors at the end of the hadron epoch, would go on to dominate the universe until the present.

After the condensation and decoupling of nucleons, at around 10^{-4} s, the universe entered the **lepton epoch**, when particles associated with the weak interaction ruled the cosmos. Leptons are the lighter fermions, the electrons, muons, and neutrinos.[5] Electrons have negative charge and are stable, muons have positive charge and are unstable, while neutrinos are, as their name indicates, neutral; they are also stable. Like quarks, leptons appear to be true point particles, genuinely "elementary." In contrast to hadrons, leptons do not interact through the strong force but do take part in the weak interaction. The weak force determines various particle phenomena, such as the decay of a free neutron, as well as beta decay, a nuclear process in which a neutron is converted into a proton and an electron, with the electron ejected from the nucleus as a "beta particle."

Neutrinos participate in interactions involving the weak force and are unaffected by the strong interaction. Their electric neutrality renders them immune to electromagnetic forces. Their mass is currently controversial, but they may well be massless, in which case they are also scarcely affected even by gravity. The feebleness of the interactions of neutrinos with ordinary matter makes them exceedingly difficult to detect. At each moment, a fantastic number of neutrinos courses through your body, originating from the Sun and from the general cosmic background. Given the average number of protons in a human body, the expected rate of interactions of your body with a neutrino from the Sun is approximately once every 72 years. The rest of the time neutrinos constantly pass through you, like visible photons through a pane of glass.

As cosmic time advanced and the temperature continued to fall, particles of less and less mass dominated. Early in the lepton epoch, heavy leptons, such as the muon, were created. As the temperature dropped, production of muons essentially ceased, but muon-antimuon annihilations continued, so electrons became dominant. The universe consisted primarily of a soup of photons, neutrinos, electrons, and positrons, with the relatively small density of leftover protons and neutrons from the unified and hadron epochs. These sparse hadrons interacted with the leptons according to such reactions as

$$\bar{\nu} + p \rightleftharpoons e^+ + n$$

and

$$\nu + n \rightleftharpoons e^- + p.$$

The first equation states that an antineutrino (ν is the symbol for a neutrino, and the overbar indicates the antiparticle) reacts with a proton (p) to create a positron (e^+) and a neutron (n); or, conversely, the positron and neutron can react to create a proton and an antineutrino. (The double-pointed arrow indicates that the reaction can proceed in either direction.) The second equation, similarly, says that a neutrino and a neutron can react to form an electron and a proton, or vice versa. These reactions must occur as indicated, because of the requirement to conserve certain properties of the particles, such as electric charge and baryon number; that is, the neutron cannot react with the antineutrino to form an electron. At high temperatures these two reactions together produced approximately equal numbers of protons and neutrons, but as the temperature continued to fall, around $t = 0.1$ s, the small difference in the masses of the proton and neutron began to have an effect. The neutron is somewhat more massive than the proton. At temperatures well below the threshold of either particle, the reaction which produces a proton from a neutron is thus slightly more energetically favorable than is its counterpart, which produces a neutron from a proton. Hence the interactions of the nucleons (the protons and neutrons) with the leptons led to a much larger number of protons than of neutrons, even though the hadron epoch ended with essentially equal numbers of both. The ratio of neutrons to protons continued to drop until the end of the lepton epoch.

Approximately 1 s after the big bang, when the temperature had fallen to 10^{10} K, the density dropped enough that neutrinos no longer interacted sufficiently with other particles to remain in thermal equilibrium, and the neutrinos streamed freely from the background stew. These neutrinos continue to travel through the universe today; if they are massless, they are redshifting in energy.[6] This neutrino background is analogous to the CBR, but it is impossible to detect the cosmic neutrino background with current technology. Neutrinos interact too weakly with other forms of matter; they are very difficult to see even when they have high energies, and the lower their energy, the less they interact. By now these relic neutrinos would have very low energy indeed. If we *could* detect them someday, they would provide invaluable insights into the conditions only a few seconds after the big bang.

When the temperature fell below the threshold of roughly 5×10^9 K for the creation of electrons, at $t \simeq 14$ s, the lepton epoch ended, fixing the ratio of protons to neutrons. This ratio is measured today to be approximately 14% neutrons to 86% protons; at this point in the history of the universe, it held significant consequences for the subsequent formation of atoms. Almost all the leptons annihilated, leaving only enough electrons to balance the protons. The last burst of electron-positron annihilation added energy to the photons, raising their temperature somewhat, but not affecting the neutrinos, which had previously gone their own way. Because of this, the temperature of the photons at the end of the lepton epoch was 40% higher than the temperature of the neutrinos.

Nucleosynthesis

Approximately 180 s after the big bang, the temperature of the universe was 10^9 K. The contents of the universe consisted of free-streaming neutrinos, photons, and a relatively small abundance of massive particles that were mostly still in thermal equilibrium with the photons. The temperatures and densities were very high, but had dropped sufficiently that the nuclei of atoms could remain stable. *Nucleosynthesis*, the creation of atomic nuclei through nuclear reactions, began at this point; hence this period in the big bang is known as the **nucleosynthesis epoch**. At high temperatures and densities, neutrons and protons can fuse directly to form **deuterium** nuclei, or *deuterons*. Deuterium, also called heavy hydrogen, is the isotope of hydrogen that has one proton and one neutron in its nucleus. The number of protons determines which chemical element a given nucleus represents, but the number of neutrons affects the nuclear properties of the isotope, such as the nuclear reactions, if any, in which it will participate. Deuterium is formed by the reaction

$$n + p \rightarrow d + \gamma$$

where γ represents a photon; this reaction liberates the *binding energy* of the deuterium nucleus in the form of a photon.

Before this point in the history of the universe, any deuterons that formed were blasted apart almost immediately by the high-energy background photons, before they had the opportunity to participate in any further nuclear reactions. Once the universe had cooled to approximately 10^9 K, however, the deuterons could survive. Under the conditions prevailing early in the nucleosynthesis epoch, deuterium easily fuses with a proton, or with another deuteron, to form the helium nucleus ^{3}He, or else with a neutron to form tritium, ^{3}H. Both of these nuclei can then react with additional particles, the ^{3}He with a neutron or a deuteron, and the tritium with a proton or a deuteron, to form ^{4}He, the most common isotope of helium. Almost all the helium in the universe, including that in the Sun, was created at this stage after the big bang.[7] That any hydrogen at all survived to the nucleosynthesis epoch tells us that the early universe must have been filled with hot radiation. If no energetic photons had been present at the end of the hadron epoch, when the numbers of neutrons and protons were approximately in balance, then all the protons in the universe would have combined immediately with the neutrons, subsequently continuing to fuse on to helium, and leaving no hydrogen behind. It was this realization that led Gamow, Alpher, and Herman to propose that the early universe must have contained a billion photons per particle of matter. They also recognized that after nucleosynthesis ceased, these photons would continue to permeate the universe, redshifting to ever-lower temperatures. From this, they predicted the existence of the background radiation, more than a decade and a half before the CBR was actually discovered and fully two decades before the thermal history of the early universe began to be understood in terms of particle physics.

At the beginning of the nucleosynthesis epoch, no nuclei existed. Instead there was a mixture of free neutrons, protons, and other particles. But a free neutron is unstable; it decays into a proton and an electron. (Under laboratory conditions, this occurs with a half-life of about 10.5 minutes.) During the nucleosynthesis epoch, two processes involving neutrons were occurring: neutrons fused with protons or deuterons, while free neutrons decayed. The competition between these two phenomena controlled the abundance of helium that was able to form. The amount of helium created during the nucleosynthesis epoch is not very sensitive to the density of the matter, but depends mainly on the ratio of neutrons to protons at the beginning of this epoch. Since neutrons were already relatively rare compared to protons, and since any neutrons which did not fuse with protons decayed, single protons were left as the most abundant nucleus. (A lone proton is a hydrogen nucleus.) Even so, about 25% of all baryonic mass ended up in the form of helium by the end of the nucleosynthesis epoch, which represents a significant amount of nuclear burning. The fusing of so much hydrogen into helium did increase the temperature of the universe somewhat, but the temperature was already so high that the energy released in the fusion reactions had only a very small effect.

Hydrogen is now by far the most common element in the universe, followed by helium. Most of the rest of the elements in the universe were created in the stars, and are much less abundant. However, a few other nuclei besides hydrogen and ^{4}He emerged from the nucleosynthesis epoch, and these isotopes have important cosmological implications; they can help to constrain cosmological models, provided that we can measure the abundances that actually exist.

One important marker left over from the nucleosynthesis epoch is the deuterium abundance. The precise abundance of primordial deuterium depends *very* sensitively upon the conditions in the universe during the nucleosynthesis epoch, especially the overall baryon density. The denser the universe, the less deuterium survives from the early nuclear reactions. If we can measure the amount of deuterium present today, accounting for destruction processes, we can use this measurement to tell us the density of the universe. Like all cosmological observations, this is not an easy observation to make, but recent data have been sufficiently good to provide important constraints on the matter density in the early universe.

Most deuterium in the universe was created in the big bang, but it can be destroyed fairly easily within stars. Therefore, in order to measure the primordial abundance of deuterium, we must look for matter that has never passed through a star and that is relatively uncontaminated by any subsequent nuclear activity. Measurements of deuterium abundances have been made in the atmosphere of Jupiter, in the local interstellar medium, and in the spectra of clouds of intergalactic gas that are illuminated by light from distant quasars.[8] All of these determinations are quite consistent with one another, as astronomical data go, and give an abundance of D/H $\approx$ 1 to 4×10^{-5}, by mass, for the cosmological ratio of deuterium to hydrogen. The deuterium abundance places a limit on the density of *baryons* in the universe, since only baryons participate in the nuclear reactions that create it. However, deuterium alone sets only an upper limit to this quantity, because deuterium is so readily destroyed in stars.

Another rare species, ^{3}He, is a bit less sensitive to density, but its primordial abundance too drops off as Ω approaches unity. This isotope of helium is fairly resistant to destruction, so its observed cosmic abundance is more reliable as a measure of its primordial abundance than is the case for deuterium. Measurement of present-day ^{3}He abundances yield density estimates that are consistent with the results of the direct measurements of deuterium.

Another nucleus thought to have been produced in the big bang is the isotope of lithium ^{7}Li. Lithium in the early universe was produced in reactions such as

$$^4\text{He} + {}^3\text{H} \rightarrow {}^7\text{Li} + \gamma,$$

$$^4\text{He} + {}^3\text{He} \rightarrow {}^7\text{Be} + \gamma \rightarrow {}^7\text{Be} + e^- \rightarrow {}^7\text{Li} + \gamma.$$

This isotope can also be created in some stellar events, as well as by cosmic rays. Moreover, ^{7}Li is easily destroyed at moderate temperatures, even those in the atmosphere of the Sun. No one expected observations of ^{7}Li to be able to tell us anything about the big bang, but starting in the 1980s, it became clear that careful observations could detect this isotope in the cool atmospheres of some very old stars. From such measurements, the primordial abundance of ^{7}Li could be inferred to be ^{7}Li/H $\simeq 10^{-10}$. This is the abundance predicted for quite reasonable assumptions about the big bang. If the measurement truly indicates the primordial ^{7}Li, it represents a powerful vindication of the accuracy of the standard cosmological model.

Taken together, the synthesis of the light elements in the early universe places a fairly stringent constraint upon the total density of baryons in the cosmos. Nucleosynthesis is almost entirely controlled by the temperature of the universe, and by the ratio of neutrons to protons. Model-dependent factors, such as the expansion rate and the geometry, influence nucleosynthesis only indirectly, by affecting the cosmic time at which the universe reaches the appropriate temperatures, as well as by controlling the density of nucleons and the neutron/proton ratio at the initiation of fusion. The major limiting factor to nucleosynthesis is the neutron, since all nuclei beyond hydrogen must contain at least one neutron. The more neutrons that decay before combining with protons, the smaller the abundance of heavier nuclei; this in turn depends upon factors such as the expansion rate. The earlier fusion begins, the more neutrons are available for the construction of heavier nuclei. Density, which is a function both of expansion rate and of the geometry, determines reaction rates, for both creation and destruction of nuclei. Deuterium is particularly sensitive to the density, since it is so easily destroyed if it interacts with other particles; thus, the higher the density, the smaller the abundance of deuterium.

All the cosmological effects, as well as all the complications of nuclear and particle physics, must be taken into account if we want to compute very precise values for the abundance of helium and the other nuclei produced shortly after the big bang. Most such calculations are performed by solving detailed models with the aid of computers. The inputs include such parameters as the density of nucleons and photons, the availability of neutrons, and the probability of occurrence of a given nuclear reaction. The probabilities of various reactions are known to very good accuracy, so the output can be used with confidence to predict conditions during the nucleosynthesis epoch. The result of such calculations is a set of curves of the predicted abundance for each nucleus as a function of the density of the universe. If we superimpose the observational limits on a plot of these curves, we can determine whether there is overlap of the measurements of the primordial abundances of the different nuclei created in the big bang. (If there were no overlap, then our model would be inconsistent in some way, or some of the data must contain errors which had not been taken into account.) The overlap of the measured abundances shows us the range in which the density of the universe is permitted to lie. When we carry out this exercise, we find that the present baryonic density of the universe is very tightly constrained; it cannot be far from $\rho_B \approx 10^{-28}$ kg/m^3.

The "raw" density in baryons is not of much interest, however. In order to learn anything about the geometry of the universe from nucleosynthesis, we must compare the measured density of baryons to the critical density. That density, expressed as a fraction of the critical density, $\Omega_B = \rho_B/\rho_c$, most likely lies within the range of roughly $0.01 \lesssim \Omega_B \leq 0.2$. One major contribution to the indicated uncertainty in Ω_B is the value of the Hubble constant. When we compare the measured density to the critical density, then we are implicitly dividing by a factor of H^2 (cf. equation [11.11].) We shall henceforth follow convention and subsume our ignorance into a new parameter, h, where

$$h \equiv H_0/100, \tag{13.8}$$

with H_0 expressed in the usual units of km/s/Mpc. All things considered,

$$\Omega_B \approx 0.015h^{-2} \tag{13.9}$$

is a good working estimate.

The determination of the density from the deuterium abundance is relevant to the *total* mass density of the present universe only if most or all of the matter is in baryons. This may not be the case; it is possible that most of the mass in the universe is due to some sort of exotic particle, in which case the deuterium limit can tell us only what fraction of the total is in baryons; it places no bounds whatsoever upon the mass density due to non-baryonic matter. If our observations show that the density of the universe is greater than baryons can provide, then we must conclude that some other form of matter, which does not participate in nucleosynthesis, is present.

Interesting information can also be obtained from measurements of the primordial abundance of ^{4}He. Carrying out such a measurement calls for some care. Although most of the present ^{4}He in the universe was created in the big bang, it can also be produced in stellar nuclear reactions, so merely measuring its abundance in stars such as the Sun, which formed from the debris of stellar explosions, would not be adequate. Galaxies that have few "metals" (in astronomical usage, all elements heavier than helium), and thus have experienced relatively little stellar nucleosynthesis, will not have created much additional ^{4}He. Careful observation of such galaxies yields an abundance of between 0.22 and 0.26, by mass, for the fraction of primordial helium, relative to hydrogen.

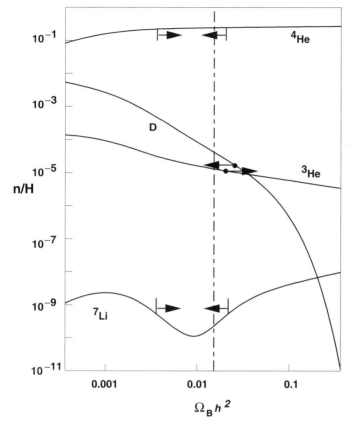

Figure 13.5 Big bang nucleosynthesis abundances expressed as a fraction of hydrogen, plotted as a function of the baryon density parameter (Ω_B) times the square of the Hubble parameter ($h = H_0/100$). The solid curves are the predicted values; the arrows indicate various upper and lower limits from observations of the abundances of these light-element isotopes. The vertical line is the "best fit" value of $\Omega_B h^2 = 0.015$. The rapid change in deuterium abundance as a function of density makes it a sensitive measure of the density parameter. (Adapted from T. Wilson and R. Rood 1994.)

As it happens, the abundance of helium helps hardly at all to pin down the geometry of the model. For helium, the higher density of a closed model, which would tend to increase the rate of fusion compared to the lower-density open model, is mostly counterbalanced by the slower expansion rate of a closed model, leading to later onset of nucleosynthesis, and hence fewer available neutrons by the time the temperature has dropped to the point that nuclei can survive. What is remarkable, however, is that the abundance of ^{4}He *can* be used to restrict the number of species of neutrinos. This is possible because the number of neutrino species affects both the expansion rate during nucleosynthesis, and also the temperature at which the ratio of neutrons to protons "freezes." The more neutrino species, the faster the expansion, and the earlier the freezing of the neutron-to-proton ratio, which results in more neutrons relative to protons. Both faster expansion and, especially, earlier "freezeout" tend to increase the production of ^{4}He. Therefore, the more neutrino species, the greater the final abundance of ^{4}He. The best estimate for the cosmic ratio of ^{4}He to hydrogen is close to 0.24 by mass. This, together with the observed abundances of deuterium and lithium, indicates that there are three species of neutrinos, which happens to be the number that has been detected. By the same token, the ^{4}He abundance must be at least near 0.23, since a lower value would require that only two species of neutrino exist, and we know of three. There are also theoretical reasons from particle physics to expect that exactly three species of neutrino should exist; thus the concordance of the helium data is comforting.

The observations of helium, deuterium, and lithium abundances show remarkable agreement with the predictions from the standard model of cosmology, independently indicating very similar numbers for various parameters that enter into the model. This consistency cannot be taken lightly, and we must be careful before we start to tinker with such success. On the other hand, the standard model is so constrained by the observations that the discovery of anything very much out of the ordinary, such as another species of neutrino, could force us to abandon the model. This is, of course, the hallmark of a scientific theory: it is falsifiable.

We conclude, then, that the simplest model of the thermal history of the universe outlined here, is quite successful at explaining the present abundances of light elements and their isotopes. There are still many open questions that address important issues in modern cosmology, some of which will be discussed in the following chapters. Nevertheless, the standard model of big bang cosmology must be regarded as a great achievement.

From Light to Dark: the End of the Radiation Era

At the end of the nucleosynthesis epoch, the universe contained massless photons and neutrinos, as well as ordinary matter in a state called a **plasma**, consisting of electrons, protons (hydrogen nuclei), helium nuclei in the form of ^{4}He (approximately 25% by mass), and traces of such light nuclei as deuterium, ^{3}He, and lithium. A plasma, in this context, is matter that is ionized, meaning that the negatively charged electrons have been stripped away from the positively charged nuclei, and both the nuclei and the electrons act as independent particles. Essentially, the temperature is sufficiently high that the electrons have too much energy, and thus are moving too fast, to be captured by the electrostatic attraction of the nuclei. The cosmic plasma was still in thermal equilibrium with the photons at this point, meaning that each of the species of matter freely interacted with the photons; the radiation and the matter both had the same characteristic temperature. Most important, the free electrons scattered the photons randomly in all directions. With the matter density still quite high, a photon could not have traveled far before encountering a free electron and being diverted from its original path; light emitted from any point would quickly end up scattered into random directions. It would have been impossible to see very far through this dense plasma; during the radiation era, the universe was opaque, like an unimaginably hot fog.

As the universe continued to expand, the temperature dropped in proportion to the scale factor. The mass density also fell, decreasing as the scale factor cubed, while the radiation energy density continued to diminish like the fourth power of

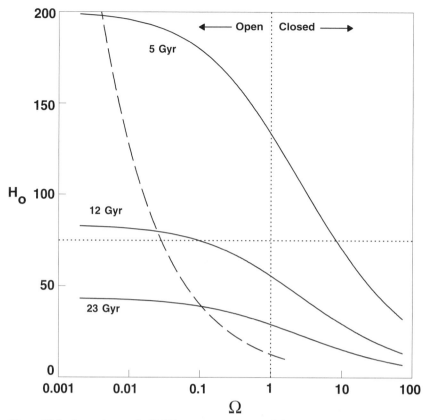

Figure 13.6 Constraints on the Hubble parameter, the age of the universe, and the baryon density limit given by big bang nucleosynthesis are combined on a plot of Hubble constant versus Ω. The solid lines indicate model parameters for universes with the indicated ages. The dashed line is $\Omega_B h^2 = 0.015$. The vertical dotted line indicates the critical density, which separates the closed and open universe models, while the horizontal dotted line marks the value $H_0 = 75$ km/s/Mpc, currently a good estimate from the latest observations. The implication is that the universe must be open if it is a standard type ($\Lambda = 0$), unless additional mass due to nonbaryonic sources (e.g., massive neutrinos) is present. However, an $\Omega = 1$ standard universe cannot be reconciled with globular cluster ages greater than 12 Gyr unless H_0 is smaller than current observations tend to indicate.

the scale factor. Although the radiation dominated the universe at the beginning, its energy density dwindled more quickly than that of matter, and eventually the matter became more important. The point of equality between the mass density and the radiation energy density occurred sometime around 10^{11} s. The time during which the matter and radiation densities were comparable can be called the **epoch of equal density**. During this interval, it is a bit more complicated to derive the scale factor as a function of time in a closed form; we will not write down the formula, as it is not particularly illuminating on its own. You would probably guess that during this epoch the scale factor was between $t^{1/2}$ and $t^{2/3}$, with the expansion occurring more like $t^{1/2}$ (characteristic of a radiation-filled universe) near the beginning of the epoch and making a smooth transition to $t^{2/3}$ (characteristic of a matter-dominated universe) by its end.

The epoch of equal density did not last long, because expansion was still quite rapid at this early stage. An important event can be attributed to this epoch, however. During the radiation era, the constant interplay between matter and radiation ensured that the plasma remained mostly smooth and homogeneous. But once the tight coupling was lost, it became possible for the matter to bunch together into clumps. These hypothetical clumps are believed to be the seeds for the structures that formed later in the universe, namely the galaxies, clusters of galaxies, and any other great agglomerations that are present. Thus this was the time of **structure formation**, a phenomenon which will be discussed in more detail in the next chapter.

After the epoch of equal density, the radiation energy density soon became negligible in determining the overall gravitation, and the universe entered the matter-dominated era, which has lasted till the present day. The next great landmark was reached when the temperature cooled to approximately 3000 K, sometime around 10^{13} s. Below this temperature, electrons are no longer moving fast enough to escape from the electric fields of the nuclei; the conditions then became suitable for most of the free electrons to be captured by the protons to form hydrogen atoms. This occurrence is known as **recombination**.[9] After recombination, very few free electrons remained to scatter photons, so the photons streamed freely through the universe. This had two effects. First, the universe became transparent; it finally became possible to see for great distances. Second, radiation and matter ceased to be in thermal equilibrium. From this point onward, the radiation would simply cool down with the expansion of the universe, with little regard for what the matter was doing.

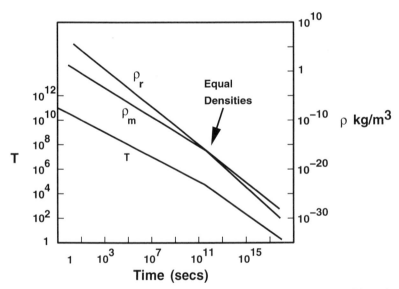

Figure 13.7 Time evolution of the temperature and densities over the history of the universe. The scale on the left is the temperature of the universe; the scale on the right is for density. The point at which the matter and radiation energy densities (ρ_m and ρ_r, respectively) are equal occurs around 10^{11} s after the big bang.

The photons that became free-streaming during the recombination epoch now constitute the CBR. Since the time of recombination, the background photons' energy has redshifted, with the peak wavelength dropping from the visible portion of the spectrum, where it was located at recombination, down to the microwave region. These are the photons first detected in 1964 by Penzias and Wilson. The present temperature of the background radiation is close to 3 K, and recombination occurred when the temperature was around 3000 K; from these facts, the redshift formula leads to the conclusion that recombination took place approximately at $z = 1000$. This redshift represents a fundamental limit on our ability to look into the past with telescopes; we will never be able see directly through the impenetrable state of the universe that existed prior to this time. The last instant that the universe was opaque is effectively the "edge" of the universe for us. This "edge" is called the *surface of last scattering*. Particle physics was over within a few seconds after the big bang; the hot, opaque plasma then dominated the universe for an interval on the order of a million years. We can only hope to understand what happened before recombination, and particularly in the first few seconds, by studying the imprint of the events which occurred then upon the universe we observe now.

The Age of Galaxies

Sometime after the epoch of equal density, the matter perturbations, which are revealed in the small temperature fluctuations seen by *COBE*, were able to pull themselves together rapidly. The details of how this process occurred are still almost complete mysteries; we shall discuss what is known in more detail in the following chapter. Regardless of how the collapse occurred, and no matter whether clusters or galaxies are the more fundamental aggregations, galaxies must have formed fairly quickly after recombination. Quasars are among the most ancient objects we can see. There is significant evidence that quasars are at least associated with galaxies, perhaps even

TABLE 13.1
Great Moments in History

Epoch	Time (secs)	Major Events
Planck	0–10^{-43}	All forces unified
GUT	10^{-43}–10^{-35}	Baryogenesis
Inflation	$\sim 10^{-37}$	Exponential increase in R
Quark	10^{-35}–10^{-6}	Universe of fundamental particles
Electroweak	$\sim 10^{-11}$	Weak and EM force decouple
Hadron	10^{-6}–10^{-4}	Matter excess frozen in
Neutrino decoupling	~ 1	Background neutrinos go free
Lepton	10^{-4}–10	Proton/neutron ratio frozen
Nucleosynthesis	~ 100	Light atomic nuclei formed
Radiation	10–10^{11}	Scale factor goes as $t^{1/2}$
Recombination	$\sim 10^{13}$	Universe becomes transparent
Matter	10^{11} to present	Galaxies, stars, life

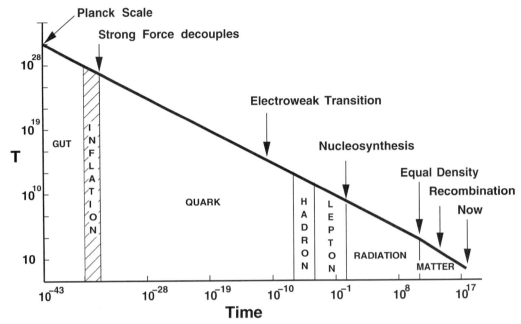

Figure 13.8 Great moments in the history of the universe. Important events and epochs are shown along a line indicating radiation temperature as a function of time.

located in the centers of early galaxies. Many recent images of very distant quasars clearly display traces of spiral arms surrounding the bright core. In other cases, especially among relatively nearby objects, quasars are members of clusters of galaxies, all with the same redshift.

There is also evidence for the very early formation of stars. High abundances of iron, even greater than in the solar neighborhood, have been found in the emissions of quasars at $z = 3.4$ and greater. Since iron originates only in stars and supernovae, this observation indicates that the quasar is an environment in which stars must have existed, strengthening the identification of at least some quasars with galaxies. The formula for the lookback time in the Einstein–de Sitter universe (cf. review question 13.12) tells us that $z = 3.4$ was only about 10% of the age of the universe. For $H_0 = 75$ km/s/Mpc, the universe would have been only about 1 billion years old when the light left a quasar at this redshift.

Even more telling is the discovery of clearly normal galaxies at great distances, corresponding to very early times in the history of the universe. An object that is most likely a galaxy is known to have a redshift of 4.25. This galaxy was formed within little more than a billion years after the big bang. Nor does this ancient galaxy seem anomalous; improved observations, especially from the *HST*, have recently found a population of comparably old galaxies. The *HST* is able to resolve galaxies at incredible distances, and some remarkable data have resulted. The most distant galaxies seem fragmentary and inchoate, as might be anticipated; anything we call a "galaxy" is, by definition, a fairly well-defined object containing highly organized structures, such as stars, globular clusters, and possibly spiral arms. A little later than the era of these primitive objects, after approximately 10% of the age of the universe, galaxies appear to have formed, although not necessarily in their final configurations.

Elliptical galaxies seem to have settled into their characteristic shapes almost immediately; an object which is nearly certain to be an elliptical, and which is apparently in its final stages of formation, has been found at $z = 3.8$. Since one of the distinguishing traits of ellipticals is a relative absence of dust, gas, and recent star formation, this finding is consistent with their present-day appearances. Ellipticals apparently underwent a rapid collapse, then settled into quiescence. Some giant ellipticals, such as the galaxy M87 near the center of the Virgo Cluster, might have continued to grow, perhaps by swallowing nearby galaxies in their tight clusters; at earlier times the galaxies were much closer together, and even in the recent universe, collisions and mergers are known to occur, especially in rich clusters that have strong gravitational attractions near their centers. Spiral galaxies seem to have led more eventful lives than did most ellipticals, especially those spirals which inhabit denser clusters. Spirals appear to have experienced intense and repeated bursts of star formation, perhaps triggered

Figure 13.9   The Hubble Deep Field image. This image was obtained by pointing the telescope into a relatively "empty" part of the sky and taking a very long exposure. Almost every smudge of light in the picture is a distant galaxy. (STScI/NASA.)

by near misses, or even outright collisions, with their neighbors. Some spirals are destroyed by such encounters, while others sprout ringed and spoked arms marked by bright young stars.

The early appearance of established galaxies presents a puzzle in modern cosmology. Existing models of galaxy formation tend first to form structures either smaller than galaxies or else considerably larger. In either case, the appearance of galaxies would require a significant lapse of time after the primordial density enhancements pulled themselves together; in the former, the smaller structures would have to merge to create larger ones, while in the latter, the larger structures would have to fragment into galaxy-sized chunks. The current models of galaxy formation have difficulty producing galaxies so early in the history of the universe, yet there they are.

The problem of structure formation is intimately tied to other issues in cosmology, including the initial conditions of the universe and the possible presence of dark matter. The prospects for genuine progress have never been better, however. Cosmology has historically been one of the most data-starved of sciences, an unfortunate circumstance which forced theory to advance almost blindly, with little help or discipline from observations. Recent advances in telescope technology, especially the refurbishment of the *HST*, have drastically changed this situation; for the time being, observations have begun to gain the upper hand on theory. The most exciting times in any science always occur when new data create an uncomfortable quandary for theory. In the last decade of the twentieth century, cosmology has suddenly found itself in such a condition, although it must also be emphasized that most of the accepted models, while certainly not perfect, still show no deeply irreconcilable differences with observations. Perhaps we are now on the verge of a truly thorough understanding of the universe.

KEY TERMS

thermal equilibrium
pair production
field
spontaneous symmetry
  breaking
Planck epoch
hadron
baryon
quark epoch
lepton epoch
plasma
recombination

matter era
quantum mechanics
graviton
Planck time
unified epoch
lepton
gluon
electroweak
  interaction
nucleosynthesis epoch
epoch of equal density

radiation era
Compton wavelength
symmetry
Planck length
grand unified theories
  (GUTs)
quark
baryogenesis
hadron epoch
deuterium
structure formation

Review Questions

1. Review the redshift formulae: Consider the universe at a redshift of $z = 2$. What is the average matter density in the universe then compared to now? What is the average radiation energy density then compared to now? What is the temperature of the background radiation then? Suppose the energy density in matter, ρc^2, is now 2000 times greater than the present energy density in radiation. At what redshift did matter and radiation have equal energy densities?

2. Describe what is meant by thermal equilibrium. What simplifications does this state of equilibrium allow us to make in our description of the early universe? What does equilibrium imply about the rates of reactions, as well as the numbers of photons and massive particles?

3. The electron has a mass of 9.11×10^{-31} kg. What is the threshold temperature, in degrees Kelvin, for an electron to be created from energy, if we can assume thermal equilibrium obtains? What is the Compton wavelength of the electron?

4. (More challenging.) At the Planck time $t = 10^{-43}$ s, the temperature is 10^{32} K. The formula for the scale factor during the radiation-dominated era can be written

$$R(t) = R_i (t/t_i)^{1/2},$$

where R_i is the Planck length and t_i is the Planck time. Assuming that this formula holds for the duration of the radiation era: (a) Compute the temperature at cosmic time $t = 10^{-33}$ s (the approximate time of the GUT symmetry breaking). (b) Compute the redshift from now back to this GUT time. (c) Compute the Hubble length at this GUT time ($c = 3 \times 10^8$ m/s).

5. What would happen if there were significant accumulations of antimatter anywhere in the universe? Why do such antimatter accumulations seem unlikely to exist? What significance does this conclusion hold for particle physics in the unified epoch?

6. Why did the lepton epoch end with more protons than neutrons in the universe?

7. What can observations of the cosmic deuterium abundance tell us about the big bang? Why is deuterium a particularly good nuclide to use for this purpose? List at least two complications that make the measurement of deuterium an uncertain proposition.

8. Characterize the following intervals in the history of the universe: matter era, radiation era, nucleosynthesis epoch, lepton epoch, hadron epoch. For each of these intervals, what were the main components of the universe, and what were the most important phenomena which occurred? What events ended each interval?

9. If a fourth species of neutrino were discovered, what might it mean for the standard hot big bang model?

10. What was the era of recombination? What happened then, and why is it significant? What is meant by the "surface of last scattering"? How does this surface limit the information we can obtain about the universe?

11. What is the connection between the early universe and the CBR?

12. Why do we believe that galaxies formed very early in the history of the universe? Use the formula for the lookback time in the flat Einstein-de Sitter model,

$$t_{lb} = \frac{2}{3H_0} \left[1 - \frac{1}{(1+z)^{3/2}} \right].$$

for redshift z to estimate, relative to recombination, when galaxies developed. How does the early appearance of galaxies create a problem for current models of their formation?

Notes

1. Specifically, the energy E of a photon with frequency ν is given by $E = h\nu$, where h is a fundamental constant of physics known as Planck's constant. See chapter 4 for further details.

2. Wave-particle duality is discussed further in chapter 16.

3. Carrier particles and their role in mediating forces are discussed briefly in chapter 4.

4. These scales are named for Max Planck, in honor of his discovery of the quantum behavior of light, for which he received the 1918 Nobel Prize in physics.

5. *Lepton* comes from the Greek *leptos*, meaning thin or small.

6. However, if they do have rest mass, these cosmic neutrinos can play a very important role in the evolution of the universe. This will be discussed in chapter 14.

7. The helium found on Earth, on the other hand, is nearly all due to the radioactive decay of atoms deep in the Earth's interior; our "primordial" helium is long gone. The Earth is too warm, and its surface gravity is too weak, to retain atmospheric helium.

8. There is deuterium in the Earth's oceans, which one day may provide the fuel for nuclear fusion reactors, but in the oceans, deuterium is 10 times more common than in the cosmos as a whole. Deuterium is chemically favored over ordinary hydrogen to form water molecules. In the very early Earth, most of the deuterium became bound into water molecules and was thus unable to escape from the atmosphere, so its Earthly abundance cannot tell us anything about the early universe.

9. Since there were no hydrogen atoms previously, they must have, in fact, simply combined for the first time; nevertheless, "recombination" is the standard terminology.

PART V

The Continuing Quest

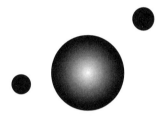

CHAPTER 14

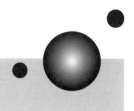

Dark Matter and Large-Scale Structure

The standard model of cosmology is very successful as a scientific theory. Until quite recently, it would have seemed unthinkable that any theory of physics, much less one so elegant and simple, would enable us to begin to understand the universe as a whole, with all its daunting complexities. The standard model makes definite predictions about observable quantities such as elemental abundances, and those predictions have been borne out to a degree that is, to some extent, surprising. The agreement with observations is so good that the standard model is the benchmark by which more sophisticated theories are evaluated. In many cases, cosmology has put rather severe constraints upon particle physics, guiding the development of our understanding of conditions we may never be able to replicate on Earth.

Of course, many questions remain to be answered. We do not yet know whether the universe is open or closed. We are not certain whether ordinary matter, or something more exotic, is the dominant constituent of the universe. At an even more fundamental level, we have assumed all along that the universe is, at its largest scales, smooth, isotropic, and homogeneous, and have found good evidence that this is true. But *why* should it be so? There is no apparent *physical* compulsion for isotropy or homogeneity. Moreover, at smaller scales, the universe most certainly is neither isotropic nor homogeneous. Galaxies are not identical; beyond that, they form clusters, and even the clusters are organized into superclusters. The better our telescopes become, the farther we can see, and the larger the structures that seem to appear. At what point does the assumed isotropy of the universe, so well measured by the *COBE* observations, begin? How can we explain the existence of structure within an isotropic universe? What can the dynamics of galaxies tell us about the matter distribution in the universe?

The standard model is fully self-consistent, and these questions do not represent a failure of the model. Rather, they stem from the assumptions that were built into it from the beginning and from the incompleteness of our knowledge. For example, the standard model *assumes* isotropy and homogeneity and thus is not able to explain how isotropy could develop. Our incomplete understanding of particle physics, and our inability to perform experiments beyond an energy scale corresponding roughly to that of the hadron epoch of the early universe, further hampers our construction of a complete cosmological model. We could abandon the effort and claim that the standard model is the best theory of the universe we can construct. This is an unsatisfying resolution, however; we have been able to advance as far as we have come in cosmology by insisting that the universe is knowable. Therefore, let us ask the difficult questions and begin to seek answers to them. Even if our first efforts may prove to be faulty, perhaps they will point the way to better explanations.

Weighing the Universe

Within the framework of the standard model, the most obvious uncertainty is the density of the universe; is it greater than, less than, or equal to the critical value? Equivalently, is the universe closed, open, or flat? How might we "weigh" the universe? In previous chapters we have seen that the density of the universe is related to the major cosmological parameters, such as the Hubble constant and the deceleration parameter. Methods that measure the mass density indirectly, by determining the overall expansion rate and structure of the universe, are called **kinematical methods**. (Figure 12.10 illustrates one such method.) The major drawback to these kinematical methods is that they require precise determinations of distance and motion over a wide range of redshift, and such data are very difficult to obtain. So far, kinematical methods have failed to pin down Ω any better than to a value between 0.01 and 2, a range that encompasses all three standard models.

Another means of measuring Ω is to determine directly the mass contained within some representative volume of the universe. The procedure is straightforward: select a volume of space, count up the galaxies within it, and multiply by the mass per galaxy. A galaxy represents a significant, localized density enhancement, perhaps by a factor of 10^5 relative to the overall cosmic density, so there may be some question as to whether the galaxies are representative of the overall matter distribution in the cosmos. However, it may certainly be the case that galaxies contain most of the matter in the universe. If this is so, we can weigh them and then average their densities over all space, in order to obtain Ω. This method is simple in principle, but unfortunately, there is no way to measure the mass of a galaxy directly. The most easily observed feature of any galaxy is its light. For nearly all galaxies, most of the light comes from stars, with some contribution from glowing regions of hot gas. Thus the amount and spectrum of the light from a galaxy gives us an estimate of the number and type of stars. A knowledge of the distribution of stars enables us to estimate the mass of the luminous matter, since we know quite accurately the mass of a star with a given luminosity. If we can then somehow obtain an estimate of the total mass of the galaxy, we can determine the ratio M/L, which is generally called the **mass-to-light ratio**.

Although only luminous matter is directly visible, all the mass in a galaxy will make itself known through its gravitational influence. We can use the gravitational interactions of galaxies, gas, and stars to infer the mass density required to produce such motions. Methods which depend on observing the dynamic interactions of stars and galaxies are known as **dynamical methods**. For example, if the galaxy is rotating, then we can exploit Kepler's laws to measure the total mass in the galaxy. Combining the rule for centripetal force with Newton's law of universal gravitation shows that a body in a Keplerian orbit at radius r from the center of the galaxy obeys

$$GM(r) = v^2 r, \tag{14.1}$$

where $M(r)$ is the total mass within radius r.

Spiral galaxies rotate, making this technique mainly suited to them. One of the characteristics of a spiral galaxy is a fairly sharp dropoff in the light distribution at some radius. If we measure the rotational velocity at this radius, we can determine the total mass within the luminous portion of the galaxy. The results for most spirals, including the Milky Way, indicate that the ratio of mass to luminosity, expressed in terms of the solar mass-to-light ratio, is approximately

$$(M/L)_{\text{spiral}} \sim 10\text{--}30 \; M_\odot / L_\odot. \tag{14.2}$$

So far we have discussed only spiral galaxies, but elliptical galaxies are also a major constituent of the universe. It is more difficult to obtain estimates of the total mass of elliptical galaxies than in the case of spirals, since by and large, ellipticals do not show any systematic rotation. One approach is to measure the *dispersion* of the velocities of stars, that is, the extent of the range of velocities about the mean velocity. We can then make use of a simple formula called the **virial theorem**. This is a statistical result for gravitating systems that relates a measurement of the velocity dispersion to the mean gravitational field. The virial theorem is somewhat similar in concept to hydrostatic equilibrium in a ball of gas, such as a star. In a star, the atoms must move fast enough (i.e., have an adequately high temperature) to generate sufficient pressure to resist gravitational collapse. In a galaxy, the stars do not collide, so there is no "gas pressure," but they still must keep moving with an average velocity high enough to avoid gravitational collapse. This statistical rule provides an estimate of the collective mass of the system. Such measurements indicate that the M/L for most elliptical galaxies is comparable to that for spirals, $M/L \sim 20 M_\odot/L_\odot$.

A mass-to-light ratio considerably greater than unity implies that most of the matter of a galaxy is much less luminous than the Sun, and the Sun is not an unusually bright star. Where, then, is this dark, or perhaps dim, matter? We are aware of the existence of *some* dark matter, specifically, the Earth and its fellow planets, but compared to the Sun the dark members of our solar system contribute negligible mass; planets alone are not likely to account for the invisible baryonic matter. Might it be found in multitudes of faint stars? The luminosity of a star on the main sequence is approximately proportional to the third power of its mass.[1] The rapid increase of luminosity with mass means that bright, massive stars are responsible for most of the light output of a galaxy; but stars of greater than about 2 $M_\odot$ are extremely rare. Because of this, low-mass stars, the *red dwarfs*, contribute disproportionately to increasing the mass-to-light ratio. Such stars are quite abundant; the Sun is actually near the upper end of the luminosity range of the relatively common stars. How can we determine whether the mass-to-light ratio for galaxies can be explained by these objects? One approach would be to study all the stars near the Sun, where even low-luminosity stars should be detectable. Such observations have shown that the median mass for nearby stars is approximately one-third the solar mass, and the overall mass-to-light ratio in the solar neighborhood is roughly

$$(M/L)_{\text{nearby}} \sim 3 \ M_{\odot}/L_{\odot}. \tag{14.3}$$

If the stars around the Sun are representative, as we believe them to be, it does not appear that dim, normal stars alone can explain the observed M/L for galaxies as a whole.

If not stars, then what might explain the mass-to-light ratios of galaxies? Some of the unseen matter could take the form of dim, compact cinders such as white dwarfs; their luminosity, at a given mass, is quite low. Neutron stars would be even better, for unless they beam as a pulsar directly toward our line of sight, they emit very little light and are nearly invisible. Best of all would be massive black holes, which could contribute a fairly large amount to the total mass without increasing the luminosity at all. The mass-to-light ratio of such dark objects would be essentially infinite! There is evidence that a realistic distribution of these stellar remnants, along with the main sequence stars, can explain the mass-to-light ratio of $M/L \sim 10 \ M_{\odot}/L_{\odot}$ that is observed in the disks of spiral galaxies. However, the *overall* M/L for spiral galaxies tends to lie more toward the middle to the upper range in equation (14.2). White dwarfs seem to be too scarce to contribute significantly to these larger ratios of mass to light; neutron stars and black hole candidates are thought to be even rarer.

Faint stars and white dwarfs are not the only way to increase M/L; there might also exist a huge population of small, substellar objects with little or no luminosity. These dimmest stars, the so-called *brown dwarfs*, would each have perhaps a few percent of the mass of the Sun but very little luminosity. From the general observation that smaller objects are more common, it might be naively expected that many of these brown dwarfs might exist. Such stars are subluminous even for their tiny mass; they could, in sufficient numbers, make a very significant contribution to the large M/L. They are very difficult to observe, precisely because they are so dim, but determining their density is quite important to cosmology. Their existence has been confirmed recently; one ultra–low–mass star, the faint companion of a star called Gliese 229, has been detected by the *HST*. (This object is shown in Fig. 5.2.) The assumed abundance of brown dwarfs has been called into question, however. Star surveys have shown an unexpected cutoff to the mass distribution of self-luminous stars. Stellar theory had predicted that the minimum mass of an object that could ignite thermonuclear fusion at its core, and thus could be defined to be a "star," should be about 0.08 $M_{\odot}$. Recent surveys instituted to look for these brown dwarfs, as well as for very low mass main sequence stars, have turned up scarcely any stars with masses less than approximately 0.2 $M_{\odot}$, certainly far fewer than had been predicted. Perhaps some phenomenon that is not taken into account in the simplest theories of star formation places an effective lower limit upon stellar masses. Further explanation of the apparent abrupt cutoff in stellar masses must await improvements in our understanding of the process of star formation.

Other small-massed objects are often referred to as "jupiters." Like the planet Jupiter, they have some significant mass, but do not shine. We have scant knowledge of the distribution of "jupiters," but such objects most likely contribute very little to the total mass of a galaxy, since even the archetypal Jupiter has only a tiny fraction of the mass of the Sun. What, then, about matter that has not assembled itself into either stars or planets? We know that spiral galaxies contain a great deal of interstellar gas and dust which, along with the stars, contributes to the $M/L \sim 10$ observed in the disk of the Milky Way. Unfortunately, estimates of the total mass of such matter fall far short of what would be required to make much of a contribution toward an $M/L \sim 30$. The conclusion is that the M/L of stars, gas, dust, and dwarfs can easily account for most of the mass-to-light ratio within a spiral's galactic disk itself; yet overall we are still missing quite a bit of mass. This is the first hint of the so-called "missing mass" problem.

From data such as mass-to-light measurements, we can construct an estimate of the portion of the mass density of the universe contributed by visible matter. Overall, estimates of luminous mass from light output and distribution indicate that

$$\Omega_L \simeq 0.005\text{--}0.01. \tag{14.4}$$

The Hubble constant does not appear in this expression, because the density obtained from studies of the luminosity of galaxies implicitly contains a distance scale; in the division to obtain the density parameter, the Hubble constant is then divided out. The result (14.4) shows clearly that luminous matter is a long way from closing the universe, and this conclusion is essentially independent of the value of the Hubble constant.

Even without the dynamical estimates, we have already seen that measurements of nuclide abundances, such as deuterium, imply a density due to baryons of

$$\Omega_B \simeq 0.015h^{-2}, \tag{14.5}$$

where $h \equiv H_0/100$ is the "ignorance parameter" defined previously. Comparing equations (14.4) and (14.5), we can immediately conclude that, for a Hubble constant in the most likely range, the overwhelming majority of the baryonic matter in the universe is not luminous. All matter that we cannot detect directly because it emits very little or no light, makes up the **dark matter** of the universe.

Observations more direct than nuclide abundances provide evidence of significant amounts of dark matter. The rotation curve (the rotation rate as a function of radius) of a spiral galaxy makes this clear. If the luminous matter accounted for all or most of the gravity, then beyond the outermost circle of stars, $M(r)$ would become constant, and orbital velocities v would decrease with increasing r according to equation (14.1). Furthermore, if essentially all the mass of a galaxy were contained within its luminous regions, then we should see a clear trend of decreasing velocities as we approach the edge, even before reaching the outlying stars. Although it becomes more difficult to track rotation curves where we cannot see any stars, there are ways in which orbital velocities beyond the edge of the main disk of a spiral galaxy can be studied. In most cases, it is possible to find stray stars and globular clusters outside the disks of very nearby spirals. Another valuable technique, still applicable only to spirals, is to observe the radio emissions of neutral hydrogen. This emission can be detected well beyond the visible edge of the disk, which immediately indicates that at least some gas surrounds typical spirals beyond the boundaries of their luminous matter. The radio output of these gas envelopes is relatively weak, however, suggesting that the gas is rather tenuous, and thus perhaps not a very significant mass contributor.

The interesting conclusion from observations to date is that for most spiral galaxies, there is no evidence for *any* decrease of orbital velocity with radius. The velocity

$v(r)$ is roughly constant, or even increases, as r increases. By equation (14.1) it follows that $M(r)$ must also continue to increase with radius, despite the dearth of visible stars. Thus we cannot determine the total mass of a typical spiral galaxy with any certainty at all. Galaxies must be surrounded by **dark halos** of unseen matter. Some evidence from rotation curves indicates that the halos of spiral galaxies may be spherical and may extend to a considerably larger radius than that of the visible galaxy.

What is in the halos of spiral galaxies that might account for all this dark matter? Is it composed of ordinary, baryonic matter, or might it be something more exotic? Extremely long exposure photographs of galaxies, both spiral and elliptical, in many cases have revealed faint light coming from greatly extended halos. This certainly confirms that at least some of the matter in massive halos is baryonic, since as far as we know, only ordinary matter can emit light. There are relatively few avenues for measuring the mass of these halos directly, but there is a very interesting technique, based upon gravitational lensing, which is capable of detecting compact objects in the halo of our own Galaxy. A gravitational lens effect can be produced by objects of any mass, no matter how puny; however, the smaller the mass, the more minute the bending of the light. Nevertheless, a compact object of even a modest mass can distort the light from a distant source in predictable ways. If a compact object were to pass in front of a background star, for instance, it would split the image into multiple images. These multiple images would be too small to resolve independently; the net result would be a temporary brightening of the background star. Such a gravitational effect is called *microlensing* in order to distinguish it from the lensing by a large object such as a galaxy, but there is no qualitative difference in the mechanism; the appellation simply refers to the type of lensing object. This class of hypothetical inhabitants of the galactic halo has been dubbed the **MACHO** (MAssive Compact Halo Object).

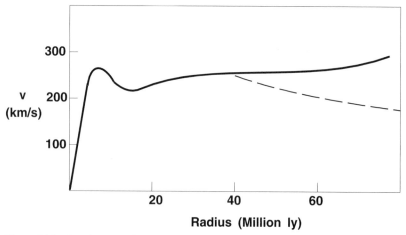

Figure 14.1 Rotation curve for a typical spiral galaxy. The dashed line corresponds to the Keplerian velocity curve that would be observed if most of the mass of the galaxy were within 40 million lt-yr. The fact that the rotation velocity continues to increase means that there is increasingly more mass present at larger galactic radii.

Although a compact object can hardly be expected to pass directly in front of any particular background star, constant monitoring of millions of background stars should find a few of these low-probability events. Several collaborations have formed, named with quaint acronyms such as OGLE and EROS, to search for microlensing phenomena. One in particular, the MACHO group, has obtained some interesting preliminary results. The MACHO project photographs the Large Magellanic Cloud, using a CCD detector array attached to a telescope in Australia. If a compact object in the halo of the Milky Way passed in front of any of the millions of stars in the Magellanic Cloud, it would produce a change in that star's brightness in accordance with the properties of a gravitational lens. Computers scan each night's photographs, looking for any changes in star brightnesses over time. So far, several lens events have been recorded. The most recent results, released early in 1996, indicate that compact objects could account for up to 50% of the mass of the halo; it is even possible that the halo could be attributed entirely to an aggregation of such objects. No particular object is specified by the data, although the observations are consistent with objects of approximately the mass of a typical white dwarf.

If these results hold up, they would tell us some interesting facts about the halo of the Milky Way. First, they would imply that the halo was mostly or entirely baryonic, since we believe that only baryons can form compact objects. Second, we would learn something about the history of the Milky Way. If the objects are truly white dwarfs, this suggests a very early, and uniform, generation of stars distributed quite differently from the newer stellar generations, which are mostly confined to the disk of the galaxy. Like all good data, the MACHO work is raising many questions, perhaps as many as it answers. Further work should help to clarify the situation.

If all the spirals have massive halos, how much might they contribute to the total density of the universe? If, as it appears, a galaxy's rotation velocity goes to a constant at large radius, then $M(r) \propto r$, from equation (14.1). Hence the total density is proportional to r^{-2}. Since the radii estimated for distant galaxies are dependent on the value of the Hubble parameter, the halo mass density scales in the same way as the critical density. The resulting dynamical estimate for the density parameter Ω is then more or less independent of the value of H_0. The best estimates of the density of the universe obtained from dynamical estimates of the mass of galaxy halos is

$$\Omega_{\text{halo}} \simeq 0.1. \qquad (14.6)$$

The techniques we have described provide estimates for individual galaxies. But what if there is substantial dark matter between the galaxies? Fortunately, we can apply similar principles to larger aggregations of matter, such as clusters of galaxies. Most galaxy clusters are almost certainly gravitationally bound; that is, their members orbit one another. For such a cluster, we can apply the virial theorem. There are still pitfalls, however. The virial theorem applies only to systems that are well approximated statistically; for a given cluster, it is not always easy to determine whether this state holds. Moreover, measurements of Doppler shifts provide only the radial velocity components, that is, the motion along our line of sight, not the full three-dimensional velocities of the galaxies; this introduces an additional uncertainty which must be taken into account. With all these caveats, and others, the results so far obtained indicate that

$$\Omega_{gc} \simeq 0.1\text{--}0.3. \tag{14.7}$$

Another method of estimating mass densities is based upon the study of large galaxy clusters that show overall infall toward their centers. The nearest such large cluster to the Milky Way is the Virgo Cluster. The Virgo Cluster is the dominant mass aggregation in our immediate neighborhood, and many smaller clusters, including our own Local Group, are falling toward it. It is possible to model this infall as a deviation from the general Hubble flow and thereby to estimate the mass of the Virgo Cluster. Such observations indicate

$$\Omega \simeq 0.1\text{--}0.2. \tag{14.8}$$

Recent work has hinted that there may be an even larger mass concentration somewhere beyond Virgo, in the direction of the constellation Hydra, toward which Virgo and all its entourage are in turn falling. If this so-called "Great Attractor" is real, the models of infall toward the Virgo Cluster are probably incorrect, as the attractor would distort the simple motions expected; this might render invalid the current estimates of the mass-to-light ratio in the Virgo Cluster. Large-scale galactic motion remains a very active area of research.

Going to larger and larger scales, we find ourselves on increasingly shaky ground to use dynamical methods. Various preliminary measurements based on techniques such as galaxy surveys, the cosmic virial theorem, and so forth, generally obtain a density parameter $\Omega \simeq 0.1\text{--}1.0$. These results must yet be regarded as tentative. It is probably safe to say, however, that none of the dynamical estimates indicate a closed, dense universe.

Summarizing all the dynamical and nuclide evidence forces us to the inescapable conclusion that most of the matter of the universe is invisible to us. The estimates from nucleosynthesis are still somewhat uncertain, and could be stretched to accommodate density ratios to perhaps approximately 0.1, if the Hubble constant turned out to be fairly small. Therefore, if we take nucleosynthesis and dynamical data at face value, and assume that all matter is baryonic, then $\Omega \simeq 0.1$. At most some 10% of it is luminous; that is, approximately 90% is dark, or nearly so, and hence unseen. The universe is open and has hyperbolic geometry, and will never recollapse.

Some of the dynamical measurements on the largest scales admit $\Omega = 1$ as a possibility. And, as we shall later discover, proponents of the inflationary cosmological model have a theoretical preference for $\Omega = 1$. If $\Omega = 1$ and all matter must be baryonic, then we would need to revise drastically our understanding of the early history of the universe. Some attempts have been made to reconcile an $\Omega_B = 1$ universe, a critical universe in which all matter is baryonic, with the nucleosynthesis calculations and the observed abundances of deuterium, ^{3}He, and ^{7}Li. For example, there is no reason, other than the usual assumption of homogeneity in the universe, that the distribution of neutrons and protons had to be completely uniform at the time of nucleosynthesis. Because neutrons have no charge, they can travel more easily than protons, which would tend to repel and scatter one another. Regions could have formed with higher-than-average neutron density. The difficulty with this scenario is that lithium is overproduced for $\Omega_B = 1$. Other nonstandard hypotheses have encountered similar difficulties in remaining within the stringent constraints of nucleosynthesis.

There is another possibility, however. If $\Omega = 1$, and we accept the constraints on baryons from the deuterium measurements, together with modern measurements which indicate that $h \sim 0.75$, then no more than 10% to 15% of the matter in the universe is baryonic, and the baryons that are visible to us constitute as little as 1% of the mass density of the universe. The majority of the mass of the universe must be nonbaryonic. Let us next examine this concept.

WIMPs

If it is true that $\Omega = 1$ and no more than 10% of the mass of the universe is baryonic, then we must hypothesize the existence of exotic forms of matter. Any nonbaryonic matter in the universe, such as the exotic particles predicted by various theories of particle physics, must remain aloof from all matter, including itself, except for its contribution to gravity; more direct interactions would have significant, and observable, effects upon the universe. Consequently, the type of particle hypothesized to account for nonbaryonic dark matter is called a **WIMP**, for *Weakly Interacting Massive Particle*.[2]

Many suggestions have been made for the identity of the possible nonbaryonic dark matter. One obvious candidate is the neutrino, which we have so far assumed to be massless. The mass of the neutrino was known to be extremely tiny from the time of its discovery; the neutrino's mass was originally postulated to be zero because there was no particular reason for the particle to have a small, yet still nonzero, mass. On the other hand, there is also no theoretical reason that the neutrino must have precisely zero mass. (This is in contrast to the photon, which must be massless for strict theoretical reasons, as experiment has verified.) As a dark matter candidate, the neutrino has the added virtue of being a particle that unquestionably exists. Neutrinos were produced in the big bang in numbers comparable to those of photons, that is, roughly a billion neutrinos per baryon. This enormous population means that even if neutrinos have a rest mass no greater than a billionth the rest mass of the proton, they would still make a large contribution to the mass density of the universe. Neutrinos alone would close the universe if their mass is about $10^{-7}h^2$ times the mass of the proton. As it happens, present experimental evidence places the upper limit on the neutrino's mass in the vicinity of this critical value. But cosmology was not the only, nor even the first, reason that physicists went hunting for the mass of the neutrino. Some of the strongest evidence for a massive neutrino comes not from any majestic theory of the structure of the universe, but right from our cosmic backyard.

According to our current understanding of nuclear theory, the Sun and other stars should emit copious amounts of neutrinos; yet only recently has "neutrino astronomy," the direct observation of these neutrinos, become barely possible. Neutrinos interact so weakly with ordinary matter that they are very difficult to detect. The first neutrino detector was built in the late 1960s, and now has taken several decades' worth of data. This detector consists of an enormous quantity of a chlorine-containing fluid commonly used in dry cleaning, buried deep underground in an old gold mine, the Homestake, in South Dakota.[3] Even though neutrino interactions are exceedingly rare, the sheer quantity of fluid in the detector means that occasionally a neutrino interacts with a chlorine atom in the fluid, resulting in a signature that can be detected. The detector must be buried in order to screen out the "ordinary" particles of cosmic rays. These particles, mostly high-energy protons, can produce events that would overwhelm and confuse the detector; however, cosmic rays cannot penetrate far into the Earth. Neutrinos, in contrast, could fly right through a sheet of lead several *light years* in thickness; a few hundred feet of rock is essentially transparent to them.

Subsequent neutrino detectors, of which several are now in operation, follow the general principles of the Homestake experiment. Some make use of the rare-earth element gallium, while others are filled with water, a substance that has the advantages of considerably less toxicity and expense than either dry-cleaning fluid or gallium. The water-based detectors use water of extreme purity. High-energy neutrinos striking electrons, and antineutrinos striking protons, lead to brief flashes of light that can be detected by photocells surrounding the fluid. Nearly all neutrino experiments are very difficult. Neutrino events are sufficiently rare that instrumental effects can be significant. A major source of error is that the shielding by the Earth is never perfect, so some spurious interaction events still occur due to cosmic rays.

The major motivation for the construction of neutrino detectors was to study neutrinos emitted by fusion processes in the core of the Sun. Ever since the Homestake experiment began to monitor the Sun nearly three decades ago, it has found that the Sun seems to produce about a third to a half as many neutrinos as theory predicts. This result has been confirmed by more than one neutrino detector, so we can be confident that the problem is not instrumental. Since other facets of stellar theory are well supported by observations of the Sun, this discrepancy led theorists to postulate that the neutrino, formerly assumed to be of strictly zero mass, may in fact have a very small mass.

The reason for this conclusion has to do with the behavior of the different members of the neutrino family. The three known species of neutrino are called the *electron neutrino*, the *tau neutrino*, and the *muon neutrino*. The electron neutrino is by far the most abundant; moreover, most detectors are sensitive only to it. As it turns out, the solar data could be explained handily if electron neutrinos converted into tau neutrinos on the journey from the Sun to the Earth, a phenomenon known as *neutrino oscillation*; for technical reasons, this could happen only if neutrinos have mass. If neutrino oscillation occurred, then a detector would be blind to the arriving tau neutrinos and we would never become aware of their passage, even though they began their trip as potentially detectable electron neutrinos.

Critical astronomical information has also been revealed by other "neutrino events." One of the most exciting observations was made in 1987, when a supernova designated SN 1987A exploded in the Large Magellanic Cloud. On February 23, 1987, a puzzling burst of neutrinos was seen over a 13 s interval by detectors in both Japan and the United States. This fusillade occurred approximately 20 hours before SN 1987A was first spotted by a telescope operator in Chile. Neutrinos and antineutrinos should be emitted in copious amounts by a supernova, as the protons and electrons are squeezed into neutrons during the collapse of the core. Because of their weak interaction with matter, the majority of the neutrinos zip into space immediately after the core collapse, whereas the photons are emitted only after a shock wave reaches the surface of the star. Hence neutrinos from SN 1987A were seen first, even though they travel at (or, perhaps, nearly at) the same speed as the photons. The detection of neutrinos from the supernova thus confirmed the basic theory of supernova explosions.

More important for cosmology, the supernova gave an upper limit on the mass of the neutrinos. If neutrinos are massless, then they must travel at the speed of light. In that event, all the neutrinos would have arrived at Earth at essentially the same time. If neutrinos have some rest mass, however, their speeds must be slightly less than c and would be spread out over a range of values; not all the neutrinos would have arrived at the same time. The actual spread of neutrino detection times for SN 1987A suggests that neutrinos have a mass less than 17 billionths the mass of the proton, less than the mass required to close the universe for Hubble constants greater than 50 km/s/Mpc.

Recent direct experimental evidence from various laboratories, such as Los Alamos National Laboratory, has suggested that the electron neutrino, the lightest and most abundant species of neutrino, might have a mass of somewhere between 0.5 and 5 billionths of the mass of the proton. This is consistent with the supernova result and again suggests that the neutrino alone cannot make $\Omega = 1$. The experimental determination of the mass of all species of neutrino has been difficult, so this tentative mass should be considered uncertain. Furthermore, only the electron neutrino has had any direct experimental limits placed upon its mass.

Although neutrinos may well prove insufficient to close the universe, their large cosmic number density means that any nonzero rest mass would go a long way toward explaining the dark matter. With even a tiny mass, the neutrino could account for the results of most of the dynamical measurements. Unfortunately, lest you decide that the problem of the dark matter is solved, massive neutrinos cannot explain everything. There are problems with explaining the origin of galaxies in a universe dominated by massive neutrinos. Smaller galaxies do not fit into the picture of baryons dancing to the gravitational tune of a distribution of massive neutrinos. In any case, the experimental limits upon the mass of the neutrino still include the possibility of zero mass. It is probably best to keep an open mind on this point, until the accumulation of better data provides firmer evidence one way or another.

Neutrinos are by no means the only possibility for a WIMP. Other, more exotic particles could contribute to the mass density of the universe without influencing the deuterium or lithium abundances produced by nucleosynthesis in the early universe. Such particles are required by various theories of particle physics; grand unified theories in particular provide no shortage of candidates. The spontaneous symmetry breaking that separates one fundamental force from another may require new massive particles called *gauge bosons*. The effect that may account for baryogenesis might be mediated by a *very* heavy particle called a *Higgs boson*, a hypothetical particle which could also play an important role in determining the masses of all particles. Unfortunately, none of these advanced theories is yet amenable to testing in the laboratory, since their elusive particles require creation energies far beyond anything we can achieve in our Earthly particle accelerators. We can only hope to examine our theories by confronting them with the evidence from the universe itself. If such WIMPs are abundant in the universe, then it may be just possible to detect their presence with experiments similar to those used for neutrinos.

If the only accomplishment of WIMPs was to provide overall mass to the universe, it would be nearly impossible to distinguish one type of particle from another. Fortunately, different species of particle produce different, observable effects in the manner of **structure formation** in the universe. This provides the opportunity to construct a theory that may be compared with observations.

Lumps in the Batter

Up to now we have modeled the universe by treating it as smooth and homogeneous, in keeping with the cosmological principle. In reality, the universe is not completely smooth; it contains stars, galaxies, clusters of galaxies, and even superclusters of galaxies. On the grandest scales, those appropriate to the Friedmann equations, these "local" density enhancements can be averaged into a smooth background. Eventually, however, the cosmologist must do better than such an approximation and must tackle this difficulty of the standard cosmology: the origin of structure in the universe.

There is a great deal of structure in the universe, far beyond those small structures upon which we live or depend. From the cosmological perspective, it is not the formation of stars and planets, which still occurs today, that is wondrous, but the formation of larger aggregations of matter. Everywhere we look, we see galaxies. The most distant ordinary galaxies may have redshifts of as much as $z = 4.5$. Many QSOs, which are thought to be associated in some way with galaxies, or perhaps protogalaxies, have redshifts greater than $z = 3$, with the currently most distant known quasars having redshifts that are pushing $z = 5$. Galaxies, then, or something compact like them, must have formed very early in the history of the universe, perhaps within a billion years after the big bang. There are at least as many galaxies in the observable universe as there are stars in the Milky Way. Galaxies may be the dominant structures in the universe.

Galaxies, like humans, are more likely to be found in the company of others than alone. Many galaxies, including our own, dwell in clusters of a few to tens of members. This Local Group of galaxies, all of which are gravitationally bound and interacting, is dominated by two large spirals, the Milky Way and the Andromeda Galaxy (the latter is also known as M31). Each of the two large spirals has several smaller satellite galaxies, which orbit their primary like moons orbit a planet. The Large and Small Magellanic Clouds, visible only from the Southern Hemisphere of Earth, are two of the larger satellites of the Milky Way. The Local Group has a few other fairly significant members, such as a small galaxy in the constellation Fornax and the modest spiral M33, as well as dozens of tiny dwarf galaxies. Many small clusters such as the Local Group exist in the universe.

The Local Group is a lonely outpost of only a few citizens, compared to the urban clustering within some galaxy clusters. The nearest large cluster is the Virgo Cluster, 18 Mpc distant, with some 250 major galaxies and a few thousand small ones. The Virgo Cluster is an example of an **irregular cluster**. Irregulars show no particular symmetry and consist mostly of spirals, with a few elliptical galaxies. Often a giant elliptical resides at the center of the cluster; the elliptical galaxy M87 occupies nearly the central spot of the Virgo Cluster. Beyond the Virgo Cluster, at a

NGC 4881
Coma Cluster

HST · WFPC2

ST Sci OPO PR95-07 · January 1995 · W. Baum (U.WA), NASA

Figure 14.2   Photo of Coma Cluster of galaxies, taken by the *Hubble Space Telescope*. The bright elliptical galaxy is NGC 4881, located on the outskirts of the cluster. (STScI/NASA.)

distance of over 50 Mpc from the Sun, is an unusually rich cluster in the constellation Coma Berenices. The Coma Cluster consists of several thousand large galaxies and an unknown number of smaller ones. It is an example of a **regular cluster**. Regular clusters are roughly spherical or ellipsoidal in overall shape and are dominated by elliptical galaxies, with few spirals. The correspondence between cluster type and typical galaxy type is intriguing but has so far yielded no firm clues to the origin of galaxy clusters.

The Local Group and several other small clusters are satellites of the Virgo Cluster. A cluster of a cluster of galaxies is called a **supercluster**. In keeping with the nomenclature, the supercluster consisting of the Virgo Cluster and its satellites is called the Local Supercluster. Many gravitationally bound superclusters are known; the organization of the universe seems to go from galaxy to cluster to supercluster. We cannot yet say which way the hierarchy travels. Did superclusters form first, then fragment into clusters, which in turn shattered into individual galaxies? This is called **top-down** structure formation. In an alternate picture, the **bottom-up** scenario, galaxies are the fundamental building block; galaxies form, then become pulled into clusters by mutual gravitational attraction, followed by the evolution of superclusters as the clusters are themselves pulled together. There are other possibilities. Did mass concentrations the size of clusters form first, which then separated into individual galaxies while attracting one another into superclusters? We have as yet no answer to this question. A complete theory of structure formation should provide an explanation for the existence of all these scales and should answer the question of which structures came first.

Counting the Galaxies

How can we discern the shapes, sizes, and distributions of these cosmos-girding structures? One way would be simply to examine maps of galaxy positions. This is not satisfactory, however, since we want a firm, quantitative measure of galaxy clustering. A more basic problem is that the human brain is so attuned to pattern recognition that it has little difficulty in discerning "patterns" in what is really just pure random noise. It has often been remarked that this must reflect an evolutionary pressure; far better to see a leopard that is not there, than to miss the one that is! Be that as it may, it renders our subjective judgements of galaxy clustering quite unreliable.

Quantification of the structures in the universe is not an easy matter. As always, there are many subtleties associated with data analysis of this kind, but the basic concepts are not difficult to visualize. Our goal is to measure the average probability that a randomly selected galaxy will lie within some fixed distance r of another galaxy; such a probability is given mathematically by a **correlation function**. Of course, we must know the distance scale in order to convert centimeters on a photograph into physical units; and if we knew the distance scale, many of our cosmological problems would be solved! Most studies of large-scale structure must thus utilize some quantity that can indicate clustering, without introducing the complication of the distance ladder.

The most obvious approach to measuring correlation functions makes use of photographs of the sky. Of course, photographs can show only the projection of the three-dimensional galaxy distribution upon the two-dimensional apparent surface of the sky. This may not seem like much with which to start, but it is possible to draw some inferences about the tendency of galaxies to cluster just from this information. Since the galaxies are photographed on the celestial sphere, without knowledge of the distance to the galaxy, their proximity on the photograph demarcates only their *angular* separation on the sky. In this case, we are interested in the probability of finding the image of another galaxy within a circle of some angular size. Of course, an angular size in the sky will correspond to some separation d on the physical photograph (or, in more modern terms, the digital image). Selecting any individual galaxy, we draw a little circle (perhaps not literally, but in a computer program) centered about that galaxy, of some specified radius d. We count the number of galaxies whose images are located on the plate within the distance d from the chosen galaxy. When this procedure is carried out for all the galaxies in a given sample, we can compute the average number of neighbors of any galaxy within that fixed distance d. By repeating such a measurement for all d, out to some reasonable limit, we obtain the average number of neighbors as a function of the distance of separation.

Clustering is not the only effect which causes two galaxy images to lie within a given distance on a plate. The probability must depend upon the total area, regardless of the clustering properties of the galaxies, since it is obvious that the bigger the area sampled, the more likely it is that we would find another galaxy somewhere within it. Moreover, even if galaxies were randomly distributed on the sky, we would find some apparent clustering merely due to the inevitable occurrences of coincidental alignments. We can, however, take those factors into account mathematically. It is possible, and not all that difficult, to compute the expected number of neighbors within any specified distance d for a random scattering of images across an area. Any clustering over and above what would be obtained from a random distribution is described by something called the *two-point correlation function*.

This sort of analysis of survey photographs of galaxies always yields a measured two-dimensional, two-point correlation function that deviates from the theoretical, purely homogeneous, function. This is the mathematical way of saying what we have already stated in words: galaxies are more likely found near other galaxies; that is, galaxies cluster. We may progress further to the *three-point correlation function*, which measures the probability that three galaxies will make a triangle of a given size. In principle, we could keep extending this process, but in practice, there is little point in going beyond the four-point correlation.

Any more quantitative analysis of the two-point (or higher) correlation function is beyond the scope of this discussion. Even so, we can gain some understanding by considering a highly simplified, "toy" problem. Figure 14.3 shows a manufactured "photograph" of galaxies on the sky, marked off into equal-area cells. From this illustration, some clustering is apparent. How is it possible to quantify that impression? First, count the number of galaxies in each cell N_i, where the index i indicates the ith cell. Next compute the average number of galaxies per cell. This average, which is just the total number of galaxies divided by the number of cells, is denoted by $\langle N \rangle$. Now we can calculate by how much each cell deviates from this average; that is, we compute $N_i - \langle N \rangle$ for each cell. A statistical measure of the extent of the clustering is given by the average of the square of this deviation, that is, the average of the quantity $(N_i - \langle N \rangle)^2$, over the entire image. If each cell had exactly the same number of galaxies, then, obviously, this quantity would be zero, and we could conclude that there was no tendency for galaxies to cluster on the scale specified by the size of the cell. At the other extreme, if all the galaxies were located in only one cell, the average deviation would reach its maximum, which would indicate strong clustering. To determine the cosmic scale of structures, then, it is necessary to repeat this calculation for many cell sizes, searching for the size at which the average deviation approaches zero, after projection effects and other biases are taken into account.

Measurement of statistical quantities for a variety of galaxy catalogs, with numerous corrections for various statistical biases, systematic errors in the data, and so forth, is nearly a minor industry among astronomers, and we shall not attempt to treat any of the details of these studies. We can state that the data conform to the law we would expect for a *clustering hierarchy* up to scales of approximately $10h^{-1}$ Mpc. From $10h^{-1}$ to around $30h^{-1}$ Mpc, the behavior changes, but still seems to indicate

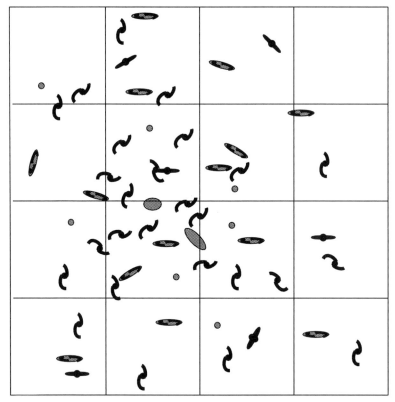

Figure 14.3 Simulation of an astronomical photograph showing galaxies. The region is divided into 16 bins. Counting galaxies within bins is the first step toward developing statistics on the tendency for galaxies to cluster.

clustering. Beyond a scale of

$$l_{cl} \sim 30h^{-1} \text{ Mpc}, \tag{14.9}$$

the correlation functions depart significantly from the hierarchical-clustering model. It cannot yet be said with certainty that the statistics go to a homogeneous, random distribution at the scale indicated by equation (14.9), but it seems we can say that we have found some limit to the typical size of structures at approximately $30h^{-1}$ Mpc.

The value of any of these statistical measurements depends upon the quality of the data. The number of galaxies is enormous, and obtaining statistics on them is hard work. As an illustration of the dedication of astronomers, C. D. Shane and C. A. Wirtanen counted a few hundred thousand galaxies in the late 1960s, by hand, from photographic plates! Fortunately, such devotion to scientific duty is no longer necessary. The availability of automatic plate-scanning equipment, which can determine the two-dimensional coordinates of a galaxy on an image, and high-speed computers, which can compute quantities such as the correlation functions, have tremendously aided the study of large-scale structures in the universe.

The next improvement involves obtaining information about the full spatial distribution of galaxies, rather than just their apparent distribution on the two-dimensional celestial sphere. From Hubble's law we know that redshift provides a measure of distance, and with better redshifts for more galaxies, it becomes possible to construct extensive three-dimensional maps of the distribution of galaxies. (The third dimension is usually plotted as redshift, not as distance, because of the uncertainties in converting redshift to distance.) Since the 1970s, the quality and quantity of redshift data have constantly improved. The quantitative analysis technique for galaxy redshift surveys is a simple extension of the two-point correlation function to three dimensions. Rather than measuring the two-dimensional *projected* distances between any two galaxies, we simply measure their three-dimensional separations, although again we must typically use redshift as the third "distance." By now, however, it should go without saying that there are considerable difficulties in determining three-dimensional structures. Even with the improvements in redshift measurements, the data are still, for the most part, insufficient to make the three-dimensional correlation functions as meaningful as we would prefer. Redshifts require spectroscopic measurements for each galaxy, and that requires much more effort than taking a simple photograph of a region of the sky. The interpretation of redshift data is further complicated by our inability to distinguish the Hubble recession from peculiar motions. At very large distances, we can safely ignore peculiar velocities, compared to recession velocities, but we also need data for closer galaxies if we are to understand the hierarchy of structures. Conversely, measuring redshifts for very distant galaxies is not easy, for the simple reason that the remote galaxies are faint, and obtaining their spectra is difficult or impossible.

There are also fundamental limitations in the samples that we may obtain. Every galaxy survey is inevitably biased in some way. A perfect survey would (accurately) measure the redshift for every galaxy, no matter how luminous, in a specified large volume of the universe. Such an ideal survey would be said to be "complete." Needless to say, this is impossible with present-day technology and may never become practical. The farther we go in redshift, the more we bias our sample toward the bright galaxies, those we can see for large distances. On the other hand, a galaxy survey which is complete, or as complete as humanly and technologically possible, for all galaxies down to a limiting brightness, cannot go to a very large redshift, although that restriction is easing as telescopes and detectors improve. Evolution effects also play a role, especially as we go to greater and greater distances; the luminosity of very young galaxies may not be well described by the behavior of nearby, and hence much older, galaxies.

Given this inherent bias in any sample, it is important to perform as many different types of surveys as possible. For example, additional information can be derived from surveys in frequency bands other than the visible. One of the largest recent catalogs of galaxies comes from data recorded by the *InfraRed Astronomical Satellite (IRAS)*. *IRAS* selected for another effect, of course, namely, infrared luminosity. A galaxy bright in the visible is not usually bright in the infrared, and, typically, vice versa. One advantage to the *IRAS* data is that the satellite could "see" through obscuring clouds of dust in the Milky Way, thus producing better coverage of the sky. It also happens that the galaxies which shine brightest in the infrared often are engaging in rather violent interactions of one form or another, such as collisions with nearby galaxies; thus they may show a few bright, strong features in their spectra, making it easy to determine their redshifts. Similar surveys have been made for radio galaxies, with similar caveats. Radio galaxies, and other radio sources, can often be "seen" for enormous distances, making them good indicators of the distribution of galaxies at the largest scales; many radio sources are nearly a Hubble length away, which for practical purposes is as far as we can see. The maps of radio sources, as well as the most distant *IRAS* galaxies, do not present any obvious structures to the eye, and such surveys are consistent with the scale of overall "homogeneity" indicated by equation (14.9).

Similar exercises can be carried out for *clusters* of galaxies, rather than the individual galaxies. The data analysis becomes somewhat more problematical in this case, partly because it is not always easy to determine the boundaries of a cluster, and partly because systematic errors, such as differing exposure times of the photographs of different parts of the sky, can cause even more acute problems than in the case of galaxy-galaxy correlation studies. The results for clusters are similar, by and large, to those for galaxies, with the striking exception that clusters seem to be more tightly correlated to one another than galaxies are to other galaxies. This is interesting, for it may hint that the distribution of *luminous* matter does not exactly follow the overall distribution of mass.

The typical scale of a supercluster is comparable to the scale at which galaxy clustering transitions toward homogeneity, approximately $30h^{-2}$ Mpc. Superclusters are about as large as a mass concentration can be and still be mutually gravitationally bound, for a universe of age 10–20 billion years. Yet recent sky surveys show structure on even larger scales. These surveys find not only clusters and superclusters, but also vast **voids** nearly empty of matter. Chains of galaxies seem to form luminous bridges between voids. We can understand the existence of bound galaxy clusters in terms of gravitational attractions occurring since the formation of galaxies, but as yet we lack a satisfactory explanation for the existence of these much larger structures. They are not incompatible with the existence of the "transition scale," as long as they do not have much effect upon the density contrast for most cells in a sample. However, their

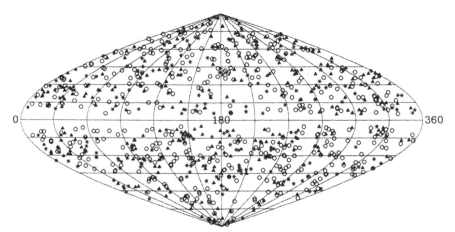

Figure 14.4 Map of the more distant *IRAS* galaxies with redshifts greater than $z = 0.02$. The empty strip in the center of the plot is due to obscuration by the plane of the Milky Way. (M. Strauss et al. 1992.)

presence does mean that we cannot say with certainty at what point the distribution of galaxies becomes fully homogeneous, as the cosmic background radiation says it must.

The newest surveys using automated techniques and dedicated equipment should be able to find the limits of structure in the universe. The Las Campanas Redshift Survey obtained over 26,000 galaxy redshifts, going as far as $z \simeq 0.2$. Although the data continue to show huge voids and filaments of galaxies, one preliminary conclusion from this survey is that the scale of homogeneity has finally been reached. Even more ambitious projects are in the works; the Sloan Digital Sky Survey will sample over a million galaxies. Such data may help to elucidate the mystery of the superclusters.

The Formation of Structure in the Universe

If the universe is so fantastically isotropic as the CBR demands, how could any overdense structures such as galaxies have formed? Gravity is always attractive, which implies that any small overdensity in a uniform background will tend to be amplified. This still requires a fluctuation in density, however, since even with gravity, a perfectly uniform density distribution will remain uniform for all time. We have already encounted this situation when considering star formation, but now we expect to form galaxies and clusters of galaxies in an *expanding* background. This means that the self-gravity of an overdense region has the additional burden of fighting against expansion as it tries to collapse.

The observation that quasars can be found as far back as redshifts of $z = 4.5$ places important constraints on any theory of structure formation. Computing the lookback time for such remote quasars shows that they existed when the universe was no more than about a billion years old. Since quasars are believed to be energetic nuclei of galaxies, this implies that galaxy formation happened promptly after recombination. Hence, no matter how clusters, superclusters, and voids may have formed, their origins must be present in the radiation-dominated era of the big bang. *Seed perturbations*, disturbances in the homogeneous background of the mass density, must have been already present very early in the universe, at least by the epoch of equal radiation and matter densities. These ripples in the early universe grew into the galaxies we see today. The ripples must have been evenly distributed everywhere, so that on the large scale, the universe was still homogeneous.

Although the amplitudes of these perturbations are uncertain, they cannot have been too large, or their traces would be clearly visible in the cosmic background radiation we observe today. How is it that we can see the ghosts of primordial structures in the present background radiation? If the CBR radiation consists only of free photons, by what mechanisms might the primordial matter perturbations have affected it? Prior to recombination, the photons and the matter interacted constantly, scattering off one another. After recombination, the photons streamed unimpeded through space, carrying the memory of their last scattering. The photons in the CBR did not cause the matter perturbations, but they did scatter off them, becoming imprinted with the scale of the fluctuations in the process. The major process by which this imprinting occurs is the *Sachs-Wolfe effect*, which is essentially gravitational redshifting and blueshifting. Photons which last scattered from a region of higher density than average must

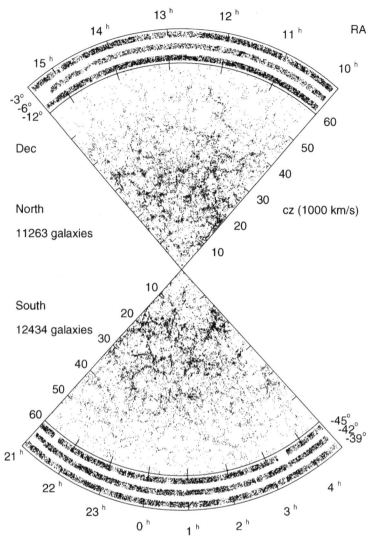

Figure 14.5 Two slices through the universe, from the Las Campanas Redshift Survey. The distance is given by redshift, which in this survey goes as high as $z = 0.2$. Galaxy clusters tend to appear as long strings of galaxies directed radially toward us because their redshifts have significant components due to the peculiar motions of the individual galaxies. This causes a distortion in the plotted appearance of the cluster. The number of galaxies in the survey decreases at large redshift, as galaxies become fainter and harder to detect. (Lin et al. 1996.)

climb out of the slightly stronger gravitational "well," and thus are redshifted, relative to the average, while photons which last scattered from a lower-density region are blueshifted. This alone would not tell us very much, except for the fact that these redshifts and blueshifts are due to a factor *other than* the cosmic expansion, and thus do *not* affect all photons equally. Consequently, the Sachs-Wolfe shifting appears to our measuring devices as a difference in temperature of the scattered photons, in comparison to the smooth background photons. This causes the temperature of the

CBR to show anisotropies, or deviations from the perfectly uniform, isotropic background temperature, from one part of the sky to another. The sizes of the primordial density anisotropies, relative to the background density, are roughly proportional to the amplitude of these temperature anisotropies.

Prior to the launch of *COBE*, astronomers had carried out extensive ground-based searches for these anisotropies, but no definite detections were made. This set an upper limit: any fluctuations present must be smaller than a very small fraction of a degree. The *COBE* satellite finally found evidence for temperature anisotropies of roughly one part in 10^5. Temperature irregularities of this size correspond approximately to density perturbations of the same order. As we shall see, the small size of these perturbations places a severe constraint upon the initial conditions for galaxy formation. The balance required to create the structures while producing no more than the observed anisotropy of the CBR is quite delicate. We seem to have encountered a requirement of special initial conditions for the big bang.

Bearing in mind the stringent limitations imposed on any hypothesis by the *COBE* data, let us begin with the straightforward assumption that galaxies are spawned from seed perturbations of ordinary matter. We assume that the early universe was nearly homogeneous, but some random density fluctuations were distributed throughout it. While the universe was radiation dominated, the energy density of the radiation prevented matter perturbations from growing. Thermal equilibrium of baryons and photons ensured that the radiation interacted strongly with the matter. Photons in slightly hotter regions diffused into slightly cooler regions in order to maintain equilibrium. As they moved, they tended to drag the matter with them, because of the tight coupling between the matter and radiation during the early epochs of the universe. Once the epoch of equal density passed, and ordinary matter controlled its own fate, gravity began to separate overdense regions and pull them to greater density, which in turn increased their self-gravity and caused them to collapse further. The history of such an overdense region should tell us how it might lead to structure formation in the present universe.

The Newtonian theory of gravitational collapse was well understood long before general relativity was developed. Newton himself realized that matter would tend to collapse under its own gravity, and he proposed in a letter to a friend that this might account for the presence of stars in the universe. The detailed theory of gravitational collapse was worked out much later than Newton's time, beginning in the nineteenth century. The full theory is complicated, but the simplest case, the collapse of a static, perfect (nonviscous) fluid, provides a good approximation for most purposes. Sir James Jeans first developed this theory, and it is the starting point for studies of gravitational instabilities.

Imagine that within a smooth distribution of matter there is a slightly overdense region surrounded by a uniform background. What will happen to this overdense region? The answer depends on the relative strengths of two forces: the gravitational force, which tries to make an overdense region collapse, and the pressure force, which resists this collapse. This is the basic idea of hydrostatic equilibrium, which we have encountered previously in our discussion of stars. If the perturbation is not too overdense, the pressure force will support it, and it will merely oscillate like an ordinary sound wave. (Sound waves represent traveling pressure perturbations in a gas or fluid.) If the density perturbation is sufficiently large, however, it grows from its own self-gravity and separates from the background to form a collapsing ball. It is possible to compute the minimum mass required for this breakaway to occur; it is called the **Jeans mass**.

We can understand the Jeans mass qualitatively without considering the equations from which it arises. It depends mainly upon the ratio of the time a sound wave takes to cross the perturbation, compared to the time for that perturbation to collapse gravitationally. If sound can cross the perturbation before the collapse can develop, then pressure waves can restore hydrostatic equilibrium. If the collapse begins before sound has a chance to communicate from one side of the overdensity to another, then gravity will win over pressure. Hence gravitational collapse favors larger clumps containing more mass, or lower temperatures resulting in a smaller sound speed, or both. Equating the gravitational energy of the perturbation, which controls its collapse, with the thermal energy, which generates the pressure by which collapse might be combatted, yields an estimate for the Jeans mass of

$$M_J = \frac{c_s^3}{G^{3/2}\rho^{1/2}}, \tag{14.10}$$

where ρ is the density of the clump, and c_s is the speed of sound.

Jeans' original calculation does not apply directly to cosmology because it does not account for an expanding background. The first relativistic calculations for an expanding background were carried out between 1946 and 1949 by E. M. Lifshitz, a Russian physicist. Remarkably, much of the classical result, in particular, the formula for the Jeans mass, is unchanged by the addition of cosmology. There are a few major relativistic effects: First, some of the quantities entering into the Jeans mass, specifically the density and sound speed, change with time due to the expansion of the universe. Consequently, the Jeans mass in a standard universe depends on time, through the scale factor R. Second, the rate of collapse of an overdensity is much slower in an expanding universe than in a static background. This makes qualitative sense; the expansion is trying to pull the matter apart while gravity tries to pull it together. The outcome of this tug-of-war still depends upon the sound-crossing time versus the collapse time, so the formula for the minimum collapsing mass is unchanged, but the speed at which a collapse can occur is affected. Third, an essentially Newtonian analysis such as that of Jeans is not applicable everywhere, but only within the Hubble sphere. This may not seem like much of a restriction because the Hubble sphere today is so huge, but keep in mind that in the early universe, the Hubble sphere was relatively small, since its radius is approximately equal to the age of the universe multiplied by the speed of light.

Lifshitz, and others after him, also worked out the details of the behavior of perturbations that are *larger* than the Hubble length at any given cosmic time. Such inhomogeneities are only wrinkles in spacetime. To these large perturbations, sound waves are irrelevant; not even light would have had time enough to cross their length over the age of the universe. They cannot collapse, because a gravitational collapse requires causal communication. The fully relativistic analysis enables us to compute the amplitude of such superhorizon perturbations at the time that the growing Hubble length "catches up" with them. It also tells us something about the nature of the perturbations. It turns out that while disturbances whose size is less than the Hubble length are essentially all of the same kind, just straightforward lumps in density, inhomogeneities with scales greater than the Hubble length can be of different types. Some of these perturbations may grow, while others die out. In principle, such information might help us to determine how the density perturbations originated in the very early universe, when nearly all inhomogeneities would have been superhorizon-scale ripples in spacetime.

The change of the cosmological Jeans mass with time in a baryon-dominated, expanding universe is sketched in Figure 14.6. Prior to recombination, the universe is dominated by photons, and matter, or at least the baryonic matter, interacts directly with these photons. For photons in thermal equilibrium, the pressure depends only upon the energy density, which in turn is a function only of temperature. The sound speed in a gas dominated by photons is a constant $c_s = c/\sqrt{3}$. Meanwhile, the density of baryons decreases in the usual way, like the cube of the scale factor. Since the Jeans mass depends upon the sound speed *divided* by the density, if the density decreases while the sound speed is constant, then the Jeans mass actually *increases* rapidly with time until recombination. It turns out, however, that before recombination the Jeans mass is much greater than the mass of baryons within the Hubble sphere at any time. Therefore, precisely those perturbations in which we are most interested, those that are potentially unstable, cannot be treated by the relatively simple Jeans analysis because their scale is greater than the Hubble length.

In a universe in which all matter is baryonic, recombination and the advent of the matter-dominated era occur at about the same time. Once the photons decouple from the matter and their energy density and pressure no longer dominate the universe, the sound speed drops precipitously and so does the Jeans mass. After recombination, the Jeans mass decreases because now the sound speed decreases with time as the matter cools and rarefies. In the baryonic universe, the Jeans mass just after decoupling is

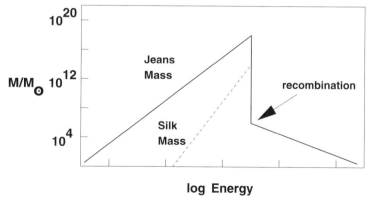

Figure 14.6   The change in the Jeans mass as a function of temperature (equivalent to an energy in this case) and hence of time, since the temperature changes in a regular manner with time in the early universe. Prior to recombination, the Jeans mass rises because the sound speed is constant while the density falls. At recombination, the sound speed drops as the photons decouple from the matter; the Jeans mass subsequently declines. The Silk mass is indicated by the dashed line for comparison.

very close to the mass of a typical globular cluster, which is at the least a suggestive coincidence.

Yet globular clusters do not seem to be fundamental building blocks for structure in the universe. The most obvious structures that we observe when we look into the cosmos are galaxies. Beyond that, we see clusters of galaxies. How can we account for these much larger structures? Perhaps the Jeans mass is not the best explanation for the conditions at recombination. Perhaps events *prior* to recombination were important. Structure formation is still a very active branch of cosmological research, and we can only give an outline of some of the more popular theories here.

The Jeans analysis, and its relativistic extension, applies to the collapse of something called a perfect fluid. A perfect fluid has many special properties, such as the absence of viscosity. Nevertheless, it is a good approximation for many purposes; it is an excellent approximation for air, for example, except for very thin layers where the air contacts a surface. It may not be such a good approximation for matter around the time of recombination. What if we allow a more general description of matter? In that case, the mathematical analysis becomes much more complex, and we can only describe some of the possibilities.

One phenomenon to consider is *photon diffusion*. Before the time of decoupling of the matter from the photons, photons were not yet freely streaming through the universe, but could only diffuse through the matter. The photons tended to have a more uniform, homogeneous distribution than the matter might have had. Since there was still coupling between photons and baryons, the photons were able to squelch perturbations in the matter, an effect known as **photon damping**. The scale of this damping is essentially determined by how far a photon can travel, on the average, before scattering from matter. The mass within a sphere defined by this length scale is called the *Silk mass*, after Joseph Silk, who developed this analysis. Perturbations with a scale smaller than the Silk mass are suppressed by photon diffusion. It is noteworthy that around decoupling (and, if the universe is entirely baryonic, recombination), the Silk mass is approximately that of a typical cluster of galaxies.

If photon damping washes out the small-scale perturbations, can we count on the remaining large-scale perturbations to form structure? For a strictly baryonic universe, this suggestion encounters the difficulty that the antics of baryons near the time of recombination are recorded in the CBR, due to the coupling between baryonic matter and the radiation up to that point. Perturbations in baryons around the time of decoupling of the photons would thus show up today as a corresponding anisotropy in the CBR. In order for galaxies to have formed from baryonic perturbations in the short time available between recombination and the appearance of the first quasars, the perturbation amplitudes at recombination would have had to be very large, much larger than allowed by the observed near-isotropy of the CBR. Cosmologists are forced to conclude that purely baryonic perturbations cannot account for the observed structure in the universe.

To this point we have been assuming that all or most of the matter in the universe is baryonic, yet many prominent theories currently assert that most of the matter is *not* baryonic. Nonbaryonic matter forces us to reconsider the issue of initial perturbations. How does such matter behave? For one thing, it can never be modeled as a perfect fluid; its interactions are too weak, and anything that can be called a "fluid" necessarily consists of particles that interact strongly with each other. Matter such as WIMPs is said to be *collisionless* because of its weak mutual interaction. It must be treated by statistical methods; such methods show that even collisionless matter causes damping of perturbations, by effects similar to photon damping. Collisionless particles can diffuse from regions of high density to regions of low density, thus smoothing the inhomogeneities. This is called **collisionless damping**, and like photon damping, its scale is determined by the mean distance traveled by a particle over a given time interval. For free-streaming, weakly interacting particles, this length is set by the average speed of the particle; faster particles create longer damping scales.

Perturbations occur with many scales, of course; perturbations on some scales will grow, while those on other scales will die out. How can we deal with all these different scales in a general way? Just as we can break light into its component wavelengths, so can we decompose any arbitrary density inhomogeneity into a spectrum of waves. In the case of the Jeans analysis, we saw that some of these waves correspond to sound waves, while others collapse. Even in the more elaborate theories, we are still interested in the behavior of the fundamental waves, or *modes*, of the perturbations. The spectrum tells us the amplitude of each mode. If we can determine the spectrum, and the evolution of each of its constituent modes, then we have all the information we need to describe the evolution of any arbitrary inhomogeneity. Unfortunately, we cannot yet directly measure the spectrum of primordial density fluctuations, so we must make a hypothesis. The simplest spectrum assumes that the ratio of the perturbation to the background density is a constant for every mode. The constant specifies the amplitude at the time when a particular mode's wavelength is the same as the Hubble length. This spectrum is not as ad hoc as it might seem; it is obtained from a well-founded assumption about the random nature of perturbations. It also has the desirable feature that it is consistent with the anisotropies seen in the CBR, and it does not contain so many strong perturbations that it would lead to excessive production of primordial black holes. It is called the **Harrison-Zel'dovich spectrum**, after the astrophysicists who independently applied it to studies of structure formation.

The Harrison-Zel'dovich spectrum requires that all perturbations have the same amplitude when their size is the same as the Hubble length. In the standard models, the Hubble length increases with time. Hence the smaller the wavelength of any particular perturbation, the earlier the time at which it equals the Hubble length. If perturbations can begin to grow at the point at which they first become smaller than the Hubble length, without being smoothed away by collisional or collisionless damping, then the small-scale perturbations are able to grow for a longer time than are the larger-scale perturbations, which fall beneath the Hubble length at later times. Thus, in the absence of damping, we should get bottom-up structure formation, with smaller structures forming first. Damping, if present and sufficiently strong, can lead to a top-down scenario, as the small-scale perturbations are quashed before they have a chance to grow.

Our study of the growth of perturbations has not addressed the origin of the fluctuations that made up the initial spectrum. What might have provided the "seeds" for the inhomogeneities? This is a major area of current research. Two ideas that have emerged depend on properties of GUTs; we will consider these ideas in more detail in the next chapter. For now, it is sufficient to know that the initial perturbations did exist: we see them directly in the *COBE* results, and we infer them indirectly through the existence of structure in the universe today.

Simulating the Universe

When we speculate about the conditions near the big bang, we must make many approximations and guesses to fill in the gaps in our knowledge. For the case of structure formation, we can assume that seed perturbations formed by some means. The subsequent behavior of these perturbations under the influence of gravity can be affected by many factors, including the initial spectrum of perturbations, whether most of the matter is collisionless or baryonic, what the mean energy was when structure formation was active, and so forth. How can we decide whether any of these speculations might be correct? One approach is to construct a model of the universe that incorporates the details of the chosen theory, evolve it forward in time, and compare the results of such a *simulation* to observations such as the redshift surveys of galaxies. We can then ask whether the outcome of a given simulation resembles the observed present-day universe. If not, then some of the basic assumptions must be wrong.

Detailed cosmological models implemented for such simulations almost always consider not the full Hubble sphere, but only a comoving volume of the universe that is large enough so as to be representative, that is, a homogeneous volume. The conclusions derived can then be extended to the entire observable universe with the help of the cosmological principle. Even with this restriction to a limited volume, most models of this type are sufficiently complex that their equations must be solved with the aid of a computer. Typical sizes for simulation boxes correspond to around 100 Mpc per side in the present universe. Into this volume are placed massive particles representing ordinary matter, and WIMPs if they are part of the model. The more particles, the better the "resolution" of the simulation. Some of the latest simulations also include the effects of gas dynamics in the ordinary matter, even though such interactions add considerably to the required computing time. The dramatic increase in computing power over the past two decades has made a significant impact on many fields of astrophysics, particularly in cosmology; greater computing capabilities mean that more particles can be followed in the simulation, and more complex interactions can be included. Both improvements lead to more realistic simulations whose results can be interpreted with greater confidence.

A simulation is begun at a time corresponding to some large redshift. Small perturbations are added to the initial smooth distribution of matter, according to the spectrum to be tested. The evolution begins and the matter particles start to interact gravitationally. The model is evolved forward in time until the present, where it can be compared with the observations. Any "structures" which form in the computer model can be treated as if they were images on an astronomical plate, and various statistics such as two-point correlation functions can be computed. These analyses determine whether the assumed perturbations, matter content, and other conditions can reproduce the observed structure in the physical universe.

Models of structure formation are classified according to the type of matter they assume to be present and dominant near the time of the decoupling of matter and photons. One of the earlier detailed models of structure formation is the **hot dark matter (HDM)** model. This scenario assumes that the dominant form of nonbaryonic matter has extremely high velocity near the time of structure formation; hence the designation "hot," since velocity can be related to temperature. The strongest candidate for the weakly interacting particle in the HDM model is a massive neutrino, as it is the only known particle that would still have a very high temperature near the epoch of equal density. Although the neutrino has mass in this picture, it is a small mass, so the particles move at speeds close to that of light. The characteristic high velocities of HDM lead to very large structures, because the collisionless damping length is long. In the neutrino-dominated HDM scenario, the first structures to form are superclusters. The neutrinos create the initial structures, and baryons fall toward the gravitational attraction of the neutrinos. Once the baryons begin to collapse, they flatten rapidly into "pancakes," which then fragment into smaller objects due to friction and heating of the baryons. The distribution of neutrinos, however, retains the large-scale structures. In the HDM scenario, then, massive neutrinos, which make up the dark matter, remain much less clustered than the baryons.

The HDM model successfully reproduces the largest-scale structures, including filaments and voids, but has difficulties with details. In order to fit the observed clustering properties of galaxies, the collapse must have occurred much too recently, at cosmological redshifts of approximately $z = 1$. This contradicts the observations of normal galaxies at much greater redshifts, perhaps as large as $z = 4$ or greater, as well as leaving little accommodation for QSOs, which are seen to redshifts as large as $z = 5$. HDM models also tend to show excessively strong clustering; maps of "galaxies" created by computer simulations of hot dark matter show tight blobs of matter and huge voids.

At the other extreme is the **cold dark matter (CDM)** model. In this scenario, the missing matter consists of a weakly interacting particle that is very massive and hence is already quite cool (i.e., has a low velocity) when it decouples from the rest of the contents of the universe; by the time of matter domination, its energy is small. Because these WIMPs have such low velocities, the collisionless damping length is extremely short, so small-scale perturbations can grow unimpeded. Thus, in the CDM model the first structures to form are smaller than galaxies. These objects then interact through gravity to form larger structures, pulling one another together and even merging. Eventually, galaxies and galaxy clusters result. This seems to be consistent with data indicating that galaxies formed quite shortly after the beginning of the matter-dominated era and only later were drawn into clusters and then superclusters. Our own Local Group may be an example; the Milky Way is certainly very old, probably about as old as any galaxy. Yet the size of the Local Group and the peculiar motions of the galaxies within it suggest that it is still forming and collapsing. Similarly, the Local Supercluster appears to be a fairly recent aggregation.

The CDM model has many attractive features. It is quite successful at reproducing the observed clustering properties of galaxies, such as density and distributions. It accounts for the formation of galactic halos as a natural outcome of the collapse of the weakly interacting particles. The simplest CDM scenarios tend to fit the observations best for open universes, which may or may not be a disadvantage to the model, depending upon which observations one chooses to believe, and one's theoretical prejudices. CDM has difficulty reproducing both the small-scale and the largest-scale structures, however. CDM alone seems incapable of providing a full explanation for structure formation.

Because the mutual failings of the two extreme models are complementary, some cosmologists have proposed that there might be a little of each kind of matter. That is, both types of dark matter are present; the HDM (perhaps massive neutrinos) accounts for the largest-scale structures, while CDM is responsible for the creation of more modest structures such as galaxies. Simulations of this type have produced some of the best results so far. While this is certainly an attractive combination, it suffers from the difficulty of requiring *two* particles for which no firm evidence yet exists, rather than only one. Moreover, there is the concern that such models have so many adjustable parameters that they might fit anything. Nevertheless, the model looks promising, and much current research has concentrated upon the "hot and cold," or *mixed* model.

It must be kept in mind that both HDM and CDM models are necessarily highly simplified, due to the limitations of computing capabilities. For one thing, most simulations have used only a few million particles. More recent simulations of CDM collapse, following the interactions of as many as 16 million particles, show better agreement with observations than did older, less-resolved calculations. Furthermore, in most simulations, only the weakly interacting particles are evolved. But what we actually observe are the baryons, and the galaxies they form. Baryons have much more complicated interactions than WIMPs; they collide, heat, behave as fluids and develop shock waves, and so forth. Without the inclusion of baryons in the simulations, however, we cannot really say where "galaxies" are.

The best way to test models of galaxy formation would be to include baryons in simulations. Some recent pioneering work by several groups has combined weakly interacting particles and baryons in the same simulation, but given the magnitude of the problem, it is probably too early to draw firm conclusions. The equations of hydrodynamics and thermodynamics for baryons have long been solved numerically on computers, but that is itself computationally challenging, requiring considerable memory and speed for realistic solutions. Adding collisionless matter magnifies the difficulties and boosts the demand on computing resources. It is very difficult to resolve all the length scales in a model containing both types of matter, from the full 100 Mpc box down to the kiloparsec scales of galaxies. An example of a recent simulation is displayed in Figure 14.7; the results are impressive, but research must continue.

As we look to increasing distances, we can see only the most unusually bright galaxies. If those galaxies are not representative of the overall matter distribution, even in baryons, then the statistical properties we derive from studying them may have little relevance to the large-scale structure of matter. To some extent, we already know that this is true. Within our Local Group are several dwarf galaxies, some of which contain scarcely more than a million or so stars. We may not even have detected all of these objects in our immediate vicinity, much less within the entire Local Group; a new member of the Local Group was found as recently as 1994, hidden behind the obscuring dust lane of the disk of the Milky Way. Seeing dim galaxies such as these at the distance of the Virgo Cluster is currently impossible. Nevertheless, if we can assume that all galaxies form a continuum, then we can allow for the existence of small galaxies. But what if bright galaxies form differently from fainter galaxies? In that case, the clustering properties of bright galaxies might be entirely different from the clustering properties of matter as a whole. We expect that the density perturbations would not have been uniform at the time of structure formation. Some regions would have been *very* overdense, compared to the mean, while others might have been quite underdense. The most overdense (and underdense) regions would have been the rarest, but bright galaxies might have preferentially formed in the anomalously dense regions. Less overdense regions might have collapsed after these peaks of density, while the regions only a little denser than the mean, which we might expect to be the most common, might have collapsed very late to form a relatively smooth background of extremely dim, or perhaps even dark, galaxies, which we cannot detect with current technology. This idea is known as **biased galaxy formation**. If this phenomenon exists, its magnitude, and how it affects our observations, are unknown at the present time.

The ideal study of structure formation in the universe would be a simulation that could include either HDM or CDM, or both, or neither. It would incorporate baryons with all their complicated interactions. All components would be fully resolved, and no unrealistic simplifications would be made. At the present time, the only computer

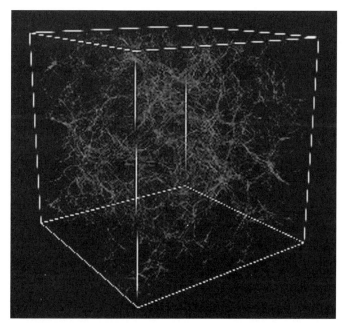

Figure 14.7 A cosmological simulation showing density at the present time for a model that includes both hot and cold dark matter, as well as baryons. In this model, cold dark matter contributes $\Omega = 0.6$; hot dark matter, $\Omega = 0.3$; and baryons, $\Omega = 0.1$. (G. Bryan and M. Norman, NCSA/University of Illinois.)

capable of carrying out such a calculation is the universe itself! It is not necessary to follow every particle and include every interaction to gain an understanding of overall behavior, but adequate resolution is still beyond the reach of our computing abilities. In the fairly near future, this might change.

The complement and arbiter to the theoretical studies is improved observations. Satellite observations have already placed strict limitations on the size of primordial temperature fluctuations. If *COBE* had restricted temperature anisotropies to a level of 10^{-6}, *all* our current theories of structure formation would have been nullified! Fortunately, the limitation turned out to be less stringent; *COBE* found anisotropies of order 10^{-5}. This is quite severe, but still allows many scenarios to survive. In fact, the only model that has difficulties under such restrictions is one in which all matter in the universe is baryonic. All-baryon models seem to be ruled out for density parameters large enough to explain the dynamical measurements because such models inevitably produce larger perturbations in the CBR than are observed. The HDM model just manages to accommodate the small anisotropies, while the CDM model fits within the upper limit established by *COBE*.

Exciting new observations continue to be made. *HST* photographs have detected primordial gas clouds from which young galaxies may be just emerging. We are on the verge of seeing the complete history of the assembly of galaxies laid out before us; yet a full explanation of their formation and clustering continues to elude us. Structures abound in the universe, ranging from humble dwarf elliptical galaxies, to huge superclusters, and beyond to the voids and galaxy filaments. Galaxies speak to us of the earliest times of the universe, but we cannot yet understand what they are saying. Only further research can clarify the mysteries of the galaxies.

KEY TERMS

kinematical method	mass-to-light ratio	dynamical method
virial theorem	dark matter	dark halo
MACHO	WIMP	structure formation
irregular cluster	regular cluster	supercluster
top-down structure	bottom-up structure	correlation function
void	Jeans mass	photon damping
collisionless damping	Harrison-Zel'dovich	hot dark matter (HDM)
cold dark matter (CDM)	spectrum	
	biased galaxy formation	

Review Questions

1. Where does most of the mass of spiral galaxies seem to be located? From what evidence do we draw such a conclusion?

2. Describe two distinct approaches to measuring the mass density of the present universe. How do these results compare with the accepted density due to baryons (ordinary matter)? To what conclusion does this lead?

3. If we live in a critical $\Omega = 1$ universe, what must make up most of its matter? Why is it not possible that such matter could be baryonic?

4. What evidence exists that neutrinos might be massive? What effect would a massive neutrino (of any species) have on cosmology?

5. Briefly describe the types and sizes of structure that exist in the universe.

6. What information does the two-point correlation function provide us? Why do astronomers still often use the projected, two-dimensional distance rather than a full three-dimensional distance for their studies of galaxy clustering?

7. Consider Figure 14.3. Compute the average number of galaxies per bin $\langle N \rangle$. Next, count the galaxies in each bin, N_i, and compute the difference $N_i - \langle N \rangle$ for each bin. Square each of these differences and find the average squared-difference for all the bins. Take the square root and divide by $\langle N \rangle$. This is the fractional fluctuation in the galaxy count. How do you think this would change if you used a smaller bin size? A larger bin size?

8. What is the Sachs-Wolfe effect, and how does it tell us about perturbations in matter?

9. Explain qualitatively the physical processes that determine the Jeans mass. Why does the Jeans mass change with time in an expanding universe?

10. Explain the difference between hot dark matter and cold dark matter. Which one corresponds to top-down and which to bottom-up structure formation, and why? Describe the strengths and weaknesses of each corresponding model of structure formation. What other possibilities might exist?

Notes

1. See chapter 5 for a discussion of the mass-luminosity relationship for main sequence stars.

2. The previously mentioned dark matter candidate named MACHO was so dubbed partly in jest, to contrast with the WIMP.

3. Legend has it that the Homestake experiment's principal investigator, Ray Davis, received a complimentary truckload of wire hangers from a grateful dry-cleaning supply company, which assumed he was opening a huge cleaning establishment.

CHAPTER 15

The Inflationary Universe

Unresolved Issues in the Standard Model

The standard hot big bang cosmological model has been a great success, providing a physical basis for understanding the observed expansion and the cosmic background radiation. Despite the many triumphs, however, there remain significant unresolved questions.

One of the most nagging of these unsolved mysteries has to do with the very homogeneity and isotropy that have been our guiding principles in the creation of cosmological models. The homogeneous and isotropic Robertson-Walker metric is an exceedingly special solution to the Einstein equations, valid for conditions of high symmetry. Perhaps we should be quite surprised that the real universe seems to be consistent with it. After all, the most general solutions to the Einstein equations, even without the cosmological constant, are neither isotropic nor homogeneous. There is also a fairly large class of known solutions called the *Bianchi models*, that are homogeneous but not isotropic. The Robertson-Walker metric, however, is *unique*; it can be *proved* to be the only solution to the Einstein equations for an isotropic, homogeneous universe. Why, out of the large number of more general solutions available, did the universe "choose" one so special? The measurements of the CBR by *COBE* showed that the universe must be isotropic to better than one part in 10^5. If this is such a special condition, why does it just so happen that the one and only universe is isotropic?

Even within the context of the standard Robertson-Walker model, there exists a full range of spherical ($\Omega > 1$) and hyperbolic ($\Omega < 1$) models, but only one very special flat model ($\Omega = 1$). Yet the universe is so close to flat that we are presently unable to determine its true geometry. If this is another special condition, why does it happen that the universe is nearly (or exactly) flat?

An examination of the shortcomings of the standard models can point the way toward a more general theory. These limitations are not the fault of the models, but of our imperfect understanding of some aspects of the universe, and we must seek to progress by honing our understanding of fundamental physics. Already one possible context for a solution to the difficulties presented by the standard models has been put forward: the inflationary universe.

The Horizon Problem

The first quandary of the standard models is how the universe even "knows" that it is homogeneous and isotropic. This is essentially a problem of causality. The present observable universe is so large that light requires billions of years to cross it; how can two points separated by so much distance resemble one another at all? You may be inclined to argue that since the observable universe started out small, it was originally easy for particles to interact, and for an overall equilibrium to prevail. A homogeneous portion of universe then simply enlarged due to the universal expansion, explaining the observed smoothness. But a careful analysis will show that this cannot be the answer in the standard models.

We have described the surface of last scattering as an impediment to our view of the universe. However, there is a more fundamental, and more important, limit to what we can see. Since the universe is of finite age, and the speed of light is finite, we cannot see all of it at any specific cosmic time. A surface that blocks light from our view is called a **horizon**. We have already encountered an example of an **event horizon** at the Schwarzschild radius of a black hole. An event horizon is a lightlike (null) surface in spacetime that forms the dividing line between those events we can see, and those we cannot. The Schwarzschild horizon occurs where gravity is so strong, as it is around a black hole, that light is trapped; light from inside the event horizon will never reach an outside observer. Another example of an event horizon is the past light cone itself; it separates the universe into a region that we can see, and regions we cannot. All the information we can obtain about the universe is carried to us, either directly or indirectly, by light; therefore, when we look into the cosmos, we are looking along our past light cone. We cannot see events that are spacelike separated from us; at any given moment, there are events that are invisible because they lie outside the light cone, in our elsewhere.

There is a different kind of horizon, of special importance to cosmology. If the universe is not infinitely old, there may exist objects whose light has not yet had time to reach us since the beginning of the universe. That is, given a time t_0 since the big bang, there exists some distance r_{max} beyond which we cannot see. Such a distance would demarcate a two-dimensional sphere in our three-dimensional space. A surface such as this sphere, beyond which all objects are invisible because their light has not yet arrived, is called a **particle horizon**. The particle horizon represents the tracing of the instantaneous light cone all the way back to time $t = 0$. Any object whose world line lies entirely outside this past light cone is beyond our particle horizon and cannot affect us. More generally, any two objects in the universe can be causally connected, and thus able to influence one another, *only* if they are within one another's particle horizon. The particle horizon changes with time; as the universe ages, more and more objects become visible to us.

In cosmological contexts, the terms "Hubble length" and "horizon" are often used interchangeably. For many applications, such as matter perturbations during the

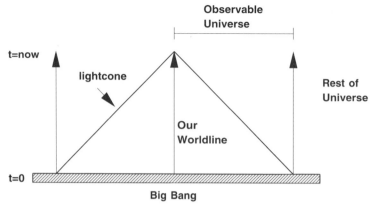

Figure 15.1 A spacetime diagram illustrating the particle horizon, the limit of the observable universe. If we trace our past light cone back to the big bang, we can find the most distant world line that was ever within our past light cone. The present distance to this world line marks the particle horizon.

epoch of galaxy formation, there is little practical difference between the two, and even astronomers often do not need to make a careful distinction. However, when we consider the very early universe, we must be more precise. The distance traveled by light through three-dimensional space over some given time interval is the *lookback distance*: the horizon distance is simply the lookback distance to the beginning of the universe. The simplest example of a particle horizon arises from the usual light cone of special relativity. In special relativity, light follows straight lightlike lines in Minkowski spacetime. We can always set coordinates for our convenience, so let us consider light traveling only along radial directions. The lightlike lines are defined by the condition that the spacetime interval is always zero, that is, $\Delta s^2 = 0$. The usual special-relativistic metric equation becomes

$$c^2 \Delta t^2 - \Delta r^2 = 0. \tag{15.1}$$

The light cone is then

$$r = \pm ct. \tag{15.2}$$

The signs merely indicate that the light can travel in either direction. This equation tells us that over a time t, light has traveled a distance r to reach us. We have used this equation, which is strictly valid only for straight light cones, when we defined the Hubble length.

The light cones in general relativity are curved by the effects of gravity, so we must use the appropriate general relativistic metric to compute them; conceptually, however, it is exactly the same as for the light cones of special relativity. Let us repeat our calculation, using the Robertson-Walker metric. We shall again consider only light traveling along a radial path to reach us, that is, we set the angular change to zero. The light follows a null geodesic specified by

$$c^2 \Delta t^2 - R(t)^2 \frac{\Delta r^2}{(1 - kr^2)} = 0. \tag{15.3}$$

Because null paths are curved in spacetime, not straight as in special relativity, we must integrate (sum) the small intervals in both space and time, in order to find the total distance that a light ray could have traveled from the big bang to the present. In this general relativistic equation for the light cone, the scale factor accounts for the expansion, and the curvature constant is present due to possible curvature in three-dimensional space. In order to evaluate the required integral, we must know the scale factor R as a function of t; that is, we must have a model of the universe. It turns out that this sum can be easily calculated for flat space ($k = 0$) with any scale factor that is of the form $R(t) = t^n$. As long as $n \neq 1$, the result is

$$r_h = \frac{ct_0}{1 - n}. \qquad (15.4)$$

For $n < 1$, this number never becomes infinite, regardless of the age of the universe; there is always a particle horizon from the very first instant. Because t increases faster than R as long as $n < 1$, the particle horizon always increases, encompassing more and more of the universe with the passage of time.

The simplest solution for the scale factor for a standard model is the matter-dominated flat Einstein–de Sitter universe; recall that $R(t) \propto t^{2/3}$ for this case. The horizon length for this important model is

$$r_h = 3ct_0 = 3c\frac{2}{3H} = \frac{2c}{H}. \qquad (15.5)$$

This calculation shows that for the Einstein–de Sitter model, the actual particle horizon is twice the Hubble radius. For example, if we can utilize the approximation that the present universe is matter dominated and still effectively flat, which should be reasonably accurate no matter what the actual curvature constant may be, then the horizon distance today is a few times 10^9 pc. The similar flat, radiation-dominated solution is characterized by $n = 1/2$, which gives a different result for the particle horizon distance.

The particle horizon length is the farthest distance to anything whose world line was ever within our past light cone; it contains all of the universe that is in any way observable to us. Not only does it limit our view of the cosmos; its consequences for causality create one of the most important unsolved puzzles of cosmology. At any given time, parts of the universe are not in causal contact with other regions. The universe became matter dominated within roughly 10^{12} s after the big bang, and this is approximately the time when structure formation began. At this time, only objects within a distance of about 10^5 lt-yr could have influenced one another. How far apart would two objects separated by this maximum distance be today? Their distance would have increased by the same amount as $R(t)$. The redshift for recombination is close to 1000, and recombination occurs at approximately the same time as the beginning of the matter era. Hence, these two objects are now about 10^8 lt-yr apart!

Around the time of last scattering, the horizon length was thus roughly 100,000 lt-yr. At the enormous distance, in time and in space, at which last scattering occurred, such a length now subtends but a small angle on the sky. Expressing this distance as an angular separation of two points on the sky, we can compute that regions of the sky now separated by more than approximately $1°$ could *not* have been in causal contact with one another around the time of recombination. For comparison, the width of the full Moon covers an angular separation of about 0.5°; and the full Moon occupies very little of the area of the sky. Yet when we measure the temperature of the CBR in different directions, we find that the deviations from perfect uniformity *over the whole sky* are barely within the sensitivity of even the *COBE* satellite. The resolution of *COBE*'s instruments was approximately 10°; that is, any two points measured by *COBE* have at least this angular separation on the sky. But when the CBR was created, today's observable universe consisted of approximately a *million* causally *disconnected* volumes!

The only processes we know by which radiation could be forced to the same temperature over a large region require communication throughout the region, and that could not have happened within the standard model. How could conditions have been so phenomenally uniform in all these isolated patches, that the temperature was everywhere the same at recombination despite the fact that the separate volumes of the universe had never been in contact with each other? It is as if there were an orchestra scattered throughout a large stadium, without a director, without music, the musicians never having met each other previously or having had any prior discussions; yet the band spontaneously began to play a piece both in tune, and in tempo. This causality riddle, namely the lack of any physical explanation for the large-scale smoothness of the universe, is often called the **horizon problem**.

The Flatness Problem

We have discussed how observations of the expansion of the universe have determined that the parameter Ω lies somewhere in the range of 0.1 to perhaps 2. Observations of primordial elements, together with the hot big bang model, indicate that the density in baryons can contribute no more than about $\Omega_B \simeq 0.1$. This may not seem particularly significant, but to an astrophysicist, "0.1" is remarkably close to "1," especially since unity, in this case, represents something special, the boundary between an open and a closed universe. Why should the density be so close to a special value? Why is it not well separated from the dividing line, perhaps $\Omega \simeq 10^{-6}$? A density so close to critical implies that the universe is very nearly flat. Since "flatness" is itself an unusual condition among the class of standard models, why "nearly flat"? Might there be some mechanism that would *require* flatness?

This is not merely a persnickety detail. The state of "nearly flat" is much more unusual than it might at first glance appear. If $\Omega = 0.1$ is not sufficiently close to unity for your persuasion, consider the situation at very early times. In a closed or open universe, Ω changes with time. At any particular time it is given by

$$\Omega = \frac{\rho}{\rho_c} = 1 + \frac{kc^2}{H^2 R^2}. \tag{15.6}$$

This shows that in a curved space, Ω evolves like $1/(H^2 R^2)$. Only in the special case of the flat universe does the time dependence vanish, and $\Omega = 1$ throughout history. For an open universe of constant negative curvature (Ω less than unity), however, Ω decreases with increasing time, since it heads toward zero as t goes to infinity. Therefore, in such a universe Ω was closer to 1 in the past and was *extremely* close to

1 during the first few seconds of the big bang. Indeed, for the present universe with Ω near 1 today, then at a cosmic time of 1 s, Ω would have differed from unity by roughly one part in 10^{16}, that is, $\Omega(t = 1 \text{ s}) \sim 1\text{--}10^{-16}$! This implies that if Ω had been already small, for instance, $\Omega \sim 0.1$, *early* in the big bang, then the universe would be obviously open, and nearly empty, today. A similar argument can be made for the closed spherical model. In order for Ω to be just barely bigger than 1 now, it would have had to be infinitesimally above 1 very early on.

If the universe were *not* very nearly flat, most likely we would not exist. If the value of Ω had been much above unity early in the history of the big bang, the universe would recollapse very quickly, perhaps even before galaxies could form, much less before life would have time to develop. If, on the other hand, the universe had been strongly open, it would have expanded so rapidly that, over a moderate interval of time, its average density would have dropped too low for structures to form. The resulting universe would be devoid of galaxies, stars, and, presumably, life. Once again, out of all possible universes, very special initial conditions seem to be required in order to create a universe capable of containing beings that could ask questions about it. The proximity of the density to the critical value is usually called the **flatness problem**.

The Structure Problem

If the universe is isotropic and homogeneous on the largest scales, how could any structures exist? In order to produce the structures we observe, the universe must be mostly, yet not completely, homogeneous; significant, although not excessively large, seed perturbations must have existed from which structure could develop. If we assume the existence of the appropriate seed perturbations, we can develop a theory, such as cold dark matter or mixed dark matter, which explains, at least potentially, the subsequent formation of structures like galaxies and galaxy clusters. But how did the seed perturbations originate? Furthermore, at the cosmic time at which the seed perturbations must have arisen, the mass of baryons within the particle horizon was much smaller than the Jeans mass appropriate to that epoch. How could any familiar physical process that depends upon communication from one region to another have created coherent inhomogeneities that could have later collapsed? Did the initial conditions somehow collude to create just the right perturbations?

Even more puzzling, how does the universe create seed perturbations that were approximately homogeneous over all of the observable universe? When we look at opposite points in the sky, we see galaxies that look similar to one another, and to all other galaxies everywhere else that we can see. The mix of galaxy types is not too different around the sky; the clusters also look similar. As far as we can tell, even the great voids and filaments of galaxies are found throughout the visible universe. Why do we not see exclusively spirals in one part of the sky and only ellipticals in another? Or why not galaxies in one region, and elsewhere something else entirely, in another region that did not communicate with the first at the time when the matter perturbations began to grow? What told all those galaxies to form in a similar way everywhere, even though in the standard model, regions of the sky separated by more than about twice the width of the full Moon could never have been in causal contact.

Once again, we are stymied in our attempts at understanding by the small size of the horizon at early cosmic times.

The improbability of forming such similar structures in an otherwise isotropic, causally disconnected universe may be called the **structure problem**. It is not a completely independent issue, but is tightly connected to the horizon problem.

The Relic Problem

Related to matter density is another puzzle that might be called the **relic problem**. GUTs and other advanced theories of particle physics predict a proliferation of massive particles. In some cases, such particles might provide just the mass density required to bring Ω to unity. Generically, however, they might be *too* massive, causing the universe not only to be closed, which does not seem to be borne out by observations, but to have already collapsed, which is definitely not supported by observations!

One such troublesome particle that arises in advanced theories of particle physics is a beast called the *magnetic monopole*. A monopole, in general, is a particle that produces a field of a particular form; for example, any charged point particle such as an electron or proton acts as an electric monopole. In classical electromagnetic theory, there exist electric monopoles (ordinary charged particles) but no magnetic monopoles. Most GUTs predict magnetic monopoles, somewhat to their embarrassment, because magnetic monopoles would have significant and observable effects on the cosmos, none of which have been seen. They would profoundly affect stellar evolution, for example, in ways that might not be pleasant for those of us who depend on one particular, stable star. Since magnetic monopoles, if they exist at all, cannot be present in any substantial density in the universe, some mechanism must dilute them.

The magnetic monopole is the prime example of an "undesirable" relic particle; there are others. If these relics exist, they would have been created too copiously in the early universe. No unequivocal detections of particles such as magnetic monopoles have ever occurred, and this places severe limits upon their possible density, at least in the neighborhood of the solar system. The simplest GUT theories predict quite a large density of magnetic monopoles in the present universe. Within the standard model, with no way to eliminate these stable particles once they have formed, they would still be traveling the cosmos, wreaking their havoc; yet we do not observe any trace of their presence.

Perhaps it is possible to circumvent this prediction with various alterations to the simple GUT theories. However, it has proved difficult to construct a GUT theory that does not predict a substantial density of monopoles. The possibility that magnetic monopoles might play an important role in the universe was first suggested long ago, in 1931, by P. A. M. Dirac, the physicist who created a special relativistic quantum theory. Dirac argued that the existence of a single magnetic monopole in the entire universe could account for the discreteness of electric charge. Of course, this was long before the formulation of modern field theories, and Dirac's monopole is not really the same creature as a GUT monopole; still, the concepts have similarities. Magnetic monopoles would also make classical electromagnetic theory perfectly symmetric between its electric and magnetic parts, a symmetry that it now lacks because it contains electric monopoles, but no corresponding magnetic monopoles. From various arguments, then, it seems that a magnetic monopole "ought" to exist. That it apparently does not, at least not in appreciable densities, must be telling us something.

The "relic problem" is not a difficulty that arises within the standard models of cosmology alone, but rather from efforts to combine the standard model with particle physics. Nevertheless, it is something that cannot quite be made to fit, and therefore invites an explanation.

The Return of the Cosmological Constant

The standard big bang model of cosmology is spectacularly successful in describing the universe we observe. Its simplicity, elegance, and predictive power are rare qualities in cosmology. Yet we have also seen that it provides no answers to some significant questions, most of which can be reduced to the issue of initial conditions. If our universe is strictly the result of steady expansion from a big bang, then it is a remarkably, perhaps extraordinarily, special universe. Perhaps that is just the way things are, and the universe is an extraordinarily low-probability special case. But we cannot draw such a conclusion on the basis of what we know now. It is possible that the physics of the Planck epoch, of which we have almost no understanding, *requires* that the initial conditions be the way they are. In such a case, "the dice were loaded," and a universe like we see around us was the necessary outcome.

Let us consider a mundane analogy. Suppose you receive a letter telling you that you have just won a lottery. You know nothing about the conditions under which the lottery was held. You do not know the odds of any particular winning combination; indeed, you do not even know that all combinations are equally likely in that particular lottery. Your only firm knowledge is that a winner, you, exists. Should you be surprised? If the odds are like those of a typical state lottery, then a fantastically improbable event has occurred. On the other hand, if the letter informing you of your good fortune bears the stamp "bulk rate," then your "win" may not be so unlikely after all. We know that state lottery winners do exist. Improbability is not the same as impossibility. But in the case of the universe, we do not know whether our universe really is so improbable. Although we know that the initial conditions seem to have been very special, we cannot yet say for certain that all initial conditions are equally likely.

Perhaps we can answer some of these questions, without recourse to an anthropocentric appeal to the anthropic principle, or to vague statements about the as-yet-impenetrable Planck epoch. One promising approach goes by the name of **inflation**, or the inflationary cosmology. This is not a completely new model, but a supplement to the earliest moments of the standard model. In the inflationary model, the universe undergoes a brief period of rapid exponential expansion early in its history. We have already encountered one solution which inflates, the de Sitter cosmology; this model contains no matter, but only a constant, repulsive cosmological constant. We know that the de Sitter model cannot be a valid description of our universe, since matter certainly exists, but it provides an example of the phenomenon. How might inflation have happened in this universe? And might there be a physical interpretation of the cosmological constant? Until now, Λ has been merely a bothersome constant in the Einstein equations, with no particular connection to reality; might we be able to understand it in terms of modern physics?

There are two ways in which to interpret the cosmological constant's role in the Einstein equations of general relativity. If it is placed on the left-hand side of the equation, as Einstein did, it enters as a geometry term, and its interpretation is somewhat abstract. However, it could equally well be placed on the right-hand side, with the terms that account for energy, pressure, and matter. When grouped with the stress-energy terms, it would represent some sort of universal energy density; specifically, it would be an energy density of space itself, a **vacuum energy** density. How could such a thing arise?

In classical physics, the energy of the vacuum is zero, but quantum mechanics changes this. According to quantum mechanics, the vacuum is not empty. This may seem to be contradictory, since a "vacuum" is *defined* classically as the state of emptiness; but the laws of quantum mechanics do not permit absolute nothingness. This strange property arises directly from one of the most important foundations of quantum mechanics, the Heisenberg **uncertainty principle**, which states that certain types of measurements cannot be made simultaneously to arbitrary precision. The more carefully we measure the momentum of a particle, for example, the less we can say about its position in space. Another such pair of measurements is time and energy. The smaller the time interval over which we measure energy, the greater the uncertainty in the energy; a perfect determination of the energy of a quantum-mechanical system would require an infinite interval of time. The relationship between energy and time has some particularly interesting consequences, since matter is just a form of energy. In the quantum universe, the "vacuum" consists of a sea of **virtual particles**, constantly being created in matter-antimatter pairs, and subsequently annihilating. These virtual particles can exist only as long as the uncertainty principle allows. They cannot be observed directly unless somehow, by some process, a virtual particle is converted into a real particle.[1]

Quantum fields are subject to the uncertainty principle. A consequence of the application of the uncertainty principle to quantum fields is that there is no such thing as a constant field. All fields undergo continual changes at the level allowed by the uncertainty principle. These random and unobservable changes are called **quantum fluctuations.** Although quantum fluctuations themselves might be unobservable, they can have physical, observable effects. An example of such a phenomenon is the *Casimir effect.* Imagine two perfectly clean metal plates in a perfect vacuum, separated by a very tiny gap. If the plates are electrically neutral, then classically there should be no force between them. However, the presence of the plates restricts the wavelengths of the quantum fluctuations permitted to exist between them, while the vacuum outside the plates has no such restriction. Thus there are fewer virtual particle pairs between the plates than are present in the region surrounding them; the vacuum between the plates is "more empty" than that outside. The consequence of this effect is a very small attractive force pushing the plates toward one another, as if there were a *negative* energy density, or negative pressure, in the region between the plates. While this may sound like science fiction, recent technology has enabled physicists to measure the Casimir force to very high accuracy, and it was found to obey the theory exactly. Quantum fluctuations are a genuine facet of nature.

Regular vacuum
All quantum wavelengths allowed

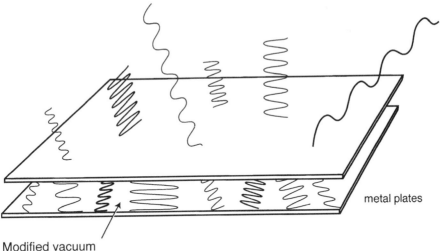

metal plates

Modified vacuum
Quantum wavelengths must fit between plates

Figure 15.2 The Casimir effect. Two metal plates in a vacuum restrict the wavelengths of the quantum fluctuations that will be possible between them. Outside the plates, the vacuum has no such restriction. The difference creates a slight attractive force, pushing the plates together.

The quantum vacuum is an active place, filled with virtual particles and fields and undergoing constant fluctuations. The vacuum energy may be exactly zero, as for the classical vacuum; but it need not be zero. If the vacuum energy is nonzero, then its corresponding energy density is fixed and is determined by the nature of the quantum fields present. Since energy density is just the energy per unit volume, when the volume of space increases, the total vacuum energy increases as well, according to $E = \mathcal{E} \times V$. Therefore, the larger the volume, the greater the total energy.

The vacuum energy density acts physically like a negative pressure. But what is the meaning of a "negative pressure"? Is that not contradictory? A conventional positive pressure results when the energy increases upon compressing a system; the increase resists further compression. This is a familiar phenomenon: squeeze a balloon, and it resists. Conversely, if a rubber balloon is stretched, or a piston is pulled outward in a cylinder, the gas pressure goes down, and the energy density decreases. A negative pressure has the opposite behavior. No matter how strange the concept of a negative pressure may seem at first, it really means only that the associated energy increases when the volume of the system increases, rather than decreasing with increasing volume as does ordinary positive pressure.

A positive pressure tends to make the universe collapse, in the nonintuitive world governed by general relativity. Would you not then expect that a negative pressure would cause it to expand? We have already encountered a quantity that causes a universal expansion. A positive (repulsive) cosmological constant means that any expansion of the universe increases the "lambda-force," which in turn pushes the universe to further expansion. Any positive cosmological constant thus behaves gravitationally like a negative pressure. But we have just identified a vacuum energy density as a source of such negative pressure. Cosmologically, then, any nonzero vacuum energy density would act exactly like a positive cosmological constant, providing a repulsive force to counteract gravity. Thus the cosmological constant may be regarded as a classical equivalent to the vacuum energy density. A model with a vacuum energy is a de Sitter cosmology, which expands exponentially and accelerates rather than decelerates.

We have justified the *possible* existence of a vacuum energy density; can we determine whether such a phenomenon is present in the modern universe? Although we cannot say for certain that $\Lambda = 0$, we know that the universe does not now have a large cosmological constant. Measurements of the expansion rate restrict the mass equivalent of the present vacuum energy density to be at most comparable to the critical density. Even with all its uncertainty, the critical density is quite small; it is comparable to about 10 atoms per cubic meter. If there *is* a nonzero vacuum energy density, then there should be some physical explanation for it, which must account both for its presence and for its small magnitude.

From the perspective of particle physics, a vacuum energy should arise from some quantum field, presumably a field involved in gravity. Attempts to obtain a naive estimate of the vacuum energy density (or cosmological constant) from considerations of spontaneous symmetry breaking would predict a value comparable to the fourth power of the *Planck energy*. The Planck energy represents a fundamental scale of physics, since it is that energy that is obtained by combining the fundamental constants of quantum mechanics (Planck's constant h), gravity (G) and relativity (c) to create a quantity with units of energy; this energy gives the scale at which gravity is thought to have decoupled from the other forces. Specifically, the Planck energy is

$$E_P = c^2 \left(\frac{hc}{2\pi G} \right)^{1/2} . \tag{15.7}$$

Working this out in MKS units, we find that the Planck energy is approximately 2×10^9 J. In units which might be more familiar, this corresponds to about 550 kw-hours, or roughly half a ton of TNT. This is an enormous energy on the scale of any elementary particle; in comparison, the rest energy of the proton is only 1.5×10^{-10} J. When the Planck energy is assigned to the quantum of a field (i.e., a particle), and the corresponding energy density of the field is computed, the fourth power of the resulting number is at least 10^{120} times greater than the observed upper limit on the vacuum energy density. This mismatch of the only known natural scale for the cosmological constant with the observational limits on its value is not understood. The smallness of the present-day vacuum energy density is a conundrum similar to the "flatness problem." The vacuum energy density (or cosmological constant) must be so tiny that it "ought" to be zero, within the limits of our ability to measure it. There is no physical explanation for why it should be zero and not a function of the fourth power of the Planck energy, or the energy associated with some other symmetry breaking; but observations of the present universe do not allow the possibility of such a large value.

We are not living today in a de Sitter universe. But is it possible that the universe might have gone through such a phase early in its history? A quantum field associated with some particle not present today would have been responsible for providing the required vacuum energy. But if it is no longer present, how would we know what it is, or anything about it? No particle accelerator we can imagine could reach the energies at which such a mysterious particle might be detected. We must rely upon theory until better data are at hand. GUTs, of which there are several varieties, predict the existence of massive particles that we cannot yet detect. These hypothetical particles play various roles within the theories; which particles are required, and of what type and mass, depends upon the particular theory.

There is, at present, no fully satisfactory model of force unification, the phenomenon which GUTs seek to explain. The electroweak unification, at least, seems to be well understood. Unfortunately, the most straightforward extension of the electroweak theory to include the strong interaction is known to be incorrect, because it predicts a lifetime for the proton that is less than the experimental lower limits. Proton decay has never been detected. This does not mean that the proton is absolutely stable; there are good reasons to believe that it may not be so. However, its expected life is certainly greater than the 10^{30} yr predicted by the simplest GUT. Therefore, we must seek a more complicated theory, and no successful one has yet emerged. Until we understand grand unification of strong and electroweak forces, we cannot make firm predictions of what particles should have existed at the energy level of the unification epoch, what their masses might be, and whether they might survive as relic particles.

A New Explanation

Some candidates exist for providing a vacuum energy density in the early universe. Many theories of symmetry breaking invoke the massive *Higgs boson*. This hypothetical boson plays a role in spontaneous symmetry breaking during the unified epoch. The Higgs boson takes part in particle interactions connected to baryogenesis, then conveniently annihilates after the epoch of unification, so would play no role in the present universe. A more troublesome grand-unified particle is the aforementioned magnetic monopole, which tends to be overproduced by GUTs. These two particles provided the original motivation for inflation, the hypothetical de Sitter phase early in the history of the universe, driven by the vacuum energy of a quantum field. Physicist Alan Guth realized in the early 1980s that if the universe did undergo a de Sitter phase during the unified epoch, then the density of magnetic monopoles, which would have formed prior to this era but not during or afterward, would have been diluted away to almost nothing by the huge increase in the volume. All that was needed was a means to induce exponential expansion. The original models of Guth and others invoked the Higgs boson to fill this role. Like all particles, the Higgs boson is associated with a field. The field in turn is associated with a **potential**, a function in spacetime that describes the energy density of the field. From the potential, we can learn where and when the value of the vacuum energy density might have been nonzero.

These concepts are awkard to express in words; here we see an example of how mathematics tremendously increases our power to understand quantitative ideas. It is far more transparent to use elementary mathematical notation when we work with the ideas behind inflation. Suppose that the field, which you will recall is just a description of a variation in space and time, is specified by some function $\phi(x, t)$.

For our illustrations, we shall allow x to stand for any spatial dimensions. (A one-dimensional example suffices for our purposes, anyway.) We do not need to know any details of the field here. All we assume is that the field is *scalar*; that is, it has a single value at each point, so that we do not require multiple components to specify the behavior of the function. All advanced theories of inflation start from a scalar field, so this is not an excessive simplification. Now we suppose that the field has an associated potential $V(\phi)$. That is, V is a function of the field quantity ϕ, and thus only indirectly does V depend upon space and time. If we wish to know the potential at any event (x, t), we first compute ϕ at that event, then insert that value into the potential function $V(\phi)$. As a concrete example, we could consider the height above sea level (as depicted on a topographic map) to be a scalar field, and the gravitational potential to be the associated potential function. In this case, the gravitational potential tells you how much energy you acquire, or expend, in moving from one height to another. In the case of the early universe, computation of the potential tells the energy density at that event.

Consider some particular event (x_i, t_i). The field, and thus the potential, have certain values at that event, specified by their respective functions. Now contemplate the region in spacetime near this event. If the potential is greater for those values of x and t that are a little different from x_i and t_i, then the field will prefer to sit where it is. If the potential is smaller in some direction, the field will tend to shift toward it. The general principle is that the field "wants" to find a minimum of its potential. The analogy of field as altitude is once again helpful. In the gravitational case, it is energetically favorable for mountains (maxima in altitude) to erode away and form a smooth, flat plain, since this would minimize the gravitational potential of the terrain. Similarly, if the region around (x_i, t_i) is very flat, as defined by the form of $V(\phi)$, then the field will not change; that is, we say it is *stable* around this event. For example, it may happen that for a certain time interval, which may be short or long depending upon the potential, the spatial variations of the energy density are very small. The field will remain stable until, perhaps after some appropriate time interval has passed, the form of the potential function changes. If the potential function changes, the field will have to adjust anew.

Now let us apply these general ideas to a scalar field in the early universe and return to the hypothetical Higgs boson as an example of a particle that is associated with a scalar field. The Higgs boson controls a symmetry breaking in the unified epoch. When the mean energy per quantum of the universe is above the scale set by the breaking of this particular symmetry, it turns out that the Higgs field is stable, with an expected, or average, value of zero, in the vicinity of some positive value of the potential, that is, $V(\phi = 0) = V_0 \neq 0$. Because this energy density is greater than zero when its allied field is zero, this state is known as the **false vacuum**. It is not an absolutely stable state, but is like the stability of a marble resting atop an inverted bowl. The marble will stay there as long as it is not disturbed, but a perturbation will cause it to roll down to the genuinely stable state on the table. The potential of the Higgs boson resembles an inverted bowl with a very flat base. The false vacuum is a *metastable* state, analogous to the base of the bowl.

While the field is in the metastable false vacuum, its vacuum energy density dominates the universe completely. As we have discussed, this energy acts like a

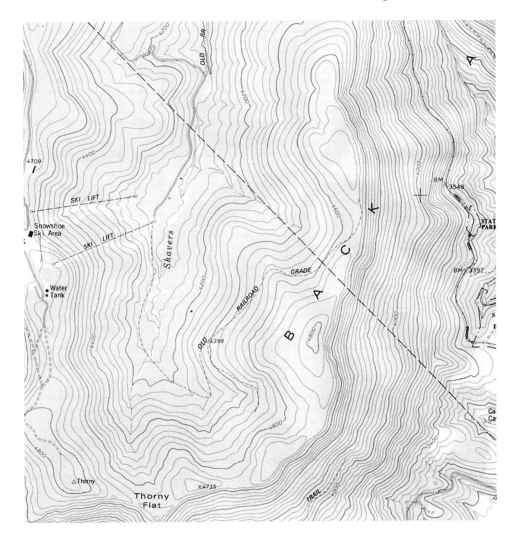

Figure 15.3 A contour map shows altitude above sea level. Each point has an elevation; this is a scalar function of position $\phi(x)$. The elevation in turn implies a gravitational potential, $V(\phi)$. It is energetically favorable for the mountains to erode into flat plains; that is, to evolve to a uniform minimum altitude with lower potential.

negative pressure, a large one in this case, which can be shown to give an equation of state

$$P = -\rho_\Lambda. \qquad (15.8)$$

The energy density in a given volume of vacuum is constant, hence ρ_Λ is constant. Inserting a constant density into the Friedmann equation (11.8) yields a term that varies as R^2, exactly as a cosmological constant does. Comparing the constant-density form of equation (11.8) with the way in which the Λ term enters equation (11.19) shows that the vacuum energy density can be equated to a cosmological constant

$$\rho_\Lambda = \frac{\Lambda}{8\pi G}.$$ (15.9)

If it happens that this vacuum energy density is large compared to conventional mass-energy densities, we obtain the familiar de Sitter solution. The flat de Sitter model expands exponentially; hence the "inflation." Any positive cosmological constant will eventually cause the universe to inflate, regardless of its source.

If we assume that the cosmological constant is a consequence of the energy density of some quantum scalar field, then we must make a connection between the potential and the cosmological constant. This relationship would vary from one field theory to another, but for the sort of theory we are taking as our example, namely, a field connected to a spontaneous symmetry breaking, we can consider the constant vacuum energy density to be specified by the value of the potential function when the field is zero. Symbolically,

$$\rho_\Lambda = V_0.$$ (15.10)

For most theories of this class, it happens that $V_0 \sim E^4$, where E is an energy that characterizes the scale of the symmetry breaking. This may provide some justification, although not an explanation, for our earlier assertion that the "natural" scale of the cosmological constant should be something like the fourth power of the Planck energy. For the particular field we have used as an example, the Higgs field, the energy in question would *not* be the Planck energy, of course; it would be the rest energy of the Higgs boson, which is still very large because the strong and electroweak interactions are unified only at very high energies. If the Higgs boson had been responsible for an inflationary phase during the unified epoch in the early universe, then the cosmological constant during that interval would have been given by $\Lambda \sim 8\pi G E_{\text{Higgs}}^4$. The energy density of the false vacuum, acting like a cosmological constant in a de Sitter universe, powers the inflation. Once the energy drops below the scale of unification, the field adjusts and drops down to a state of zero vacuum energy density, the **true vacuum**, and inflation ceases. The true vacuum is truly stable, and the field remains there indefinitely.

The Higgs boson is not gospel, however. It is an example of a particle that does play a well-defined role in a theory of grand unification, and that does have some characteristics that would have to be present in a particle which could cause inflation. Historically, it was the particle that motivated the inflationary model in the first place. But it turned out that the Higgs boson is probably not the responsible particle. The original model of inflation described above suffered in the details. Inflation occurred while the field was in the metastable false vacuum. However, for various technical reasons, nothing subsequently could eject the field from that metastable state; the inflation never stopped. This clearly contradicts our observations of the universe, at least that which we can see, so the original version of the inflationary universe cannot be correct. The model was rescued by a change in the potential, V. In this "new inflation" scenario, inflation occurs not while the field is *in* the false vacuum, but during the transition from the false to the true vacuum. Since the potential plays the role of the cosmological constant, we can see that a slow decrease during inflation will provide a simple way for the inflation to come to an end after a period of time. In other words, the "cosmological constant" changes slowly during the new inflation. To ensure that "enough" inflation occurs, the potential must be very flat so that the field carries out the transition very slowly—slowly, that is, in comparison to the characteristic rate of expansion at that time.

If such an inflation occurred, it would have happened around 10^{-37} s after the big bang and would have required approximately 10^{-32} s to complete. During this cosmic eyeblink, the scale factor would have inflated by a dizzying factor of some 10^{40}–10^{100} or even more. This more successful new inflationary scenario was first proposed independently by Andrei Linde and by Andreas Albrecht and Paul Steinhardt, and subsequently developed by many researchers. New inflation does not depend upon the presence of any particular particle, but merely requires the existence, at the appropriate stage in the history of the universe, of a particle with an extremely flat potential and a slow transition to the true vacuum. This generic particle has come to be known as the **inflaton**.

You might suspect that, since temperature drops as the inverse scale factor, inflation would cool the universe tremendously, and it does. How do we reconcile this with the good agreement between observations and the standard hot big bang model, which never undergoes such a drastic expansion? At the end of the inflationary period, the vacuum energy from the earlier false-vacuum and inflation stage is consumed in the creation of particles, which decay and ultimately convert their energy into reheating the universe. A huge energy density would have been locked up in the quantum field, so it is not artificial for the universe to be reheated back to where it should have been if inflation had not occurred, once this energy was released and converted into other, more conventional, forms of energy. Before inflation, the evolution of the universe was dominated by particle physics. After this burst of heating, the mostly classical, radiation-dominated phase of the standard models began.

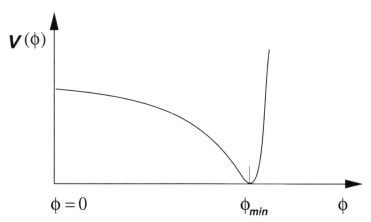

Figure 15.4   Schematic illustration of an inflationary potential. The false vacuum occurs because $V(\phi)$, which represents the vacuum energy, is nonzero when the field ϕ is zero. The field eventually transitions to the true vacuum of zero potential at some associated scale given by $\phi = \phi_{min}$. During this transition, the universe contains a large energy density due to the field, and inflation occurs.

a)

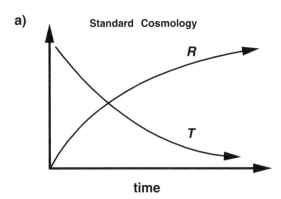

Standard Cosmology

R

T

time

Figure 15.5 Comparison of the evolution of the scale factor R and the temperature T in the standard and inflationary universes. In the inflationary model, an exponential expansion phase results in strong cooling. At the end of the inflationary phase, reheating occurs as the vacuum energy is converted into more conventional forms of energy. The universe subsequently evolves as in the standard model.

b)

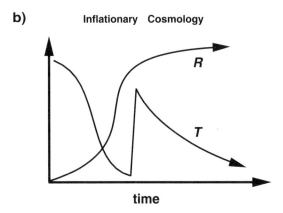

Inflationary Cosmology

R

T

time

Inflation and the Initial Conditions

Inflation achieves the goal of explaining the initial conditions of the big bang by rendering them insignificant. What we take for special initial conditions in the standard model now arise naturally as a necessary consequence of inflation. Consider the horizon problem. Because of the enormous increase in the scale factor, what is now the observable universe began from a tiny region that *was* in causal contact prior to the beginning of inflation. Thus the isotropy of the cosmic background radiation is automatically assured, and the "horizon problem" is eliminated.

Moreover, the exponential expansion means that the observable universe becomes effectively flat, regardless of its original state, because any curvature it might have possessed initially is stretched away. We can see this by considering the Friedmann equation

$$\left(\frac{\dot{R}}{R}\right)^2 = \frac{8}{3}\pi G\rho - \frac{kc^2}{R^2}. \tag{15.11}$$

During inflation the vacuum energy density is by far the dominant component of the total energy density ρ. Since the vacuum energy *density* is constant, the total vacuum energy increases significantly as the universe expands. The scale factor R grows so enormously during this interval that the curvature term drops rapidly toward zero, whereas the density term on the right-hand side stays fixed. Therefore, at the end of inflation, the observable universe effectively expands from then onward as a flat ($k = 0$) universe, regardless of the original value of k. This effect can be visualized by imagining a balloon blown up far beyond its normal proportions, until the surface is stretched so much that any small neighborhood around a point on that surface will appear flat. Such a simple analogy should not be taken excessively literally, of course. The inflated universe is "stretched" by such an enormous, unimaginable factor that any relict curvature has a scale that is far, far greater than the Hubble length. In an inflationary universe, there is no "flatness problem," since the observable universe would always end up indistinguishable from a flat universe. The critical density is not a special condition in this model but a natural consequence of particle physics in the early universe.

Furthermore, any relic particles created before inflation, such as magnetic monopoles, are exponentially diluted away, their density diminishing to insignificance. This means that baryogenesis must occur *after* inflation, or else the baryon density would also have dropped effectively to zero, and galaxies as we see them would never have formed. However, after the universe reheats, baryogenesis proceeds exactly as in the standard model, so this is not a problem for inflationary scenarios.

But in solving these problems, have we aggravated the structure problem? If the universe now can stretch away any inhomogeneities, where are the seeds for the formation of structure? A leading possibility is that the perturbations originated in the quantum fluctuations of a field associated with some kind of elementary particle.

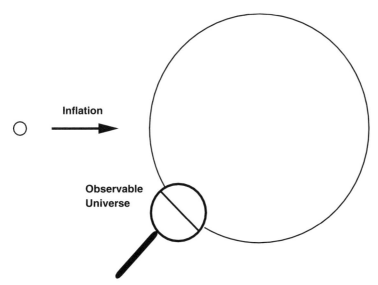

Figure 15.6 A small universe undergoes exponential inflation by a huge factor. Any initial curvature is enormously stretched. Our observable universe, a tiny bit of this inflated structure (here magnified for view) has no measurable curvature within the Hubble length. Thus, in the inflation picture it is natural that we observe a flat universe.

In the inflation scenario, the seed perturbations for structure formation are created during the inflation itself, from quantum fluctuations in the inflating field. Like other lengths, the sizes of these fluctuations are stretched by the enormous increase in the scale factor, until they become large enough to generate macroscopic inhomogeneities. It turns out that this process naturally leads to the Harrison-Zel'dovich spectrum of perturbations. This is a very pleasing result: this spectrum was proposed long before inflationary models were developed because it had properties that seemed to fit the observed structures well and because it required minimal assumptions about the nature of the perturbations. Its appearance in a well-motivated model provides a physical explanation for what had previously been justified primarily on mathematical and empirical grounds.

Quantum fluctuations are not the only candidate for seeding the universe. When a symmetry in the fundamental forces of nature is broken, it is as if the universe undergoes a phase transition, such as the freezing of water into ice. What if a phase transition occurs irregularly? If different portions of an ice cube freeze at different rates, the interior of the cube will be divided by many fine planes, separating one region from another. Each separate portion froze uniformly, and the planes are discontinuities where the rates of freezing did not match. A very similar phenomenon might have happened as the universe went through a phase transition associated with some kind of symmetry breaking. Discontinuities called **cosmic strings** might have formed. These cosmic strings can move, like the lines between the tiles of a shuffle puzzle as you shift a tile from one position to another. As they move through the universe, they might attract a wake of matter, which would form the initial density perturbations. Cosmic strings are exotic, much more so than quantum fluctuations. The particular kind of phase transition that creates them may not have occurred. If they existed, they would have some observable consequences, beyond creating matter perturbations. They tend to decay into gravitational radiation, which, if sufficiently sensitive detectors could be built, would be observable. If any long-lived strings still existed today, they could cause gravitational lensing and discontinuities in the temperature of the CBR. No observational evidence exists for cosmic strings; neither have the observations completely ruled out their presence. However, future observations with greater sensitivity may soon rule out cosmic strings as an *important* constituent of the cosmos.

Chaotic Inflation

Inflation wipes away the memory of any initial conditions and ensures that the observable universe will be flat and smooth, regardless of how it began, while also providing the necessary quantum seed perturbations from which to form structure. But if our entire observable "universe" began as a tiny patch within a larger big bang, does that mean that our "universe" is not all there is? Are there other "universes"? Inflation will occur only in patches where the vacuum energy of the false vacuum is dominant; this condition is not guaranteed to hold over the entire universe during the unified epoch. The inflating field would not likely be constant everywhere.

The most general case is **chaotic inflation**, an idea originally proposed by Andrei Linde.

In the chaotic inflation model, quantum fluctuations in a primordial field cause some portions of the universe to inflate, while other regions do not. In some cases, the inflation begins, but falters. In this scenario, the inflating patches rapidly form "child" universes that are attached by wormholes to the "mother" universe. The wormhole effectively cuts off communication between the "child" and "mother" universes. Our observable universe might be only one of many noncommunicating universes, some of which have inflated, and appear flat, and others of which retain the original initial curvature! Presumably, this process goes on into the indefinite future and continues into the indefinite past.

Chaotic inflation offers an intriguing interpretation of the anthropic principle. It is not particularly significant that conditions seem to be remarkably right in our universe for life to have formed. Each child universe in the chaotic inflation model could well have its own set of physical conditions. When there are many universes from which to choose, only those in which the correct conditions prevailed would have given rise to life forms who could ask questions about their environment. In such a case, marveling at the apparent wonder of our existence is like speculating about the astonishingly unlikely set of genes that any human possesses. The specific set of genes that makes up any given individual is almost infinitely improbable, from a strictly statistical point of view; yet no one is surprised about the existence of any particular person. This idea of multiple, possibly infinite, child universes seems like a natural way out of the "specialness" problem. However, the special conditions of our universe do not constitute a "proof" of the chaotic-inflation model.

We can think of this in terms of a game of "straight" poker, in which each player is dealt five cards and must play them as they are. For our purposes, let us define a "hand" as some combination of five cards from a deck of 52. The number of such hands is enormous, more than two and a half million. The probability of receiving *any* hand of five specific cards is equal, approximately one in two and a half million; but often there is more than one way to achieve a certain scoring combination. For example, there are many ways in which a pair, or even three of a kind, could occur in a hand, but only four royal flushes (ten, jack, queen, king, and ace of one suit) are possible; hence the probability of a royal flush is roughly four in two and a half million. The "gambler's fallacy" is the assumption that a long sequence of poor hands means that a good hand is somehow "due"; most of us know, at least rationally, that this is not the case. Now turn this around; suppose that you visit a poker game at which the first thing you see is a player laying down a royal flush. No doubt you would be amazed. If you then concluded that a large number of hands must have previously been played, you would have fallen into the *inverse gambler's fallacy*. The occurrence of an improbable event does not imply that any previous trials need have taken place at all. Similarly, the winner of a state lottery may never have played before. Thus the apparent specialness of our universe does not in itself *require* that other universes exist; this would be the inverse gambler's fallacy again. In cosmology, unlike poker, we do not even know yet what the odds are.

The chaotic inflationary model would be difficult or impossible to test observationally, since the individual "child universes" cannot communicate. Guth and others have suggested that it might be possible to create miniature quantum bubbles in the laboratory, but this experiment would seem to be extremely difficult (and perhaps dangerous). Indirect tests might be possible, if ever the inflaton is detected, so that its associated potential can be determined and compared with theory. For now, however, chaotic inflation is another interesting speculation which has, as yet, little observational justification.

Testing Inflation

Something as seemingly bizarre as inflation prompts the suspicion that cosmologists have ventured beyond the scientific and testable, into the purely speculative. However, inflation can be put to the test. After all, Guth's original proposal for inflation was found to be unworkable, and hence was falsifiable. Physicists continue to work toward GUTs, quantum gravity, and theories of the high-energy state of the early universe. A direct consequence of such a new theory may very well be something like inflation. Even in its present form, inflation makes some predictions about the subsequent evolution of the universe, and these can be tested against observations.

Further data from *COBE*, which was just becoming available early in 1996, may provide a means of testing one fairly universal aspect of nearly all inflationary models. Inflationary models produce a particular spectrum of primordial seed perturbations. Improved data for temperature anisotropies might show whether they correspond to "standard" perturbation models or whether something such as cosmic strings might provide a better explanation for seed perturbations. Since inflationary models predict a Harrison-Zel'dovich spectrum of perturbations, whereas "defects" such as cosmic strings would generate a different spectrum, it is possible to distinguish these two hypotheses. This test does not generally favor one inflationary model over another, but it might provide some observational evidence for inflation, or something similar to it. This program will also be aided by more complete redshift surveys of galaxies. Such surveys will give us an increasingly accurate description of perturbation structures on length scales smaller than those detectable by *COBE* and its successors.

Another obvious consequence of the standard inflation picture is that $\Omega = 1$. However, most measurements of Ω still tend to point toward $0.1 \leq \Omega \leq 0.3$. Many of the dynamical estimates of the cosmic matter density, taken together with models of big bang nucleosynthesis, indicate a universe that is open and consists of approximately 10%–20% baryons. If theories such as inflation, which seek to eliminate the "flatness problem" by *requiring* a perfectly flat universe, are to be relevant, then the dynamical measurements are still missing most of the matter of the universe.

How can we reconcile the observations with the prediction of $\Omega = 1$ from inflation? There are several possibilities. First and simplest is the possibility that for whatever reason, the density parameter of the universe simply *is* $\Omega \simeq 0.2$. This cannot be discounted at the present time, but it contributes little to our understanding of how the initial conditions might have set us up for such a state. Even if the inflation scenario could be modified to accommodate a lower Ω, much of the motivation for it would be lost with the return of special, finely tuned conditions.

A second possibility is that we have underestimated Ω because clusters of galaxies are still simply not big enough to detect the overall matter density of the universe. The size of the Virgo Cluster is approximately 20 Mpc, still small on cosmic scales. Some dynamical methods that utilize larger scales have obtained results with Ω near or at unity. However, at the present time these methods are much less firmly founded, both theoretically and observationally, so we cannot discard the Virgo and similar results just yet.

A third possibility is that we are overlooking the main distribution of matter. It is clear that luminous matter is not distributed uniformly, but is clumped into galaxies, and these galaxies themselves tend to clump into clusters. Dynamical methods depend upon our observations of the gravitational effects of the dark matter upon the visible. There is, however, no a priori reason for dark and visible matter to clump in the same manner. Moreover, the dynamical methods can measure only the *deviation* from the background. If there is a uniform background distribution, it would not contribute to local gravitational dynamics on the scale of galaxies or galaxy clusters. If this is the case, the dynamical and nucleosynthesis estimates of Ω would suggest the intriguing possibility that baryonic matter, whether dark or luminous, aggregates together, while most of the undetected matter, whatever it might be, makes up the smooth overall matter distribution of the universe.

More troubling to the inflationary models are recent measurements of the Hubble constant. Many recent observations have been quite consistent at producing a number in the range of $65 \lesssim H_0 \lesssim 80$ km/s/Mpc. A Hubble constant greater than 65 km/s/Mpc would be a very serious blow to standard inflationary models. Inflationary scenarios generally demand that $\Omega = 1$, but for the newer values of the Hubble constant, the corresponding Einstein–de Sitter universe is too young to accommodate the ages of the globular clusters, which have been fairly reliably dated to approximately 14 billion years old. The epochs in which particle physics and inflation would have occurred are extremely short; afterward, the inflationary universe would continue to evolve as a classical Einstein–de Sitter ($k = 0$) model, which has an age of two-thirds the Hubble time. A Hubble constant of 75 km/s/Mpc corresponds to a Hubble time of 15 billion years. The age of an Einstein–de Sitter model with this Hubble time would be only 10 billion years, much younger than the apparent ages of the globular clusters.

Some theorists have speculated that observables such as H_0 might vary from region to region within the universe; thus our "local" measurements might show a high value of H_0, whereas the overall Hubble constant could be considerably lower; indeed, a few measurements on larger scales show $H_0 \simeq 50$ km/s/Mpc. While this is one possible way out of the current age problem, in the absence of any compelling evidence it must be regarded as suspect, since it tends to violate at least the spirit, if not the letter, of the cosmological principle. In any case, no final verdict has yet been returned on the Hubble constant, and the higher values for H_0 are still tentative. Cosmology has encountered age problems before, which have generally been resolved by better data and better models. There are still too many uncertainties for any firm conclusions about the viability of the Einstein–de Sitter model to be drawn, with or without inflation. But it may be that in the very near future we will have a fairly well determined value of H_0, which we will simply have to accept.

Some possibilities remain for explaining the low Ω data even within the context of a flat model. Inflation is not incompatible with the presence of a cosmological constant from another, possibly classical, source. Inflation or no, the cosmological constant alone can account for a flat universe with subcritical density. If there is a permanent cosmological constant, we exchange one special situation, a matter density that "ought" to be unity, for another, a cosmological constant that is extremely small, yet nonzero. Or it could be that the universe *is* open, with a density ratio of roughly 0.2, and the answer to the apparent special conditions is to be found elsewhere. On the other hand, it is difficult to understand how the universe could have become so incredibly uniform unless *something* like inflation occurred.

Even assuming that better data will settle the "Ω problem," does inflation solve all our cosmological problems? Unfortunately, it does not; there are still mysteries in cosmology. The model does not explain why the present-day vacuum energy density, or equivalently the cosmological constant, is so small. Inflation also does not yet seem to fit comfortably within any known scenario of particle physics. Originally, the Higgs boson was thought to be the particle responsible for inflation. However, the limitation on the amplitudes of seed perturbations, as set by the isotropy of the CBR, requires an extremely flat potential, which seems to be a special requirement of its own. Further investigation showed that a potential which could be associated with a Higgs boson was not sufficiently flat to cause inflation without also disrupting the CBR. Several other potentials have been suggested, most of which have some justification in particle physics, but none of which corresponds to a particle that has any role other than to produce inflation. Replacing one set of ad hoc requirements, special initial conditions, with another, a particle that seems to have nothing else to do with particle physics, at least not yet, does not seem all that satisfying.

It is mysteries such as these that drive progress. Work will continue; perhaps someday a theory will come along that will explain all the data within a fully developed combination of particle physics and general relativity. Unfortunately for the curious among us, no such theory is in sight at present, but a breakthrough is always possible. For the time being, inflation must be regarded as a promising and interesting suggestion that as yet cannot provide all the consistent answers we seek. No matter what, however, we cannot base our models on theoretical prejudice or aesthetic appeal alone. All sciences, including cosmology, must ultimately be founded on empirical evidence. Better data will guide theoretical progress, and from improved theory we will achieve greater understanding of the origins of our universe.

KEY TERMS

horizon	event horizon	particle horizon
horizon problem	flatness problem	structure problem
relic problem	inflation	vacuum energy
uncertainty principle	virtual particle	quantum fluctuation
potential	false vacuum	true vacuum
inflaton	cosmic string	chaotic inflation

Review Questions

1. Explain the difference between a *particle horizon* and an *event horizon*.

2. (More challenging.) Write down the formula for the horizon length during the radiation era. Compute the horizon length at cosmic time $t = 10^{-37}$ s. Consider two particles separated by this distance at this cosmic time. How far apart would they be, assuming only standard cosmic expansion, at recombination, $t = 10^{13}$ s? Assume that the radiation-dominated formula for the scale factor still holds. Use meters as your unit for this problem.

3. Discuss the horizon, flatness, structure, and relic problems. Do they constitute genuine "problems" for the standard model, in the sense of being inconsistent with its assumptions? In your opinion, which of these problems is most troublesome? Why?

4. What effect would a negative pressure have had in the early universe? How might such a phenomenon have been produced? What connection might this have to the cosmological constant?

5. Explain qualitatively what a "potential" is. What is the essential property that a potential must have for inflation to occur? Illustrate your answer with a sketch.

6. Distinguish the true vacuum and the false vacuum. Which is important for inflation? Why?

7. How does the inflationary model account for the formation of structure? What features of inflation are particularly appealing in this regard?

8. If the universe inflated enormously, what would happen to the temperature of any preinflation constituents? What is the source of the energy for "reheating" at the end of inflation?

9. Discuss how inflation solves the horizon, flatness, and relic problems. What problems does inflation itself introduce?

10. Discuss "chaotic inflation." What intriguing interpretation of the anthropic principle does it offer?

Note

1. We have already examined Hawking radiation, one example of such a process.

CHAPTER 16

The Edge of Time

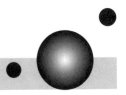

> Common sense is the collection of prejudices acquired by age eighteen.
>
> —ALBERT EINSTEIN

In our journey through the universe, we have encountered many wonders. Once we leave behind the bounds of our Earthly velocities and distances, we realize that our cozy intuition is often wrong, our common sense does not apply. Special relativity seemed so strange at its introduction that even some of the most distinguished scientists of the day refused to accept it. General relativity had an easier time, but quickly gained such a formidable reputation that most people became convinced that none but the brightest genius could hope to grasp its concepts. Yet both these theories are elegant and straightforward in their fundamental ideas; confusion occurs because they demand a way of thinking that is so at odds with our everyday experience and intuition. Nevertheless, the special and general theories of relativity are at their cores pure classical physics, the appropriate extensions of Newtonian physics to spacetime itself. For most of our story, we have been concerned with astronomically sized objects and with scales, of both time and distance, that are enormous even in comparison to our solar system; therefore, we have easily been able to remain within the realm of classical physics. Yet every now and then, the shadowy world of quantum mechanics has intruded even into our modern version of Newton's clockwork.

Relativity and quantum mechanics were the two great triumphs of twentieth-century physics. Both were developed during the first 30 years of the century. Both are spectacularly successful within their respective domains. Quantum mechanics governs the world at the smallest scales, the level of particles, atoms, and molecules, while general relativity, as a theory of gravity, rules the largest scales, from stars and planets to that of the universe itself. Low-energy quantum mechanics, as well as special relativity, boast ample experimental verification from the laboratory. General relativity is much more difficult to test experimentally, so its empirical foundation rests upon its success at explaining and predicting certain observed astronomical phenomena.

Nevertheless, every such test has produced results completely consistent with the predictions of general relativity.

Gravity is by far the weakest force in the universe; in the hydrogen atom, the electromagnetic force between the proton and the electron is about 10^{40} times as great as the gravitational force between them. This is fairly representative of the difference in scales between the quantum and gravitational realms, and accounts for our ability, through most of our study of cosmology, to separate the two theories without ambiguity. Yet they must inevitably meet. Near a singularity, the curvature of spacetime must be so great that the scale of gravity becomes comparable to that of the other fundamental forces. To describe such a state, we must find a theory of **quantum gravity**. Moreover, quantum mechanics has already been applied to the explanation of the other three forces, the electromagnetic force and the strong and weak interactions; should not gravity be similar? It might seem as though the challenge of developing quantum gravity should not be so great. After all, *special* relativity and quantum mechanics were united in the 1920s by the British physicist Paul A. M. Dirac. The most significant result of Dirac's theory was its requirement that antiparticles exist, a prediction that was confirmed in 1932 by the discovery of the positron (the anti-electron). The Dirac theory is now well established as the special relativistic quantum mechanics. More than 70 years later, however, general relativity has still not been successfully incorporated into a consistent quantum formulation.

We are able to take into account the quantum-mechanical nature of matter in white dwarfs and neutron stars because we understand the behavior of matter under the pressures and densities encountered in these objects. For a black hole, we have no such understanding. The center of a black hole marks a singularity in spacetime, where classical general relativity must break down. In the collapse to an infinitely dense singularity, the physics of gravity necessarily enters the quantum realm of the microscopic. We simply do not know how to describe the properties of matter under such extreme conditions, but we know that at some point, quantum mechanics *must* play a role. Without a theory of quantized gravity, we cannot know what lurks at the center of a black hole. We do not even need to go all the way to the singularity to find quantum effects associated with black holes; Hawking radiation shows that the very strong gravitational field near the hole's event horizon has predictable quantum consequences. What other phenomena might we discover from a full theory of quantum gravity? At present, we cannot say.

Another meeting point of general relativity and quantum mechanics is the very beginning of the universe, the big bang itself. The Planck time marks the limit of our ability to speak at all about the evolution of the universe. Yet the Planck epoch may be crucial to our understanding of some of our most fundamental questions about the universe, such as how perturbations arose and whether the initial conditions are restricted in some way, or indeed whether they matter at all.

The absence of a quantized theory of gravity is not due to lack of effort by theorists. Many proposals have been put forward, and some progress toward this goal has been made. There as yet exists no complete theory, although the various suggestions may hold pieces of the answer. But before we provide some flavor of the attempts at the unification of gravity and quantum mechanics, we must first discuss a few of the basic ideas of standard quantum theory.

Particles and Waves

Although it is a quite modern theory, quantum mechanics has its origins in an old question: Is light a particle, or is it a wave? Newton was an early advocate of the corpuscular theory of light, although he recognized that the data were insufficient to decide the issue. However, the observation of wave phenomena such as interference and diffraction, as well as the development of the electromagnetic theory of light in the nineteenth century, seemed to answer the question most convincingly in favor of the wave. And there things might have remained, had it not been for the problem of explaining the blackbody spectrum. Max Planck found he was able to do so by hypothesizing that blackbodies can emit light only in discrete amounts, or *quanta*, with energies proportional to the frequency of the light.[1] In 1905 Einstein applied this idea to the *photoelectric effect*, in which a light beam shining upon a metal plate causes an electrical current to flow. (The explanation for this effect won Einstein his Nobel Prize; relativity was thought to be too exotic for the Nobel Prize, which specifies service to humanity.) Einstein showed that light has a particle alter ego, the photon. When a photon of sufficient energy strikes the metal surface, it ejects an electron; these liberated electrons constitute the observed electrical current. The photoelectric effect cannot be explained if light is considered to be a wave; it can be understood only if light occurs in the form of discrete photons. On the other hand, refraction and diffraction cannot be understood if light behaves as a particle; for these phenomena, light must be a wave.

In quantum mechanics, the blending of particles and waves extends to everything. Not only does light behave both as a particle and as a wave, but also electrons, protons, atoms, molecules, and, by extension, even macroscopic objects have both a particle and a wavelike nature. This insight is due to Louis DeBroglie, who proposed in his doctoral thesis in 1924 that a particle could be described by a wave whose wavelength was determined by its momentum. The duality of particle and wave is one of the most counterintuitive ideas of quantum mechanics. Surely, it might seem, an entity should be either a wave or a particle, but not both. However, according to quantum mechanics, *no* experiment can be devised in which both wave and particle behavior simultaneously appear. Although the same entity may sometimes exhibit wavelike properties, and at other times seems to be a particle, only one such manifestation can be observed at a time. This has proved to be true in all experimental tests so far. Under most circumstances, what we envision as an "elementary particle," such as a proton or an electron, will behave as a particle; but under some conditions, an electron or proton will behave as a wave. It is more unusual to see the wave behavior of the electron, for example, than for light, because the wavelength of light is not small compared to "reasonably" sized objects, whereas the wavelength of the electron is very short. As a specific example, green light has a wavelength of about 500 nanometers (nm), whereas the wavelength of an electron is only of the order of 0.2 nm, about the size of an atom. Light will therefore exhibit wave behavior under many everyday conditions, such as when interacting with air molecules, whereas an electron will show its wave nature only under more unusual circumstances.

The small wavelength of the electron is exploited by the electron microscope. The resolving power of a microscope, that is, its ability to distinguish two close points, is inversely related to the wavelength of the probe. The shorter the wavelength, the greater the resolving power. Not only do electrons have a very short wavelength, but since they are charged and respond to electromagnetic forces, an electron beam can be focused by magnets, just as a beam of visible light can be focused by lenses. In this, electrons have a distinct advantage over X-rays, the electromagnetic radiation of comparable wavelength, because X-rays cannot be focused by conventional lenses. In the electron microscope, a beam of electrons is accelerated through an evacuated tube toward the specimen on the stage. Electrons striking the specimen scatter from it, creating an interference pattern that can be refocused at the objective into an image. Electron microscopes are available in several designs and are widely used in research, as well as in the manufacturing of certain items such as semiconductor devices. Electron waves are sufficiently real that an industry has been built up around them.

Quantum mechanics is the physical theory that accommodates this particle-wave duality. Ordinary, nonrelativistic quantum mechanics is based upon the **Schrödinger equation**, an equation which describes the behavior of an entity called the **wavefunction**. The interpretation of the wavefunction is still not fully unambiguous. The wavefunction must not be regarded as *the* "wave of the particle." Instead, a very successful and useful interpretation, used every day by physicists working in many subfields, is that the wavefunction describes the *probability distribution* of properties of the system to which it corresponds. A set of attributes, such as energy, momentum, position, and so forth, make up the **quantum state** of a particle; the wavefunction specifies the probability of the particle's being in a certain state.

The fact that only probabilities, and not absolute certainties, can be assigned to states is ultimately a consequence of the wave nature of particles; this lies at the heart of Heisenberg's famous uncertainty principle. For example, a wave fills space, and therefore its "position" cannot be unambiguously determined; hence the location of the corresponding particle is uncertain. We can, however, compute the most likely positions for the particle at a given time by means of the Schrödinger equation. If we measure the position of the particle to greater and greater precision, we find that we can say less and less about its momentum, because for a wave, it is impossible to know both those quantities to arbitrary precision at the same time. To "squeeze" a wave into a perfectly located position, we cannot use a "monochromatic," or single-frequency wave; a perfectly monochromatic wave fills all space. In order to localize a wave, we must add together, or *superpose*, many such monochromatic waves. As we add more and more frequencies, however, we find that the momentum of the wave, which is related to its frequency, is less and less determined. Perfect localization requires an infinite superposition of frequencies, and the momentum becomes completely undefined.

Quantum mechanics is by its very nature a statistical theory. The probabilities that a measurement of some variable, such as momentum, position, energy, spin, and so forth, will yield a certain value, can be computed from the wavefunction, but the behavior of a particle is fundamentally unknowable. The limit of our knowledge is defined by *Planck's constant*, which in MKS units has the value 6.6×10^{-34} Js. It is the small size of this number which means that we do not see quantum effects in our everyday lives. Yet how do we leap from such a strange, probabilistic realm to the deterministic classical world? If a measurement of some quantity is performed on an ensemble, that is, a large number of identical systems, all permitted values will be obtained, but the most probable result, called the *expectation value*, will represent the average behavior. It is impossible to predict how long any single atom of uranium will exist before decaying; but the expectation value of the lifetime, measured over a large sample of identical uranium atoms, determines the half-life of the isotope. As the size of the ensemble grows, the expectation value begins to behave more and more like a classical variable; however, no clear-cut demarcation exists at which "quantum" crosses over automatically to "classical."

The wavefunction itself cannot be observed experimentally; only the probabilities computed from it can be observed. It is unclear whether the wavefunction has any physical reality of its own. In the standard interpretation of quantum mechanics, the wavefunction serves merely to define a probability distribution, and only this probability distribution is connected to reality. It is possible to formulate a consistent theory of quantum mechanics, such as that of David Bohm, which differs from standard theory mainly in that the wavefunction does have an objective existence. However, under all conditions achievable on Earth, the predictions of Bohm's theory are identical to those of the standard theory.[2] We shall not discuss this interesting digression further, but mention it only to show that the interpretation of quantum mechanics is still undecided.

The standard view of the wavefunction is called the **Copenhagen interpretation**, because it was formulated by the Danish physicist Niels Bohr. According to Bohr, the wavefunction is merely a mathematical formality that characterizes our state of knowledge about a system; the wavefunction tells us everything that it is possible to know about the system. In the Copenhagen interpretation, there is a demarcation between the *system* and the *observer*.[3] Prior to an observation, the wavefunction evolves according to the Schrödinger equation. The variable to be observed, such as momentum, spin, or energy, is described by the probabilities that it will take various values. Each possible value corresponds to a state of the wavefunction. Until a measurement is performed, that variable literally *has no* value, but the wavefunction represents a **superposition of states**, the combination of all possible outcomes for a measurement of that variable. Only when an interaction occurs that demands a particular value for some quantity—for example, a measurement is performed—does the observed variable take on a specific value, that which was measured. This rather odd phenomenon is called the **collapse of the wavefunction**. The act of observing causes the wavefunction to assume a state that was previously only a potentiality.

The *two-slit experiment* demonstrates these concepts. In this experiment, a metal plate with two slits, comparable in thickness to the wavelength of light, is set up perpendicular to the path of a light beam. A screen on the other side of the plate registers the arrival of the light. What is recorded is not two small points of light corresponding to the images of the slits. Instead, the light waves emerging from each slit interfere with one other, reinforcing in some places and cancelling in others, to produce an *interference pattern*, bands of alternating light and dark along the screen. The wave behavior here is very similar to the interference patterns resulting when the waves from two sources overlap on the surface of an otherwise still lake.

What if we replace the light source with a source of electrons? This time, we arrange for a single electron to be emitted behind the metal plate and aimed toward

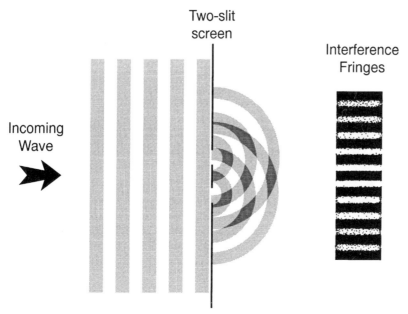

Figure 16.1 Interference fringes produced by a wave passing through a two-slit screen and striking a detector. Each slit acts like a source for waves. Where the waves from the two slits reinforce, they produce a bright fringe; mutual cancellation produces dark fringes.

it by the attraction of a charged object on the other side. Beyond the plate, we place some kind of detector which we have covered with a phosphorescent material, such as that which coats the display screen of a television set or computer monitor. The impinging of an electron is recorded by a bright glow where it strikes the phosphor. By "observing" the electron, the phosphorescent screen collapses the wavefunction of the electron.

If we repeat this experiment many times, we will find a distribution of positions for the electrons, each occurring with a certain frequency, and all of which can be computed from the wavefunction. After a large number of such measurements of individual electrons, we will find that the locations and repetitions of their measured positions will eventually build into a continuous pattern of interference fringes. The fringes are found to be exactly such as would be produced by a wave which passes through both slits and recombines on the other side. This result is obtained regardless of whether the electrons pass through the slits in a continuous beam or are emitted as individual electrons separated by arbitrarily long intervals of time. More than that, if we placed an appropriate "wave" detector beyond the screen, we would find an interference pattern no matter how few electrons are admitted into the apparatus; even a *single* electron produces interference fringes! Such fringes could be produced by one solitary electron only if it passed through *both* slits and interfered with itself. Even if this may seem too bizarre to be possible, it has been confirmed experimentally. When electrons pass through both slits, an interference pattern is obtained; we observe the wave aspect of the electron. Something fundamental about the electron's nature is wavelike.

Suppose you contrived to find out through which slit a single electron "really" passed. You place at each slit a detector that registers the passage of each electron but allows it to proceed through the slit. Instead of interference fringes at the screen, you will discover two distinct regions of electron impact points, one for each slit. The act of detecting the electron at one or the other slit collapses the wavefunction. With the electrons localized to one slit or the other, the wave interference that produces fringes is lost.[4]

The picture of an electron passing through two separate slits at once is so counter to common sense that it can be quite disturbing. It may be tempting to grant to the invisible world of electrons and photons some strange, almost eerie, properties, but to fall back on the comforting assurance that the "real world" behaves more logically. But in principle, quantum mechanics describes the behavior of *all* matter, just as, in principle, special relativity is the correct theory of dynamics at all speeds. It is simply that the wavelength corresponding to any macroscopic object is so tiny that quantum effects are unobservable. The quantum wavelength of a thrown baseball, for example, is of the order of 10^{-35} m.

The Tale of Schrödinger's Cat

Acceptance of the governance of quantum mechanics over the macroscopic world implies that the quandaries raised by quantum theory carry over to that world as well. If a system unobserved is indeterminate, how does the observer's act of measurement introduce determinacy? We tend to think of the "system" as microscopic, and the observer as macroscopic. The prejudice is that "large" objects, such as experimental apparatus, computers, scientists, and the like, must be clearly distinguishable from the bizarre world of barely imaginable particles. But what if the system and the observer are of comparable scale? If quantum mechanics applies to all systems, then quantum effects must in some way control even familiar classical objects; yet quantum behavior seems impossible for "everyday" entities.

The best-known illustration of this paradox has become part of the folklore of physics, the tale of *Schrödinger's cat*. Suppose a cat were placed into a closed box that is completely isolated from its surroundings. This box contains an elaborate and diabolical contraption. If an atom of some radioactive element decays, the emitted alpha particle trips a Geiger counter, which, by a prearranged switching mechanism, causes a vial to be broken and a poisonous gas to be released into the box. The fate of the cat is tied to a probabilistic quantum effect, specifically, the decay of an atom. Within the box, there is no paradox: the Geiger counter trips if it observes the decay, and the cat dies or lives accordingly. But consider the system from the point of view of someone outside. The entire setup of box, atom, Geiger counter, and cat is, in principle, a quantum system described by some complicated wavefunction. In any time interval, all that is known is that there is some probability that the atom will decay; so while the cat is in the box, unobserved, it is unknown whether it is alive or dead. Adhering strictly to the Copenhagen interpretation forces us to conclude that while the cat is in the box, no measurement has been made, and therefore the cat is neither alive nor dead, or else is both alive and dead; that is, the cat +

box system represents a superposition of the states "alive" and "dead." When the outside observer opens the box to observe the state of the cat, it becomes alive or dead at that moment, according to the probability that the Geiger counter had been tripped.

This is all completely in accord with the laws of quantum mechanics, yet the conclusion seems nonsensical. What might explain the apparent discrepancy between quantum mechanics and well-established common-sense notions that a living being must be either alive or dead, but not both? One suggestion is that a cat in a box is a macroscopic object that is composed of a very large number of microscopic quantum objects, that is, its atoms and molecules, which collude to create the classical behavior we observe. The collective quantum state of such a system would be extraordinarily complex. Writing the "wavefunction of the cat + box system" would be an impossible undertaking, at present. But is it impossible in principle? Perhaps Schrödinger's equation simply does not apply to such an assemblage, an attitude adopted by many pioneers of the theory, including Schrödinger himself. But this line of argument begs the question, for no clear delimitation has yet been found for the point at which Schrödinger's equation breaks down in the transition from the microscopic to the macroscopic world.

No literal experiments have ever been carried out with cats, of course. However, many experiments have been performed with great precision upon microscopic entities; the predictions of quantum mechanics and the Copenhagen interpretation are invariably borne out. It is difficult to reconcile Schrödinger's cat with the "classical" picture of a cat which is, at any moment, either alive or dead, with a certain probability of its death occurring at any time. It might seem quite reasonable to presume that the states "alive" and "dead" are not quantum states and thus are not subject to superposition. Yet if quantum mechanics ultimately underlies our macroscopic reality, it must have some validity for apparently classical objects. Perhaps the wavefunction of the cat is so complicated that it is impossible to observe any quantum superposition of states. Or it may be that we cannot prepare a box that is truly so isolated from the rest of the universe that the quantum state describing the total system (cat, radioactive atom, Geiger counter, and poison) can evolve undisturbed by the "outside" world. Any infringement upon the box by the state of the supposedly external observer might constitute a "measurement," which would collapse the wavefunction of the cat. However, none of these alternatives seems to resolve the fundamental paradoxes inherent in the so-called **measurement problem**. What happens in a "measurement" to cause such a drastic change in the evolution of the system? The collapse of the wavefunction is not described by the Schrödinger equation, but is overlaid upon it as part of the Copenhagen interpretation. When an observation of some variable occurs, the system abruptly ceases to obey the Schrödinger equation; all subsequent measurements of that quantity will continue to yield the same result, as long as the system is not otherwise altered. The act of measurement seems to impose reality upon a previously unknowable state; but if that is so, what is "reality"?

The collapse of the wavefunction is one of the most vexatious problems of quantum mechanics. Philosophically, most physicists agree that it is at best uncomfortable. Einstein hated it, and his attitude influenced his own efforts to find a unified theory of quantum mechanics and gravity. Much effort is still devoted to analyzing the philosophical underpinnings of quantum mechanics. For operational purposes, however, most physicists set aside aesthetic worries and use the formal theory of quantum mechanics to make detailed calculations. The predictions of quantum mechanics, including its generalization to include electromagnetics, the theory called *quantum electrodynamics*, have been verified to an astounding precision by experiment. We cannot object too much to quantum mechanics on philosophical grounds, then, as it unquestionably describes something very fundamental and deep about the workings of the universe. Perhaps eventually a better interpretation will be found that will clarify these issues.

But there is a more serious objection to the collapse of the wavefunction if we seek to apply quantum mechanics to cosmology. The universe is, by definition, everything observable. The observer is part of the universe. In the Copenhagen interpretation, the collapse of the wavefunction depends upon a clear separation between the observer and the system observed, a distinction which is, obviously, untenable in cosmology. We shall have to put aside this concern for now, however, as it as yet unresolved, and forge ahead.

Quantum Cosmology

A direct approach to quantum gravity is to attempt to make a generalization from the Schrödinger equation in one great leap and to write an equation for the universe as a whole. This is the method that is usually called **quantum cosmology**. How can quantum mechanics, which treats microscopic particles, be extended to the universe as a whole? In quantum mechanics, the Schrödinger equation describes the space and time behavior of the wavefunction for a system, such as a particle, that has some energy, both kinetic and potential. Can this idea be extended to cosmology? The Friedmann equation for $\dot{R}^2$ (eq. [11.8]) plays a role much like that of an equation for the energy of a system. This equation can be transformed into a quantum-mechanical equation for the evolution of the scale factor. There is no unique way in which to convert the Friedmann equation into a quantum equation; however, one of the best-known and most widely applied versions was developed by John Wheeler and Bryce DeWitt. Their quantum cosmological equation is now known as the **Wheeler-DeWitt equation**.

We shall not write down this equation here, as it is far beyond our scope, but we shall describe some of its consequences. The Wheeler-DeWitt equation requires that the four-dimensional spacetime of general relativity be broken into a three-dimensional, purely spacelike, surface, and a timelike curve. These, of course, may be identified with "space" and "time" respectively. The decomposition is not unique for a given spacetime, however, but depends upon the choice of coordinates, as required by general relativity. The Wheeler-DeWitt equation then describes the evolution of a quantity that bears the rather grandiose title of the **wavefunction of the universe**. Remarkably enough, this quantity is a function of the three-dimensional geometry of the universe and of the matter-energy field; it contains no explicit dependence upon time. Certainly the wavefunction of the universe can change, but what we think of as "time" may not be the standard by which we should measure that change.

It is also unclear whether the split into a spacelike surface and a timelike curve is the correct one to consider. Other possibilities exist and have been studied, some of which may be more promising. Quantum cosmology as a program may be a good approach, but the correct formulation may be lacking. Nevertheless, some interesting results can be obtained. One of the easier cases to examine is that of a universe which, classically, corresponds to a variant of the de Sitter solution. This model has spherical spatial geometry and contains a cosmological constant (or, equivalently, a constant vacuum energy density), but no other matter. It is most convenient to study the quantum mechanics of a closed universe; open universes create troublesome extra terms in some of the integrals required, and it is not clear how to handle some of these terms. Moreover, the spherical geometry turns out to be particularly appropriate for quantum cosmology not only because it is highly symmetric, and finite, but also because the resulting equations resemble those for a situation familiar to most physicists: that of a particle moving in a potential with a deep minimum, or "well," in ordinary quantum mechanics.

It happens that the solution for the scale factor of the spherical de Sitter model is a function that has both an exponentially increasing and an exponentially decreasing part. The original de Sitter flat-space model selected only the exponentially increasing solution; the corresponding spherical geometry admits only a solution with both an expanding and a contracting part. Such a function stretches from $t = -\infty$, at which point $R = \infty$; it contracts as time approaches zero, passes through its minimum size, as specified by the cosmological constant, at $t = 0$, and then expands forever. The minimum size, however, is not zero; this is not a big bang solution, and therefore it contradicts our observations. But the classical solution is not the whole story.

In quantum mechanics, the wavefunction is not confined strictly to the region allowed to it classically. The wavefunction spills over into "forbidden" regions, although its amplitude becomes small in such regions. Quantum mechanically, there is a nonzero probability that a particle may be found in a region where it could not be located classically. This phenomenon is called *quantum tunneling*, and it is exploited in some scientific instruments in use today. In quantum cosmology, the "particle" is the entire universe. Therefore, even though this spherical de Sitter solution cannot classically start from $R = 0$, quantum mechanically, it can do so; the universe can tunnel from the state $R = 0$. Beyond the classical minimum value allowed for R, the universe ceases behaving in a quantum-like manner and can continue to evolve classically. This little marvel is a simple consequence of solving the Wheeler-DeWitt equation; the quarrel comes in what to do next.

The Wheeler-DeWitt equation describes only the evolution of the wavefunction; it tells us nothing about the appropriate *boundary conditions*. Different assumptions about these conditions can produce drastically different behaviors. For example, A. Vilenkin has argued that the correct condition should be an *outgoing* wave, that is, a wave which expands in all directions from $R = 0$, while S. Hawking and J. B. Hartle thought that the appropriate boundary condition should be that there be *no* boundary; the solution should contain both expanding and contracting functions. It is impossible to resolve which, if either, is correct; there simply are no data to help us make such a decision.

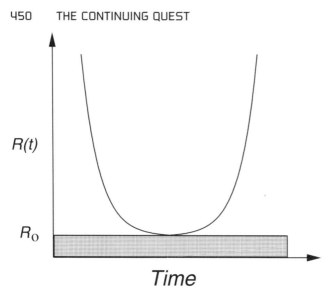

Figure 16.2   The spherical de Sitter solution. The shadowing indicates the classically forbidden region. Quantum mechanics permits the universe to tunnel from $R = 0$, resuming classical behavior at $R = R_0$.

Neither does the Wheeler-DeWitt equation address the issue of initial conditions. Different assumptions once again yield different behaviors. It may be that inflation wipes out the initial conditions anyway. It may be that quantum cosmology *creates* the appropriate conditions for inflation to begin. It may be that quantum cosmology sets up initial conditions that would snuff out inflation. It is completely unknown.

What *is* the "wavefunction of the universe," anyway? It sounds very pretentious, particularly when one considers that it arises from a drastic simplification for the universe, a quantum analog of the Friedmann equation. But, despite the necessary oversimplifications, quantum cosmology is beginning to outline important questions that will have to be answered someday in a more complete model. Although the present universe can certainly be said to be evolving in a manner consistent with purely classical equations, it must nevertheless obey the laws of quantum mechanics. Quantum physics must become increasingly important as we probe back to the Planck time and would be important again should the universe end in a big crunch. But since the universe is apparently a unique thing, how can its "wavefunction" be given a probabilistic interpretation? And what is the meaning of time within this picture? Currently these questions are mysteries.

The Nature of Time

What is *time*? This question has troubled philosophers and scientists throughout humanity's history. Our intuitions say that time is different from space. In space we can travel in any direction, limited only by the capabilities of our modes of transportation. Time, in contrast, seems to be a one-way street, moving inexorably from the past to the future. We remember the past, but can only guess about the future. The past is fixed, unchangeable. The future is indeterminate, mutable, unpredictable. That is, at least, how we perceive time. But physics takes a quite different view, one that is not easily reconciled with our experience of time.

We have already seen this with our study of relativity. In special relativity, time and space are joined into spacetime. Does special relativity imply that time and space are fully equivalent? Not really; time enters into the metric with a different sign from that of the spatial dimensions, and this is a distinguishing factor, although we have learned that we cannot pin absolute labels onto the passing of time. A particular definition of time depends upon the frame of the observer. Only *proper time*, the time measured by a clock at rest on a given world line, is invariant. One of the most counterintuitive consequences of the blending of space and time is that simultaneity is not invariant in special relativity, but depends upon the observer: one observer's future may be another observer's past. We insisted upon the preservation of cause and effect, however; an effect could never precede its cause in any frame. The merging of space and time into spacetime in special relativity implies that a world line is not something that creeps forward at some rate, revealing reality as time passes. A world line is an entity in the spacetime. Its future and past are *already there*.

General relativity preserves the basic spacetime of special relativity, with the extension in the general theory that the measurement of time, as well as space, depends not only upon velocity, but also upon what masses happen to be in the vicinity. An extreme example is the interior of a black hole, where time and space (as defined by the external observer) seem to exchange roles; going forward in time means falling toward the center. But in general relativity as in special, a world line is a path in the spacetime, determined once and for all time by the equations of the theory.

What about the cosmos itself? We have noted previously that the Friedmann models include a good cosmic time, which is conveniently defined as that proper time kept by an observer at rest with respect to the universe as a whole. But how can we define such an observer during the Planck epoch? Where time and space are themselves subject to uncertainty, we must seek a way of describing spacetime events that is free of the arbitrariness of coordinates. It is difficult to know how to begin such a task.

Perhaps time is the wrong marker. Perhaps what we call "time" is merely a labeling convention, one that happens to correspond to something more fundamental. What about the scale factor, which is related to the temperature of the universe? In our standard solutions, the scale factor, and hence the temperature, is not a steady function of "cosmic time." Intervals marked by equal changes in the temperature will correspond to very different intervals of cosmic time. In units of this "temperature time" the elapsed interval, that is, the change in temperature, from recombination till "Now" is less than the elapsed change from the beginning to the end of the lepton epoch.[5] As an extreme example, if we push "temperature time" all the way to the big bang, the temperature goes to infinity when cosmic time goes to zero. In temperature units, the big bang is in the infinite past!

In an open universe, the temperature goes to zero at infinite cosmic time, and temperature and cosmic time always travel in *opposite* directions. If the universe is closed, on the other hand, there is an infinite "temperature time" in the future, at some finite cosmic time. A closed universe also has the property, not shared by the open (or flat) universe, of being finite in both cosmic time and in space. In this case, the beginning and the end of the universe are nothing special, just two events in the four-geometry. Some cosmologists have argued for this picture on aesthetic grounds; but as we have seen, such a picture has little observational support, and no particular theoretical justification other than its pleasing symmetry.

If we are looking for clues to a physical basis for the flow of time, however, perhaps we are on the right track with temperature. All of the formal theories of physics are time symmetric. It makes no difference whether time travels backward or forward; nothing changes in the equations if we substitute $-t$ for t. Even quantum mechanics makes no distinction between "past" and "future"; the collapse of the wavefunction complicates the picture, but whether that is merely an interpretation of the act of measurement, or represents a genuine time asymmetry, is currently a matter of debate. The Schrödinger equation itself is as time symmetric as Newton's laws. In all of physics, there is only one fundamental law that has a definite time preference: the second law of thermodynamics, which states that entropy increases with time. Newton's laws, Einstein's laws, quantum mechanics—all are invariant under a time reversal. Only the second law of thermodynamics proclaims that any process has a direction. How does this fit into the rest of physics?

The second law of thermodynamics is an *empirical* statement, based ultimately on observations of steam engines in the nineteenth century. Thermodynamics could exist without it. In its modern form, the second law declares that *the entropy of a closed system never decreases, but either remains the same, or increases*. What is **entropy**? We have stated previously that entropy is a measure of the *disorder* of a system. This is certainly true, but more precisely, entropy is related to the total number of macroscopically indistinguishable states that a system can occupy. The more states available, the higher the entropy. For example, the entropy of the air in a room is related to the number of ways in which the air molecules, given the count of molecules present and the total energy available to them, can be physically arranged in space, including rotational freedom of the molecules, such that the macroscopic characteristics of the air are identical. A little thought should convince you that there are far more ways to arrange molecules that will result, on the large scale, in an even distribution throughout the room, than there are arrangements in which the molecules are clumped in one corner. The evenly distributed state has high entropy (many possible equivalent states), whereas the clumped molecules have low entropy (few possible states). This is why entropy is a measure of disorder; there are generally far more disordered states available than there are ordered states. According to the second law, entropy at best remains the same, and, in general, increases. Air clumped in the corner of a room, perhaps by means of a piston, will, as soon as the opportunity presents itself, find the state of maximum entropy allowed to it under the circumstances. The only processes in which entropy remains the same are *reversible* processes, which are idealizations that do not occur in nature on the macroscopic level. Real macroscopic processes are *irreversible*; entropy increases.

Irreversibility seems to be intimately related to the direction of time. For instance, a glass tumbler is in a highly ordered state. If it falls and shatters, it enters a more disordered state. Broken shards of glass never spontaneously reassemble themselves into a tumbler; the only way to recreate the tumbler is to melt the fragments and start anew. The second law applies only to *closed* systems, those in which no energy enters or exits, so local exceptions to the rule of increasing entropy can always be found. As in the example of the recreated tumbler, entropy can be decreased locally by the expenditure of energy. You maintain your highly ordered state only at the cost of enormous consumption of food energy. Your automobile converts fuel into mechanical work, specifically the ordered motions of the pistons and wheels, by extracting the chemical energy of the gasoline.

The universe is certainly a closed system; thus in any process, the entropy of the universe as a whole increases. The gasoline that was burned is changed in its composition and disappears forever as various combustion products, all of which are much less capable of conversion into work. Entropy increases. You eat, and most of the energy in your food is spent to maintain your body temperature; only a fraction goes into driving biochemical processes, while the rest is radiated away into the atmosphere as waste heat. Entropy increases. You die, and the ordered molecules break down into simpler, more disorganized constituents. (They would do so spontaneously, without the aid of bacteria, over a long enough time interval; the bacteria speed up the process and use the energy *they* extract for their own battle against entropy.)

It is not energy which makes the world go 'round. Energy is conserved, after all. The chemical energy released by the burning of gasoline or of your food is converted into various forms. Some goes into work; driving pistons and turning wheels, or moving your muscles. Some is dissipated by friction; the wheels or your feet must overcome friction in order to move, generating heat in the process. Some of the energy goes into maintaining a low-entropy state, such as storing your memories in your brain. Some is released as waste heat; through the exhaust of the car, or from your skin. But the total amount of energy, in all forms, is conserved. Not so entropy; entropy increases. From these examples we can also see that another definition of entropy is related to the *capacity to do work*, where *work* is defined strictly in physics as the exertion of a force over a distance to produce a motion. A higher-entropy state has much less capacity to do work than does a lower-entropy state. For example, it is easy to see that "uniform heat" is a higher-entropy form of energy than, say, kinetic energy. A car moving along at high speed has a large amount of energy. If that energy is dissipated into the brake pads, it is distributed into random motions in a huge number of separate molecules. Potentially useful kinetic energy is now spread out in essentially useless, random molecular motions.

Our sense of time moving forward is associated with the change from a state of lower entropy to one of higher entropy. This is why it is so easy to distinguish a motion picture running forward from one that is running backward. Broken glass shards fly together and reassemble themselves into a glass that leaps back onto a table. Crumpled automobiles back away from shattered brick walls, reassembling themselves, and the wall, in a series of highly coordinated movements. We all know that processes in which order spontaneously increases never occur, so a film shown in reverse strikes us as amusing. In our world, energy must be expended for order to increase; in most natural processes, disorder is created. The **arrow of time** is determined by the inexorable increase in entropy, an increase that is seen in all macroscopic occurrences.

However, there is still a mystery to this. Even with the second law of thermodynamics, there is time symmetry; a system that is disordered today is likely to have been disordered yesterday. That is, if the equations of physics are run backward, a room full of random air molecules does not revert to an ordered, low-entropy state, but remains in a random, disordered, high-entropy state. To have a sense of the arrow of time, we must start from a low-entropy state. What provides a past and a future in the universe is that it *began* in a state of low entropy; this makes the past distinguishable from the future. Thus the arrow of time is not due to the second law itself. It is due to the *initial conditions*.

What is the entropy of the universe? One measure of this quantity is the number of photons. There are perhaps 10^{80} baryons in our visible universe, and about 10^9 photons per baryon. This produces a figure of 10^{89} for the entropy of the CBR and would represent almost all the entropy in the universe, were it not for gravity, and more specifically black holes. Black holes are not completely black; they emit Hawking radiation. Since this radiation is blackbody, a temperature can be assigned to it, and hence to the black hole itself.[6] The association of a temperature with a black hole leads to a full theory of black hole thermodynamics, from which an entropy can be derived. The entropy of a black hole is found to be proportional to its surface area; the larger the hole, the greater its entropy. Because the surface area is proportional to the mass squared, it follows that the entropy per unit mass increases with mass. From such calculations, it turns out that black holes are the most entropy-laden objects in the universe. If the entire estimated mass of the observable universe collapsed to a black hole, the entropy associated with that black hole would be 10^{123}, a number that is beyond any genuine comprehension, and that dwarfs the modest, by comparison, value of 10^{89} of the current universe.

One of the distinguishing features of the black hole is its very strong tidal force. The particles created via Hawking radiation obtain their energy from this tidal force; this suggests that the huge entropy of a black hole is somehow tied up with the tidal forces. Physicist Roger Penrose has extended this idea to *any* spacetime geometry and associated an entropy with the tidal force, which can be computed in a straightforward way from the metric of the spacetime. When we apply this to the standard models, we find that the geometries of these spacetimes have zero tidal force at the big bang. This is quite distinct from the singularity of a black hole, which has, in contrast, infinite tidal force. Thus the singularity at the beginning of the universe is quite different from that which is to be found in a black hole. Indeed, if the universe is closed, the final collapse will be like a black hole and will have enormously high entropy, quite a different state from which the cosmos emerged. This alone suggests that the so-called *cyclic model*, in which the universe begins anew following a big crunch, is not likely.

Even an open or flat universe will end in a state of relatively high entropy, due to the increase in entropy as stars burn out and galaxies fade away. This is a remarkable arrangement. Why did the universe begin in such a low-entropy state? Since, as we have asserted, systems seek the state in which their entropy is maximized, we can assume that the "natural" state of the universe is the aforementioned black hole. This leads us to the conclusion that our universe is "special," in that it actually began from zero entropy, to one part in 10^{123}. This implies a specialness of the initial conditions to an almost incomprehensible degree. No physical theory known at present is able to account for this phenomenon; the resolution, if any, may be buried in the Planck epoch.

The physics of the Planck epoch is inextricably tied to quantum gravity; it was during the Planck epoch that the conditions were set which resulted in this state of extreme low entropy. Penrose has argued, on this basis, that the theory of quantum gravity must be a time-asymmetric theory. In this viewpoint, quantum gravity necessarily requires that initial singularities, such as the big bang, be smooth, low-entropy singularities. Such an argument suggests that the apparently special initial conditions are actually part of the laws of physics and that the relentless march of time is a direct consequence of quantum gravity. Of course, this is only a prescription for a theory, not the theory itself. Still, it is fascinating to contemplate that our perception of the arrow of time might be telling us about the nature of the big bang spacetime singularity, from which the macroscopic universe was spawned.

Time Travel and Many Universes

Time is special. Not only does it have a *preferred* direction, but that direction seems inviolable. This does not seem to be demanded by the second law of thermodynamics, with its apparently bland, yet profound, statement about entropy; the second law would seem to require only that time tend to track the increase in entropy. Can we travel backward in time? All our experience denies this. If it were possible, some severe paradoxes would result. The **grandfather paradox** is one of the best-known of these conundrums. What if a perverse time-traveler visited his own grandfather while he was a baby in his crib, and killed the infant? Would the time traveler disappear, since without his grandfather he could not have eventually come into existence? What if a more benign time-traveler went back in time in order to prevent World War II? Before he left, the world contained the horrors of that war and its aftermath. If the time traveler succeeded in his beneficent mission, would the people killed as a result of the war suddenly return to life? This seems nonsensical.

We have seen that in classical general relativity, any world line of a material particle must be timelike. However, world lines do not evolve; they are the complete four-dimensional histories of the particle. Each point, or event, on the world line represents a particular place at a particular time. If we are to find world lines that allow time travel, that is, world lines for which the future lies in the past, we must search for a *closed timelike path* in spacetime. Such a world line is timelike at every event, yet still forms a closed loop. The standard cosmological solutions and classical black hole solutions do not allow closed timelike curves. But these are by no means the only solutions to Einstein's equations; might some solutions permit these unusual world lines? A new class of closed timelike curves associated with wormholes, discovered by Kip Thorne and collaborators, may seem to be a realization of the dreams of science-fiction writers, but such "time machines" occur only under extremely special conditions, and probably could not be traversed by any real particle. A true world line is infinitesimally thin; any extended particle describes a world tube in spacetime. The world tubes found by Thorne are certainly too narrow for macroscopic particles and may not even be traversable by elementary particles. They also require a preexisting wormhole that is maintained in a very special state.

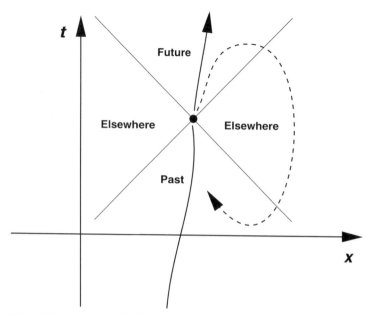

Figure 16.3 In special relativity, a particle's world line must follow a timelike trajectory. Time travel (*dashed line*) requires a spacelike trajectory. The question is, does general relativity allow time travel along a timelike world line? This would permit a "closed timelike curve."

One known solution to Einstein's equations that does freely admit closed timelike curves is the *Gödel solution*, found by the German mathematician Kurt Gödel in 1949. The term in Einstein's equations that describes the distribution of stress-energy, which acts as the source of the gravitation, is called, sensibly enough, the *source term*. The Gödel solution is obtained from the same source term as the Einstein static solution, but with different boundary conditions. This vexed Einstein, for it meant that his theory did not perfectly embody Mach's Principle; if Mach's Principle were strictly upheld, a source distribution would uniquely determine the metric. Gödel's solution proved that this was not the case. In retrospect, it is not surprising, however; the Einstein equations are *nonlinear*, and it is a well-known property of nonlinear equations that more than one solution may be obtained for the same source. Only linear equations, in which no variable appears to any power other than the first, can guarantee a unique form of the solution for a fixed source.

In the Gödel solution, it is possible to travel into one's own past, already a remarkable result. The solution has other curious properties; for one thing, it rotates. What does this mean, that a universe "rotates"? With respect to what? To understand this, we must consider the motion of a test particle. Suppose we send a rocket toward a distant galaxy along an inertial world line. For simplicity, we shall assume that there are no gravitational influences on the rocket other than the galaxy; it simply coasts toward its target. The rocket follows a geodesic, which would, in flat space, correspond to a straight line; eventually the rocket arrives at the galaxy. In the Gödel universe, however, the rocket's path spirals away from the galaxy at which it was aimed.

The Gödel universe does not seem to have much to do with the universe in which we live. No observations have ever detected an overall rotation of the universe, and certainly closed timelike curves are an uncomfortable property at best. The solution demonstrates, however, that the boundary conditions seem to be the way in which Mach's principle is incorporated into general relativity. Those boundary conditions, in the real universe, may specify that classical closed timelike curves are not allowed. This is not a certainty, however.

Might quantum mechanics have anything to say about this? In quantum mechanics, it turns out that the uncertainty principle permits particles to travel backward in time, so long as causality is not violated. It has even been suggested that antiparticles can be interpreted as particles that are traveling backward in time! Such an interpretation cannot, however, apply to macroscopic objects. But what if we are wrong, and closed timelike curves do exist? The Robertson-Walker metric is certainly merely an approximation to the universe, valid only on the largest scales; locally, another metric must apply, one which may be greatly complicated, and of unknown form. Perhaps such a metric would allow a more varied structure, including closed timelike paths, than do the simple metrics with which we are familiar. If that is true, or else if we find that quantum mechanics somehow permits us to get around the classical limitations, what *would* happen if you went back into time and killed your grandfather in his cradle?

One possible resolution to the contradictions of time travel is provided by an alternative view of the wavefunction, an exotic conjecture called the **many-worlds interpretation**. This interpretation of quantum mechanics, originally proposed by Hugh Everett, was developed to deal with the "measurement problem" in quantum cosmology. The Copenhagen interpretation depends upon a distinction between the observer and the system, a distinction which cannot be maintained in quantum cosmology. In the many-worlds interpretation, an infinite number of universes exist. These are not the usual kind of "parallel universes" of science fiction, nor are they the "child universes" of the chaotic inflation model. They represent the set of universes in which all possible outcomes of all quantum process occur. When a measurement is made, no "collapse of the wavefunction" takes place; rather, the probability of obtaining a given outcome is proportional to the number of universes in which that result is obtained. The issue of the meaning of the "collapse of the wavefunction" is avoided by requiring all possibilities to occur. As an illustration, return to Schrödinger's cat. When the box is opened, the universe splits into those in which the cat is alive, and those in which it is dead. After the measurement, universes which were once indistinguishable can now be distinguished by whether the cat jumps from the box or not.

The many-worlds interpretation solves not only the "measurement problem" but the "grandfather paradox" as well. If a time-traveler murders his grandfather in the cradle, it merely means that now there are distinguishable universes. In one, the time traveler is never born. The universe containing the murderer and the dead infant continues along its own path. In the other universe, the murderer disappears the moment he travels into the past. The objection that the time-traveler and his grandfather are macroscopic, classical objects is inapplicable in this case, because ultimately quantum mechanics must apply to the universe as a whole if we are to solve the "measurement problem," and therefore there is no such thing, strictly speaking, as classical behavior.

But if "I" am present in multiple, perhaps infinite, universes, then what are we to make of consciousness? It seems continuous; we remember a past that appears to occur in some linear way. Are there multiple consciousnesses of the same apparent individual, none of which can communicate with the others and each of which regards itself as a single entity? It may be strange to think of Schrödinger's cat as neither alive nor dead, but is it any more satisfying to suppose that in some universes the cat is alive, while in others it is dead? On the other hand, we have repeatedly stressed that the universe is not bound by our "intuition" or by our sense of aesthetics. If the many-worlds interpretation, or some variant of it, turns out to be the only way in which to fit gravity into quantum mechanics, then we must adjust our "common-sense" beliefs accordingly.

Whither Physics?

Is there any hope that someday we might be able to understand all of these mysteries? It is difficult to know because, at the moment, we are not even sure we are asking meaningful questions. What answers might lie in the immediate future, as more is learned at the frontiers of physics? We had already concluded that we cannot make sense of the beginning of the universe without a theory of quantum gravity, but now we must realize that it goes deeper than that. Quantum mechanics apparently not only accounts for what goes on during the Planck epoch, but also fundamentally determines the state of the universe, including the arrow of time. Hawking radiation and the entropy of black holes provide tantalizing hints of the wonders that a complete theory of quantum gravity will reveal.

It appears, then, that the arrow of time, which manifests itself in the physics we know only in the form of the second law of thermodynamics, is probably a consequence of quantum gravity. Might quantum gravity determine other things as well? Quantum mechanics is involved with setting the masses of particles; although no complete unified theories have been worked out, masses of some particles can already be computed from theory. The basic mechanism is that of spontaneous symmetry breaking. When a force decouples from the others, certain particles acquire mass, although this phenomenon is still incompletely understood. The more fundamental of the dual roles of mass in physics is that of "gravitational charge," with inertia related via the equivalence principle, or whatever underlying quantum principle determines it. It seems likely, then, that quantum gravity will be found to play a major role in determining the masses of elementary particles.

What of other fundamental constants, such as the gravitational constant or Planck's constant? Some approaches to quantum gravity find these constants from the presence of *hidden dimensions*. It seems that even four-dimensional spacetime might not be big enough. The reasons for introducing further dimensions have to do with consistency of various particle models and are too technical for us to consider here. All these dimensions are spatial in nature, so time remains as a "special" dimension. These speculated extra dimensions are certainly unobservable at present, since they are *compactified*; that is, they are curled into sizes of the order of the Planck length, 10^{-35} m. If the universe began with many such spatial dimensions, and for some reason three of them underwent a protoinflationary expansion, all the others would be left behind as compactified dimensions. These additional dimensions influence our four-dimensional spacetime in that the "correct" physics is that of the higher-dimensional theory. Our world, and the fundamental constants we find to govern it, are obtained in this view by integrating over the hidden dimensions. During the Planck epoch, however, all spatial dimensions might have made comparable contributions, as all would have had the same scale during that interval.

Theories that require hidden dimensions represent a different approach to particle physics, and its unification with gravity, from what we have considered so far. Quantum cosmology is inherently *geometrical*, following the spirit of Einstein's formulation of general relativity. Gravity follows from geometry. Many distinguished physicists have become convinced, however, that the geometrical viewpoint is not leading anywhere, and may even have hindered our progress because it has blinded us to other possibilities. In the alternative approach, particle physics and the unification of forces is fundamental. Gravity arises much like the other forces of nature: the exchange of gravitons accounts for gravity, rather than the curvature of spacetime. The question still unanswered is why gravity nevertheless *appears* to be so geometrical. This is not necessarily a natural consequence of the newer theories, although further development of the models should illuminate the issue. Perhaps the apparent geometrical nature of gravity is simply an approximation to its true nature, which has yet to be uncovered.

One of the most promising, or at least most interesting, of the nongeometrical approaches is **superstring theory**. Superstrings must not be confused with cosmological strings; they are completely different phenomena. In superstring theory, the fundamental building blocks of nature are extremely tiny, but finite, strings and loops. These are quantum objects and their vibrations and interactions occur in a quantum manner. At the level of the Planck scale of lengths, times, and energies, spacetime is not continuous but consists of a "foam" of oscillating and interacting loops and strings. Ultimately, time may not be the linear, smooth function we perceive it to be, but a shifting froth. Superstring theory is the only known theory that unifies all four forces of nature in a finite and self-consistent way. This does not guarantee that the theory is actually a good description of nature, of course. Superstring theory is complex and difficult to understand even for specialists. Nevertheless, it represents a great achievement and provides hope that a unified theory of particle physics and of the four fundamental forces is within our reach.

The Ultimate Question

Throughout most of our study of the universe, we have carefully avoided broaching the question of how the universe began. We have discovered that we are able to describe the evolution of the universe, with considerable success, down to approximately 0.01 s after the beginning. More than that, we have at least some good ideas, and promising hypotheses, to understand the universe to as little as 10^{-43} s after the big bang. If we try to push to times earlier than the Planck time, however, our confidence evaporates, and our ability to say much, beyond the vaguest expressions of our belief in the ultimate unity of the fundamentals of nature, disintegrates. Unless and until we achieve an understanding of the Planck epoch, it is hardly more than scientific bravado even to speculate about the origin of the universe. It may well be that this question is beyond science. Still, we cannot resist. After all, it was not so long ago that the nucleosynthesis epoch, which we now believe we understand quite clearly, was thought to be unreachable by any scientific models. We can contemplate this ultimate question, keeping firmly in mind that any hypothesizing about the origin of the universe can as yet be no better than educated guesswork, and may prove to be nonsense.

The universe is quantum mechanical at its heart. About this there is no question. We still do not understand how best to interpret quantum mechanics, nor do we understand how classical gravity, which works so well for the largest scales of space and time, fits into this indeterminate universe in which we live. Yet if quantum mechanics ultimately governs the life of the universe, is it possible that it might have been responsible for its beginning as well? Quantum mechanics permits the creation of something from nothingness within the universe we observe today. Even galaxies might trace their ancestry to quantum fluctuations in a quantum field. Might quantum mechanics explain the very origin of the universe itself? If a galaxy could begin as a fluctuation, why not the cosmos as a whole?

Perhaps the fundamental reality is a foam of superstrings, or else, as a newer theory suggests, of quantized units of multidimensional spacetime with scales of the order of the Planck length and time. Perhaps a fluctuation occurred in at least one region of this foam, causing three spatial dimensions to expand enormously in a kind of inflation; the result would have been a "universe" that was dominated by an apparently four-dimensional spacetime, and that then continued to evolve according to the theories we have discussed. This is an intriguing hypothesis, and it has some support from known physics, but as we still have almost no understanding of the Planck epoch, it must remain an interesting idea which may be all dressed up, but as yet has no place to go.

Quantum gravity is one of the great frontiers of physics at the end of the twentieth century. When it is achieved, it will certainly answer many of our questions about the universe—but perhaps not all. More than once has the end of physics been pronounced, when it was thought that everything that could be known, was known. Science is a process of successive approximations. Philosophers sometimes argue that Truth is ultimately unknowable, that our scientific models can never be more than our best description of our own experiences. This may well be correct, but science has enabled us to develop a coherent, self-consistent view of the universe that, in some limit, must show us at least the shape of Truth. We began our exploration of cosmology with the ancient myths and ideas that placed humanity at the center of an unfathomable cosmos. Much of what we have accomplished in the past four centuries has relentlessly removed humanity from any favored place in the universe and has changed the physical universe from a capricious, mysterious realm to a domain that obeys laws we can, at least in some sense, understand. Although Earth is not the center of the universe, it is representative—all points are central. Thus the conditions required for our existence, the state of nature that has permitted us to form, evolve, and ask these questions, is a state present throughout the universe. At the end of our inquiry we find that, despite what we have learned, we return to those same questions the ancients asked: Why are we here? Why is the universe here? We are a little closer to answering these questions; perhaps, as the ancients suspected, the answers are linked.

KEY TERMS

quantum gravity
quantum state
collapse of the
   wavefunction
wavefunction of the
   universe
grandfather paradox

Schrödinger equation
Copenhagen
   interpretation
measurement problem
entropy
many-worlds
   interpretation

wavefunction
superposition of states
quantum cosmology
arrow of time
superstring theory

Review Questions

1. What theory is prominent by its absence in our search for the understanding of very extreme conditions? Why do we need such a theory?

2. What is meant by the term "collapse of the wavefunction"? How is this understood within the "Copenhagen interpretation"? How is it treated in the "many-worlds interpretation"? How does Schrödinger's cat illustrate these concepts?

3. Why is the spherical ($k = 1$) de Sitter solution not an acceptable model within classical physics? What quantum effect occurs to make the spherical de Sitter model compatible with observations?

4. What does it mean to say that Einstein's equations or the Schrödinger Equation do not distinguish between time running forward or backward? Which law(s) of physics do make a distinction?

5. Define and discuss the concept of *entropy*. What is its apparent importance to the arrow of time?

6. Give an example of a macroscopic phenomenon that looks the same whether time runs forward or backward. Give an example of a macroscopic phenomenon that is not time symmetric. What distinguishes the two cases?

7. Give a brief example of the grandfather paradox of time travel. How would you resolve this paradox? Many science fiction stories have centered around travel into the past. If you are familiar with an example, how did the author deal with the subject?

8. Explain how the "many-worlds interpretation" of quantum mechanics can accommodate the grandfather paradox. Discuss the "measurement problem" and explain how the many-worlds interpretation can account for this as well. Do you think that the many-worlds interpretation offers any testable predictions? Is it falsifiable?

9. What is meant by the geometrical nature of general relativity? What is the alternative interpretation of gravity?

10. What does "superstring theory" aim to achieve? Like many newer theories, superstring theory requires hidden dimensions; why can't we observe them? If they exist, how do they affect the observable universe?

Notes

1. The constant of proportionality, symbolized by the letter h, is now known as Planck's constant.

2. Any alternative version of quantum mechanics must reproduce the considerable experimental success of the standard theory in order to be acceptable, in accordance with our usual rules for scientific theories.

3. The term "observer" has the regrettable property of suggesting a conscious entity doing the observing. This is not the case: an "observer" is anything that interacts with the system.

4. This phenomenon occurs whether you personally read out the detector data or not. Again, "measurement" or "observation" does not imply human involvement.

5. Figure 13.8 illustrates "temperature time."

6. Black holes of sizes that are likely to exist are extremely cold; the temperature of a solar-mass black hole is only 10^{-7} K, a remarkably low number, since 0 K is absolute zero, the lowest temperature that can be achieved. In fact, according to the principle called the third law of thermodynamics, absolute zero cannot be attained, but only approached arbitrarily closely.

APPENDIX A

Some Useful Numbers

Speed of light	$c = 3.00 \times 10^8$ m/s
Gravitational constant	$G = 6.67 \times 10^{-11}$ N m^2/kg^2
Boltzmann's constant	$k = 1.38 \times 10^{-23}$ J/K
Planck's constant	$h = 6.63 \times 10^{-34}$ J s
Proton mass	$m_p = 1.67 \times 10^{-27}$ kg
Electron mass	$m_e = 9.11 \times 10^{-31}$ kg
Mass of Earth	$M_\oplus = 5.98 \times 10^{24}$ kg
Radius of Earth	$R_\oplus = 6.37 \times 10^6$ m
Mass of Sun	$M_\odot = 1.99 \times 10^{30}$ kg
Radius of Sun	$R_\odot = 6.96 \times 10^8$ m
Astronomical Unit	AU $= 1.50 \times 10^{11}$ m
Light year	lt-yr $= 9.46 \times 10^{15}$ m
Parsec	pc $= 3.08 \times 10^{16}$ m

APPENDIX B

Scientific Notation

Astronomy, even more so than most other sciences, demands the use of very large and very small numbers. A convenient notation is essential for dealing with such numbers; specifically, the standard generally known as *scientific notation*. A number expressed in this convention has the form

$$N.F \times 10^b,$$

where N is between zero and nine, F is any fractional part, and b is the *exponent*. The number $N.F$ is called the *mantissa*. The exponent simply indicates how many times its *base*, 10 in this case, is to be multiplied by itself. For example, 100 is 10×10. Ten is multiplied by itself twice; therefore,

$$100 = 10^2.$$

In scientific notation, we can write the number 100 as

$$1 \times 10^2,$$

although the leading "1" is usually regarded as optional.

Similarly, we can write

$$10^0 = 1, \quad 10^1 = 10, \quad 10^2 = 100, \quad 10^3 = 1000, \quad \text{etc.}$$

Let us attempt a more general number. Suppose we wish to write 33,500 in scientific notation. We move the decimal place until we achieve a number between zero and 9, counting the number of places we have shifted it. For this case, we must move the decimal point four places to the left. The number of places moved indicates the power of 10 by which we must multiply the mantissa. It should be clear that $33,500 = 3.35 \times 10,000$; thus we need only write 10,000 in exponent form to obtain

$$33,500 = 3.35 \times 10^4.$$

So much for very large numbers; what about the small ones? Negative exponents indicate values less than one:

$$10^{-1} = 0.1, \quad 10^{-2} = 0.01, \quad 10^{-3} = 0.001, \quad 10^{-4} = 0.0001, \quad \text{etc.}$$

Note that in general $10^{-x} = \frac{1}{10^x}$.

Multiplication and division are easy with scientific notation. Since 10^3 means $10 \times 10 \times 10$ and 10^2 means 10×10, it follows that $10^3 \times 10^2$ is equal to 10^5, or

$10 \times 10 \times 10 \times 10 \times 10$. In general:

$$\text{multiplication: } 10^a \times 10^b = 10^{a+b};$$

$$\text{division: } 10^a \div 10^b = 10^{a-b};$$

$$\text{exponentiation: } (10^a)^b = 10^{a \times b};$$

$$\text{taking roots: } \sqrt[b]{10^a} = 10^{a \div b}.$$

The appropriate operation must also be performed upon the two mantissas. If the result is not between zero and 9, it may be adjusted, with a corresponding change in the exponent. For example:

$$\text{multiplication: } 5 \times 10^4 \times 7 \times 10^6 = 35 \times 10^{10} \text{ or } 3.5 \times 10^{11};$$

$$\text{division: } 8 \times 10^{10} \div 4 \times 10^5 = 2 \times 10^5;$$

$$\text{another example: } 8 \times 10^{10} \div 4 \times 10^{20} = 2 \times 10^{-10};$$

$$\text{exponentiation: } (3 \times 10^3)^4 = 3^4 \times 10^{12} = 81 \times 10^{12} \text{ or } 8.1 \times 10^{13};$$

$$\text{taking roots: } \sqrt[3]{2.7 \times 10^{10}} = \sqrt[3]{27 \times 10^9} = 3 \times 10^3.$$

Scientific notation is almost essential for numbers that are very large or very small. It would be difficult to write a number such as Avogadro's number, 6.023×10^{23}, without the help of scientific notation. It would nearly as difficult to follow a long string of "million million million millions" if we were to attempt to write such a number in words. This number is so large as to be far beyond those which occur in most human activities, and thus it is not easy to conceptualize. Yet it is an important constant of nature, describing the number of atoms in a standard quantity of any chemical substance. Scientific notation makes such numbers manageable to humans. Of course, large and small numbers create no difficulties for Nature!

APPENDIX C

Units

The *metric system* is used for nearly all scientific purposes. Either of two combinations of units are standard: the "cgs" (centimeter, gram, second) or "MKS" (meter, kilogram, second) system. That is, if a length is given in centimeters, then any mass should be expressed in grams, whereas if the length is given in meters, the mass should be measured in kilograms. The unit of time known as the second is used with either choice. Units that adhere to this convention are *consistent* and can be combined mathematically. Use of units from both systems produces a *mixed* result; this is to be discouraged, although some common quantities are given in such units.

Special prefixes indicate powers of 10 by which units might be multiplied:

Number	Prefix	Abbreviation	Meaning	Example
10^3	kilo	k	thousand (10^3)	kilogram, kilometer
10^6	mega	M	million (10^6)	megaparsec, megaton
10^9	giga	G	billion (10^9)	gigayear, gigawatt
10^{-3}	milli	m	thousandth (10^{-3})	millimeter, milliamp(ere)
10^{-6}	micro	μ	millionth (10^{-6})	micrometer, microsecond
10^{-9}	nano	n	billionth (10^{-9})	nanometer, nanosecond

Because metric units were developed with reference to Earth (the meter was originally intended to be one ten-millionth of the distance from the North Pole to the Equator), they are much too small when applied to astronomical distances and masses; fortunately, scientific notation permits their use. However, astronomers often make use of "natural" units, such as the light year, the parsec, and the solar mass. These units have been defined in the text and are specified in terms of the MKS system in Appendix A. "Natural" units are often much more convenient for cosmological quantities, even though they sometimes result in oddly inconsistent units, such as the km/s/Mpc usually quoted for the Hubble constant.

Units of any sort are arbitrary, of course. They simply form a set of standards to which we can refer the measurement of physical quantities. The metric units for the most important physical quantities are given below.

1. **Time:** The basic unit of time is the second (s). Larger aggregates are the minute, the hour, the day (86,400 s), and the year (3.16×10^7 s).

2. **Length:** The units of length are the *meter* and the *centimeter*. The *kilometer* is often used, but remember that it is not a consistent unit within either the cgs or MKS system. (One kilometer is approximately equal to 0.6 of a mile).
Several special units of length are also used in astronomy:

   i. The *Astronomical Unit (A.U.):* This is defined as the distance between the Earth and the Sun and is approximately 1.5×10^{13} cm.

   ii. The *light year (lt-yr):* This is defined to be the distance that light travels in 1 year. It is approximately 9.5×10^{17} cm. Note that the light year is a unit of distance, not of time.

   iii. The *parsec (pc):* This is about 3.25 lt-yr; it is widely used as a convenient measure of distances between stars. For intergalactic distances, the megaparsec is most common.

3. **Velocity:** Standard units of velocity are meters per second (m/s or m s^{-1}) or centimeters per second (cm/s or cm s^{-1}). For astronomers a more natural choice is often kilometers per second, 1 km/s $\sim$ 2200 miles per hour. For example, the velocity of light is 300,000 km/s.

4. **Mass:** The units of mass are the gram (g) and the kilogram (kg = 10^3 g). For astronomy, we may also use units of Earth mass ($M_{\oplus} = 5.98 \times 10^{24}$ kg) or more usually solar mass ($M_{\odot} = 1.99 \times 10^{30}$ kg). In scientific usage, 1 g is the mass of 1 cubic centimeter (1 cm^3) of pure water under standard conditions of temperature and pressure. (In the Système Internationale of units used in most countries, however, the gram is one thousandth of the kilogram, where the kilogram is defined as the mass of a reference block of an alloy of platinum and iridium, kept in a vault in Paris.)

5. **Density:** Density is the amount of mass present in a chosen unit of volume. For example, the density of water is one gram per cubic centimeter (1 g/cm^3 or 1 g cm^{-3}).

6. **Temperature:** Scientists usually measure temperature in *degrees Kelvin*, or occasionally in degrees Celsius (centigrade). The size of the unit of each scale is the same, but the Kelvin scale locates its zero point at the thermodynamic standard called "absolute zero." At absolute zero, all molecules are as stationary as quantum mechanics allows. Absolute zero corresponds to -273 C or to -459 F. The modern Celsius scale sets its zero at the freezing point of water ($+273$ K); both positive and negative values are possible. The centigrade or Celsius scale came to be used in science because it is based upon the phase changes of water, a common substance easily purified and measured in the laboratory. We will avoid the Fahrenheit scale, which has smaller units and a rather arbitrary zero point; its zero was based upon a particular combination of ice, salt, and water that is not so easily reproduced consistently in the laboratory.

7. **Angular measure:** As the name implies, these are units for the measurements of angles. There are 360 *degrees* (360°) in a full circle, 60 *minutes* ($60' = 1°$) of arc in one degree, and 60 *seconds* of arc (1 arcsecond) ($60'' = 1'$) in every minute. The Big Dipper (Ursa Major) occupies about 20° on the sky; the full moon subtends

30 arcminutes; the eye can barely distinguish two objects separated by 1 arcminute. The angular size of a dime seen at a distance of 2 km is approximately 1 arcsecond.

8. **Force:** The standard units of force are the *dyne* (cgs) and the *Newton* (MKS). The dyne is 1g cm per second squared, whereas the Newton is 1kg m per second squared. We often refer to the "weight" of something, but strictly, the unit of weight is a force. The kilogram is a unit of mass, *not* a unit of weight, although it is casually employed for weight in most countries that utilize the metric system; for some reason, the Newton is not widely used in everyday life. Near the surface of the Earth, however, there is little practical distinction between mass and weight.

9. **Energy and power:** The standard units of energy are the *erg* (cgs) or the *Joule* (MKS). The Joule is much larger than the erg, with $1 \text{ J} = 10^7$ ergs. One Joule corresponds to the energy obtained from dropping 1 kg (e.g., a small bag of sugar) from a height of 10 cm (close to 4 inches). The more familiar unit, the *watt*, is actually a unit of **power**, which is the rate of energy production or release per second. Specifically, 1 W is equal to 1 J per second. Thus a 100 W light bulb expends 100 J of energy every second. The hybrid unit *kilowatt-hour*, which appears on most electric bills in the United States, consists of a unit of power multiplied by a unit of time, and thus is itself a unit of energy; customers are billed for the total energy consumed over some interval of time, not for "power" *per se*. In astronomy, a natural unit of power is the luminosity of the Sun, $1 \ L_\odot = 3.9 \times 10^{26} \text{ W} = 3.9 \times 10^{33} \text{ ergs/s}$.

Units Conversions

Many people find units conversions intimidating. One way to prevent mistakes is to remember that the symbols for units behave exactly like algebraic variables. It is possible to write an equation containing units, and to cancel like (and *only* like) units appropriately. For example, suppose it is desired to convert kilometers into miles. We have the fundamental equation that

$$0.61 \text{ mile} = 1 \text{ km.}$$

This is an equality, so we may write

$$\frac{0.61 \text{ mile}}{1 \text{ km}} = 1$$

or its reciprocal

$$\frac{1 \text{ km}}{0.61 \text{ mile}} = 1.$$

Algebraically, we may always multiply by "1" in some form. Suppose we wish to know

$$56 \text{ miles} = ? \text{ km.}$$

If we always keep in mind that we must manipulate the problem into the form such that the units we wish to eliminate will cancel algebraically, we can see that we obtain

$$56 \text{ miles} \times \frac{1 \text{ km}}{0.61 \text{ mile}} = 91.8 \text{ km}.$$

This method works even when compound units are to be converted. For instance, suppose we must convert a density from g/cm^3 to kg/m^3. As a specific example, the density of water is 1 g/cm^3; what is its density expressed in kg/m^3?

We begin with the equalities

$$1000 \text{ g} = 1 \text{ kg},$$

$$100 \text{ cm} = 1 \text{ m}.$$

From the second of these, we find that

$$100^3 \text{ cm}^3 = 1^3 \text{ m}^3$$

or

$$10^6 \text{ cm}^3 = 1 \text{ m}^3.$$

Thus

$$\frac{1 \text{ g}}{\text{cm}^3} \times \frac{1 \text{ kg}}{10^3 \text{ g}} \times \frac{10^6 \text{ cm}^3}{1 \text{ m}^3} = 1000 \text{ kg/m}^3.$$

Hence the density of water is 1000 kg/m^3.

Glossary

absolute zero: The lowest possible temperature, attained when a system is at its minimum possible energy. The Kelvin temperature scale sets its zero point at absolute zero ($-273.15°$ on the Celsius scale, and $-434.07°$ on the Fahrenheit scale).

absorption spectrum: A spectrum consisting of dark lines superimposed over a continuum spectrum, created when a cooler gas absorbs photons from a hotter continuum source.

acceleration: The rate of increase of velocity with time.

accretion disk: A disk of gas that accumulates around a center of gravitational attraction, such as a white dwarf, neutron star, or black hole. As the gas spirals in, it becomes hot and emits light or even X-radiation.

active galactic nucleus (AGN): An unusually bright galactic nucleus whose light is not due to starlight.

active galaxy: A galaxy whose energy output is anomalously high. About 1% of galaxies are active. Most contain an AGN at their cores.

amplitude: *See* wave amplitude.

angular size: The angle subtended by an object on the sky. For example, the angular size of the moon is 30 arcminutes.

anthropic principle: The observation that, since we exist, the conditions of the universe must be such as to permit life to exist.

anthropocentrism: The belief that humans are central to the universe.

anthropomorphism: The projection of human attributes onto nonhuman entities such as animals, the planets, or the universe as a whole.

antimatter: Particles with certain properties opposite to those of matter. Each matter particle has a corresponding *antiparticle*. The antiparticle has exactly the same mass and opposite electric charge as its partner. When a particle and its antiparticle collide, both are annihilated and converted into photons. (*See also* baryogenesis.)

arrow of time: The direction, apparently inviolable, of the "flow" of time that distinguishes the past from the future.

Astronomical Unit (AU): The mean distance from the Earth to the Sun.

astronomy: The study of the contents of the universe beyond Earth.

atom: The smallest component of matter that retains its chemical properties. An atom consists of a nucleus and at least one electron.

atomic number: The number of protons present in the nucleus of an atom. This determines its elemental identity.

baryogenesis: The creation of matter in excess of antimatter in the early universe. Only the relatively few unmatched matter particles survived to make up all subsequent structures.

baryon: A fermionic particle consisting of three quarks. The most important baryons are the proton and the neutron.

baryon number conservation: The principle that the number of baryons must remain the same in any nuclear reaction.

biased galaxy formation: The theory that bright galaxies form preferentially from anomalously overdense perturbations in the early universe.

big bang: The state of extremely high (classically, infinite) density and temperature from which the universe began expanding.

big crunch: The state of extremely high density and temperature into which a closed universe will recollapse in the distant future.

Birkhoff's theorem: A theorem of general relativity that states that all spherical gravitational fields, whether from a star or from a black hole, are indistinguishable at large distances. A consequence of this is that purely radial changes in a spherical star do not affect its external gravitational field.

blackbody: A perfectly absorbing (and perfectly emitting) body.

blackbody radiation: A special case of thermal radiation, emitted by a blackbody and characterized by thermal equilibrium of the photons. A blackbody spectrum is completely determined by the temperature of the emitter.

black hole: An object that is maximally gravitationally collapsed, and from which not even light can escape.

black hole thermodynamics: The theory that permits a temperature and an entropy to be defined for black holes.

blueshift: A shift in the frequency of a photon toward higher energy.

boost factor: The quantity $\Gamma = 1/\sqrt{1 - v^2/c^2}$ in the special theory of relativity that relates measurements in two inertial frames.

boson: A class of elementary particles whose spin is an integer multiple of a fundamental quantized value. The major function of bosons is to mediate the fundamental forces. The best-known boson is the photon.

bottom-up structure formation: The idea that small structures, perhaps galaxies or even smaller substructures, form first in the universe, followed later by larger structures.

brown dwarf: A substellar object that is below the minimum mass required for nuclear fusion reactions to occur in its core.

carrier boson: A particle that carries one of the fundamental forces between other interacting particles. For example, the carrier boson for the electromagnetic force is the photon.

CBR: Cosmic background radiation.

CDM: Cold dark matter.

Cepheid variable: A type of variable star whose period of variation is tightly related to its intrinsic luminosity.

Chandrasekhar limit: The maximum mass, approximately 1.4 $M_\odot$, above which an object cannot support itself by electron degeneracy pressure; hence, the maximum mass of a white dwarf.

chaotic inflation: A model in which many distinct universes form from different regions of a "mother" universe, with some inflating and others perhaps not.

charge: The fundamental property of a particle that causes it to be affected by the electromagnetic force.

closed universe: A standard universe with a spherical three-dimensional spatial geometry. Such a universe is finite in both space and time, and recollapses.

cold dark matter model: A model of structure formation in which an exotic particle whose energy is low at the time it decouples from other matter is responsible for structure formation.

collapse of the wavefunction: In the Copenhagen interpretation of quantum mechanics, the result of an act of measurement, in which the potentialities inherent in the quantum wavefunction take on a specific value, namely, that which is measured.

collisionless damping: The tendency of weakly interacting (collisionless) matter to smooth out gravitational perturbations by freely streaming from overdense to underdense regions.

comoving coordinates: Coordinates fixed with respect to the overall Hubble flow of the universe, so that they do not change as the universe expands.

Compton wavelength: The quantum wavelength of a particle with a highly relativistic velocity.

conservation of angular momentum: The principle that the angular momentum of a system (the momentum of rotation about a point) remains the same as long as no external torque acts.

conservation of energy: The principle that the total energy of a closed system never changes, that energy is only converted from one form to another. This principle must be enlarged under special relativity to include mass-energy.

conservation of matter: The principle that matter is neither created nor destroyed. This principle is only approximately true, since special relativity shows that matter and energy are equivalent and interconvertible.

conservation of momentum: The principle that the linear momentum of a system (in Newtonian mechanics, mass times velocity) remains the same as long as no external force acts.

consistent: The property possessed by a scientific theory when it contains and extends an earlier well-supported theory; for example, general relativity is consistent with Newtonian gravity.

coordinate singularity: A location at which a particular coordinate system fails, such as the Schwarzschild metric coordinates at the Schwarzschild radius of a black hole.

coordinates: Quantities that provide references for locations in space and time.

Copenhagen interpretation: In quantum mechanics, the interpretation of the wavefunction as a description of the probabilities that the state of the system will take on different values.

Copernican principle: The principle that Earth is not the center of the universe.

Copernican revolution: The revolution in thought resulting from the acceptance of the heliocentric model of the solar system.

correlation function: A mathematical expression of the probability that two quantities are related. In cosmology, the correlation function indicates the probability that galaxies will be found within a particular distance of one another, thus providing a quantitative measure of the clustering of galaxies (or of clusters).

cosmic background radiation: The blackbody radiation, now mostly in the microwave band, which consists of relic photons left over from the very hot, early phase of the big bang.

cosmic censorship: The principle that singularities are never "naked," that is, do not occur unless surrounded by a shielding event horizon.

cosmic distance ladder: The methods by which increasing distance is measured in the cosmos. Each method depends on a more secure technique (or "rung") used for smaller distances.

cosmic strings: Long, stringlike concentrations of matter-energy that may have formed during symmetry breaking in the first moments of the big bang. If they exist, they would be candidates for the seed perturbations of structure formation.

cosmic time: A time coordinate that can be defined for all frames in a homogeneous metric, representing the proper time of observers at rest with respect to the Hubble flow. In a big bang model, this coordinate marks the time elapsed since the singularity.

cosmological constant: A constant introduced into Einstein's field equations of general relativity in order to provide a supplement to gravity. If positive (repulsive), it counteracts gravity, while if negative (attractive), it augments gravity. It can be interpreted physically as an energy density associated with space itself.

cosmological principle: The principle that there is no center to the universe, that is, that the universe is everywhere isotropic on the largest scales, from which it follows that it is also homogeneous.

cosmological redshift: A redshift caused by the expansion of space.

cosmology: The study of the origin, evolution, and behavior of the universe as a whole.

critical density: The density that just stops the expansion of space, after infinite cosmic time has elapsed. In the standard models, the critical density requires that the spatial geometry be flat.

crucial experiment: An experiment that has the power to decide between two competing theories.

curvature constant: A constant (k) appearing in the Robertson-Walker metric that determines the curvature of the spatial geometry of the universe.

dark halos: Massive, nonluminous matter of unknown kind that surrounds and envelopes galaxies.

dark matter: Matter that is invisible to us because it emits little or no light. As much as 90%–99% of the mass of the universe may be dark.

data: The outcome of a set of measurements from which inferences may be drawn, theories constructed, and so forth.

de Sitter model: A model of the universe that contains no matter but only a positive cosmological constant. It expands exponentially forever.

deceleration parameter: A parameter (q) that denotes the rate of change with time of the Hubble constant.

density parameter: The ratio of the actual mass density of the universe to the critical density. Also called omega (Ω).

deuterium: An isotope of hydrogen whose nucleus contains one proton and one neutron.

distance ladder: *See* cosmic distance ladder.

dynamical method: A method of measuring the mass of a galaxy, cluster, or even the universe which makes use of the gravitational interactions of two or more bodies.

Doppler effect: The change in frequency of a wave (light, sound, etc.) due to the relative motion of source and receiver.

Einstein–de Sitter model: The flat ($k = 0$), pressureless standard model of the universe.

electromagnetic force: The force between charged particles, which accounts for electricity and magnetism. One of the four fundamental forces of nature, it is carried by photons and is responsible for all observed macroscopic forces, except for gravity.

electromagnetic spectrum: The full range of light wavelengths or frequencies, from low-energy radio waves to high-energy gamma rays.

electron: An elementary lepton with a negative charge. One of the components of atoms, the electrons determine the chemical properties of an element.

electron degeneracy: A condition of matter in which all quantum states available to the electrons are filled.

electron degeneracy pressure: A form of pressure arising from electron degeneracy; the electrons resist being forced closer together because of the exclusion principle.

electroweak interaction: The unified electromagnetic and weak forces. Also called the electroweak force.

element: A particular type of atom, with specific atomic number and chemical properties. The smallest unit into which matter may be broken by chemical means.

ellipse: A geometric figure generated by keeping the sum of the distance from two fixed points (the foci) constant.

elliptical galaxy: A galaxy whose shape is roughly spheroidal or ellipsoidal. Most ellipticals contain little dust or gas and show no evidence of recent star formation.

elsewhere: That region of spacetime outside the light cone at a given event. Events that are elsewhere from each other are mutually unobservable and cannot be causally connected.

emission distance: The distance to the source of light at the time the light was emitted.

emission spectrum: A spectrum consisting of bright lines, created when a hot gas emits photons characteristic of the elements of which the gas is composed.

energy: The capacity to perform work.

entropy: A quantitative measure of the disorder of a system. The greater the disorder, the higher the entropy.

equal density epoch: That interval in the early history of the universe when the gravitational contributions of matter and radiation were approximately equal.

equilibrium: A balance in the rates of opposing processes, such as emission and absorption of photons, creation and destruction of matter, and so on.

equivalence principle: The complete equality of gravitational and inertial mass, gravity and acceleration, and the identification of freefalling frames with inertial frames. (*See also* weak equivalence principle and strong equivalence principle.)

ergosphere: The region of a rotating Kerr black hole between the static surface and the event horizon.

escape velocity: The minimum velocity required to escape to infinity from the gravitational field of an object.

Euclidean geometry: Flat geometry based upon the geometric axioms of Euclid.

event: A point in four-dimensional spacetime; a location in both space and time.

event horizon: A surface that divides spacetime into two regions: that which can be observed and that which cannot. The Schwarzschild radius of a nonrotating black hole is an event horizon.

exclusion principle: The property that fermions of the same type that can interact with each other cannot simultaneously occupy the same quantum state.

experiment: A controlled trial for the purpose of collecting data about a specific phenomenon.

explanatory power: The ability of a scientific hypothesis to account for known data.

extinction: In astronomy, the removal of light from a beam by whatever means, such as absorption and scattering.

false vacuum: A metastable state in which a quantum field is zero, but its corresponding vacuum energy density is not zero.

falsifiable: The property of a scientific hypothesis that it is possible to perform an experiment that would disprove, or falsify, the hypothesis.

fermion: A class of elementary particles whose spin is a half-integer multiple of a fundamental quantized value. Fermions make up matter. The best-known fermions are protons, neutrons, electrons, and neutrinos. Fermions obey the exclusion principle.

field: A mathematical representation of a quantity describing its variations in space or time, or both.

fission: The splitting of a heavy atomic nucleus into two or more lighter nuclei.

flat geometry: Geometry in which the curvature is zero; ordinary Euclidean geometry.

flatness problem: The observed fact that the geometry of the universe is very nearly flat, a very special condition, without an explanation of why it should be flat.

flat universe: A model whose three-dimensional spatial geometry is flat.

flux: The amount of some quantity, usually energy, crossing a unit area per unit time.

force: An action such as a push or pull which produces an acceleration.

frame of reference: The coordinate system to which a particular observer refers his or her measurements.

free fall: Unrestrained motion under the influence of a gravitational field.

frequency: *See* wave frequency.

Friedmann equation: The equation that describes the evolution of the cosmological scale factor of the Robertson-Walker metric.

Friedmann models: A class of cosmological models that are isotropic and homogeneous, contain a specified matter-energy density, conserve matter, and admit no cosmological constant. Also called standard models.

fundamental forces: The four forces (strong, weak, electromagnetic, and gravitational) that account for all interactions of matter.

fusion: The joining of two or more lighter elements to create a heavier nucleus.

future: Those events that could be influenced by a given event.

galaxy cluster: A group of galaxies that are mutually gravitationally bound.

Galilean relativity: The transformation from one inertial frame of reference to another in the limit of very small velocities and very weak gravitational fields.

gauge boson: *See* carrier boson.

geocentric: Taking Earth to be the center; for example, of the solar system.

geodesic: In geometry, that path between two points/events which is an extremum in length. In some geometries, such as Euclidean, the geodesics are the shortest paths, whereas in others, such as in the spacetime geometries appropriate to general relativity, the geodesics are the longest paths.

globular clusters: Aggregations of approximately 100,000 stars that orbit many galaxies. Globular clusters contain almost exclusively very ancient stars and are thought to be among the oldest structures in the universe.

gluon: A hypothetical particle that binds quarks together into hadrons.

grand unified theories: A class of theories that seeks to explain the unification of the strong, weak, and electromagnetic forces.

grandfather paradox: The contradictory idea that a time traveler could kill her grandfather while he is an infant in his crib, thus preventing the traveler's own birth.

gravitational constant: A fundamental constant of nature, G, which determines the strength of the gravitational interaction.

gravitational lens: A massive object that causes light to bend and focus due to its general relativistic effect upon the spacetime near it.

gravitational radiation: The emission of gravitational waves by the creation of a gravitational field that changes in time. Also: the waves (*see* gravitational wave) so radiated.

gravitational redshift: A shift in the frequency of a photon to lower energy as it climbs out of a gravitational field.

gravitational wave: A propagating ripple of spacetime curvature that travels at the speed of light.

graviton: A hypothetical massless boson that is the carrier of the gravitational force.

gravity: The weakest of the four fundamental forces; that force which creates the mutual attraction of masses.

Great Attractor: A proposed mass concentration beyond the Hydra-Centaur galaxy cluster, toward which the Virgo Cluster and its entourage may be falling.

GUTs: Grand unified theories.

hadron: A class of particles that participates in the strong interaction. Hadrons consist of those particles (baryons, mesons) that are composed of quarks.

hadron epoch: That interval in the early history of the universe after the quarks had condensed into hadrons and before the temperature dropped below the threshold temperature for protons.

half-life: The interval of time required for half of a sample of a radioactive material to decay.

Harrison-Zel'dovich spectrum: A proposed spectrum for the matter perturbations in the early universe that gave rise to the observed structure. The Harrison-Zel'dovich spectrum is scale free; that is, perturbations of all sizes behave in the same way.

Hawking radiation: Emission of particles, mostly photons, near the event horizon of black holes due to the quantum creation of particles from the gravitational energy of the black hole.

HDM: Hot dark matter.

heat: A form of energy related to the random motions of the particles (atoms, molecules, etc.) that make up an object.

heat death: The fate of the flat and open universe models in which the temperature drops toward zero, stars die out, black holes evaporate from Hawking radiation, entropy increases, and no further energy is available for any physical processes.

heliocentric: Taking the Sun to be the center; for example, of the solar system.

Hertzsprung-Russell (H-R) diagram: A plot of magnitude (related to luminosity) versus color (related to surface temperature) for a large number of stars, such as the members of a cluster. The points are always found to lie close to a small number of curves, with each curve appropriate to stars at a specific stage of their lives, such as the main sequence for hydrogen-fusing stars.

Higgs boson: A hypothetical particle which plays an important role in grand unified theories. The Higgs boson would be associated with processes leading to baryogenesis and might play a role in endowing all particles with mass.

homogeneity: The property of a geometry in which all points are equivalent.

horizon: Any surface that demarcates events which can be seen from those that cannot be seen.

horizon problem: The conflict between the observed high uniformity of the cosmic background radiation and the fact that regions of the sky separated by an angular size of more than approximately $1°$ could not have been in causal contact at the time of recombination.

hot dark matter model: A model of structure formation in which a particle whose energy is high at the time it decouples from other matter is responsible for the origin of large-scale structure.

Hubble constant: The constant of proportionality (H) between recession velocity and distance in the Hubble law. It is not strictly a constant, because it can change with time over the history of the universe.

Hubble flow: The separation of galaxies due to the expansion of space, not to their individual gravitational interactions.

Hubble law: The relationship between recession velocity and distance, $v = H\ell$, for an isotropic expanding universe.

Hubble length: The distance traveled by light along a straight geodesic in one Hubble time, $r_H = ct_H$.

Hubble sphere: A sphere centered about any arbitrary point whose radius is the Hubble length. The center of the Hubble sphere is not a "center" to the universe because each point has its own Hubble sphere. The Hubble sphere approximately defines that portion of the universe that is observable from the specified point at a specified time.

Hubble time: The inverse of the Hubble constant. The Hubble time, also called the Hubble age or the Hubble period, provides an estimate for the age of the universe.

hydrostatic equilibrium: The balance between gravity and gas pressure in an object such as a star.

hyperbolic geometry: A geometry that has negative constant curvature. Hyperbolic geometries cannot be fully visualized because a two-dimensional hyperbolic geometry cannot be embedded in the three-dimensional Euclidean space. However, the lowest point of a saddle, that point at which curvature goes both "uphill" and "downhill," provides a local representation.

hypothesis: A proposed explanation for an observed phenomenon. In science, a valid hypothesis must be based upon data and must be subject to testing.

ideal gas: A gas in which the mutual interactions of the gas particles are negligible, except for their momentary collisions.

ideal gas law: The formula that relates temperature, pressure, and volume for an ideal gas. Nearly all real gases obey the ideal gas law to very high temperatures and pressures, even those found in the interiors of stars.

IMF: Initial mass function.

inertia: That property of an object which resists changes in its state of motion.

inertial force: A force arising from the acceleration of an observer's frame of reference.

inertial motion: Motion free of any force, that is, constant velocity motion.

inertial observer: An observer occupying an inertial frame of reference.

inertial reference frame: A reference frame in which a free particle experiences no force.

inflation: A period of exponential increase in the scale factor due to a nonzero vacuum energy density; inflation occurs very early in the history of the universe in certain cosmological models.

inflaton: The generic name of the unidentified particle that may be responsible for an episode of inflation in the very early universe.

initial mass function: The theoretical function describing the number of stars for each given mass that will be produced in an episode of star formation.

interference: The interaction of two waves in which their amplitudes are reinforced or cancelled.

interference fringes: A pattern of alternating reinforcement and destruction caused by the interference of two or more waves.

interferometer: A device that carries out some measurement by detecting wave interference.

interstellar medium: Gas, dust, bits of ice, and so on, which fill the space between the stars. Nearly all of the interstellar medium is hydrogen and helium gas, with hydrogen most abundant.

invariance: The property of remaining unchanged under a transformation of the frame of reference or the coordinate system.

ion: An atom that has gained or lost an electron and thereby acquired an electric charge. (Charged molecules are called radicals, not ions.)

irregular cluster: A cluster of galaxies with no particular shape. Irregular clusters often contain many spiral galaxies.

irregular galaxy: A galaxy with an ill-defined, irregular shape. Many irregulars are interacting or even colliding with other galaxies, which may account for their disorganized appearance.

isotope: One of the forms in which an element occurs. One isotope differs from another by having different numbers of neutrons in its nucleus. The number of protons determines the elemental identity of an atom, but the total number of nucleons affects properties such as radioactivity or stability, the types of nuclear reactions, if any, in which the isotope will participate, and so forth.

isotropy: The property of a geometry of being the same in all directions.

Jeans instability: A gravitational instability that occurs if the collapse time for an overdense region is shorter than the time for sound waves to cross it.

Jeans mass: The minimum mass that can collapse from the Jeans instability for given conditions of temperature, pressure, and so on.

Kepler's laws: The three laws of planetary motion discovered by Johann Kepler.

Kerr metric: The metric that describes the spacetime around a rotating black hole.

kinematical method: A method of measuring the mass density of the universe indirectly, by means of overall parameters of the universe such as its expansion rate. Kinematic methods exploit the fact that expansion rate, deceleration parameter, density, and curvature are not completely independent quantities but are related by the Friedmann equation, possibly extended to include a cosmological constant.

kinetic energy: The energy associated with macroscopic motion. In Newtonian mechanics, the kinetic energy is equal to $\frac{1}{2}mv^2$.

law: In scientific usage, a theory that has become particularly well confirmed and well established.

law of inertia: Another name for Newton's first law.

law of universal gravitation: Newton's mathematical formulation of the law of attraction between two masses, $F_g = GM_1M_2/R^2$.

Lemaître model: The cosmological model developed by Georges Lemaître, which contains a positive cosmological constant, uniform matter density, and spherical spatial geometry.

length contraction: An apparent contraction of the length of an object in motion relative to a given observer, caused by the Lorentz transformation from one frame to another. Sometimes called the Lorentz contraction.

lepton: A member of a class of fermionic particles that do not participate in the strong interaction. The best-known lepton is the electron.

lepton epoch: The interval in the early history of the universe when leptons dominated.

light cone: The surface representing all possible paths of light which could arrive at or depart from a particular event.

lightlike: *See* null.

light year (lt-yr): A measure of distance equal to that traveled by light in 1 year.

line radiation: Radiation of a particular wavelength produced by an electron moving from one orbital to another of lower energy. (*See also* emission spectrum.)

Local Group: The small cluster of galaxies of which our Galaxy and the Andromeda Galaxy are prominent members.

long-range force: A force that does not become equal to zero within any finite distance. The long-range forces are gravity and electromagnetism, both of which decrease as R^{-2} with increasing distance R.

lookback time: The time required for light to travel from an emitting object to the receiver.

Lorentz contraction: *See* length contraction.

Lorentz transformation: The transformation, valid for all relative velocities, which describes how to relate coordinates and observations in one inertial frame to those in another such frame.

luminiferous ether: A supposed medium for the transmission of light. The concept was rendered superfluous by the special theory of relativity early in the twentieth century.

luminosity: The total power output of an object in the form of light. (Sometimes extended to include all forms of energy.)

luminosity distance: The inferred distance to an astronomical object derived by comparing its observed brightness to its presumed total luminosity.

Mach's principle: The principle, elucidated by Ernst Mach, that the distribution of matter in the universe determines local inertial frames.

MACHO (massive compact halo object): Any object such as a white dwarf, neutron star, or black hole that could account for some or all of the dark matter in the halos of galaxies.

magnetic monopole: A hypothetical particle representing one unit of magnetic "charge." Although required by grand unified and other theories, no magnetic monopole has been unequivocally observed.

main sequence: The curve on a Hertzsprung-Russell diagram along which stable hydrogen-fusing stars lie.

many-worlds interpretation: An interpretation of the measurement problem in quantum mechanics which holds that each act of measurement causes the universe to split into noncommunicating, parallel, quantum entities.

mass: That property of an object that causes it to resist changes in its state of motion; also, that property that generates gravitational attraction.

mass-to-light ratio: The ratio of the total mass of a luminous aggregate of matter in solar masses to its total luminosity in solar luminosities.

matter-dominated era: The epoch of the universe, lasting from approximately the time of recombination until the present, during which the energy density of radiation is negligible in determining the overall gravitational field of the universe, and the density of matter is dominant.

measurement problem: The name for the enigma of how a measurement changes a quantum system into a definite state from one that evolves according to the probabilistic Schrödinger equation.

mechanics: The science of motion.

metals: In astronomy, all elements heavier than helium, regardless of whether they are chemically "metals" or not.

metric coefficient: The functions in the metric that multiply with the coordinate differentials (e.g., Δx) to convert these differentials into physical distances.

metric equation: The expression that describes how to compute the distance between two infinitesimally separated points (or events) in a given geometry. Also called simply the "metric."

microlensing: Gravitational lensing by relatively small objects such as stars or stellar remnants.

Milky Way: The name of our Galaxy. Also the name given to the band of diffuse light seen in the night sky that originates in the disk of our Galaxy.

Minkowski spacetime: The geometrically flat, four-dimensional spacetime appropriate to special relativity.

model: A hypothesis or group of related hypotheses that describes and clarifies a natural phenomenon, entity, and so forth.

myth: A narrative intended to explain or justify the beliefs of a people. The term usually suggests a lack of historical and factual basis.

nebula: A cloud of gas in space.

neutrino: Any of three species of very weakly interacting lepton with an extremely small, possibly zero, mass.

neutron: A charge-neutral hadron which is one of the two particles that make up the nuclei of atoms. Neutrons are unstable outside the nucleus, but stable within it.

neutron degeneracy: A condition of matter in which electrons and protons are crushed together to form neutrons, and all quantum states available to the neutrons are filled.

neutron degeneracy pressure: A form of pressure that arises from neutron degeneracy, when the neutrons cannot be forced further together because of the exclusion principle.

neutron star: A compact dead star supported by neutron degeneracy pressure.

Newton's first law: The law of motion that states that an object in a state of uniform motion will remain in that state unless acted upon by an external force.

Newton's second law: The law of motion that states that the net applied force on an object produces an acceleration in proportion to the mass, $F = ma$.

Newton's third law: The law of motion that states that if A exerts a force on B, then B will exert an equal and oppositely directed force on A. For every action, there is an equal and opposite reaction.

no-hair theorem: The theorem that the gravitational field of a black hole is entirely determined by only its mass, angular momentum, and electric charge (if any).

nova: An abrupt, very bright flare-up of a star. Most likely due to the accumulation of hydrogen from a companion upon the surface of a white dwarf. The pressure and temperature grow in the accreted matter until a thermonuclear explosion occurs.

nuclear forces: Two of the fundamental forces, or interactions, the strong interaction and the weak interaction. Not necessarily confined exclusively to the nucleus, despite the name. The strong interaction not only holds nucleons together in the nucleus, but also binds quarks into hadrons. The weak interaction is involved in some nuclear processes such as radioactivity, but also causes free neutrons to decay.

nucleon: Either of the two baryons, the proton and the neutron, which form the nuclei of atoms.

nucleosynthesis: The process by which nuclear reactions produce the various elements of the periodic table.

nucleosynthesis epoch: The interval in the early history of the universe when helium, as well as traces of a few other light isotopes, was created.

nucleus: The central region of an atom, which gives it its elemental identity.

null: Of a spacetime interval, capable of being traversed only by a massless particle such as a photon. A null or lightlike spacetime interval is zero.

Occam's razor: The principle that all other things being equal, the simplest explanation is preferred.

Olbers' paradox: The fact that the night sky is dark even though in an infinite universe with stars that live forever, the night sky would be as bright as the surface of a star. The paradox disappears when it is realized that stars do not live forever. In the modern big bang model, expansion of the universe also plays a role in making the sky dark at night by redshifting the cosmic background radiation to a band well below the visible.

omega: *See* density parameter.

open universe: A standard universe that expands forever and is infinite in space and time, although it begins with a big bang. Sometimes (as in this text) applied strictly to the hyperbolic standard model, although both the hyperbolic and flat models are open in the sense of expanding forever.

pair production: The creation of a particle and its antiparticle from some form of energy, such as photons.

parallax: The apparent shift in the position of an object, such as a star, due to the changing vantage point of the observer. Astronomical parallax is caused by the orbital motion of Earth.

parsec (pc): The distance of an object with a parallax angle of exactly 1 arcsecond. Corresponds to approximately 3.26 lt-yr.

particle horizon: A surface beyond which we cannot see because the light from farther objects has not had time to reach us over the age of the universe.

past: Those events that could have influenced a given event.

peculiar velocity: The unique velocity of an object such as a galaxy, due to its individual gravitational interactions with other objects and not due to the general cosmological recession.

perfect cosmological principle: The principle that the universe is unchanging, homogeneous in time as well as in space. Refuted by the direct observation that the oldest objects in the universe are not like those in our immediate surroundings.

periodic table: A tabulation of the elements in increasing order of atomic number.

photon: A boson that is the particle of electromagnetic radiation (light). The photon is also the carrier particle of the electromagnetic force.

photon damping: The tendency of photons in the early universe to smooth out inhomogeneities in matter with which they are in thermal equilibrium.

photon sphere: The radius around a black hole at which light paths are gravitationally bent into a circle, thus causing the photons to orbit the hole.

Planck epoch: The epoch from the beginning of the universe until the Planck time. Very little is known about this interval, although probably all four fundamental forces were united.

Planck length: The Hubble length of the universe at the Planck time, approximately 10^{-33} cm.

Planck time: The cosmic time near the beginning of the universe, 10^{-43} s, at which classical gravity gained control of the universe as a whole.

Planck's constant: A fundamental constant of physics, h, which sets the scale of quantum-mechanical effects.

plasma: A gas in which many or most of the atoms are ionized.

Population I: Second-generation and younger stars, such as the Sun.

Population II: An older generation of stars, such as are found in globular clusters.

positron: The antimatter partner of the electron.

potential: In physics, a mathematical function which describes the energy density of a field.

potential energy: The energy possessed by something by virtue of its location in a field, such as its position in a gravitational field.

predictive power: The ability of a hypothesis or model to predict as yet unobserved effects. This provides an important means of testing a hypothesis.

primordial elements: Those elements and isotopes formed in the big bang: specifically hydrogen, helium (both ^{3}He and ^{4}He), most deuterium and tritium, and some lithium (^{7}Li).

principle of causality: The principle that a cause must precede its effect in time.

principle of reciprocity: The principle in special relativity that two inertial frames will observe exactly the same phenomena when each observes the other; for example, each will see lengths in the other frame to be contracted by the same amount. This follows directly from the *relativity principle*. This principle applies only to comparisons between inertial frames.

proper length: The length of an object measured in its own rest frame.

proper time: The time interval between two events as measured in the rest frame in which those events occurred. Numerically equal to the invariant spacetime interval.

proton: A hadron that is one of the two particles that make up atomic nuclei. The proton is the least-massive baryon. Its absolute stability is uncertain, but its half-life is at least 10^{31} yr.

pulsar: A rotating neutron star that emits regular, periodic bursts of radio emissions.

quantum: The smallest unit of some quantity.

quantum cosmology: A theory that attempts to describe the evolution of the universe in quantum-mechanical terms.

quantum fluctuations: The small variations that must be present in a quantum field due to the uncertainty principle.

quantum gravity: A unification of gravity and quantum field theory, not yet achieved.

quantum mechanics: The theory that describes the behavior of the very small, such as molecules, atoms, and subatomic particles. Spectacularly successful at explaining experimental data, but gravity cannot yet be made to fit within the theory.

quantum state: A particular configuration of the quantum properties, such as energy, spin, momentum, charge, and so on, that define a particular system.

quark: One of the six fundamental particles that make up hadrons.

quark epoch: The interval in the early universe during which quarks were unconfined in hadrons, and dominant.

quasar: An object that emits an extremely large luminosity from a small region. Invariably found only at large redshifts and hence distances. Also called *quasistellar objects* or *QSO*s.

quasi-stellar object (QSO): *See* quasar.

radiation: The emission of particles or energy. Also the particle or energy so emitted.

radiation-dominated era: The epoch in the history of the universe, lasting from the big bang until approximately the time of recombination, during which the energy density of radiation controlled the gravity of the cosmos.

radioactive dating: The determination of the age of a sample by the measurement of the ratio of the decay products to the precursor, for one or more radioactive isotopes. Radioactive dating is possible because each unstable isotope has a well-defined half-life.

radioactivity: Emission of particles or gamma rays from the nucleus of an atom.

reception distance: The distance of the source of light at the time the light is received.

recombination: The formation of atoms in the early universe, which occurred when the temperature became sufficiently low that free electrons could no longer overcome the electrostatic attraction of the nuclei and were captured.

red dwarf: A small, dim, low-mass main-sequence star.

red giant: A star near the end of its life, which is fusing heavier elements in its core and has a greatly expanded outer layer.

redshift: A shift in the frequency of a photon toward lower energy.

regular cluster: A cluster of galaxies with a relatively smooth, approximately spherical shape. Most regular clusters are dominated by elliptical galaxies.

relativity: The rules relating observations in one inertial frame of reference to the observations of the same phenomenon in another inertial frame of reference. Casually applied only to Einstein's special theory of relativity, but a more general term.

relativity principle: The postulate of the special theory of relativity which states that the laws of physics are the same in all inertial frames of reference.

relevant: Of a scientific hypothesis, the requirement that it must be directly related to the phenomenon it seeks to explain.

relic problem: The unresolved issue in standard cosmology in which various theories of particle physics would invariably produce massive particles that are not observed.

rest energy: The energy corresponding to the rest mass according to $E_0 = m_0 c^2$.

rest mass: The mass of an object measured in its own rest frame. An important invariant quantity.

retrograde motion: The apparent reversal in the motion of a planet across the sky relative to the background stars, caused by Earth passing it or being passed by it.

Riemannian geometry: A generalized geometry that has the property of being locally flat; that is, in a sufficiently small region, a Riemannian geometry can be approximated by a Euclidean or Minkowski geometry.

Robertson-Walker metric: The metric that describes an isotropic and homogeneous cosmological spacetime.

Sachs-Wolfe effect: The scattering of photons from perturbations in the early universe. Photons that last interacted with an overdense region suffer a gravitational redshift, whereas those that last scattered from an underdense region are blueshifted.

scale factor: The quantity that describes how lengths change in the expanding (or contracting) universe.

Schrödinger equation: The equation that describes the evolution of a nonrelativistic wavefunction.

Schwarzschild radius: The radius of the event horizon of a nonrotating black hole of mass M, equal to $2GM/c^2$.

scientific method: An investigative approach in which data are gathered, a hypothesis is formulated to explain the data, and further experiments are performed to test the hypothesis.

second law of thermodynamics: The law that states that the entropy of a closed system always increases or at best remains the same in any process.

Silk mass: The minimum mass of a perturbation that can collapse gravitationally in the presence of photon damping.

simplicity: Of a scientific hypothesis, the principle that a proposed explanation must not be unnecessarily complicated.

simultaneity: The coincidence of the time coordinate of two events; the observation that two occurrences take place at the same time. Simultaneity is not invariant but depends upon the reference frame of the observer.

singularity: In classical general relativity, a location at which physical quantities such as density become infinite.

solar luminosity ($L_\odot$): The energy output or luminosity of the Sun, used as a standard in astronomy.

solar mass ($M_\odot$): The mass of the Sun, used as a standard in astronomy.

spacelike: Of a spacetime interval: incapable of being connected by anything which travels at or below the speed of light *in vacuo*. A spacelike spacetime interval is an imaginary number.

spacetime: The geometry that merges space and time coordinates.

spacetime diagram: A depiction of spacetime, usually including time and only one spatial dimension.

spacetime interval: The invariant distance in spacetime between two events, as specified by the metric equation.

speed: The magnitude of the velocity.

speed of light: The finite speed at which light travels. Unless otherwise stated, usually refers to the fundamental constant c, the speed of light in a perfect vacuum.

spherical geometry: A geometry that has positive constant curvature.

spiral galaxy: A galaxy that shows spiral arms, resembling a glowing pinwheel. Spirals typically contain a spheroidal nuclear bulge surrounded by a flat disk of stars, dust, and gas through which the spirals are threaded. The spirals themselves are delineated by bright young stars and probably represent density waves traveling through the disk.

spontaneous symmetry breaking: The loss of symmetry that causes fundamental forces to become distinguishable. In most theories, this occurs in the early universe when the temperature becomes low enough that the different energy scales of the different forces become important.

standard candle: An object of known intrinsic luminosity, useful in the measurement of luminosity distances.

standard models: A class of big bang cosmological models that are generated with the minimum set of assumptions, namely that the cosmological principle holds and the cosmological constant is zero. Sometimes also called Friedmann models.

star: A self-luminous object held together by its own self-gravity. Often refers to those objects that generate energy from nuclear reactions occurring at their cores, but may also be applied to stellar remnants such as neutron stars.

static surface: The surface surrounding a Kerr black hole at which even light cannot resist being dragged along in the direction of the rotation of the hole.

steady state model: A cosmological model that obeys the perfect cosmological principle. Generally applied to specific models that contain a cosmological constant generated by the regular creation of matter.

stellar parallax: *See* parallax.

strong equivalence principle: The principle that *all* physical laws, not just those of mechanics, are the same in all inertial and freely falling frames of reference.

strong interaction: The fundamental force that binds quarks into hadrons and holds nucleons together in atomic nuclei. Sometimes called the strong force or the strong nuclear force.

structure formation: The development of organized structures in the universe, ranging from galaxies to clusters to superclusters, and possibly beyond to huge filaments and voids.

structure problem: The incompletely resolved difficulty of explaining the origin of structure, representing local inhomogeneities, in a universe that is isotropic and homogeneous on the largest scales.

Sunyaev-Zel'dovich effect: The distortion in the cosmic background radiation due to the interaction of photons of the CBR with hot gas in distant galaxy clusters.

supercluster: A cluster of galaxy clusters.

supernova: The explosive death of a star. Type I supernovae occur when a white dwarf accumulates too much gas from a companion upon its surface, causing it to exceed the Chandrasekhar limit. Type II supernovae occur when a massive star has reached the endpoint of nuclear fusion and can no longer support itself. In both cases, the result is a catastrophic gravitational collapse and an explosion so violent that elements heavier than iron are created. Any remaining core becomes a neutron star or a black hole.

superposition of states: In quantum mechanics, the description of an unobserved system in terms of the probabilities of all possible states.

superstring theory: A theory in which the ultimate reality is quantum strings and loops, whose vibrations are associated with what we call particles.

symmetry: The property under which some quantity does not change when certain attributes, such as spatial location, time, rotation, and so forth, vary.

temperature: A measure of the average kinetic energy of random motion of the constituents (e.g., molecules, atoms, photons) of a system.

testable: Of a hypothesis, capable of being tested because it makes a specific prediction. Similar to *falsifiable*.

theory: In scientific usage, a hypothesis or related group of hypotheses that have become well established.

thermal equilibrium: A state in which energy is equally distributed among all particles, and all the statistical properties of the particles can be described by a single parameter, the temperature.

thermal radiation: Radiation emitted by any object with a temperature greater than absolute zero. A thermal spectrum occurs because some of the heat energy of the object is converted into photons. In general, a thermal spectrum depends not only upon the temperature, but also upon the composition of the object, its shape, its heat capacity, and so forth. Compare *blackbody radiation.*

thought experiments: Experiments that could be performed in principle, but might be very difficult in practice, and whose outcome can be predicted by pure logic. Often used to develop the consequences of a theory, so that more practical phenomena can be predicted and put to actual experimental tests.

tidal force: In Newtonian gravity, the net force on an extended body due to a difference in gravitational force from one region of the body to another. In general relativity, a force arising when nearby geodesics diverge in spacetime, because the world lines of all parts of an extended body cannot travel along a single geodesic.

time dilation: An apparent decrease in the rate of the flow of time (i.e., the ticking of a clock) in a frame moving relative to a given observer, determined by the Lorentz transformation from one frame to the other.

timelike: Of a spacetime interval: capable of being connected by anything that travels below the speed of light *in vacuo.* World lines of particles with mass follow timelike paths through spacetime.

top-down structure formation: The formation of large structures, such as galaxy superclusters or perhaps even the vast filaments and voids, prior to the formation of smaller structures such as individual galaxies.

true vacuum: A stable state in which a quantum field is zero and the corresponding potential is also zero; that is, the vacuum energy density is zero.

Tully-Fisher relationship: An empirical relationship between the width of the 21 cm line of hydrogen emission from spiral galaxies and the luminosity of the galaxy. The relationship arises because more luminous galaxies have a larger mass; a greater mass increases the rotation rate, and a faster rotation causes a broader line; the precise calibration must be determined observationally.

turnoff mass: The mass of the largest star in a cluster that is still on the main sequence. The age at which a star moves from the main sequence to the red giant phase depends almost entirely upon its mass and chemical composition, with more massive stars leaving the main sequence earlier. The stars in a cluster all formed at essentially the same time and have similar chemical composition, so the turnoff mass can be used to determine the age of the cluster.

uncertainty principle: The principle of quantum mechanics which states that the values of both members of a certain pairs of variables, such as position and momentum or energy and time interval, cannot be determined simultaneously to arbitrary precision. For example, the more precisely the momentum of a particle is measured, the less determined is its position. The uncertainty in the values of energy and time interval permits the quantum creation of virtual particles from the vacuum.

unified epoch: That interval in the early history of the universe when three of the four fundamental forces, the strong and weak interactions and the electromagnetic force, were unified.

uniform motion: Motion at a constant velocity. The state of rest is a special case of uniform motion.

universe: That which contains and subsumes all the laws of nature, and everything subject to those laws; the sum of all that exists physically, including matter, energy, physical laws, space, and time.

vacuum energy: The energy associated with space itself, whether from a classical cosmological constant or from quantum fluctuations.

vector: A mathematical entity that has direction as well as magnitude. Important physical quantities represented by vectors include velocity, acceleration, and force. A vector changes whenever either its direction or its magnitude change.

velocity: The rate of change of position with time. Velocity is speed of motion along with direction of motion.

Virgo Cluster: A nearby irregular cluster of galaxies located in the constellation Virgo. The distance to the Virgo Cluster is an important rung in the distance ladder.

virial theorem: A statistical result that relates the mean gravitational field of a cluster to the dispersion of the velocities of the members of the cluster.

virtual particles: Particles that exist only as permitted by the uncertainty principle.

void: In astronomy, a huge region of space that is unusually empty of galaxies. Recent research has shown that voids are not entirely empty, but they are underdense and contain far fewer bright galaxies than average.

wave: A propagating disturbance that transmits energy from one point to another without physically transporting the oscillating quantity.

wave amplitude: The size of the departure from the average of the quantity that supports the wave.

wave frequency: The number of wave crests that pass a fixed point in a fixed interval of time.

wavefunction: The quantity that obeys the Schrödinger equation. In the Copenhagen interpretation of quantum mechanics, the wavefunction is a mathematical entity which describes the probabilities that the quantum system will assume any of several possible states upon a measurement.

wavefunction of the universe: A wavefunction that treats the scale factor as a quantum variable and describes its evolution in quantum, rather than classical general relativistic, terms.

wavelength: The distance from one crest of a wave to the next.

weak equivalence principle: The principle that the laws of mechanics are the same in inertial and free-falling frames of reference. This implies that gravitational mass and inertial mass are equivalent.

weak interaction: The fundamental force that accounts for some particle interactions, such as beta decay, the decay of free neutrons, neutrino interactions, and so forth. Sometimes called the weak force, or the weak nuclear force.

weight: The gravitational force experienced by an object. Usually refers to the gravitational attraction due to a large object, such as a planet, upon smaller objects at or near its surface.

white dwarf: A compact stellar remnant supported by electron degeneracy pressure and shining only by the diffusion of light from its interior. White dwarfs cool slowly; if the universe exists long enough they will all cool into nonluminous black dwarfs.

WIMP (weakly interacting massive particle): A particle with a nonzero mass that participates only in the weak interaction.

work: In physics, a compound of the force exerted with the displacement produced.

world line: The path of a particle in spacetime.

world tube: The set of world lines traced by all the particles of which an extended body is composed.

Bibliography

Mythology, Philosophy, and History References

Beier, Ulli. *The Origin of Life and Death: African Creation Myths*. London: Heinemann Educational Books Ltd., 1966.

Birch, Cyril. *Chinese Myths and Fantasies*. New York: Henry Z. Walck, Inc., 1961.

Brundage, Burr Cartwright. *The Fifth Sun: Aztec Gods, Aztec World*. Austin: University of Texas Press, 1979.

Campbell, Joseph. *The Power of Myth*. New York: Doubleday, 1988.

Copi, Irving M. *Introduction to Logic*, Fourth edition. New York: MacMillan Publishing Company Inc., 1972.

Hetheringon, Noriss S., editor. *Cosmology: Historical, Literary, Philosophical, Religious, and Scientific Perspectives*. New York: Garland Publishing, Inc., 1993.

Koestler, Arthur. *The Sleepwalkers*. New York: MacMillan, 1968.

Pais, Abraham. *'Subtle Is the Lord': The Science and the Life of Albert Einstein*. Oxford: Oxford University Press, 1982.

Popper, Karl. *The Logic of Scientific Discovery*. London: Hutchinson, 1959.

Taylor, Colin F., editor. *Native American Myths and Legends*. New York: Smithmark Publishers Inc., 1994.

Tuchmann, Barbara W. *The March of Folly*. New York: Ballantine Books, 1984.

Walls, Jan, and Yvonne Walls. *Classical Chinese Myths*. Hong Kong: Joint Publishing Company, 1984.

Popular-Level Books on Cosmology and Related Subjects

Barrow, John. *The Origin of the Universe*. New York: Basic Books, 1994.

Cosmology +1: Readings from Scientific American. San Francisco: W. H. Freeman, 1977.

Ferris, Timothy. *Coming of Age in the Milky Way*. New York: William Morrow and Company, Inc., 1988.

Hawking, Stephen. *A Brief History of Time*. Toronto: Bantam Books, 1988.

Penrose, Roger. *The Emperor's New Mind*. Oxford: Oxford University Press, 1989.

Silk, Joseph. *The Big Bang*. Revised and updated edition. New York: W. H. Freeman and Company, 1989.

Silk, Joseph. *A Short History of the Universe*. Scientific American Library. New York: W. H. Freeman and Company, 1994.

Thorne, Kip S. *Black Holes and Time Warps: Einstein's Outrageous Legacy*. New York: W. W. Norton, 1994.

Weinberg, Steven. *Dreams of a Final Theory*. New York: Pantheon Books, 1992.

Weinberg, Steven. *The First Three Minutes*. Updated edition. New York: Basic Books, 1988.

Zuckerman, Ben, and Matthew Malkan. *The Origin and Evolution of the Universe*. Boston: Jones and Bartlett, 1996.

Magazine Articles

Albert, David Z. "Bohm's Alternative to Quantum Mechanics." *Scientific American*, 270, No. 5, 58. May 1994.

Brashear, Ronald S., Donald E. Osterbrock, and Joel A. Gwinn. "Edwin Hubble and the Expanding Universe." *Scientific American*, 269, No. 1. July 1993.

Deutsch, David, and Michael Lockwood. "The Quantum Physics of Time Travel." *Scientific American*, 270, No. 3, 68. March 1994.

DeWitt, Bryce. "Quantum Cosmology." *Scientific American*, 249, No. 6, 112. December 1983.

Halliwell, Jonathan J. "Quantum Cosmology and the Creation of the Universe." *Scientific American*, 265, No. 6, 76. December 1991.

Gould, Stephen Jay. "The Evolution of Life on Earth." *Scientific American*, 271, No. 4, pp. 84–91. October 1994.

Kirschner, Robert. "The Earth's Elements." *Scientific American*, 271, No. 4, pp. 58–67. October 1994.

McCloskey, Michael. "Intuitive Physics." *Scientific American*, 248, No. 4, pp. 122–130. April 1983.

Peebles, P. James, *et al.* "The Evolution of the Universe." *Scientific American*, 271, No. 4, pp. 52–57. October 1994.

Schramm, David N. "Dark Matter and the Origin of Cosmic Structure." *Sky and Telescope*, 88, pp. 28–35, October 1994.

Weinberg, Steven. "Life in the Universe." *Scientific American*, 271, No. 4, pp. 44–51. October 1994.

Introductory Texts on Astronomy, Cosmology, and Physics

There are dozens of excellent introductory texts on general astronomy and physics. We list here only a few.

Chaisson, Eric, and Steve McMillan. *Astronomy Today*. Englewood Cliffs, New Jersey: Prentice Hall, 1993.

Halliday, David, and Robert Resnick. *Physics*. Third edition. New York: John Wiley and Sons Inc., 1978.

Harrison, Edward R. *Cosmology: The Science of the Universe*. Cambridge: Cambridge University Press, 1981.

Kaufmann, William J. *Universe*. Fourth edition. New York: W. H. Freeman and Company, 1994.

Kuhn, Karl F. *In Quest of the Universe*. Second edition. Minneapolis: West Publishing Company, 1994.

Advanced Books and Texts

Adler, Ronald, Maurice Bazin, and Menahem Schiffer. *Introduction to General Relativity*. Second edition. New York: McGraw-Hill Book Company, 1975.

Allen, C. W. *Astrophysical Quantities*. Third edition. Dover: Athlone Press, 1976.

Bernstein, Jeremy. *An Introduction to Cosmology*. Englewood Cliffs, New Jersey: Prentice Hall, 1995.

Einstein, Albert, *et al. The Principle of Relativity*. New York: Dover, 1952.

Hawking, Steven, and W. Israel, eds. *General Relativity: An Einstein Centenary Survey*. New York: Cambridge University Press, 1979.

Kolb, Edward W., and Michael S. Turner. *The Early Universe*. Redwood City, California: Addison Wesley Publishing Company, 1990.

Mandolesi, N., and N. Vittorio, eds. *The Cosmic Microwave Background: 25 Years Later*. Dordrecht: Kluwer Academic, 1990.

Merzbacher, Eugen. *Quantum Mechanics*. Second edition. New York: John Wiley and Sons, Inc., 1970.

Misner, Charles W., Kip S. Thorne, and John Archibald Wheeler. *Gravitation*. San Francisco: W. H. Freeman and Co., 1973.

Peebles, P. J. E. *Principles of Physical Cosmology*. Princeton, New Jersey: Princeton University Press, 1993.

Perkins, Donald H. *Introduction to High Energy Physics*. Second edition. Reading, Massachusetts: Addison-Wesley, 1982.

Rees, M. *Perspectives in Physical Cosmology*. Cambridge: Cambridge University Press, 1995.

Rindler, Wolfgang. *Essential Relativity*. Second edition. New York: Springer-Verlag, 1977.

Rowan-Robertson, Michael. *Cosmology*. Oxford: University of Oxford Press, 1977.

Schwarzschild, Martin. *Structure and Evolution of the Stars*. New York: Dover Publications Inc., reprinted from Princeton University Press, 1958.

Shapiro, Stuart L., and Saul A. Teukolsky. *Black Holes, White Dwarfs, and Neutron Stars: The Physics of Compact Objects*. New York: John Wiley and Sons, 1983.

Vangioni-Flam, E., M. Casse, J. Audouze, and J. Tran Thuanh Van, eds. *Astrophysical Ages and Dating Methods*. Gif-sur-Yvette, France : Editions Frontiéres, 1990.

Weinberg, Steven. *Gravitation and Cosmology: Principles and Applications of the General Theory of Relativity*. New York: John Wiley and Sons, 1972.

Technical Journal Articles

Bolte, M., and C. J. Hogan, 1995. "Conflict over the age of the Universe." *Nature*, 376, 399.

Coles, P., and G. Ellis, 1994. "The case for an open Universe." *Nature*, 370, 609.

Coulson, D., P. Lerreira, P. Graham, and N. Turok, 1994. "Microwave anisotropies from cosmic defects." *Nature*, 368, 27.

Dunlop, J. S., *et al.* 1994. "Detection of a large mass of dust in a radio galaxy at $z = 3.8$." *Nature*, 370, 347.

Elston, R., K. L. Thompson, and G. J. Hill, 1994. "Detection of strong iron emission from quasars at redshift $z > 3$." *Nature*, 367, 250.

Freedman, W. L., *et al.,* 1994, "Distance to the Virgo cluster galaxy M100 from *Hubble Space Telescope* observations of Cepheids." *Nature*, 371, 757.

Fukugita, M., C. J. Hogan, and P. J. E. Peebles, 1996. "The history of the galaxies." *Nature*, 381, 489.

Gott, R., *et al.,* 1974. "An unbound universe." *Astrophys. J.*, 194, 543.

Hoskin, Michael A., 1976. "The 'Great Debate': What really happened." *J. Hist. Astron.*, 7, 169.

Johnson, H. L., and W. W. Morgan 1953. "Fundamental stellar photometry for standards of spectral type on the revised system of the Yerkes Spectral Atlas." *Astrophys. J.*, 117, 313.

Johnson, H. L., and A. R. Sandage 1956. "Three color photometry in the globular cluster M3." *Astrophys. J.*, 124, 379.

Lin, H., *et al.*, 1996. "The Power Spectrum of Galaxy Clustering in the Las Campanas Redshift Survey." *Astrophys. J.*, 471, 617.

Mather, J. C., *et al.*, 1990. "A preliminary measurement of the cosmic microwave background radiation spectrum by the *Cosmic Background Explorer* satellite." *Astrophys. J*, 354, L37.

Pierce, M. J., *et al.*, 1994. "The Hubble Constant and Virgo cluster distance from observations of Cepheid variables." *Nature*, 371, 385.

Roth, K. C., D. M. Meyer, and I. Hawkins, 1993. "Interstellar Cyanogen and the temperature of the cosmic microwave background radiation." *Astrophys. J.*, 413, L67.

Songaila, A., *et al.*, 1994. "Measurement of microwave background temperature at a redshift $z = 1.776$." *Nature*, 371, 43.

Strauss, M. A., *et al.*, 1992, "A Redshift Survey of *IRAS* Galaxies. IV. The Galaxy Distribution and the Inferred Density Field." *Astrophys. J.*, 385, 421.

Tanaka, Y., *et al.*, 1995. "Gravitationally redshifted emission implying an accretion disk and massive black hole in the active galaxy MCG$-6 - 30 - 15$." *Nature,* 375, 659.

Taylor, J., and J. Weisberg, 1989. "Further experimental tests of relativistic gravity using the binary pulsar PSR $1913 + 16$." *Astrophys. J.*, 345, 434.

Wilson, T. L., and R. T. Rood, 1994. "Abundances in the Interstellar Medium." *Ann. Rev. Astron. Astrophys*, 32, 191.

Index

494